MATERIALS FOR
ARCHITECTS AND BUILDERS

Materials for Architects and Builders provides a clear and concise introduction to the broad range of materials used within the construction industry and covers the essential details of their manufacture, key physical properties, specification and uses.

Understanding the basics of materials is a crucial part of undergraduate and diploma construction or architecture-related courses, and this established textbook helps the reader to do just that with the help of colour photographs and clear diagrams throughout.

This new edition has been completely revised and updated to include the latest developments in materials research, new images, appropriate technologies and relevant legislation. The ecological effects of building construction and lifetime use remain an important focus, and this new edition includes a wide range of energy-saving building components.

Arthur Lyons was formerly Head of Quality, principal lecturer and Teacher Fellow for building materials in the Leicester School of Architecture, Faculty of Art, Design & Humanities, De Montfort University, UK. He was a lecturer in building materials within schools of architecture and surveying for thirty-five years, and is now an established writer on construction materials.

MATERIALS FOR ARCHITECTS AND BUILDERS

Fifth Edition

ARTHUR LYONS

Routledge
Taylor & Francis Group

LONDON AND NEW YORK

First edition published 1997
by Arnold and reprinted by Butterworth-Heinemann in 2002
Second edition 2003
Third edition 2007
Fourth edition 2010

This fifth edition published 2014
by Routledge
2 Park Square, Milton Park, Abingdon, Oxon OX14 4RN

and by Routledge
711 Third Avenue, New York, NY 10017

Routledge is an imprint of the Taylor & Francis Group, an informa business

British Library Cataloguing in Publication Data
A catalogue record for this book is available from the British Library

Library of Congress Cataloging-in-Publication Data
Lyons, Arthur.
Materials for architects and builders/Arthur Lyons. – Fifth edition.
   pages cm
   Includes bibliographical references and index.
   1. Building materials. I. Title.
   TA403.L95 2014
   691–dc23   2013050714

ISBN13: 9780415704977 (pbk)
ISBN13: 9781315768748 (ebk)

Typeset in Minion and Futura
by Florence Production Ltd. Stoodleigh, Devon EX16 9PN

CONTENTS

———

ABOUT THE AUTHOR

Dr Arthur Lyons, author of texts on building materials, was formerly Head of Quality, principal lecturer and Teacher Fellow for building materials in the Leicester School of Architecture, Faculty of Art and Design, De Montfort University, Leicester, UK. He was educated at Trinity Hall Cambridge, Warwick and Leicester Universities in the fields of natural sciences and polymer science, and has a postgraduate diploma in architectural building conservation. He was a lecturer in building materials within schools of architecture and surveying for thirty-five years. In recognition of his services to architects and architecture, Arthur Lyons was honoured with life membership of the Leicestershire and Rutland Society of Architects and he is a Fellow of the Higher Education Academy. He retains his active interest in architecture through liaison with the local society of architects and the Leicester School of Architecture of De Montfort University, where he is an honorary visiting researcher. In addition to this text, Arthur Lyons has written chapters in the *ICE Manual of Construction Materials* (2009, Institution of Civil Engineers); the *Metric Handbook – Planning and Design Data* (4th edn, 2012, Routledge); and the *Construction Materials Reference Book* (2013, Routledge). Arthur Lyons also authored the first and current second edition of *The Architecture of the Universities of Leicester*, published by Anchorprint.

PREFACE TO FIFTH EDITION

Materials for Architects and Builders is written as an introductory text to inform students at undergraduate degree and national diploma level of the relevant visual and physical properties of the widest range of building materials. The fifth edition has been significantly enhanced by the addition of more colour images, illustrating the materials and in many cases their use in new buildings of architectural merit. The text embraces the broad environmental issues with sections on energy-saving and recycled materials. The chapter on sustainability reflects the current debate on climate change and governmental action to reduce carbon emissions and ameliorate global warming. There are 18 chapters covering the wide range of materials under standard headings. Each chapter describes the manufacture, salient properties and typical uses of the various materials, with the aim of ensuring their appropriate application within awareness of their ecological impact.

European Standards are taking over from the previous British Standards, and the European Norms have now been published for most key materials. Generally, this has led to an increase in the number of relevant standards for building materials. However, in some cases, both the British and European Standards are current, and are therefore included in the text and references.

New and rediscovered old materials, where they are becoming well integrated into standard building processes, are described, together with innovative products yet to receive general acceptance. Other materials no longer in use are generally disregarded, except where increased concern for environmental issues has created renewed interest. The use of chemical terminology is kept to the minimum required to understand each subject area, and is only significantly used within the context of the structure of plastics. Tabulated data is restricted to an informative level appropriate to student use. An extensive bibliography and listed sources of technical information are provided at the end of each chapter to facilitate direct reference where necessary.

The text is well illustrated with over 300 line drawings and colour photographs, showing the production, appearance and appropriate use of materials, but it is not intended to describe construction details, as these are illustrated in the standard texts on building construction. Environmental concerns including energy-conscious design and the effects of fire are automatically considered as part of the broader understanding of the various materials.

The text is essential reading for honours and foundation degree, BTEC and advanced GNVQ students of architecture, building, surveying and construction, and those studying within the broad range of built environment subjects, who wish to understand the principles relating to the appropriate use of construction materials.

Arthur Lyons
January 2014

ACKNOWLEDGEMENTS

I acknowledge the support of the University Library and the Leicester School of Architecture, Faculty of Art, Design and Humanities, De Montfort University, Leicester. I wish to thank my wife Susan for her participation and support during the production of this work; also my daughters Claire and Elizabeth for their constant encouragement. I am indebted to the numerous manufacturers of building materials for their trade literature and for permissions to reproduce their published data and diagrams. I am grateful to building owners, architectural practices and their photographers for the inclusion of the photographs; to Her Majesty's Stationery Office, the Building Research Establishment, the British Standards Institute and trade associations for the inclusion of their material.

I should like to thank the following organisations for giving permission to use illustrations:

Aircrete Products Association (Fig. 2.3)

AkzoNobel Group (Fig. 15.5)

Ancon Building Products (Fig. 1.18)

Angle Ring Company Ltd (Fig. 5.10)

Architectural Ceramics (Fig. 8.10)

Aurubis (Figs 5.25 and 5.29)

British Cement Association (Figs 3.5, 3.8 and 3.23)

British Fenestration Rating Council (Fig. 7.18)

British Standards Institute (Figs 2.9 and 5.33). Permission to reproduce extracts from BS EN 771 Part 1: 2003 and BS 6915: 2001 is granted by BSI. British Standards may be obtained in PDF or hard copy formats from the BSI online shop (www.bsi group.com/Shop) or by contacting BSI Customer Services for hard copies only (Tel: +44 (0) 20 8996 9001, Email: cservice@bsigroup.com)

Building Research Establishment (Figs 2.3, 4.14 and 9.19); photographs from GBG 58, Digest 476 and IP 10/01, reproduced by permission of IHS

Construction Resources (Fig. 4.39)

Copper Development Association (Figs 5.28 and 5.29)

Durisol UK (Fig. 2.11)

Edenhall Ltd (Fig. 1.27)

EH Smith (Fig. 8.8)

Forticrete Ltd (Fig. 2.8)

Glass Block Technology (Fig. 7.6)

Hanson Brick Ltd (Fig. 1.5)

Ibstock Brick Ltd (Figs 1.3, 1.9, 1.10, 1.12, 1.15, 1.21 and 2.10)

James & Son Ltd (Fig. 11.8)

Johnson-Tiles (Figs 8.6 and 8.9)

Lead Contractors Association (Figs 5.32 and 5.34)

Lead Sheet Association (Fig. 5.31)

Lignacite Ltd (Fig. 2.7)

London Legacy Development Corporation (Figs 4.19 and 5.3)

Marshalls plc (Fig. 2.16)

Metal Cladding and Roofing Manufacturers Association (Fig. 5.15)

Metra Non-ferrous Metals Ltd and Rheinzinc (Fig. 5.36)

Monodraught (Figs 16.7 and 16.8)

Natural Stone Products Ltd (Fig. 9.14)

NBT Thermoplan System (Fig. 2.9)

NCS – Natural Colour System® © property of and used on licence from NCS Colour AB, Stockholm 2014. References to NCS® © in this publication are used with permission from the NCS Colour AB (Fig. 15.4)

Norman and Underwood (Fig. 5.30)

Pilkington plc (Figs 7.7, 7.10, 7.12 and 7.25); images are reproduced by permission of Pilkington plc

Pyrobel (Fig. 7.16)
Rimex Metals Group (Fig. 5.18)
Ruberoid Building Products (Fig. 6.3)
Securiglass Company Ltd (Fig. 7.14)
Smith of Derby (Fig. 11.2)
Solar Century (www.solarcentury.com) (Fig. 16.3)
Stancliffe Stone (Figs 9.1, 9.4, 9.5 and 9.9)
Steel Construction Institute (Figs 5.7 and 5.12)
Stone Federation of Great Britain (Fig. 9.7)
Tata Steel (Figs 5.2, 5.4–5.6, 5.11 and 5.13)
Tecu Consulting KME UK (Figs 5.26 and 5.27)
TRADA Technology Ltd (Figs 4.14 and 4.20)
Trent Concrete Ltd (Figs 1.23, 3.20, 3.21, 9.22, 11.5 and 11.6)
Weinerberger UK (Figs 1.22, 1.24 and 1.25)
Zinc Development Association (Fig. 5.36)

I wish to thank Dr Richard Leese and Mr Mike Taylor of the Mineral Products Association for their advice relating to cement and concrete; also Jon Castleman of Norman & Underwood for advice relating to lead and metal roofing systems.

The text uses the generic names for building materials and components wherever possible. However, in a few cases, products are so specific that registered trade names are necessarily used. In these cases the trade names are italicised in the text.

INTRODUCTION

Specific information relating to the materials described in each chapter is given at the end of the appropriate section; however, the following are sources of general information relating to construction materials.

- Building Regulations, including current Amendments and Approved Documents to 2013
- RIBA Office Library and Barbour Index
- Building Research Establishment (BRE) publications
- Trade association publications
- Trade exhibitions
- Trade literature
- Architecture and built environment journals
- British Board of Agrément certificates
- British Standards
- European Standards
- Eurocodes
- The Construction Information Service

European Standards (EN) have been published for a wide range of materials. A full European Standard, known in the UK as BS EN, is mandatory and overrules any conflicting previous British Standard which must be withdrawn. Prior to full publication, the draft European Standards are coded pr EN and are available for comment, but not implementation. BRE Information Paper IP 3/99 (1999) identifies the issues relating to the adoption in the UK of the structural Eurocodes.

The Construction Products Regulation (2013) makes CE marking of all products defined by harmonised European Product Standards mandatory. This is designed to ensure a unified set of rules to underpin the performance characteristics of construction products across the European Union. Under the Construction Products Regulation, technical specifications are defined in either the appropriate European Standards (ENs) or in a relevant European Assessment Document. Defined performance characteristics include mechanical properties, stability, reaction to fire, health and safety issues, energy and sustainability. Within the European Standards (ENs) issues relating to harmonised product standards including CE marking are defined in the Annex ZA.

The Building Research Establishment (BRE) publishes informative and authoritative material on a wide range of subjects relating to construction. Trade associations and manufacturers produce promotional literature and websites relating to their particular area of interest within the building industry. Architecture and building journals give news of innovations and illustrate their realisation in quality construction. Much literature has recently been presented, including from governmental organisations, in respect of the need to reduce energy consumption within the built environment sector to ameliorate the effects of global warming and climate change.

Information for this text has been obtained from a wide selection of sources to produce a student text with an overview of the production, nature and properties of the diverse range of building materials. New individual products and modifications to existing products frequently enter the market; some materials become unavailable. Detailed information and particularly current technical data relating to any specific product for specification purposes should therefore be obtained directly from the manufacturers or suppliers, and cross-checked against current standards and regulations.

ABBREVIATIONS

General

AAC	autoclaved aerated concrete
ABS	acrylonitrile butadiene styrene
AC	aggressive chemical (environment)
ACD	approved construction details
ACEC	aggressive chemical environment for concrete
APAO	atactic poly α-olefin
APM	additional protective measures
APP	atactic polypropylene
AR	alkali-resistant
ASR	alkali-silica reaction
BER	building emission rate
BFRC	British Fenestration Rating Council
BRE	Building Research Establishment
BREEAM	BRE environmental assessment method
BS	British Standard
CAC	calcium aluminate cement
CAD	computer-aided design
CCA	chromated copper arsenate
CEN	European committee for standardisation
CFCs	chlorofluorocarbons
CG	cellular glass
CIGS	copper indium gallium selenide
CIS	copper indium selenide
CL	air lime
CLT	cross-laminated timber
CMYK	cyan magenta yellow black
CO_2e	carbon dioxide equivalent emissions
COSHH	control of substances hazardous to health
CPE	chlorinated polyethylene
CPR	construction products regulation
CPVC	chlorinated polyvinyl chloride
CS	calcium silicate
CSA	Canadian Standards Association
CSM	chlorosulphonated polyethylene
DC	design chemical (class)
DC	direct current
DD	draft for development
DER	door energy rating
DFEE	dwelling fabric energy efficiency
DPC	damp-proof course
DPM	damp-proof membrane
DR	dezincification-resistant
DRF	durability of reaction to fire
DS	design sulphate (class)
DSA	design stage assessment
DSER	door set energy rating
DZR	dezincification-resistant
EN	Euronorm
ENV	Euronorm pre-standard
EP	expanded perlite
EPC	energy performance certificate
EPDM	ethylene propylene diene monomer
EPR	ethylene propylene rubber
EPS	expanded polystyrene
ETFE	ethylene tetrafluorethylene copolymer
EV	exfoliated vermiculite
EVA	ethylene vinyl acetate
EVOH	ethyl vinyl alcohol copolymer
FEES	fabric energy efficiency standards
FEF	flexible elastomeric foam
FL	formulated lime
FPA	flexible polypropylene alloy
FRP	fibre-reinforced polymer
FSC	Forest Stewardship Council
GGBS	ground granulated blast furnace slag
GRC	glassfibre-reinforced concrete
GRG	glassfibre-reinforced gypsum
GRP	glassfibre-reinforced plastic or polyester
GS	general structural (timber)

HAC	high alumina cement		PHA	partially halogenated alkane
HACC	high alumina cement concrete		PIB	polisobutylene
HB	hardboard		PIR	polyisocyanurate foam
HCFCs	hydrochlorofluorocarbons		PMMA	polymethyl methacrylate
HD	high density		PP	polypropylene
HDPE	high-density polythene		pr EN	draft Euronorm
HIP	home information pack		PS	polystyrene
HL	hydraulic lime		PTFE	polytetrafluroethylene
HLS	hue lightness saturation		PUR	rigid polyurethane foam
HS	structural tropical hardwood		PV	photovoltaic
ICB	insulating corkboard		PVA	polyvinyl acetate
ICF	insulating concrete formwork		PVB	polyvinyl butyral
IGU	insulating glass unit		PVC	polyvinyl chloride (plasticised)
LA	low alkali (cement)		PVC-U	polyvinyl chloride (unplasticised)
LD	low density		PVC-UE	extruded polyvinyl chloride
LDPE	low-density polythene		PVDF	polyvinylidene fluoride
LED	light-emitting diode		RBM	reinforced bitumen membrane
LRV	light reflectance value		RGB	red green blue
LVL	laminated veneer lumber		SAP	standard assessment procedure
MAC	minor additional constituents		SB	softboard
MAF	movement accommodation factor		SBEM	simplified building energy model
MB	mediumboard		SBS	styrene butadiene styrene
MDF	medium-density fibreboard		SCC	self-compacting concrete
MF	melamine formaldehyde		SFI	sustainable forest initiative
MMC	modern methods of construction		Sg	specific gravity
MPa	mega Pascal		SIP	structural insulated panel
MTCS	Malaysian Timber Certification Scheme		SS	special structural (timber)
MW	mineral wool		ST	standard (concrete mix)
NA	National Annex (British Standard)		SuDS	sustainable drainage systems
NAOBL	National Windspeed Database		T	tolerance (class)
NCS	Natural Colour System®		TER	target emission rate
NHL	non-hydraulic lime		TFEE	target fabric energy efficiency
ODP	ozone depletion potential		TFS	thin film silicon
OPC	ordinary Portland cement		TH	temperate hardwood
OSB	oriented strand board		THF	tetrahydro furan
PAR	processed all round		TMT	thermally modified timber
PAS	publicly available specification		TPE	thermoplastic elastomer
PB	polybutylene		TPO	thermoplastic polyolefin
PBAC	polystyrene-bead aggregate cement		TRADA	Timber Research and Development Association
PC	polycarbonate		TRM	total relative movement
PCM	phase change materials		UF	urea formaldehyde
PCS	post-construction stage assessment		UHPC	ultra-high performance concrete
PD	published document		UV	ultraviolet (light)
PE	polyethylene		VET	vinyl ethylene terpolymer
PEF	polyethylene foam		VIP	vacuum insulation panel
PEFC	programme for the endorsement of forest certification		VOC	volatile organic compounds
PET	polyethylene terephthalate		WER	window energy rating
PEX	cross-linked polyethylene		WF	wood fibre
PF	phenolic foam/phenol formaldehyde		WPC	wood plastic composite
PFA	pulverised fuel ash		WRAP	net waste tool

WW	wood wool
XLAM	cross-laminated timber
XPS	extruded polystyrene

Units

dB	decibel
GPa	giga Pascal (1 GPa = 1000 MPa)
MPa	mega Pascal (1 MPa = 1 N/mm^2)
μm	micron (10^{-6}m)
nm	nanometre (10^{-9}m)

Chemical symbols

Al	aluminium
As	arsenic
C	carbon
Ca	calcium
Cd	cadmium
Cl	chlorine
Cr	chromium
Cu	copper
F	fluorine
Fe	iron
Ga	gallium
In	indium
Mn	manganese
Mo	molybdenum
N	nitrogen
Ni	nickel
O	oxygen
S	sulphur
Se	selenium
Si	silicon
Sn	tin
Te	tellurium
Ti	titanium
Zn	zinc

Cement notation

C_2S	dicalcium silicate
C_3S	tricalcium silicate
C_3A	tricalcium aluminate
C_4AF	tetracalciumaluminoferrite

BRICKS AND BRICKWORK

CONTENTS

Introduction

Originally, bricks were hand-moulded from moist clay and then sun-baked, as is still the practice in certain arid climates. The firing of clay bricks dates back well over 5000 years, and is now a sophisticated and highly controlled manufacturing process; yet the principle of burning clay, to convert it from its natural plastic state into a dimensionally stable, durable, low-maintenance ceramic material, remains unchanged.

The quarrying of clay and brick manufacture are high-energy processes, which involve the emission of considerable quantities of carbon dioxide and other pollutants including sulphur dioxide. The extraction

of clay also has long-term environmental effects, although in some areas former clay pits have now been converted into bird sanctuaries or put to recreational use. However, well-constructed brickwork has a long life with low maintenance, and although the use of Portland cement mortar prevents the recycling of individual bricks, the crushed material is frequently recycled as aggregate in further construction.

The elegant cathedral at Evry near Paris (Fig. 1.1), designed by Mario Botta, illustrates the modern use of brickwork. The cathedral of Saint Corbinian, built with 670,000 bricks, was dedicated in 1997. The

such as fireclay, according to their source. The largest UK manufacturer uses the Lower Oxford clays of Bedfordshire, Buckinghamshire and Cambridgeshire to produce the *Fletton* brick. This clay contains some carbonaceous content that reduces the amount of fuel required to burn the bricks, lowering cost and producing a rather porous structure. Other particularly characteristic bricks are the strongly coloured *Staffordshire Blues* and *Accrington Reds* from clays containing high iron content and the yellow *London stocks* from the Essex and Kent chalky clays with lower iron content.

Brick manufacturers are endeavouring to reduce the environmental impact of production by improving energy efficiency of the firing process and the incorporation of recycled materials such as quarry fines, pulverised fuel ash, ground granulated blast furnace slag and waste glass into the extracted clay. The majority of clay bricks manufactured in the UK qualify for the maximum number of credits in the 'Materials' category of the 'Code for Sustainable Homes'. External walls built with clay bricks are generally classified as A+ in 'The Green Guide to Specification'.

SIZE

Within Europe, the dimensions of clay masonry units (BS EN 771–1: 2011) have not been standardised, but in the UK, the standard metric brick referred to in the National Annex (informative) to BS EN 771–1: 2011 is 215 × 102.5 × 65 mm, although this size is not a specified requirement. Dimensions are given in the order, length, width and height, respectively. These UK dimensions match those in BS 4729: 2005, which relates to special shapes and sizes of bricks. The standard brick weighs between 2 and 4 kg, and is easily held in one hand. The length (215 mm) is equal to twice its width (102.5 mm) plus one standard 10 mm joint and three times its height (65 mm) plus two standard joints (Fig. 1.2).

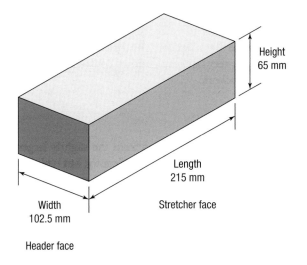

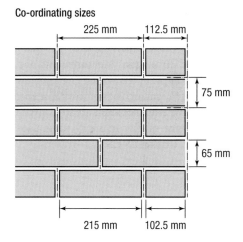

Co-ordinating sizes

Work sizes (norminal 10 mm joint)

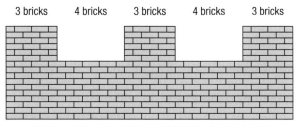

Brickwork dimensioned to 'work bricks' (no broken bond)

Fig. 1.2 Brick and co-ordinating sizes

Table 1.1 Tolerances on brick sizes to BS EN 771–1: 2011

	HD type clay brick dimensions (mm)	Limits for the deviation (±) from work size of mean of ten units (mm)					Maximum range of work size of mean of ten units (mm)				
		T1	T1+	T2	T2+	Tm	R1	R1+	R2	R2+	Rm
Length	215	6	6	4	4	*	9	9	4	4	*
Width	102.5	4	4	3	3	*	6	6	3	3	*
Height	65	3	1	2	1	*	5	1	2	1	*

Notes:

The reduced height tolerances for categories T1+, T2+, R1+ and R2+ are relevant when thin layer mortars are specified.

* Limits for Tm and Rm are as declared by the manufacturer (these may be wider or closer than the other categories).

The building industry modular coordination system (BS 6750: 1986) is based on the module (M) of 100 mm and multimodules of 3M, 6M, 12M, 15M, 30M and 60M. For metric brickwork, the base unit is 3M or 300 mm. Thus four courses of 65 mm brickwork with joints give a vertical height of 300 mm, and four stretchers with joints coordinate to 900 mm.

Table 1.1 illustrates the two types of dimensional tolerance limits set for clay masonry units including the metric brick, which relate to the square root of the work size dimension. Measurements are based on a random sample of ten bricks. The calculation based on the use of the square root of work size ensures that the dimensional tolerance limits are appropriate for the wide range in size of clay masonry units used within the European Union (BS EN 771–1: 2011).

Tolerances

Mean value

Tolerance limits are set for the difference between the stated work size (e.g. 215, 102.5 and 65 mm) and the measured mean from the samples, for each of the three brick dimensions (length, width and height). These are categorised as T1, T2 and Tm where Tm is a tolerance quoted by the manufacturer.

T1   ± 0.40 √(work size dimension) mm
     or 3 mm if greater
T2   ± 0.25 √(work size dimension) mm
     or 2 mm if greater
Tm   deviation in mm declared by the manufacturer

Range

The maximum range of size for any dimension is designated by categories R1, R2 and Rm.

R1   0.6 √(work size dimension) mm
R2   0.3 √(work size dimension) mm
Rm   range in mm declared by the manufacturer

There is no direct correlation between the limits on mean value (T) and those for the range (R); thus, a brick conforming to category T2 may be within the wider range R1. Category R2 bricks may only be required for very tight dimensional control, as in short runs of brickwork.

Alternative sizes

The metric standard evolved from the slightly larger Imperial sizes, which varied significantly, but typically were 9 × 4⅜ × 2⅞ in (228 × 110 × 73 mm) or 8⅝ × 4⅛ × 2⅝ in (219 × 105 × 68 mm). Some manufacturers offer a range of bricks to full Imperial dimensions, or alternatively to an appropriate height (e.g. 50, 68, 70, 73, 75 or 80 mm) for bonding in to Imperial brickwork for restoration and conservation work.

The 1970s also saw the introduction of metric modular bricks with coordination sizes of either 200 or 300 mm in length, 100 mm wide and either 75 or 100 mm in height. These bricks have now been replaced by a range of linear bricks up to 490 mm in length, which give the architect opportunities for increasing the horizontal emphasis within traditional brickwork (Fig. 1.3). Sizes include 327 × 50 or 65 mm, 440 × 50 or 65 mm and 490 × 50 or 65 mm with standard widths of 102 mm. A range of colours and textures is available for normal, quarter or third bonding, or alternatively stack bonding in non-load-bearing situations.

Fig. 1.3 Linear bricks. *Photographs:* Courtesy of Ibstock Brick Ltd

MANUFACTURE OF CLAY BRICKS

There are five main processes in the manufacture of clay bricks:

- extraction of the raw material;
- forming processes;
- drying;
- firing;
- packaging and distribution.

Extraction of the raw material

The process begins with the extraction of the raw material from the quarry and its transportation to the works, by conveyor belt or road transport. Topsoil and unsuitable overburden is removed first and used for site reclamation after the usable clay is removed.

The raw material is screened to remove any rocks, then ground into fine powder by a series of crushers and rollers with further screening to remove any oversize particles. Small quantities of pigments or other clays may be blended in at this stage to produce various colour effects; for example, manganese dioxide will produce an almost black brick and fireclay gives a teak brown effect. Occasionally, coke breeze is added into the clay as a source of fuel for the firing process. Finally, depending on the subsequent brick-forming process, up to 25% water may be added to give the required plasticity.

Forming processes

Handmade bricks

The handmade process involves the throwing of a suitably sized clot of wet clay into a wooden mould on a bench. The surplus clay is struck off with a framed wire and the green brick removed. The bricks produced are irregular in shape with soft arrises and interestingly folded surfaces. Two variations of the process are pallet moulding and slop moulding.

In pallet moulding, a stock board, the size of the bed face of the brick, is fixed to the bench. The mould fits loosely over the stock board, and is adjusted in height to give appropriate thickness to the green brick. The mould and board are sanded to ease removal of the green brick, which is produced with a *frog* or depression on one face. In the case of slop moulding, the stock mould is placed directly on the bench, and is usually wetted rather than sanded to allow removal of the green brick which, unlike the pallet moulded brick, is smooth on both bed faces (Fig. 1.4).

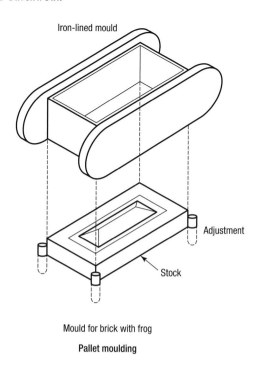

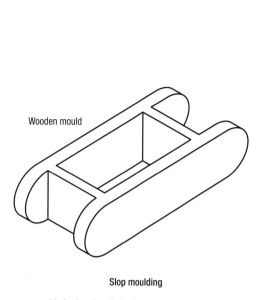

Fig. 1.4 Moulds for handmade bricks

Soft mud process

The handmade process has now been largely auto-mated, with the clay being mechanically thrown into pre-sanded moulds; the excess clay is then removed and the bricks released from the mould. These *soft mud* process bricks retain much of the individuality associated with true handmade bricks, but at a lower cost.

Pressed bricks

In the *semi-dry* process used for *Fletton* bricks the appropriate quantity of clay is subjected to a sequence of four pressings within steel moulds to produce the green brick. These bricks usually have a deep frog on one bed face. For facing bricks, texturing on both headers and one stretcher may be applied by a series of rollers. A water spray to moisten the surface, followed by a blast of a sand/pigment mixture, produces the sand-faced finish.

With clays that require a slightly higher water content for moulding, the *stiff plastic* process is used in which brick-size clots of clay are forced into the moulds. A single press is then required to form the brick. Engineering bricks made by this process often have shallow frogs on both bed faces. In all cases the size of the mould is calculated to allow for the anticipated drying and firing shrinkage.

Extruded wire-cut bricks

In this process clay with a water content of up to 25% is fed into a screw extruder which consolidates the clay and extracts the air. The clay is forced through a die and forms a continuous column with dimensions equal to the length and width of a green brick (Fig. 1.5). The surface may then be textured or sanded, before the clay column is cut into brick units by a series of wires. The bed faces of wire-cut bricks often show the drag marks where the wires have cut through the extruded clay. Perforated wire-cut bricks are produced by the incorporation of rods or tines between the screw extruder and the die. The perforations save clay and allow for a more uniform drying and firing of the bricks without significant loss of strength. Thermal perform-ance is not significantly improved by the incorporation of voids.

Drying

To prevent cracking and distortion during the firing process, green bricks produced from wet clays must be allowed to dry out and shrink. Shrinkage is typically 10% on each dimension depending upon the moisture content. The green bricks, laid in an open chequerwork pattern to ensure a uniform loss of moisture, are stacked in, or passed through, drying chambers which

Fig. 1.5 Extruding wire cut bricks

are warmed with the waste heat from the firing process. Drying temperatures and humidity levels are carefully controlled to ensure shrinkage without distortion.

Firing

Both intermittent and continuous kilns are used for firing bricks. The former is a batch process in which the single kiln is loaded, fired, cooled and unloaded. In continuous kilns, the firing process is always active; either the green bricks are moved through a fixed firing zone, or the fire is gradually moved around a series of interconnecting chambers to the unfired bricks. Both continuous systems are more energy efficient than the intermittent processes. Generally, for large-scale production, the continuous tunnel kiln (Fig. 1.6) and the Hoffman kiln (Fig. 1.7) are used.

Clamps and intermittent gas-fired kilns are used for the more specialised products. Depending on the composition of the clay and the nature of the desired product, firing temperatures are set to sinter or vitrify the clay. Colour variations called *kiss-marks* occur where bricks were in contact with each other within the kiln and are particularly noticeable on *Flettons*.

Tunnel kiln
In the tunnel kiln process the bricks are loaded 10 to 14 high on kiln cars which are moved progressively through the preheating, firing and cooling zones. A carefully controlled temperature profile within the kiln and an appropriate kiln car speed ensure that the green bricks are correctly fired with the minimum use of fuel, usually natural gas. The maximum firing temperature within the range 940–1200°C depends on the clay, but is normally around 1050°C, with an average kiln time of three days. The oxygen content within the atmosphere of the kiln will affect the colour of the brick products. Typically a high temperature and low oxygen content are used in the manufacture of blue bricks. A higher oxygen content will turn any iron oxide within the clay red.

Hoffman kiln
Introduced in 1858, the Hoffman kiln is a continuous kiln in which the fire is transferred around a series of chambers which can be interconnected by the opening of dampers. There may be 12, 16 or 24 chambers, although 16 is usual. The chambers are filled with typically 100,000 green bricks. The chambers in front of the fire, as it moves around, are preheated, then firing takes place (960–1000°C), followed by cooling, unloading and resetting of the next load. The sequence moves on one chamber per day, with three days of burning. The usual fuel is natural gas, although low-grade coal and landfill methane are used by some

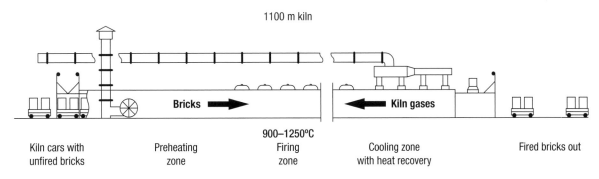

Fig. 1.6 Tunnel kiln

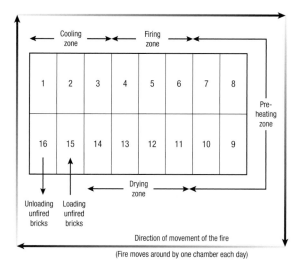

Fig. 1.7 Hoffman kiln plan

manufacturers. Hoffmann kilns are being phased out of production in the UK.

Intermittent gas-fired kilns
Intermittent gas-fired kilns are frequently used for firing smaller loads, particularly *specials*. In one system, green bricks are stacked onto a concrete base and a mobile kiln is lowered over the bricks for the firing process. The firing conditions can be accurately controlled to match those within continuous kilns.

Clamps
The basis of clamp firing is the inclusion of coke breeze into the clay, which then acts as the major source of energy during the firing process. In the traditional process, alternate layers of unfired bricks and additional coke breeze or wood are stacked up and then sealed over with waste bricks and clay. The clamp is then ignited with kindling material and allowed to burn for two to five weeks. After firing, the bricks are hand selected because of their variability from under- to over-fired. Currently some handmade bricks are manufactured in gas-fired clamps which give a fully controlled firing process but still produce bricks with the characteristic dark patches on their surfaces due to the burnt breeze content.

Packaging and distribution

Damaged or cracked bricks are removed prior to packing. Most bricks are now banded and shrink-wrapped into packs of between 300 and 500, for easy transportation by fork-lift truck and specialist road vehicles. Special shapes are frequently shrink-wrapped onto wooden pallets.

SPECIFICATION OF CLAY BRICKS

To specify a particular brick it is necessary to define certain key criteria which relate to form, durability and appearance. The European Standard BS EN 771–1: 2011 requires an extensive minimum description for masonry units, including the European Standard number and date (e.g. BS EN 771–1: 2011), the type of unit (e.g. high density – HD), dimensions and tolerances from mean value, configuration (e.g. a solid or frogged brick), compressive strength and freeze/thaw resistance. As well, depending on the particular end use, additional description may be required. This may, as appropriate, include dry density, dimensional tolerance range, water absorption, thermal properties, active soluble salt content, moisture movement, reaction to fire and vapour permeability.

Within the building industry the classification usually also includes some traditional descriptions:

- place of origin and particular name (e.g. Staffordshire smooth blue);
- clay composition (e.g. Gault, Weald or Lower Oxford Clay, Etruria Marl, Keuper Marl [Mercian Mudstones] or shale);
- variety – typical use (e.g. Class A engineering, common or facing);
- type – form and manufacturing process (e.g. solid, frogged, wire-cut);
- appearance – colour and surface texture (e.g. coral red rustic).

Variety

Bricks may be described as common, facing or engineering.

Common bricks
Common bricks have no visual finish, and are therefore usually used for general building work especially where the brickwork is to be rendered, plastered or will be unseen in the finished work.

Facing bricks
Facing bricks are manufactured and selected to give an attractive finish. The particular colour, which may be uniform or multicoloured, results from the blend

Table 1.2 Properties of traditional UK engineering and DPC bricks (NA to BS EN 771–1: 2011)

Performance characteristic	Clay engineering bricks	
	Class A	Class B
Minimum compressive strength (MPa)	125	75
Maximum water absorption (% by mass)	4.5	7.0
(also when used as DPC units)	(and DPC1)	(and DPC2)
Freeze/thaw resistance category	F2	F2
Active soluble salts content category	S2	S2

of clay used and the firing conditions. In addition, the surface may be smooth, textured or sand-faced as required. A slightly distressed appearance, similar to that associated with reclaimed bricks, is obtained by tumbling either unfired or fired bricks within a rotating drum. Facing bricks are used for most visual brickwork where a pleasing and durable finish is required.

Engineering bricks

Engineering bricks are dense and vitreous, with specific load-bearing characteristics and low water absorption. The National Annex NA (informative) to BS EN 771–1: 2011 relates the properties of the two classes (A and B) of clay engineering bricks to their minimum compressive strengths, maximum percentage water absorption, freeze/thaw resistance and soluble salt content (Table 1.2). Engineering bricks are used to support heavy loads, and also in positions where the effects of impact damage, water absorption or chemical attack

need to be minimised. They are generally *reds* or *blues* and more expensive than other machine-made facing bricks because of their higher firing temperature.

Type

Type refers to the form of the brick and defines whether it is solid, frogged, cellular, perforated or of a special shape (Figs 1.8 and 1.9). Bricks may be frogged on one or both bed faces; perforations may be few and large or many and small. Cellular bricks have cavities closed at one end. Keyed bricks are used to give a good bond to plaster or cement rendering. Because of the wide range of variation within brick types, the manufacturer is required to give details of the orientation and percentage of perforations in all cases.

For maximum strength, weather resistance and sound insulation, bricks should be laid with the frogs uppermost so that they are completely filled with mortar; with double-frogged bricks the deeper frog should be uppermost. However, for cheapness, speed and possibly minimisation of the dead weight of construction, frogged bricks are frequently laid frog down. Inevitably this leads to a resultant reduction in their load-bearing capacity.

Standard specials

Increasingly, *specials* (special shapes) are being used to enhance the architectural quality of brickwork. British Standard BS 4729: 2005 illustrates the range of standard specials, which can normally be made to order to match standard bricks (Fig. 1.9).

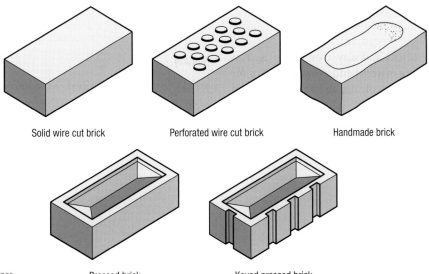

Fig. 1.8 Brick types

Solid wire cut brick · Perforated wire cut brick · Handmade brick · Pressed brick · Keyed pressed brick

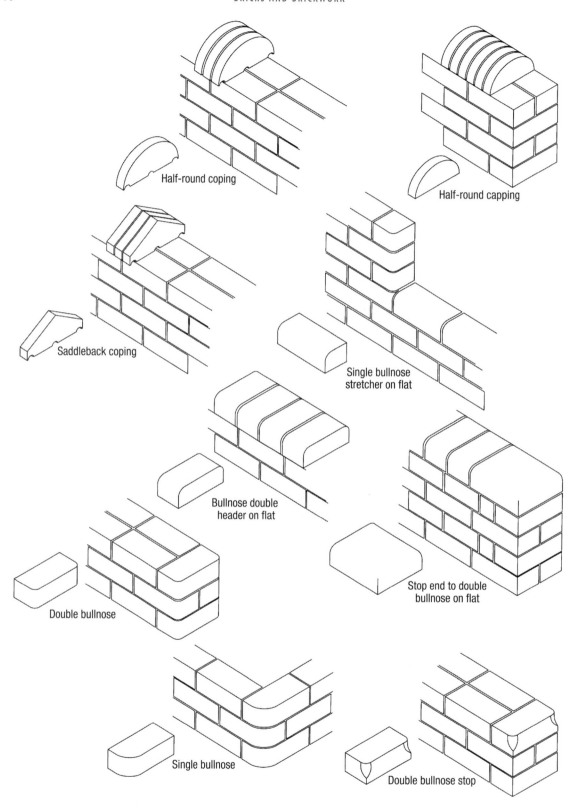

Half-round coping

Half-round capping

Saddleback coping

Single bullnose
stretcher on flat

Bullnose double
header on flat

Stop end to double
bullnose on flat

Double bullnose

Single bullnose

Double bullnose stop

Fig. 1.9 Specials. *Photograph*: Courtesy of Ibstock Brick Ltd

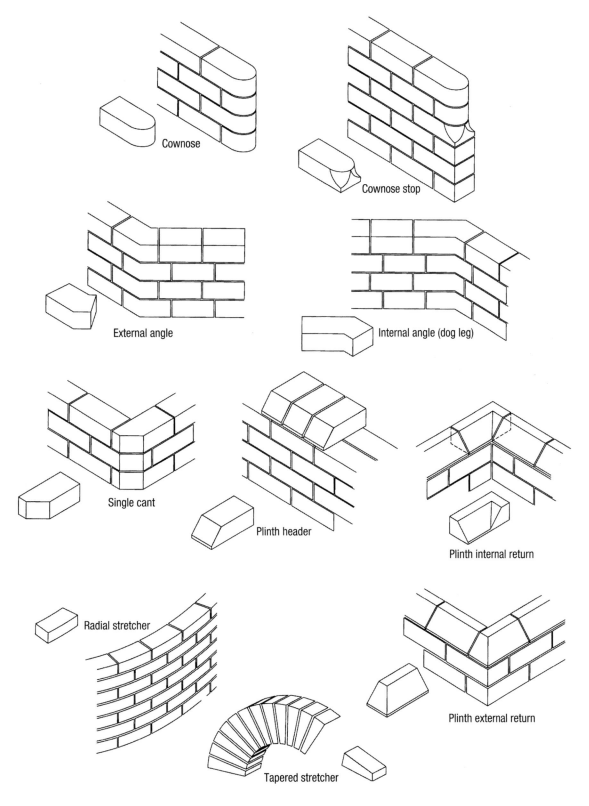

Cownose

Cownose stop

External angle

Internal angle (dog leg)

Single cant

Plinth header

Plinth internal return

Radial stretcher

Tapered stretcher

Plinth external return

Fig. 1.9 *(continued)*

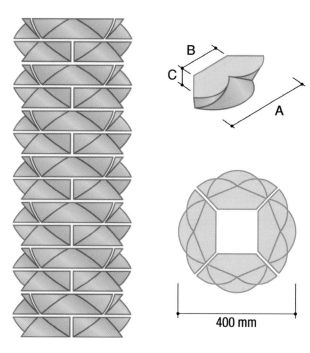

Fig. 1.9 *(continued)*

Designation of standard specials:

Angle and cant bricks
Bullnose bricks
Copings and cappings
Plinth bricks
Arch bricks
Radial bricks
Soldier bricks
Cuboid bricks
Bonding bricks
Brick slips

Manufacturers also frequently make purpose-made specials (*special specials*) to the particular requirements of the architect or builder. Inevitably, delivery on specials takes longer than for ordinary bricks, and their separate firing frequently leads to some colour variation between the specials and the standard bricks, even where the clay used is identical. The more complex specials are handmade, usually in specially shaped stock moulds, although some can be made by modifying standard bricks before firing. The range of shapes includes copings and cappings (for parapets and freestanding walls), bullnose (for corner details, e.g. window and door reveals), plinths (for corbelling details and cills), cants (for turning angles), arches and brick slips (to mask reinforced concrete lintels, etc.).

Special bricks are also manufactured by cutting standard bricks, then, if necessary, bonding the pieces with epoxy resins. This has the advantage of ensuring an exact colour match to the standard bricks. Many brick slips, dog leg and arch voussoir sets (bricks to create an arch) are produced by this method. Specialist brick-cutting companies also supply bricks for conservation projects as in the refurbishment of the elegant façade of St Pancras International railway station in London. Some manufacturers offer bat and bird boxes as special brick eco-habitats for building into standard masonry.

APPEARANCE

The colour range of bricks manufactured in the UK is extensive. The colours range from light buffs, greys and yellows through pastel pink to strong reds, blues, browns and deep blue/black, depending mainly on the clay and the firing conditions, but also on the addition of pigments to the clay or the application of a sand facing. Colours may be uniform, varied over the surface of individual bricks or varied from brick to brick. The brick forms vary from precise to those with rounded arrises; textures range from smooth and sanded to textured and deeply folded, depending on the forming process (Fig. 1.10).

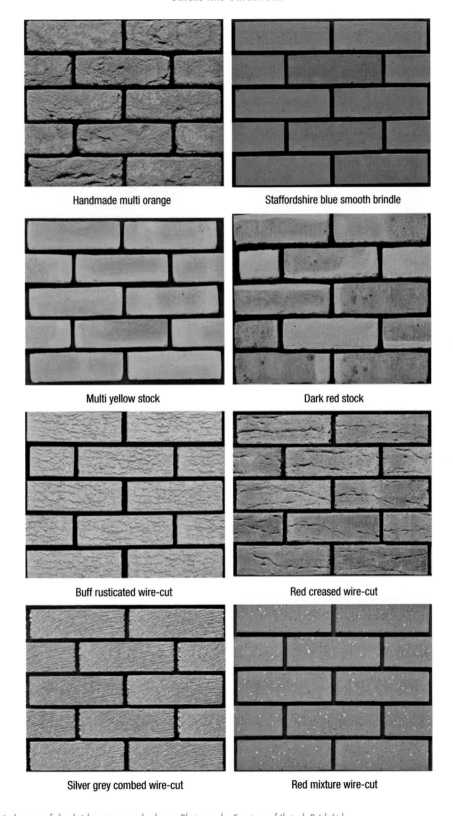

Handmade multi orange   Staffordshire blue smooth brindle

Multi yellow stock   Dark red stock

Buff rusticated wire-cut   Red creased wire-cut

Silver grey combed wire-cut   Red mixture wire-cut

Fig. 1.10 Typical range of clay brick textures and colours. *Photographs*: Courtesy of Ibstock Brick Ltd

Fig. 1.11 Handmade Hoskins Bromley Red bricks.
Photograph: Arthur Lyons

In view of the variability of bricks from batch to batch it is essential that they should be well mixed, preferably at the factory before palleting, or, failing this, on site. If this is not done sufficiently well, accidental colour banding will appear as the brickwork proceeds. Sand-faced bricks are liable to surface damage on handling, which exposes the underlying colour of the brick. Chipping of the arrises on bricks with *through colour* is visually less detrimental. Where rainwater run-off is an important factor (e.g. on cills and copings), smooth rather than heavily rusticated bricks should be used, as the latter would saturate and stain. Handmade bricks with deep surface folds (Fig. 1.11) should be laid frog up so that the creases or *smiles* tend to shed the rainwater from the face of the brickwork.

Glazed bricks, available in a wide range of intense colours, are sometimes used for their strong aesthetic effect (Fig. 1.12) or resistance to graffiti. They are either manufactured in a two-stage process, which involves the initial firing of the green brick to the *biscuit* stage, followed by the application of a *slip* glaze and a second firing. In the alternative one-stage process, the glaze is applied before a single firing. Manufacturers offer a standard range of colours or the option to specify from the RAL colour range. Bricks may be fully or partially glazed and special shapes are normally available to order.

The visual acceptability of facing bricks and the quality of the bricklaying would normally be assessed on site by the construction of a reference panel of at least 1 m² to the standard PAS 70: 2003, using randomly selected bricks with examples of any colour banding,

the proposed bonding, mortar and jointing. All subsequent brick deliveries and constructed brickwork should then be checked against the reference panel.

DURABILITY

Frost resistance

High-density (HD) clay masonry units including bricks are classified into one of the three categories, F2, F1 and F0 according to their frost resistance within a standardised freezing test (Table 1.3). Only category F2 bricks are totally resistant to repeated freezing and thawing when in a saturated condition. Category F1 bricks are durable, except when subjected to repeated freezing and thawing under saturated conditions. Therefore, category F1 bricks should not be used in highly exposed situations such as below damp-proof courses, for parapets or brick-on-edge copings, but they are suitable for external walls which are protected from saturation by appropriate detailing. Category F0 bricks must only be used where they are subject to passive exposure, as when protected by cladding or used internally.

Soluble salt content

The soluble salt content of high-density (HD) clay masonry units including bricks is defined by three categories: low (S2), normal (S1) and no limits (S0) (Table 1.3). Both the S2 and S1 categories have defined maximum limits for sodium/potassium and magnesium salt contents. The soluble salts derive from the original clay or from the products of combustion during the firing process. Soluble salts can cause efflorescence and soluble sulphates may migrate from the bricks into the mortar or any rendering, causing it

Table 1.3 Designation of freeze/thaw resistance and active soluble salts content for clay bricks (HD Units to BS EN 771: 2011)

Durability designation	Freeze/thaw resistance
F2	masonry subjected to severe exposure
F1	masonry subjected to moderate exposure
F0	masonry subjected to passive exposure
	Active soluble salts content
S2	sodium/potassium 0.06%, magnesium 0.03%
S1	sodium/potassium 0.17%, magnesium 0.08%
S0	no requirement

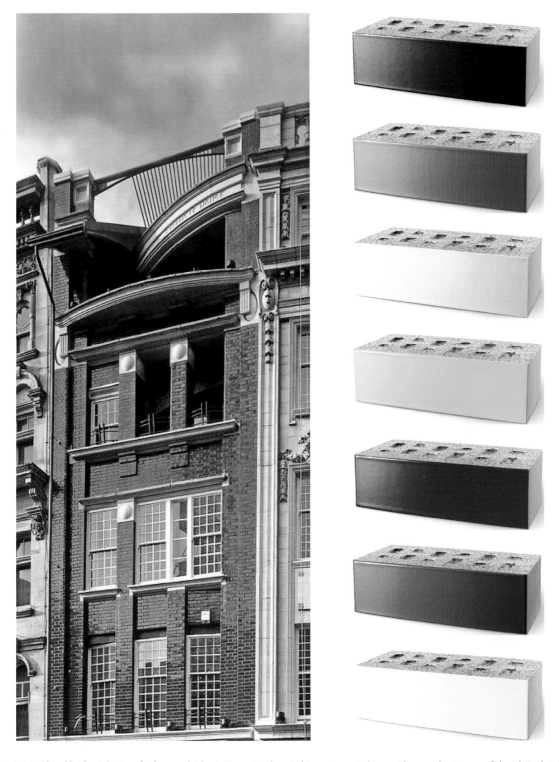

Fig. 1.12 Glazed bricks. Selection of colours and Atlantic House, London. *Architects*: Proun Architects. *Photographs*: Courtesy of Ibstock Brick Ltd

to expand and deteriorate by sulphate attack. If used in an exposed situation, S1 and S0 category bricks should be bonded with sulphate-resisting cement mortar.

Efflorescence

Efflorescence sometimes appears as a white deposit on the surface of new brickwork (Fig. 1.13). It is caused by moisture carrying salts from inside the bricks and mortar to the surface where the water evaporates leaving the crystalline salts. Under most conditions it disappears without deleterious effect within one year. In exposed brickwork that is constantly subjected to a cycle of wetting and drying, efflorescence may occur at any time; further, a buildup and expansion of crystalline salts under the surface (*crypto-efflorescence*) may cause the face of the brickwork to crumble or spall.

Fig. 1.13 Efflorescence

Staining

The surface of brickwork may be stained by cement during the building process, or by lime leaching out of the fresh mortar (Fig. 1.14). In either case the excess should be brushed and washed off, without saturating the brickwork.

Fig. 1.14 Lime leaching on brickwork

PHYSICAL PROPERTIES

Compressive strength

High density (HD) clay bricks are available with a range of compressive strengths from around 5 MPa to well over 100 MPa. The criteria for general use, damp-proof courses and engineering use are set out in Table 1.2 (page 9).

To determine the crushing strength of bricks, both bed faces are ground down until flat and parallel. The bricks are then crushed dry without filling the voids or frogs. Where frogs are to be laid upwards and filled in the construction, the crushing strength (MPa) is based on the net bearing area. Where frogs are to be laid downwards, the manufacturers give guidance on the compressive strength to be used. For solid and perforated bricks the compressive strength is based on the gross area of the brick bed face.

Water absorption and suction

The level of water absorption is critical when bricks are to be used for damp-proof courses, or as engineering bricks. The standard BS EN 771–1: 2011 requires the water absorption value of HD-type masonry units to be used as a DPC to be stated but no threshold is specified as this is left to national decision. For the UK, appropriate limits are shown in Table 1.2, although generally absorption ranges from 1% to 35%. Suction rates are now quoted by most brick manufacturers, as high values can adversely affect the bricklaying process. Bricks with high suction rates absorb water rapidly from the mortar, making it insufficiently plastic to allow for repositioning of the bricks as the work proceeds. Generally low- or medium-suction rates (1.0–2.0 kg/m² per min) are advantageous. In warm weather, high-suction-rate bricks may be wetted in clean water before laying, but any excess water will cause the brick to float on the mortar bed and also increase the risk of subsequent efflorescence and staining.

Moisture and thermal movement

After the firing process, bricks absorb moisture from the atmosphere and expand irreversibly, up to a maximum of 0.1%. It is therefore recommended that bricks should not be used for at least two weeks after firing (although it is now recognised that this irreversible process may continue at a decreasing rate for 20 years). Subsequent moisture and thermal movements are largely reversible, and movement joints allowing for a 1 mm movement per 1 m of brickwork should be allowed, typically at 10–12 m centres and at a maximum of 15 m, in restrained walls. In addition, allowing for the compressibility of the filler, movement joints at 12 m centres should be about 16 mm wide (PD 6697: 2010). Unrestrained or lightly restrained walls should have movement joints at 7–8 m centres. Horizontal movement joints should be at approximately 12 m intervals, as the vertical movement is of the same order as movement in the horizontal direction. In external unreinforced non-load-bearing walls a maximum spacing of 15 m between vertical movement joints is permitted (NA to BS EN 1996–2: 2006).

For many buildings the necessary movement joints can be made inconspicuous by careful detailing or featured as part of the design. Appropriate locations for movement joints would be where differing structural forms adjoin, such as abutments between walls and columns or where the height or thickness of a wall changes; alternatively, at design details such as brickwork returns, re-entrant corners or the recesses for downpipes. In expansion joints, fillers such as cellular polythene, polyurethane or foam rubber should be used, as these are easily compressible. Pointing should be with a flexible sealing compound such as two-part polysulphide.

Typical reversible moisture movement = 0.02%
Typical reversible thermal movement = 0.03%
Thermal movement = 5–8×10^{-6} deg C^{-1}

Thermal conductivity

The thermal conductivity of brickwork is dependent upon its density and moisture content but generally clay bricks are poor thermal insulators, so brick density has only a moderate effect on the overall U-value of a composite wall. Brick manufacturers quote thermal conductivities at a standard 5% moisture content for exposed brickwork, and may also give the 1% moisture content figure for protected brickwork.

Using 2000 g/m³ bricks with an average thermal conductivity of 0.96 W/m K, a typical partial fill cavity system is:

102.5 mm fairfaced brickwork ($\lambda = 0.96$ W/m K)
5 mm clear cavity
100 mm insulation ($\lambda = 0.022$ W/m K)
100 mm lightweight blockwork ($\lambda = 0.15$ W/m K)
13 mm plasterboard ($\lambda = 0.16$ W/m K)

Giving a U-value of 0.18 W/m² K

(A more accurate determination of U-value, with allowances for the effects of wall ties and mortar joints, can be calculated according to BRE BR 443: 2006 using worked examples in BRE FB 42: 2012.)

The thermal conductivity of clay bricks at 5% moisture content typically ranges from 0.60 to 1.95 W/m K.

Fire resistance

Clay brickwork generally offers excellent fire resistance by retaining its stability, integrity and insulating properties. Bricks with less than 1% organic material are automatically categorised as Euroclass A1 with respect to reaction to fire.

Acoustic properties

Good-quality brickwork is an effective barrier to airborne sound, provided that there are no voids through the mortar for the passage of sound. All masonry joints should be sealed and bricks laid with filled frogs to achieve the necessary mass per unit area and avoid air pathways.

At the junction between a cavity blockwork separating wall and an external brick and blockwork wall, if the external cavity is not fully filled with thermal insulation, then the separating wall cavity must be closed with a flexible cavity stop to reduce sound transmission sufficiently to comply with the Building Regulations Part E performance requirements.

Impact sound absorption by brickwork over the normal frequency range is fairly low and further decreased by the application of dense plaster or paint. However, the application of acoustic plasters or the addition of an independent panel of plasterboard backed by absorbent material improves impact sound insulation.

QUALITY CONTROL

To meet the consistent standards of quality required by clients, many brick manufacturers are now operating quality assurance systems (pr EN ISO 9001: 2014 and BS EN ISO 9004: 2009). These require manufacturers to document all their operational procedures and set out standards to which products must adhere. Quality is controlled by a combination of an internal self-monitoring system and two to four independent spot-check reviews per year. Both the content of the technical literature and the products themselves are subjected to this scrutiny.

In addition, BRE document BES 6001: 2009 together with the British Standard BS EN ISO 14001: 2004 set the criteria for the responsible sourcing of construction materials which are followed by major brick manufacturers.

Unfired clay bricks

Unfired clay bricks for internal non-load-bearing applications are produced from clays which would be less suitable for standard fired bricks. Earth bricks require only low energy input for drying and have high potential recyclability. They have the advantage of inhibiting condensation and regulating the relative humidity of the internal environment. Traditionally, unfired bricks are laid with clay or moderately hydraulic lime mortar. However, for thinner walls (105 mm) sodium silicate or lignosulphonate-stabilised clay-based mortars provide higher strength bonds. Unfired brick or block walls may be left exposed internally or finished with vapour-permeable renders or plasters such as clay or lime. Suitable finishes are lime wash or vapour-permeable paint. Careful detailing is required for any exposed areas of unfired clay masonry. Offcuts from construction can be soaked in water to produce clay mortar or be discarded to the landscape. Products include vertically perforated bricks with keyed or smooth finishes to sizes $220 \times 105 \times 67$ mm and $220 \times 105 \times 133$ mm (Fig. 1.15). Compressive strengths are normally in the range 2.0–4.0 MPa for dry densities between 1700 and 2200 kg/m^3.

The thermal conductivity of unfired clay bricks ranges from 0.5 to 1.0 W/m K depending upon the degree of compression in manufacture.

Reclaimed clay bricks

Reclaimed bricks are often selected for aesthetic reasons, but their appearance is not a guarantee of durability. In particular, frost resistance is uncertain and it is not valid to test a sample to classify the whole consignment. Imperial sizes may vary considerably and some material may be contaminated with sulphates, or be liable to efflorescence from absorbed soluble salts. However, the strength and water absorption properties of reclaimed bricks are usually appropriate for domestic-scale construction. Reclaimed paving bricks that have previously been exposed to frost will normally be durable, but walling bricks may not be durable when used as pavers.

Fig. 1.15 Unfired clay brick and block. *Photographs*: Courtesy of Ibstock Brick Ltd

Recently some buildings have been constructed using lime mortar, specifically with a view to the potential recycling of the metric bricks at the end of the useful life of the construction. Lime mortar is significantly easier to clean from bricks than modern Portland cement mortar.

Of the estimated 2500 million bricks demolished each year in the UK only about 5% are reclaimed and nearly half are crushed and used as hardcore fill.

Brickwork

CLAY BRICKWORK

The bonding, mortar colour and joint profile have a significant visual effect upon brickwork. The overall effect can be to emphasise as a feature or reduce to a minimum the impact of the bonding mortar on the bricks. In addition, the use of polychromatic brickwork with complementary or contrasting colours for quoins, reveals, banding and even graphic designs can have a dramatic effect upon the appearance of a building. The three-dimensional effects of decorative dentil courses and projecting corbelled features offer the designer further opportunities to exploit the effects of light and shade. Normally, a projection of 10–15 mm is sufficient for the visual effect without causing increased susceptibility to staining or frost damage. Curved brickwork constructed in stretcher bond shows faceting and the overhang effect, which is particularly accentuated in oblique light. With small-radii curvatures, the necessary change of bonding pattern to header bond can also be a visual feature, as an alternative to the use of curved-radius bricks. In all brickwork the vertical joints (perpends) should be plumb.

The *Gothic Revival* exterior of the Queens Building, De Montfort University, Leicester (Fig. 1.16) illustrates the visual effects of polychromatic brickwork and voussoir specials. The energy-efficient building maximises the use of natural lighting, heating and ventilation, using massive masonry walls to reduce peak temperatures. The mortar, which matches the external coral-red brickwork, reduces the visual impact of the individual bricks, giving the effect of planes rather than walls. This is relieved by the colour and shadow effects of the polychromatic and corbelled features, which are incorporated into the ventilation grilles and towers. The special bricks, cill details and banding are picked out in a deeper cadmium red and silver buff to contrast with the characteristic Leicestershire red-brick colouring.

Mortars

The mortar in brickwork is required to give a bearing for the bricks and to act as a sealant between them. Mortars should be weaker than the individual bricks, to ensure that any subsequent movement does not cause visible cracking of the bricks, although too weak a mix would adversely affect durability of the brickwork. Mortar mixes are based on blends of either cement/lime/sand, masonry cement/sand or cement/sand with plasticiser. Suitable alternatives to Portland cement are listed in Table 1.4 to PD 6678: 2005. When the mix is gauged by volume an allowance has to be made for bulking of damp sand.

Fig. 1.16 Decorative brickwork – Queens Building, De Montfort University, Leicester. *Architects*: Short Ford & Associates. *Photograph*: Arthur Lyons

For *designed* masonry mortars to BS EN 998–2: 2010, the composition and manufacturing methods are chosen by the producer to achieve the specified properties (e.g. compressive strength category M5,

Table 1.4 Appropriate cements for mortars (PD 6678: 2005)

Cement	Standard	Cement notation
Portland cement	BS EN 197–1	CEM I
Sulphate resisting Portland cement	BS EN 197–1	CEM I–SR
Portland slag cement	BS EN 197–1	CEM II/A-S or II/B-S
Portland fly ash cement	BS EN 197–1	CEM II/A-V or II/B-V
Portland limestone cement	BS EN 197–1	CEM II/A –LL(L)
Masonry cement (inorganic filler other than lime)	BS EN 413–1	Class MC
Masonry cement (lime)	BS EN 413–1	Class MC

Table 1.5). *Prescribed* masonry mortars are manufactured from a predetermined combination of constituents (e.g. cement 15%, lime 10%, aggregates 75% – by volume). *General purpose* masonry mortar (G) is without special characteristics; *thin layer* masonry mortar (T) has a prescribed maximum aggregate size and *lightweight* masonry mortar (L) has a dry hardened density equal to or less than 1300 kg/m^3.

Table 1.5 Mortar classes

Class	M1	M2.5	M5	M10	M15	M20	Md
Compressive Strength MPa	1	2.5	5	10	15	20	d

Note: d is a compressive strength greater than 20 MPa as a multiple of 5 declared by the manufacturer.

In the repointing of old brickwork it is particularly important to match the porosity of the brick to the water-retention characteristics of the mortar. This prevents excessive loss of water from the mortar before hydration occurs, which may then cause the pointing to crumble.

The use of lime mortar, as in the Building Research Establishment environmental building in Garston, Watford, will allow for the ultimate reuse of the bricks at the end of the building's life cycle. The recycling of bricks is not possible, except as rubble, when strong Portland cement mortar is used.

Sands for mortars are normally graded to BS EN 13139: 2013 into categories designated by a pair of sieve sizes d/D which define the lower and upper size limits in mm, respectively. The majority of the particle size distribution should lie between the stated limits. The preferred grades are 0/1 mm, 0/2 mm, 0/4 mm, 0/8 mm, 2/4 mm and 2/8 mm. Typically between 85 and 99% of the sand should pass through the larger sieve limit, and between 0 and 20% should pass through the smaller sieve size limit. The grades with more fines (63 micron or less) require more cement to achieve the same strength and durability as the equivalent mortars mixed with a lower fines content.

Ideally, brickwork should be designed to ensure the minimal cutting of bricks and built with a uniform joint width and vertical alignment of the joints (perpends). During construction, brickwork should be kept clean and protected from rain and frost. This reduces the risk of frost damage, patchiness and efflorescence. Brickwork may be rendered externally or plastered internally if sufficient mechanical key is provided by appropriate jointing or the use of keyed bricks. For repointing existing brickwork, it is necessary to match carefully the mortar sand and to use lime mortar where it was used in the original construction.

Bonding

Figure 1.17 illustrates the effects of bonding. The stretcher bond is standard for cavity walls and normally a half-lap bond is used, but an increase in horizontal

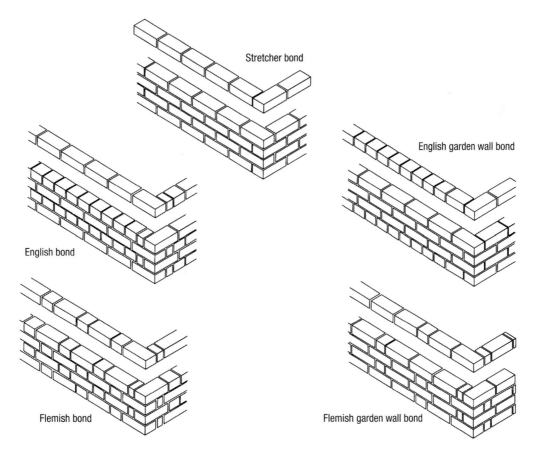

Fig. 1.17 Brick bonding

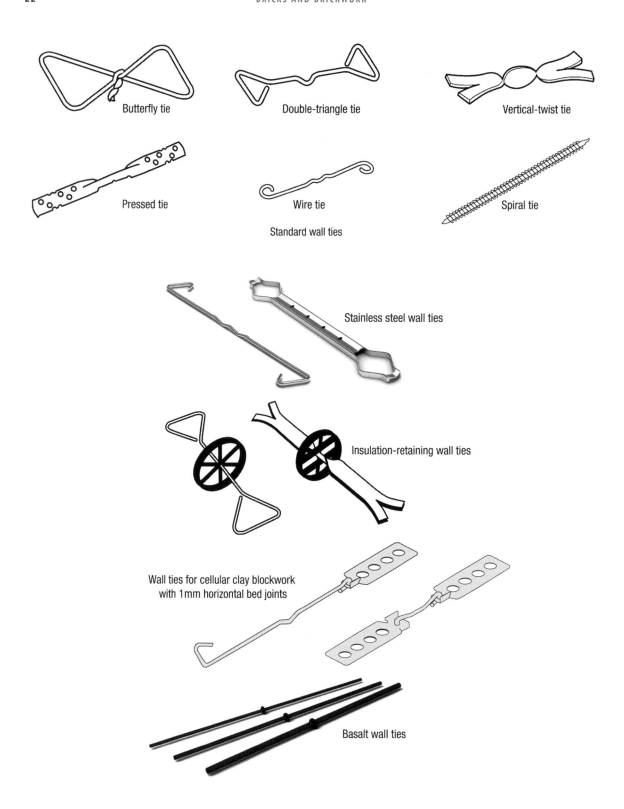

Butterfly tie

Double-triangle tie

Vertical-twist tie

Pressed tie

Wire tie

Spiral tie

Standard wall ties

Stainless steel wall ties

Insulation-retaining wall ties

Wall ties for cellular clay blockwork
with 1mm horizontal bed joints

Basalt wall ties

Fig. 1.18 Wall ties – Standard, stainless steel and basalt ties. *Photographs*: Courtesy of Ancon Building Products

Table 1.6 Wall ties to PD 6697: 2010

Types	Least leaf thickness (mm)	Nominal cavity width (mm)	Tie length (mm)
Types 1, 2, 3 or 4	75	75 or less	200
depending on	90	76 to 100	225
design loading	90	101 to 125	250
and cavity width	90	126 to 150	275
	90	151 to 175	300
	90	176 to 300	cavity width + 125 mm

Notes:

Type 1 – Masonry: Heavy duty. Suitable for most masonry construction except where large differential movements are anticipated.

Type 2 – Masonry: General purpose. Suitable for construction up to 15 m above ground level, when built of double leaves of similar thicknesses in the range 90–150 mm. Maximum basic wind speed of 31 m/sec.

Type 3 – Masonry: Basic. As Type 2 but for lower basic winds speeds of 27 m/sec maximum.

Type 4 – Masonry: Light duty. Suitable for dwellings up to 10 m in height, when built of double leaves of similar stiffness and thicknesses in the range 90–150 mm. Suitable for most internal separating cavity walls.

– All specifications are subject to verification by calculation.

– A minimum embedded length of 50 mm is required in each leaf.

– The minimum number of wall ties per unit area should be calculated according to BS EN 1996–1-1: 2005.

emphasis can be achieved by the less standard quarter or third bond. In conservation work it may be necessary to use half bricks (snap headers) to match the appearance of bonding in solid brick walls. For one-brick-thick walls more variations are possible; most typical are the English and Flemish bonds. The equivalent English and Flemish garden wall bonds, which have more stretchers, are primarily used for one-brick-thick walls where the reduced number of headers makes it easier to build both sides fairfaced. Panels of herringbone brickwork (raking bond), or dog tooth and dentil courses, as in Victorian brickwork, can generate interesting features.

Wall ties manufactured from galvanised steel, stainless steel and polypropylene are described in BS EN 845–1: 2013 (Fig. 1.18). They should be installed in cavity wall construction according to PD 6697: 2010 drip down and level, or sloping down ($\leqslant 25°$) towards the outer leaf, with embedment to a depth of 50 mm in both leaves. Wall ties are categorised as Types 1 to 4 to PD 6697: 2010 according to their strength and appropriate use, and lengths are dependent upon the cavity width (Table 1.6). Stainless steel ties are specified for all housing. An estimated minimum service life of 60 years is recommended by the publication DD 140–2: 1987. Low thermal conductivity wall ties ($\lambda = 0.7$ W/m K) are available, manufactured in pultruded composite basalt

fibre and epoxy resin, for cavities from 50 to 300 mm. (Stainless steel has a high thermal conductivity, $\lambda = 16$ W/m K.) Masonry-to-masonry ties are usually installed at a rate of 2.5 ties/m² – typically at 900 mm horizontal and 450 mm vertical centres.

Where mortar bed-joints do not coordinate between masonry leaves, slope-tolerant cavity wall ties must be used. In partially filled cavities, the wall ties should clip the insulation cavity batts to the inner leaf. In all cases the cavity, insulation and ties should be kept clear of mortar droppings and other residues by using a protective board. Asymmetric wall ties are used for fixing masonry to timber or thin-joint aircrete blockwork. Movement-tolerant wall ties bend, or slide within a slot system fixed to one leaf of the masonry. Screw-in wall ties are available for tying new masonry to existing walls.

Coloured mortars

Mortar colour has a profound effect on the overall appearance of the brickwork, as with stretcher bond and a standard 10 mm joint the mortar accounts for 17% of the brickwork surface area. A wide range of light-fast coloured mortars is available which can be used to match or contrast with the bricks, thus highlighting the bricks as units or creating a unity within the brickwork.

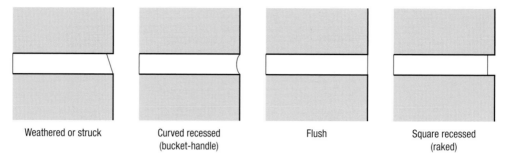

| Weathered or struck | Curved recessed (bucket-handle) | Flush | Square recessed (raked) |

Fig. 1.19 Joint profiles

The coloured mortars contain inert pigments, which are factory blended to a tight specification to ensure close colour matching between batches. Occasionally, black mortars may bloom due to lime migration to the surface. Coloured mortars can be used creatively to enhance the visual impact of the brickwork and even create designs on sections of otherwise monochromatic brickwork. The quantity of pigment should not exceed 10% by weight of the cement.

Mortar colours may also be modified by the use of stains after curing; however, such applications only penetrate 2 mm into the surface, and therefore tend to be used more for remedial work. Through-body colours are generally more durable than surface applications.

Joint profiles

The standard range of joint profiles is illustrated in Figure 1.19. It is important that the main criteria should be the shedding of water to prevent excessive saturation of the masonry, which could then deteriorate. Normally the brickwork is jointed as the construction proceeds. This is the cheapest and best method, as it gives the least

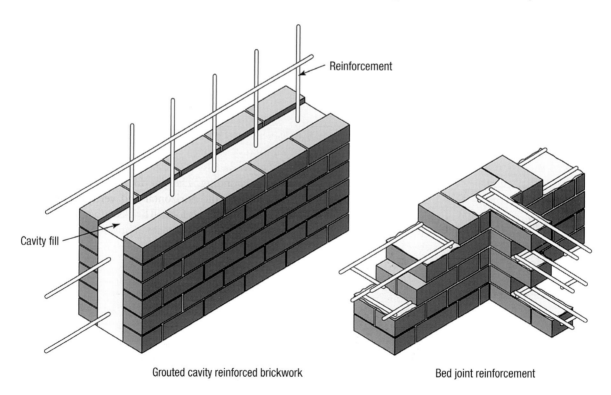

Grouted cavity reinforced brickwork   Bed joint reinforcement

Fig. 1.20 Reinforced brickwork

disturbance to the mortar bed. Pointing involves the raking out of the *green* mortar to a depth of 13–20 mm, followed by refilling the joint with fresh mortar. This is only appropriate when the desired visual effect cannot be obtained directly by jointing; for example, when a complex pattern of coloured mortar joints is required for aesthetic reasons.

The square recessed (raked) joints articulate the brickwork by featuring the joint, but these should only be used with durable (F2, S2), high-absorption bricks under sheltered conditions; furthermore, the recess should be limited to a maximum depth of 6 mm. The struck or weathered joint also accentuates the light and shade of the brickwork, while, as a tooled joint, offering good weather resistance in all grades of exposure. If the visual effect of the joint is to be diminished the flush joint may be used, but the curved recessed (bucket handle) joint, which is compressed by tooling, offers better appearance and weathering properties. No mortar should be allowed to smear the brickwork, as it is difficult to remove subsequently without the use of dilute acid or pressure jets of water.

Reinforced brickwork

Reinforcement may be introduced vertically or horizontally into brickwork (Fig. 1.20). Bed-joint reinforcement, usually austenitic stainless steel, should be completely surrounded by mortar with a minimum cover of 20 mm. For continuity in long walls, sections of reinforcement should be sufficiently end lapped. Vertical reinforcement is possible in the cavity or in pocket-type walls, where the void spaces are formed in the brickwork, then reinforcement and concrete is introduced after the masonry is completed. Care should be taken in the use of vibrators to compact the concrete within new masonry.

Decorative brickwork

Tile-bricks replicate the visual effect of hung tiles. A profiled edge on each brick overhangs the brick course below, creating the tiled effect. The bricks are laid in 10 mm mortar which is cut back 15 mm from the front face, using only a 1 mm vertical joint so that the tiles appear to butt joint. A range of colours with smooth and sand-faced finishes is available with matching specials for corner returns. The brickwork must be completed in courses ahead of the blockwork to ensure that internal vertical joints are sealed. The tile-bricks are typically used for gable-ends on housing.

Profiled bricks (Fig. 1.21), which coordinate with standard brickwork, can create features with enhanced shadow effects. Shapes include sawtooth, spheres, pyramids and wave effects in a variety of colours and finishes to create feature panels or to be interspaced within standard brickwork.

Plaques, motifs, murals and sophisticated sculptures (Fig. 1.22) can be manufactured to individual designs both for new buildings and for the renovation or refurbishment of Victorian *terracotta*. The designs are carved as a bas-relief in soft, solid through-colour brickwork or moulded in the unfired clay in relatively small units and joined on site with a matching mortar. For repetitive units, the clay is shaped in an appropriate wooden mould. Relief depths of 10–30 mm give shadow and contrast sufficient for most sculptural effects to be seen, although the viewing distances and angles must be considered. For large brickwork sculptures, the whole unit may be built in green bricks, with allowances made for the mortar joints and drying contraction. The design is then carved, numbered, dismantled, fired and reassembled on site.

Thin-bed masonry

The use of thin-bed masonry, with joints of between 2 and 6 mm, significantly reduces the visual effect of the mortar joints from 17% in 10 mm joint standard brickwork, to only 8% in 4 mm joints. This effect is further enhanced by the use of glue-mortars which are applied to create a recessed joint. Thus the joint becomes only a shade line and the visual effect of the wall is totally determined by the colour and texture of the bricks. Because the glue-mortar is stronger than traditional mortar and has tensile properties, the brickwork patterns are not constrained to standard stretcher bonding. The glue-mortar is applied in two lines to both the horizontal and vertical joints, and therefore solid or perforated bricks rather than frogged bricks are most appropriate. Thin-bed masonry wall ties and special aramid bed-joint reinforcement are used as appropriate.

A totally joint-free appearance can be produced by using traditionally laid specially shaped bricks (204 × 100 × 60 mm). The upper bed face has a 10 mm recess to be filled with standard mortar. Thin open perpends allow for some ventilation of the cavity. A uniform colour, uninterrupted by mortar joints, is produced.

Fig. 1.21 Profiled bricks. *Photographs:* Courtesy of Ibstock Brick Ltd

Preassembled brickwork

The use of preassembled brickwork supported on reinforced concrete or steel frames offers the builder a potentially higher level of quality control and increased speed of construction on site (Fig. 1.23). It also offers the scope to create complex details and forms, such as long, low arches, that would be expensive or impossible in traditional brick construction. Specialist manufacturers produce large complete brick-clad precast concrete panels either with whole bricks or brick slips. Typically the rear faces of brick slips are drilled at an angle, then stainless steel rods are inserted and fixed with resin adhesive. The brick slips are laid out with spacers within the panel mould, prior to the addition of steel reinforcement and concrete. Finally, the brick slips are pointed up, giving the appearance of normal brickwork.

Fig. 1.22 Decorative carved brickwork. *Photograph:* Courtesy of Wienerberger UK

Fig. 1.23 Preassembled brickwork. *Photograph:* Courtesy of Trent Concrete Ltd

Factory-made brickwork panels of solid wall dimensions and single-storey height, incorporating openings and ducts for windows, doors and electrical installations, enable fast erection on site for domestic-scale construction. One manufacturer supplies bay window units constructed on steel frames for easy installation on site. The panels can either be pre-pointed or left unfinished for subsequent pointing on site.

In addition to very large preassembled brickwork units, manufacturers also produce smaller components, some of which can be lifted by hand. Units include soldier and dentil courses, arches, quoins and chimney stacks for incorporation into standard brickwork.

Brick cladding systems

Brick slip and brick tile cladding systems, designed to have the appearance and durability of traditional brickwork, offer a significantly reduced installation time compared to traditional brick construction. In one system, external walls are built with aerated concrete blockwork and faced with an extruded polystyrene insulation panel to which 16 mm brick slips are applied onto the pre-formed grid, giving the appearance of standard external leaf brickwork. The polystyrene grid panels have an overlap to ensure horizontal joints are watertight and are tongued and grooved to interlock vertically. Adhesive is applied to the polystyrene and the brick slips are pushed into place with the appropriate horizontal spacing. Mortar is applied either with a pointing gun or a mortar bag and tooled to the required joint profile.

An alternative system uses a plastic-coated galvanised steel profile fixed to the structural wall (Fig. 1.24). The specially shaped brick tiles then clip into the steel system with appropriate vertical joint spacing. Mortar (typically a 1 : 1 : 6 mix) is applied with a pointing gun and smoothed off to the required profile, usually bucket handle. A range of special tiles is manufactured to produce dados, plinths, cills and external returns, giving the appearance of traditional brickwork. Because the brickwork is non-structural, a range of bond patterns including stack, quarter and diagonal is optional. This type of prefabrication offers the potential for increased off-site construction work, and some manufacturers supply pre-formed brick-tile panels ready for fixing on site.

Brick slip systems with integral insulation are used for upgrading poor-quality masonry construction where the aesthetic of traditional brickwork is required.

Fig. 1.24 Brick cladding system. *Photograph*: Courtesy of Corium, Wienerberger UK

One system uses combined adhesive and mechanically fixed mineral wool insulation followed by 20 mm brick slips fixed with mortar adhesive. Matching corner and reveal slips together with cement pointing complete the traditional brickwork appearance.

Some manufacturers produce directly a wide colour range of 15 mm brick slips.

CLAY BRICK PAVING

Many clay brick manufacturers produce a range of plain and chamfered paving bricks together with a matching range of paver accessories. Bricks for flexible paving are frequently nibbed to set the spacing correctly. The material offers a human scale to large areas of hard landscape, especially if creative use is made of pattern and colour. Typical patterns (Fig. 1.25) include herringbone, running bond, stack bond, basket-weave, and the use of borders and bands. However, it should be noted that not all paver sizes coordinate for herringbone and basket-weave designs. Profiled brick designs include decorative diamond and chocolate-bar patterns, and pedestrian-management texturing. The paving bricks may be laid on a hard base with mortar joints or alternatively on a flexible base with fine sand brushed between the pavers. Edge restraint is necessary to prevent lateral spread of the units.

The British Standard BS EN 1344: 2013 stipulates minimum paver thicknesses of 40 mm and 30 mm for flexible and rigid construction, respectively. However, 50 mm pavers are generally used for flexible laying and 60 mm pavers are necessary when subjected to substantial vehicular traffic (BS 7533–1: 2001). Table 1.7 shows the standard sizes. Clay pavers are classified on the basis of freeze/thaw resistance. Pavers with designation FP0 are unsuitable for saturated freezing conditions, while pavers designated FP100 may be used under freeze/thaw conditions. The standard BS EN 1344: 2013 classifies pavers into five categories (T0 to T4) of transverse breaking strength, with the lowest category T0 being appropriate only for rigid construction. Slip resistance for the unpolished pavers is categorised as high, moderate, low or extremely low. This factor needs to be considered particularly for potentially wet conditions to ensure safe pedestrian and traffic use. The standard BS 7533 (Parts 1, 2, 3, 9 and 13) describes the design for heavy- and light-duty pavements, and flexible, rigid and permeable pavements in clay blocks, respectively.

Table 1.7 Standard work sizes for pavers

Length (mm)	Width (mm)	Thickness (mm)
215	102.5	50
215	102.5	65
210	105	50
210	105	65
200	100	50
200	100	65

Calcium silicate bricks

Calcium silicate bricks, also known as sandlime or flintlime bricks, were first produced commercially in Germany in 1894, and then in the UK in 1905. Initially their use was confined to common brick applications, but in the 1950s their durability for foundations was exploited. Subsequently a full range of load-bearing strength classes and attractive facings were produced. However, the last UK manufacturer of calcium silicate bricks closed down in 2010 and supplies must now be imported. The European Standard BS EN 771–2: 2011 specifies the full range of calcium silicate masonry units, and calcium silicate bricks are described within the National Annex of this standard.

Fig. 1.25 Typical range of clay pavers and hard landscape at Birmingham. *Photographs*: Courtesy of Wienerberger UK

Fig. 1.25 *(continued)*

Fig. 1.25 *(continued)*

SIZE

The UK work size for calcium silicate bricks is 215 ×
102.5 × 65 mm. Generally, calcium silicate bricks are
more accurate in form and size than fired clay bricks.
The dimensional tolerances for calcium silicate
bricks defined in the standard BS EN 771–2: 2011 are
generally ±2 mm on each dimension, except for
thin-layer mortar construction when a maximum of
only ±1 mm tolerance is permitted on the height.

MANUFACTURE OF CALCIUM SILICATE MASONRY UNITS (SANDLIME AND FLINTLIME BRICKS)

The raw materials are silica sand (approximately
90%), hydrated lime, crushed flint, colouring pigments
and water. The addition of colouring pigments or
crushed-flint aggregate to the standard components
or the application of texturing to the brick surface
gives the full product range of smooth, rustic and
textured finishes. The appropriately proportioned
blend is pressed into brick units and subjected to steam
pressure for 4 to 15 hours at 180°C in an autoclave.

APPEARANCE

The manufacturing process results in accurate shapes
and dimensions, and, with the untextured calcium
silicate bricks, a smooth finish. The colour range is
extensive, from white and pastel shades through to
deep reds, blues, browns, greens and yellows.

SPECIFICATION OF CALCIUM SILICATE BRICKS

Types

Both solid and frogged calcium silicate bricks have
been produced, including matching specials to BS 4729:
2005.

Durability

Calcium silicate bricks have good frost resistance,
but should not be exposed repeatedly to strong salt
solutions, acids or industrial effluent containing magnesium or ammonium sulphates. The bricks have a
negligible salt content and therefore efflorescence,
and sulphate attack on the mortar, cannot arise from
within the bricks. The bricks are themselves resistant
to sulphate attack and may therefore be used below

ground with a suitable sulphate-resisting cement mortar. However, calcium silicate bricks should not be used as pavers where winter salting may be expected.

PHYSICAL PROPERTIES

Compressive strength

The British Standard BS EN 771–2: 2011 defines the range of compressive strength classes, as shown in Table 1.8.

Table 1.8 Minimum compressive strength for calcium silicate bricks

Compressive strength class	Normalised compressive strength (MPa)
5	5.0
7.5	7.5
10	10.0
15	15.0
20	20.0
25	25.0
30	30.0
35	35.0
40	40.0
45	45.0
50	50.0
60	60.0
75	75.0

Weight

Most standard calcium silicate bricks weigh between 2.4 and 3.0 kg, with densities in the range 1700–2200 kg/m^3. However, densities of calcium silicate masonry units can range from below 500 to above 2800 kg/m^3.

Water absorption

Water absorption is usually in the range 8–15% by weight and typically 12%.

Moisture and thermal movement

Unlike clay bricks, which expand after firing, calcium silicate bricks contract. Reversible moisture movement is greater for calcium silicate bricks than for clay bricks, so expansion joints must be provided at intervals of no

more than a maximum of 9.0 m (NA to BS EN 1996–2: 2006). Such movement joints should not be bridged by rigid materials. Generally a weak mortar mix should be used (e.g. 1 : 2 : 9 cement : lime : sand), except below damp-proof course level (DPC) and for copings, to prevent visible cracking of either the mortar or the bricks.

Typical reversible moisture movement = ± 0.05%
Typical reversible thermal movement = ± 0.05%
Thermal movement = 8–14 × 10^{-6} deg C^{-1}

Thermal conductivity

The thermal conductivity of calcium silicate bricks is equivalent to that of clay bricks of similar densities, and ranges from 0.6 W/m K (Class 20) to 1.3 W/m K (Class 40).

Fire resistance

Calcium silicate bricks (with less than 1% organic material) are designated Euroclass A1 with respect to reaction to fire.

Acoustic properties

Acoustic properties are related to mass and are therefore the same as for clay bricks of equivalent density.

CALCIUM SILICATE BRICKWORK

Calcium silicate bricks have been particularly popular for their light-reflecting properties, for example, in light wells or atria. Their smooth, crisp appearance with a non-abrasive surface is particularly appropriate for some interior finishes and also forms an appropriate base for painted finishes.

The interior of the Queens Building of De Montfort University, Leicester (Fig. 1.26) illustrates the effective use of calcium silicate brickwork in creating a light internal space. Incorporated within the ivory Flemish-bond brickwork are restrained bands of polychromatic features and robust articulation of obtuse-angle quoins.

Fig. 1.26 Polychromatic calcium silicate brickwork – Queens Building, De Montfort University, Leicester. *Architects*: Short Ford & Associates.
Photographs: Lens-based media – De Montfort University

Concrete bricks

Developments in the use of iron oxide pigments have produced a wide range of colour-stable quality concrete brick products. Currently concrete bricks are competitively priced and hold approximately 4% of the facing brick market share.

SIZE

The British Standard BS 6073–2: 2008 does not refer to standard concrete sizes for bricks, but only to aggregate concrete coursing units as set out in Table 1.9. However, the National Annex NA (informative) of the current BS EN 771–3: 2011 standard still refers to UK work sizes of bricks, including the unit 215 × 103 × 65 mm as for clay bricks.

Due to their manufacturing process, concrete bricks can be made to close tolerances (Table 1.10), so accurate alignment is easy to achieve on site. Half-brick walls can readily be built fairfaced on both sides.

Table 1.9 Standard work sizes for aggregate concrete coursing units (BS 6073–2: 2008)

Length (mm)	Width (mm)	Height (mm)
190	90	65
190	90	90
215	100	65
290	90	90
440	90	65
440	100	65
440	90	140
440	100	140

Table 1.10 Tolerances on concrete blockwork sizes (BS EN 771–3: 2011)

Tolerance category	D1	D2	D3	D4
Length	+3, −5	+1, −3	+1, −3	+1,−3
Width	+3, −5	+1, −3	+1, −3	+1,−3
Height	+3, −5	+2,−2	+1.5, −1.5	+1, −1

Notes:
− Tolerances do not apply to surfaces manufactured to be non-planar.
− Tolerance category D1 is used for most common applications.
− Tolerance category D4 is appropriate for use with thin layer mortar (3 mm joints).

MANUFACTURE OF CONCRETE BRICKS

Concrete bricks are manufactured from blended dense aggregates (e.g. crushed limestone and sand) together with cement under high pressure in steel moulds. Up to 8% of appropriately blended iron oxide pigments, depending on the tone and depth of colour required, are added to coat the cement particles which will then form the solid matrix with the aggregate. The use of coloured aggregates also increases the colour range. The accurate manufacturing process produces bricks that have clean arrises.

APPEARANCE

A wide range of colours, including multicolours, is available, from red, buff and yellow to green, blue, brown and black (Fig. 1.27). Surfaces range from smooth and rumbled to simulated natural stone, including those characteristic of handmade and textured clay bricks. Because of the wide range of pigments used in the manufacturing process, it is possible to match effectively new concrete bricks to old and weathered clay or calcium silicate bricks for the refurbishment or extension of old buildings.

SPECIFICATION OF CONCRETE BRICKS

Types

Concrete bricks may be solid, perforated or frogged, also common or facing according to the manufacturer. 'Facing masonry units' may or may not be left exposed externally, while 'exposed masonry units' offer visual qualities without further protection. Compressive strengths and densities are manufactured to client requirements. A normal range of specials to BS 4729: 2005 is produced, although as with clay and calcium silicate bricks, a longer delivery time must be anticipated. The manufacturer's reference, the crushing strength, the dimensions and the brick type must be clearly identified with each package of concrete bricks. High-compressive-strength concrete units should be used below ground where significant sulphate levels are present according to the classification given in the BRE Special Digest 1, *Concrete in aggressive ground* (2005).

Durability

Concrete bricks are resistant to frost and are therefore usable in all normal levels of exposure. Like all concrete products, they harden and increase in strength with age.

Exterior detail

Interior detail

Fig. 1.1 Brick construction – Evry Cathedral, Essonne, France. *Architect:* Mario Botta. *Photographs:* Arthur Lyons

building exhibits fine detailing both internally and externally. Externally the cylindrical form rises to a circle of trees. Internally the altar is surmounted by a corbelled structure leading one's view upwards to the central rooflight. Three-dimensional internal brickwork is finely detailed to generate the desired acoustic response.

Clay bricks

The European Standard BS EN 771–1: 2011 refers only to clay masonry units. However, the National Annex to BS EN 771–1: 2011 refers to bricks as currently understood within the UK construction industry, and described in this chapter.

The wide range of clays suitable for brick making in the UK gives a diversity to the products available. This variety is further increased by the effects of blending clays, various forming processes, the application of surface finishes and the adjustment of firing conditions. In the early twentieth century most areas had their own brickworks with characteristic products; however, ease of road transportation and continuing amalgamations within the industry have left a reduced number of major producers and only a few small independent works. Most UK bricks are defined as high-density (HD) fired-clay masonry units with a gross dry density greater than 1000 kg/m³. The European Standard BS EN 771–1: 2011 also refers to low-density (LD) fired-clay masonry units and these blocks for unprotected masonry are described in Chapter 2, together with the larger high-density units.

The main constituents of brick-making clays are silica (sand) and alumina, but with varying quantities of chalk, lime, iron oxide and other minor constituents

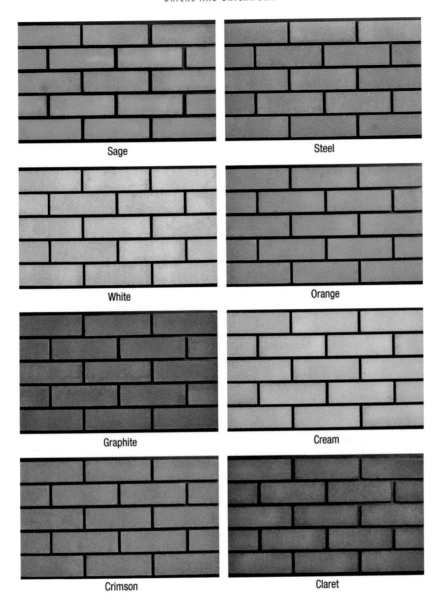

Fig. 1.27 Typical range of concrete bricks. *Photographs:* Courtesy of Edenhall Ltd

As with calcium silicate bricks, they may be made free from soluble salts and thus free from efflorescence. Concrete bricks should not be used where industrial effluents or acids are present.

PHYSICAL PROPERTIES

Weight and compressive strength

The standard brick weighs typically 3.2 kg depending on its composition. The compressive strengths of concrete masonry units are specified according to Table 1.11, within the range 2.9–40.0 MPa (BS 6073–2: 2008).

Water absorption

Water absorption is typically 8%, but engineering quality bricks average less than 7% after 24 hours' cold immersion, and are suitable for aggressive conditions such as retaining walls, below damp-proof course level and for inspection chambers.

Moisture and thermal movement

Concrete bricks have a typical drying shrinkage of 0.04%, with a maximum of 0.06%. Moisture and thermal movements are greater than those for calcium silicate bricks and movement joints should be at 5–6 m centres. Because of their moisture movement, prior to laying, concrete bricks should not be wetted to overcome excessive suction, but the water retentivity of the mortar should be adjusted accordingly. Brick stacks should be protected on site from rain, frost and snow.

Thermal conductivity

The thermal conductivities of concrete bricks are equivalent to those of clay and calcium silicate bricks of similar densities. Partially filled cavities, maintaining a clear cavity, are recommended to prevent water penetration to the inner leaf.

The thermal conductivity of concrete bricks ranges between 0.99 and 1.8 W/m K.

Fire resistance

Concrete masonry units with less than 1% of organic material are classified as Euroclass A1 with respect to reaction to fire.

Acoustic properties

Dense concrete bricks are suitable for the reduction of airborne sound transmission. On a weight basis, they are equivalent to clay and calcium silicate bricks.

Table 1.11 Typical strengths of concrete units (BS 6073–2: 2008)

Compressive strength (MPa)
2.9
3.6
7.3
8.7
10.4
17.5
22.5
30.0
40.0

Note:
Strengths 10.4 to 40.0 MPa inclusive apply only to aggregate concrete units and not aircrete.

CONCRETE BRICKWORK

With the wide range of colour and texture options offered by concrete brick manufacturers, it is frequently difficult to distinguish visually, except at close quarters, between concrete and clay brickwork. The visual effects of using coloured mortars and various jointing details are as for clay bricks, but for exposed situations the use of raked joints is not recommended.

Mortarless brick system

The non-structural cladding system is based on interlocking S-section concrete bricks, with appropriate special units for 45° and 90° internal and external angles as well as a PVC-U section as a base starter track. The visual effect is that of standard brickwork without the mortar joints. The system is suitable for cladding steel or timber frame buildings up to 18 m in height. It may also be used on conventional masonry or insulated concrete formwork (ICF) construction. The system has resistance to moisture penetration but is not airtight. A range of colours from buff, beige and brown through to red is available. The horizontal joints are slightly chamfered to give an impression of standard brickwork coursing.

References

FURTHER READING

Allbury, K., Franklin, E. and Anderson, J. 2013: *Environmental impact of brick, stone and concrete.* Bracknell: IHS BRE Press.

Arya, C. 2009: *Design of structural elements. Concrete, steelwork, masonry and timber. Designs to British Standards and Eurocodes.* Abingdon: Taylor and Francis.

Brick Development Association. 2001: *Use of traditional lime mortars in modern brickwork. Properties of bricks and mortar generally.* No. 1.3. Windsor: BDA.

Brick Development Association. 2001: *Observations on the use of reclaimed clay bricks. Properties of bricks and mortar generally.* No. 1.4. Windsor: BDA.

Brick Development Association. 2011: *Brick. Building a sustainable resource for the future.* Windsor: BDA.

Brick Development Association. 2012: *The BDA guide to successful brickwork,* 4th edn. Oxford: Butterworth Heinemann.

Brick Development Association. 2012: *Brick industry sustainability strategy report.* Windsor: BDA.

British Cement Association. 2005: *BCA guide to materials for mortar*. Camberley: British Cement Association.

Campbell, J.W.P. 2003: *Brick: A world history*. London: Thames and Hudson.

English Heritage. 2012: *Mortars, renders and plasters (Practical building conservation)*. Farnham: Ashgate Publishing.

English Heritage. 2013: *Earth, brick and terracotta (Practical building conservation)*. Farnham: Ashgate Publishing.

Greeno, R. and Chudley, R. 2012: *Building construction handbook*. Abingdon: Routledge.

Hammett, M. 2003: *Brickwork and paving for house and garden*. Marlborough: Crowood.

Institution of Structural Engineers. 2008: *Manual for the design of plain masonry in building structures to Eurocode 6*. London: ISE.

Key, M. 2009: *Sustainable masonry construction*. EP 99. Bracknell: IHS BRE Press.

Kreh, D. 2007: *Masonry skills*, 6th edn. Nantwich: Delmar.

McKenzie, W.M.C. 2001: *Design of structural masonry*. London: Palgrave.

Mortar Industry Association. 2005: *Efflorescence and bloom on masonry*. Data Sheet 8. London: Mortar Industry Association.

Thorpe, M. and Hodge, J.C. 2010: *Brickwork level 2 for CAA construction diploma and NVQs*. Oxford: Butterworth Heinemann.

Weston, R. 2008: *Materials, form and architecture*. London: Laurence King Publishing.

STANDARDS

BS 743: 1970
Materials for damp-proof courses.

BS 4729: 2005
Clay and calcium silicate bricks of special shapes and sizes. Recommendations.

BS 6073
Precast concrete masonry units:

Part 2: 2008	Guide for specifying precast concrete masonry units.

BS 6100
Building and civil engineering vocabulary:

Part 0: 2002	Introduction.
Part 6: 2008	Construction parts.

BS 6515: 1984
Specification for polythene damp-proof courses for masonry.

BS 6750: 1986
Specification for modular co-ordination in building.

BS 7533

Parts 1–13	Pavements constructed of clay, natural stone or concrete pavers.

BS 8000
Workmanship on building sites:

Part 0: 2014	Introduction and general principles.
Part 3: 2001	Code of practice for masonry.

BS 8103
Structural design of low rise buildings:

Part 1: 1985	Existing traditional cavity construction.
Part 2: 2013	Code of practice for masonry walls for housing.

BS 8215: 1991
Code of practice for design and installation of damp-proof courses in masonry construction.

BS EN 413–1: 2011
Masonry cement. Composition, specifications and conformity.

BS EN 771
Specification for masonry units:

Part 1: 2011	Clay masonry units.
Part 2: 2011	Calcium silicate masonry units.
Part 3: 2011	Aggregate concrete masonry units.
Part 4: 2011	Autoclaved aerated concrete masonry units.
Part 5: 2011	Manufactured stone masonry units.

BS EN 772
Methods of test for masonry units:

Part 1: 2011	Determination of compressive strength.
Part 18: 2011	Determination of freeze/thaw resistance of calcium silicate masonry units.
DD CEN/TS Part 22: 2006	Determination of freeze/thaw resistance of clay masonry units.

BS EN 845
Specification for ancillary components for masonry:

Part 1: 2013	Wall ties, tension straps, hangers and brackets.
Part 2: 2013	Lintels.

Part 3: 2013   Bed joint reinforcement of steel meshwork.

BS EN 934–3: 2009

Admixtures for masonry mortar.

BS EN 998–2: 2010

Specification for mortar for masonry. Masonry mortar.

BS EN 1015

Methods of test for mortar for masonry.

BS EN 1052

Methods of test of masonry.

BS EN 1344: 2013

Clay pavers. Requirements and test methods.

BS EN 1365–1: 2012

Fire resistance tests for loadbearing elements. Walls.

BS EN 1996

Eurocode 6: Design of masonry structures:

Part 1.1: 2005 Rules for reinforced and unreinforced masonry.

Part 1.2: 2005 Structural fire design.

Part 2: 2006   Design considerations, selection of materials and execution of masonry.

Part 3: 2006   Simplified calculation methods for unreinforced masonry structures.

NA Part 2: 2006 UK National Annex to Eurocode 6: Design of masonry structures. Design considerations, selection of materials and execution of masonry.

pr EN ISO 9001: 2014

Quality management systems. Requirements.

BS EN ISO 9004: 2009

Managing for the sustained success of an organization. A quality management approach.

BS EN 13139: 2013

Aggregates for mortar.

BS EN ISO 14001: 2004

Environmental management systems. Requirements with guidance for use.

PAS 70: 2003

HD clay bricks. Guide to appearance and site measured dimensions and tolerance.

DD 140–2: 1987

Wall ties. Recommendations for design of wall ties.

PD 6678: 2005

Guide to the specification of masonry mortar.

PD 6682–3: 2003

Aggregates for mortar.

PD 6697: 2010

Recommendations for the design of masonry structures to BS EN 1996–1-1.

BUILDING RESEARCH ESTABLISHMENT PUBLICATIONS

BRE Digests

BRE Digest 329: 2000

Installing wall ties in existing construction.

BRE Digest 441: 1999

Clay bricks and clay brick masonry (Parts 1 and 2).

BRE Digest 460: 2001

Bricks, blocks and masonry made from aggregate concrete (Parts 1 and 2).

BRE Digest 461: 2001

Corrosion of metal components in walls.

BRE Digest 487: 2004

Structural fire engineering design: Materials and behaviour. Masonry (Part 3).

BRE Digest 502: 2007

Principles of masonry conservation management.

BRE Special digests

BRE SD1: 2005

Concrete in aggressive ground.

BRE SD4: 2007

Masonry walls and beam and block floors. U-values and building regulations.

BRE Good building guides

BRE GBG 41: 2000

Installing wall ties.

BRE GBG 44: 2000

Insulating masonry cavity walls (Parts 1 and 2).

BRE GBG 50: 2002

Insulating solid masonry walls.

BRE GBG 58: 2003

Thin layer mortar masonry.

BRE GBG 62: 2004

Retro-installation of bed joint reinforcement in masonry.

BRE GBG 66: 2005
　　Building masonry with lime-based bedding
　　mortars.

BRE Information papers

BRE IP 8/08
　　Determining the minimum thermal resistance of
　　cavity closers.
BRE IP 6/09
　　Framework standard for the responsible sourcing
　　of construction products (BES 6001).
BRE IP 16/11
　　Unfired clay masonry. An introduction to low-
　　impact building materials.

BRE Reports

BES 6001 Issue 2.0: 2009
　　Framework standard for the responsible sourcing
　　of construction products.
BR 443: 2006
　　Conventions for U-value calculations.
FB 42: 2012
　　U-value conventions in practice. Worked
　　examples using BR 443.

BRICK DEVELOPMENT ASSOCIATION PUBLICATIONS

Design note

BDA Design Note 7: 2011
　　Brickwork durability.

Brick information sheet

Brick Information Sheet 6.4: 2001
　　Tinting of brickwork – colour variation.

ADVISORY ORGANISATIONS

Brick Development Association, The Building Centre,
　　26 Store Street, London WC1E 7BT, UK (0207 323
　　7030).
International Concrete Brick Association, PD Bricks,
　　Danygraig Road, Risca, Newport, Gwent NP1 6DP,
　　UK (01633 612671).
Mortar Industry Association, Gillingham House, 38–44
　　Gillingham Street, London SW1V 1HU, UK (0207
　　963 8000).

2

BLOCKS AND BLOCKWORK

CONTENTS

Introduction

The variety of commercially available concrete blocks (masonry units) is extensive, from dense through to lightweight, offering a range of load-bearing strength, sound and thermal insulation properties. Where visual blockwork is required, either internally or externally, fairfaced blocks offer a selection of textures and colours at a different visual scale compared to that associated with traditional brickwork. Externally, visual concrete blockwork weathers well, providing adequate attention is given to the quality of the material and rainwater run-off detailing. Blockwork has considerable economic advantages over brickwork in respect of speed of construction, particularly as the lightweight blocks can be lifted in one hand.

While clay blocks are used extensively for masonry construction on the continent of Europe, until recently there had been modest demand from the building industry within the UK. However, both fired and unfired clay blocks are now readily available. Gypsum blocks may be used for internal non-load-bearing partitions and the internal insulation of walls. Inverted reinforced concrete T-beams with concrete or clay block infill is a standard form of domestic-scale floor construction.

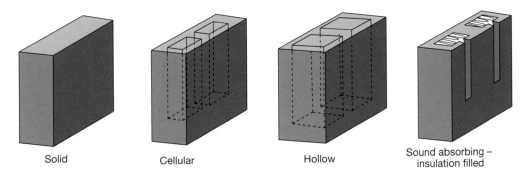

Solid   Cellular   Hollow   Sound absorbing –
                                insulation filled

Fig. 2.1 Types of concrete blocks

Concrete paving blocks, which offer opportunities for creative hard landscaping with their diversity of form and colour, are widely used for town pedestrian precincts and individual house driveways. Concrete interlocking blocks with planting are used to create environmental walls.

Concrete blocks

TYPES AND SIZES

Concrete blocks (masonry units) are defined as solid, cellular or hollow, as illustrated in Figure 2.1.

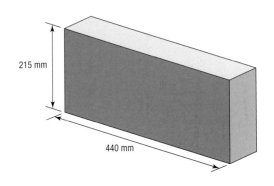

215 mm

440 mm

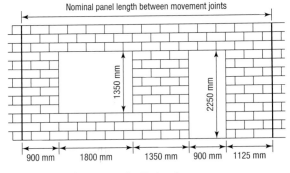

Nominal panel length between movement joints

1350 mm

2250 mm

900 mm   1800 mm   1350 mm   900 mm   1125 mm

Fig. 2.2 Co-ordinating sizes for blockwork

Concrete blocks are manufactured to various work face dimensions in an extensive range of widths, offering a wide choice of load-bearing capacity and level of insulation. Manufacturer's work size dimensions should be indicated as length, width and height, in that order, to BS EN 771–3: 2011 and BS 6073–2: 2008.

The standard work face size, which coordinates to three courses of metric brickwork allowing for 10 mm mortar joints, is 440×215 mm (Fig. 2.2), but the other sizes in Table 2.1 are marketed for aesthetic and constructional reasons. For example, narrow bands of a different colour may be used as visual features within fairfaced blockwork, and foundation wall blocks are normally laid flat. The use of thin-joint masonry offers speedier construction, especially when using large format blocks (Fig. 2.3), which are approximately equivalent in size to two standard units. However, blocks heavier than 20 kg should not be lifted by a single person, as this can potentially lead to injury. Within the 20 kg limit are 100 mm-wide aircrete blocks with face dimensions 610×375 mm for speedy construction using the thin-joint system.

The European Standard BS EN 771–3: 2011 describes a wide range of aggregate concrete masonry units incorporating either dense or lightweight aggregates. Under the European Standard, the minimum description for concrete blocks includes the European Standard number and date (e.g. BS EN 771–3: 2011), the type of unit (e.g. common or facing), work size dimensions and tolerance category, configuration (e.g. solid or with voids) and compressive strength. Depending on the particular end use, additional description may also be required. This may, as appropriate, include surface finish, net and gross dry density, coordinating size, thermal properties and moisture movement. Tolerance limits for regular-shaped blocks are defined at four levels in Table 2.2. Compressive strengths of concrete masonry units are

Table 2.1 Typical work sizes and strengths for concrete blocks to BS 6073–2: 2008

	Length (mm)	Width (mm)	Height (mm)
Aggregate concrete blocks			
Coursing blocks	190	90	65
	190	90	90
	215	100	65
	290	90	90
	440	90	65
	440	100	65
	440	90	140
	440	100	140
Standard blocks	390	90	190
	390	100	190
	390	140	190
	390	190	190
	440	75	215
	440	90	215
	440	100	215
	440	140	215
	440	190	215
	440	215	215
Aircrete concrete blocks			
Coursing blocks	215	90 to 150	65
	215	90 to 150	70
Standard blocks	440	50 to 350	215
	610	50 to 350	215
	620	50 to 350	215

Notes:
- Not all sizes are produced by all manufacturers and other sizes may be available.
- Widths quoted by certain manufacturers include: 50, 70, 75, 90, 100, 115, 125, 130, 140, 150, 190, 200, 215, 255, 265, 275, 300 and 355 mm.
- Foundation blocks with face sizes of typically 300×250, 300×275, 320×280, 350×250, 350×310, 620×140 or 620×215 mm are laid flat.
- Beam and block floor rectangular units are usually 100 mm thick, with face dimensions of 440×215, 440×350, 440×540, 620×215, 610×350 or 620×430 mm.
- Common crushing strengths to BS 6073–2: 2008 are 2.9, 3.6, 7.3, 8.7, 10.4, 17.5, 22.5, 30.0 and 40.0 MPa, but some manufacturers supply addition intermediate strengths (e.g. 4.2 MPa).

Fig. 2.3 Thin-joint masonry using large format blocks. *Photograph*: Reproduced by permission of IHS and courtesy of Aircrete Products Association

Table 2.2 Limit of tolerances on aggregate concrete block sizes

Tolerance category	D1	D2	D3	D4
Length (mm)	+3	+1	+1	+1
	−5	−3	−3	−3
Width (mm)	+3	+1	+1	+1
	−5	−3	−3	−3
Height (mm)	+3	±2	±1.5	±1.0
	−5			

Notes:
- BS 6073: 2008 states that tolerance categories D3 and D4 are intended for use with thin layer mortar joint systems. Therefore most units used within the UK conform to tolerance categories D1 and D2.
- Closer tolerances may be declared by the manufacturer.

Table 2.3 Limit of tolerances on autoclaved aerated concrete block sizes

	Standard joints of general purpose and lightweight mortar	Thin layer mortar joints	
	GPLM	TLMA	TLMB
Length (mm)	−5 to +3	±3	±1.5
Width (mm)	±3	±2	±1.5
Height (mm)	−5 to +3	±2	±1.0

Notes:
- For autoclaved aerated concrete units of category TLMB, the maximum deviation from flatness of bed faces and plane parallelism of bed faces is ≤ 1.0 mm in each case.
- Closer tolerances may be declared by the manufacturer.

classified as Category I or Category II. Category I units have a tighter control with only a 5% risk of the units not achieving the declared compressive strength.

The European Standard BS EN 771–4: 2011 gives the specification for autoclaved aerated concrete (AAC) masonry units. The maximum size of units within the standard is 1500 mm length × 600 mm width × 1000 mm height. The tolerance limits on the dimensions are defined in Table 2.3, and are dependent on whether the units are to be erected with standard or thin-layer mortar joints. The standard manufacturer's description for AAC masonry units must include the European Standard number and date (e.g. BS EN 771–4: 2011), dimensions and tolerances, compressive strength (Category I or II, as for concrete units) and dry density.

Further description for specific purposes may include durability, configuration (e.g. perforations or tongued and grooved jointing system) and intended use.

MANUFACTURE

Dense concrete blocks, which may be hollow, cellular or solid in form, are manufactured from natural dense aggregates including crushed granite, limestone and gravel. Medium and lightweight concrete blocks are manufactured incorporating a wide range of aggregates including expanded clay, expanded blast furnace slag, sintered ash and pumice. Concrete is cast into moulds, vibrated and cured. Most autoclaved aerated (aircrete) concrete blocks are formed by the addition of aluminium powder to a fine mix of sand, lime, fly ash (pulverised fuel ash) and Portland cement. The hydrogen gas generated by the dissolution of the metal powder produces a non-interconnecting cellular structure. The process is accelerated by pressure steam curing in an autoclave (Fig. 2.4). Standard blocks, typically natural grey or buff in colour, are usually shrink-wrapped for delivery. Different grades of blocks are usually identified by scratch marks or colour codes.

PROPERTIES

Density and strength

The British Standard BS 6073–2: 2008 lists common compressive strengths of 2.9, 3.6, 7.3, 8.7, 10.4, 17.5, 22.5, 30.0 and 40.0 MPa for the range of aircrete and aggregate concrete blocks. However, the majority of

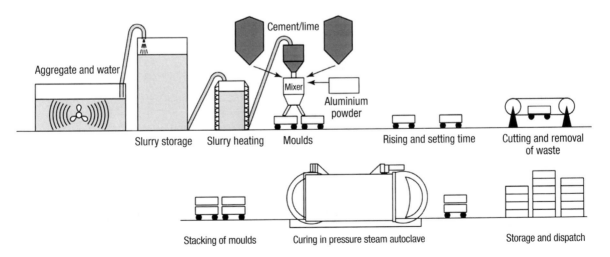

Fig. 2.4 Manufacture of aerated blocks

Table 2.4 Typical relationship between density and thermal conductivity for concrete blocks

Nominal density (kg/m³)	2200	2000	1800	1600	1400	1200	1000	900	800	750	700	600	500	460	420
Typical thermal conductivity (W/m K)	1.5	1.2	0.83	0.63	0.47	0.36	0.27	0.24	0.20	0.19	0.17	0.15	0.12	0.11	0.10

Note:
Blocks of differing compositions may vary significantly from these average figures and manufacturers' data should be used.

concrete blocks fall in the range from 2.8 to 30 MPa, with associated densities of 420–2200 kg/m³ and thermal conductivities from 0.10 to 1.5 W/m K at 3% moisture content (Table 2.4). Drying shrinkages are typically in the range 0.03–0.05%. Autoclaved aerated concrete masonry units typically have densities within the range 300–1000 kg/m³ (BS EN 711–4: 2011). Some manufacturers colour code or mark blocks to facilitate on-site recognition of their particular physical properties.

Durability

Dense concrete blocks and certain aerated lightweight blocks are resistant to freeze/thaw conditions below damp-proof course (DPC) level. However, some lightweight concrete blocks, with less than 7 MPa crushing strength, should not be used below DPC level, except for the inner skin of cavity construction.

Fixability

Aerated and lightweight concrete blocks offer a good background for fixings. For light loads, nails to a depth of 50 mm are sufficient. For heavier loads, wall plugs and proprietary fixings are necessary. Fixings should avoid the edges of the blocks.

Thermal insulation

The Building Regulations Approved Document Part L (2013 revision) requires new dwellings (Part L1A) to conform to the Target Fabric Energy Efficiency (TFEE) and the Target Emission Rate (TER) calculated for a notional building of the same size and shape as the one to be constructed. If the building is constructed entirely to the notional dwelling specifications it will meet the carbon dioxide and fabric energy-efficiency targets. The Dwelling Fabric Energy Efficiency (DFEE) and Dwelling Emission Rate (DER) calculations must be based on the dwelling as designed and finally checked

against the dwelling as constructed. Calculations are based on the Standard Assessment Procedure (SAP 2012).

Notional U-value specifications for new dwellings:

Element	U-value (W/m² K)
Walls	0.18
Floors	0.13
Roofs	0.13
Party walls	0.00
Windows, doors and roof lights	1.40 (whole window value)

However, variations from the notional specifications are permitted, proving that no element falls to a lower specification than the limiting fabric parameters, and that the overall energy performance is not compromised.

Limiting fabric parameters for new dwellings (providing that overall performance is maintained):

Element	U-value (W/m² K)
Walls	0.30
Floors	0.25
Roofs	0.20
Party walls	0.20
Windows, doors and roof lights	2.00

The limiting fabric parameters are absolute, and flexing is likely to permit only small easing of the elemental requirements. For example, the use of triple glazing (U-value 0.9 W/m² K) might be balanced against a slightly relaxed wall insulation level (U-value 0.22 W/m² K), depending on the full range of factors, including the exact dwelling plot.

The maximum space heating and cooling energy demand, the Target Fabric Energy Efficiency (TFEE), is expressed in units of kilowatt hours of energy per m² floor area per year (kWh/m²/year). The carbon emissions performance, Target Emission Rate (TER), is expressed as the mass of CO_2 in units of kg per m² per year.

New buildings other than dwellings (Part L2A) are required to conform to their relevant Target Emission Rate (TER). Emission rates are calculated using the Simplified Building Energy Model (SBEM), or other approved tool, for the total useful floor area.

Notional U-value specifications for new buildings other than dwellings:

Element	U-value (W/m² K)
Walls	0.26
Floors	0.22
Roofs	0.18
Windows	1.60
Air permeability (≤ 250 m²)	5.0 m³/hr/m² at 50 Pa (except where top lit)
Air permeability (≥ 250 m²)	3.0 m³/hr/m² at 50 Pa (except where top lit)

However, trade-offs between elements are permitted, proving that no element falls to a lower specification than the limiting fabric parameters.

Limiting fabric parameters for new buildings in England other than dwellings (provided that overall performance is maintained):

Element	U-value (W/m² K)
Walls	0.35
Floors	0.25
Roofs	0.25
Windows, doors and roof lights	2.20
High usage doors	3.50
Vehicle doors	1.50
Air permeability	10.0 m³/hr/m² at 50 Pa

For replacements and extensions to existing dwellings (Building Regulations Approved Document Part L1B)

and to existing buildings other than dwellings (Part L2B), a U-value of 0.28 W/m² K is the standard for new exposed walls.

The following material combinations achieve a wall U-value of 0.18 W/m² K (Fig. 2.5).

Partially filled cavity:

> 102.5 mm fairfaced brickwork outer leaf
> (λ = 0.84 W/m K)
> 50 mm clear cavity
> 90 mm foil-faced polyurethane foam
> (λ = 0.022 W/m K)
> 100 mm lightweight blocks (λ = 0.15 W/m K)
> 12.5 mm plasterboard on dabs (λ = 0.16 W/m K).

Fully filled cavity:

> 102.5 mm fairfaced brickwork outer leaf
> (λ = 0.84 W/m K)
> 180 mm full-fill cavity of blown mineral wool
> (λ = 0.038 W/m K)
> 100 mm lightweight blocks (λ = 0.15 W/m K)
> 13 mm dense plaster (λ = 0.5 W/m K).

Solid wall:

> 16 mm external render (λ = 1.0 W/m K)
> 80 mm phenolic foam insulation
> (λ = 0.023 W/m K)
> 215 mm high-performance lightweight blocks
> (λ = 0.11 W/m K)
> 12.5 mm plasterboard on dabs
> (λ = 0.16 W/m K).

For domestic construction, Approved Construction Details (ACDs) should be used to ensure compliance with the thermal and sound requirements of the Building Regulations. Calculated linear transmittance values may then be used in the DER and DFEE rate calculations. Building performance often falls below the predicted levels because of thermal bridges.

Phase change material blocks

Phase change materials (PCMs) incorporated into aerated concrete blocks offer some additional thermal stability to the internal environment by absorbing excessive summer heat, which is then released during the cooler periods. This phase change at 26°C effectively increases the thermal capacity of the lightweight blocks.

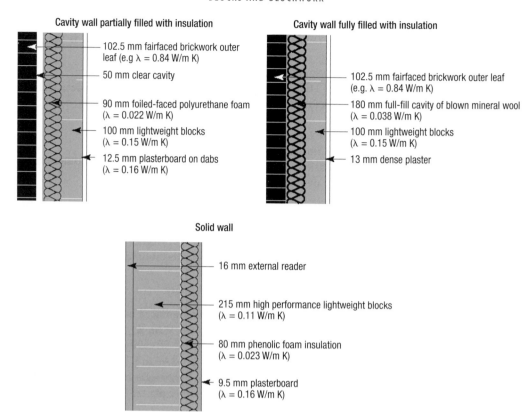

Fig. 2.5 Typical blockwork construction achieving U-values of at least 0.18 W/m² K

One manufacturer colour codes the phase change material blocks green for easy identification. The phase change material is described in Chapter 12 (page 388).

Fire resistance

Concrete block construction offers good fire resistance. Solid 100 mm load-bearing walls can give up to 120 minutes' fire protection. Dense, lightweight and autoclaved aerated concrete blocks with less than 1% organic material are automatically categorised as non-combustible Class A1 with respect to fire.

Sound insulation

The Building Regulations 2010 Approved Document E recognises the need to provide adequate sound insulation both between and within dwellings, as well as between rooms in hostels, hotels and residential accommodation. The regulations require minimum airborne sound insulation of 45 $D_{nT,w} + C_{tr}$ dB for separating walls and 40 R_w dB for internal bedroom or

WC walls. The passage of airborne sound depends on the density and porosity of the material. The use of *Robust Details* or Pre-Completion Testing is required to demonstrate compliance.

The following alternative systems should perform to the required airborne insulation standard for separating walls of new build dwellings:

> 13 mm plaster or cement render with plaster skim (10 kg/m³)
> 100 mm dense (1850–2300 kg/m³) or lightweight (1350–1600 kg/m³) solid blockwork
> 75 mm clear cavity only linked by Type A wall ties to the required spacing
> 100 mm dense (1850–2300 kg/m³) or lightweight (1350–1600 kg/m³) solid blockwork
> 13 mm plaster or cement render with plaster skim (10 kg/m³).

These alternatives perform to the required standard only if there are no air leaks within the construction, all joints are filled, and the cavities are kept clear except for the approved wall ties and optional mineral wool insulation (maximum 40 kg/m^3). Chasing out on opposite sides of the construction must be staggered, and all voids fully filled with mortar.

Sound absorption

The majority of standard concrete blocks with hard surfaces are highly reflective to sound, thus creating long reverberation times within building enclosures. Acoustic absorbing concrete blocks are manufactured either with a specially designed hollow core geometry and appropriate aggregate mix or with a slot on the profiled exposed face which admits sound into the central cavity (Fig. 2.1). In the latter case, the void space contains a sound-absorbing fibrous material to dissipate the incident sound, reducing reverberation effects. Acoustic control blocks in fairfaced concrete are suitable for use in swimming pools, sports halls, industrial buildings and auditoria.

SPECIALS

Most manufacturers of blocks produce a range of *specials* to match their standard ranges. Quoins, cavity closers, splayed cills, flush or projecting copings, lintel units, bullnose ends and radius blocks are generally available, and other specials can be made to order (Fig. 2.6). The use of specials in fairfaced blockwork can greatly enhance visual qualities. Matching full-length lintels may incorporate dummy joints and should bear on to full, not cut, blocks.

FAIRFACED BLOCKS

Fairfaced concrete blocks are available in a wide range of colours from white, through buff, sandstone, yellow, to pink, blue, green and black. Frequently the colour is *all through*, although some blocks have an applied surface colour. Blocks may be uniform in colour or flecked. Textures range from polished, smooth and weathered (sand- or shot-blasted) to striated and split face, the latter intended to give a random variability associated more with natural stone. New finishes incorporate crushed sea shells or recycled glass which is either reflective or glows in the dark (Fig. 2.7).

Glazed masonry units are manufactured by the application of a thermosetting material to one or more faces of lightweight concrete blocks, which are then heat-treated to cure the finish. The glazed blocks are available in an extensive range of durable bright colours

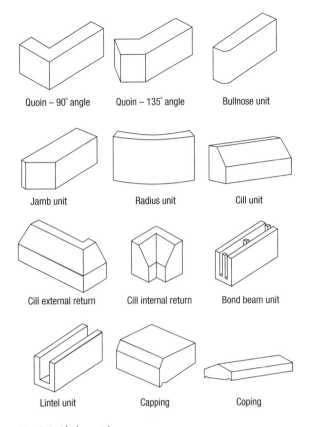

Fig. 2.6 Block specials

Quoin – 90° angle Quoin – 135° angle Bullnose unit

Jamb unit Radius unit Cill unit

Cill external return Cill internal return Bond beam unit

Lintel unit Capping Coping

and are suitable for interior or exterior use (Fig. 2.8). Where required, profiled blocks to individual designs can be glazed by this system. Most manufacturers produce a range of specials to coordinate with their standard fairfaced blocks, although, as with special bricks, they may be manufactured from a different batch of mix, and this may give rise to slight variations. In specific cases, such as individual lintel blocks, specials are made by cutting standard blocks to ensure exact colour matching.

Clay blocks

FIRED-CLAY BLOCKS

Masonry clay honeycomb-insulating blocks may be used as a single skin for external load-bearing construction as an alternative to standard cavity construction. These fired-clay honeycomb blocks combine structural strength, insulation and, when externally

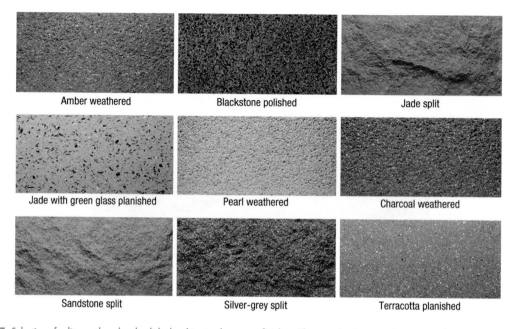

Fig. 2.7 Selection of split, weathered and polished architectural masonry finishes. *Photographs:* Courtesy of Lignacite Ltd

rendered, moisture protection. The internal surface is normally finished directly with gypsum plaster. Block sizes for monolithic construction typically range from 247 mm to 373 mm in length and 249 mm in height for thin joint mortar construction. Wall block widths are 240, 300, 365, 425 and 490 mm. A 425 mm wide block ($\lambda = 0.11$ W/m K) with 60 mm wood fibre-board insulation ($\lambda = 0.043$), two-coat external render and 15 mm internal plaster finish has a U-value of 0.18 W/m^2 K.

For internal walls, blocks are 308 to 400 mm in length and range in width from 100 to 240 mm. Horizontal joints require either 10 mm of a lightweight mortar or a thin joint mortar, but the vertical joint edges, if tongued and grooved, remain dry. The British Standard BS EN 771–1: 2011 illustrates a selection of high-density (HD) vertically perforated units and a range of low-density (LD) fired-clay masonry units. The LD units may be vertically or horizontally per-forated, with butt jointing, mortar pockets or a tongue and groove system (Fig. 2.9). Special blocks are manu-factured for corners, reveals, lintels, etc., but individual blocks may also be cut.

Fairfaced fired-clay blocks, as illustrated in Figure 2.10, offer an alternative to traditional brickwork. They are manufactured to natural, riven or textured finish in a range of colours including terracotta red, ochre, buff and blue, also to high gloss or satin finish in strong or pastel shades. Where used as infill rather than load-bearing, alternative bonding is possible including stack bond. Typical work sizes, depending on the manu-facturer, are 215 × 215 mm, 327 × 215 mm, 327 × 140 mm, 440 × 215 mm, 390 × 240 mm, 390 × 190 mm and 490 × 190 mm with widths of 90 and 102 mm. A standard 10 mm mortar joint is appropriate, which may match or contrast with the block colour.

UNFIRED-CLAY BLOCKS

Unfired blocks manufactured from clay and sometimes incorporating straw may be used for non-load-bearing partition walls. Blocks (typically 500 × 250 mm and 450 × 225 mm × 100 mm wide) may be tongued and grooved or square edged. Sodium silicate or ligno-sulphonate-stabilised clay-based mortars are necessary for their added strength over natural clay- or lime-based mortar. Blocks are easily cut to create architectural features, and are usually finished with a skim coat of clay plaster, although they may be painted directly. Internal walls are sufficiently strong to support shelving and other fixtures. Unfired-clay block walls are recyclable or biodegradable, and have the advantage of absorbing odours and stabilising internal humidity and tem-perature by their natural absorption and release of moisture and heat. A 100 mm-thick wall gives a 45 dB sound reduction and 90 minutes' fire resistance. (The thermal conductivity of perforated unfired-clay blocks is typically 0.24 W/m K.)

Fig. 2.8 Glazed blocks and Astra glazed masonry units at Redcar Primary Care Trust. *Photographs*: Courtesy of Forticrete Ltd

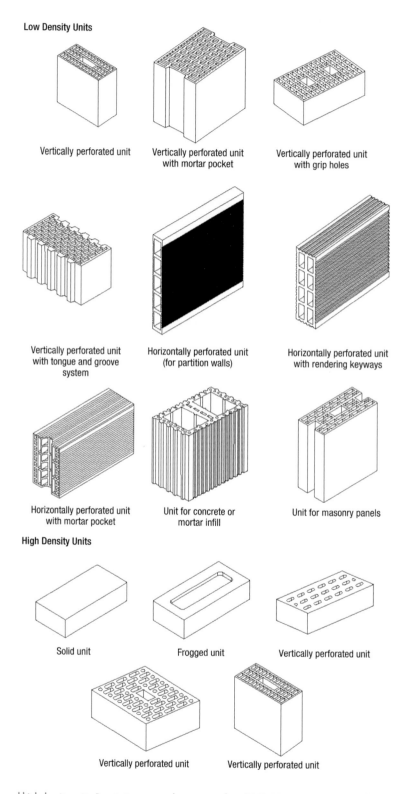

Fig. 2.9 Low density and high density units. Permission to reproduce extracts from BS EN 771–1: 2011 is granted by the British Standards Institute

Fig. 2.9 *(continued)* Hollow clay blocks. *Photograph*: Courtesy of NBT ThermoPlan System

Non-clay blocks

GYPSUM BLOCKS

Gypsum blocks are available with densities ranging from 600 to 1500 kg/m³, in thicknesses from 50 to 100 mm. The maximum thickness to BS EN 12859: 2011 is 150 mm. The preferred face dimensions are 666 mm in length and 500 mm in height, with a maximum length of 1000 mm. Gypsum blocks are classified by density and water absorption.

Gypsum blocks – density class:

Low density (L)	600–800 kg/m³
Medium density (M)	800–1100 kg/m³
High density (H)	1100–1500 kg/m³

Fig. 2.10 Fairfaced blockwork – IDP Offices, Glasgow. *Architects*: IDP. *Photograph*: Courtesy of Ibstock Brick Ltd

Gypsum blocks – water absorption class:

H3	>	5%
H2	≤	5%
H1	≤	2.5%

The standard BS EN 15318: 2007 details the sound insulation properties of gypsum block partitions in relation to wall thickness and block density. Gypsum blocks may be used as non-load-bearing partitions and internal insulation of walls. They are assembled with gypsum-based adhesives as specified in BS EN 12860: 2001.

HEMP BLOCKS

Hemp blocks are manufactured using hemp blended with a lime and cement binder. Hemp, as a hygroscopic material (Chapter 3, page 67), affords some degree of humidity control. Standard blocks with dimensions of 440 × 100 × 215 mm have a compressive strength of 3 MPa and a thermal conductivity of 0.36 W/m K. Structural hemp blocks are normally laid on a standard 10 mm bed of lime-based mortar. Blocks require protection from adverse weather conditions before and immediately after laying. Lime-based renders are recommended for external and internal finishes.

INSULATED CONCRETE FORM (ICF) BLOCKS

Insulated concrete form blocks are hollow units, partially filled with mineral wool or foam insulation, but with vertical voids for steel reinforcement and poured concrete (Fig. 2.11). Blocks made of cement-bound wood chip can be stacked in staggered dry bond

Fig. 2.11 Insulated concrete forms. *Photograph:* Courtesy of Durisol UK – The Sustainable Walling System

to create the formwork which is then filled with reinforced concrete. Alternative ICF systems based on blocks or panels of expanded polystyrene are described in Chapter 3 (page 86).

CALCIUM SILICATE BLOCKS

Large format calcium silicate blocks 1000 × 545 and 645 mm, with thicknesses ranging from 100 to 300 mm, are imported for the swift construction of internal walls using thin bed mortar. Lifting gear is necessary to locate the large blocks which are manufactured with the appropriate lifting points.

Blockwork

FAIRFACED CONCRETE BLOCKWORK

Within fairfaced blockwork, an appropriate choice of size is important for both coordination and visual scale. While blocks may be cut with a masonry cutter, the addition of small pieces of block, or the widening of perpends over the 10 mm standard, is unacceptable. The insertion of a thin *jumper* course at floor or lintel height may be a useful feature in adjusting the coursing. Curved blockwork may be constructed from standard blocks, the permissible curvature being dependent on the block size. The oversail between alternate courses should not normally exceed 4 mm in fairfaced work. If the internal radius is exposed, the perpends can be maintained at 10 mm with uncut blocks, but if the external radius is exposed, the blocks will require cutting on a splay. For tighter curves *specials* will be required.

THIN-JOINT MASONRY SYSTEMS

Thin-joint blockwork may be constructed with mortar joints of only 2–3 mm, provided that the aircrete or equivalent blocks have been manufactured to fine tolerances and on-site workmanship is good. The special rapid-setting mortar sets typically within 60 minutes and the full bond strength is achieved after only two hours, allowing more courses to be laid each day. In the case of brick and block cavity construction, the inner leaf is built first, providing a weatherproof enclosure as quickly as possible. The outer skin of brickwork can subsequently be built up, using wall ties fixed to the face, either screwed or hammered into the completed blockwork. Bed joints in thin-layer mortar blockwork do not necessarily coordinate with those of the brickwork, so conventional cavity wall ties can only be used if they are slope-tolerant.

Usually, inner leaf construction commences with a line of 440 × 215 mm standard-height blocks, with normal bedding mortar to compensate for variations in the foundation level, followed by the larger 440 or 620 mm × 430 mm-high blocks, which should weigh less that 20 kg for repeated lifting by one operative. Heavier blocks require mechanical lifting or two-person handling. Thin-joint mortars, consisting of polymer-modified 1 : 2 cement : sand mix with water-retaining and workability admixtures, are factory pre-mixed and require only the addition of water, preferably mixed in with an electrically powered plasterer's whisk. The mortar is applied with a proprietary applicator of the appropriate width or through a pumped system to achieve a uniform thickness of typically 2 mm. Work should only proceed at temperatures above 5°C.

The main advantages of thin-joint systems over traditional 10 mm joint blockwork are:

- increased productivity allowing storey-height inner leaves to be completed in one day;
- up to 10% improved thermal performance due to reduced thermal bridging by the mortar;
- improved airtightness of the construction;
- the accuracy of the wall which allows internal thin-coat sprayed plaster finishes to be used;
- higher quality of construction and less wastage of mortar.

The acoustic properties of thin-joint mortar walls differ slightly from those of walls constructed with 10 mm mortar joints. Resistance to low-frequency noise is slightly enhanced, while resistance to high-frequency sound is slightly reduced.

Completed thin-joint blockwork acts as a monolithic slab, which, if unrestrained, may crack at the weaker points, such as near openings. To avoid this, the block units should be laid dry to avoid shrinkage and bed joint reinforcement (typically 1.25 mm stainless or galvanised steel welded mesh or 1.5 mm GRP mesh) should be appropriately positioned. Larger structures require movement joints at 6 m centres.

Certain extruded multi-perforated clay and calcium silicate blocks are designed for use with thin mortar bed joints and dry interlocking vertical joints. One system of clay blocks requires only horizontal proprietary adhesive joints of 1 mm applied with a special roller

tool, as the units are ground to exact dimensions after firing. Blocks may be used for inner and/or outer leaf cavity construction or for solid walls. While this reduces the initial construction time, exposed sides of the units subsequently require plaster or cement render to minimise heat loss by air leakage. Typical block sizes are 300 × 224 mm and 248 × 249 mm with widths of 100, 140, 190 and 365 mm.

BOND

A running half-block bond is standard, but this may be reduced to a quarter bond for aesthetic reasons. Blockwork may incorporate banding of concrete bricks, but because of differences in thermal and moisture movement, it is inadvisable to mix clay bricks with concrete blocks. Horizontal and vertical stack bond and more sophisticated variations, such as basket-weave bond, may be used for infill panels within framed structures (Fig. 2.12). Such panels will require reinforcement within alternate horizontal bed joints, to compensate for the lack of normal bonding.

REINFORCEMENT

Blockwork will require bed-joint reinforcement above and below openings where it is inappropriate to divide the blockwork into panels, with movement joints at the ends of the lintels. Bed-joint reinforcement would be inserted into two bed joints above and below such openings (Fig. 2.13). Cover to bed reinforcement should be at least 25 mm on the external faces and 13 mm on the internal faces. Combined vertical and horizontal reinforcement may be incorporated into hollow blockwork where demanded by the calculated stresses. Typical situations would be within retaining basement walls, and large infill panels to a framed structure.

MOVEMENT CONTROL

Concrete blockwork is subject to greater movements than equivalent brickwork masonry. Therefore, the location and form of the movement joints requires greater design-detail consideration, to ensure that inevitable movements are directed to the required locations and do not cause unsightly stepped cracking or fracture of individual blocks. Ideally, such movement joints should be located at intersecting walls, or other points of structural discontinuity, such as columns. In addition, movement joints are required at changes

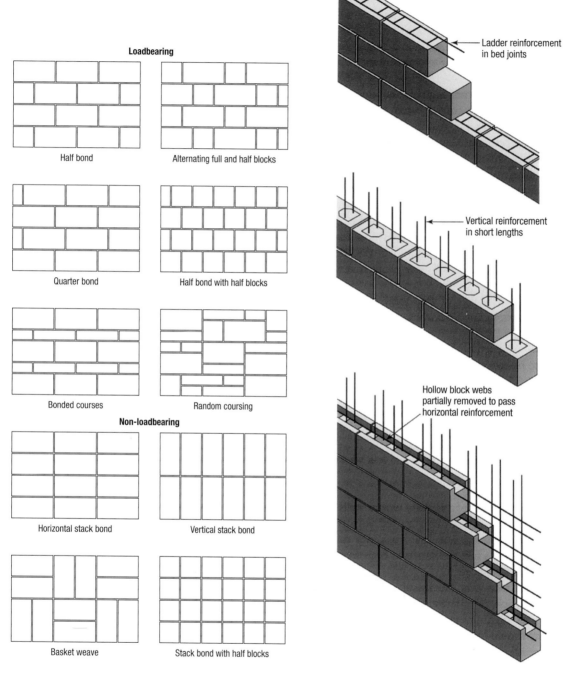

Loadbearing

Half bond

Alternating full and half blocks

Quarter bond

Half bond with half blocks

Bonded courses

Random coursing

Non-loadbearing

Horizontal stack bond

Vertical stack bond

Basket weave

Stack bond with half blocks

Ladder reinforcement
in bed joints

Vertical reinforcement
in short lengths

Hollow block webs
partially removed to pass
horizontal reinforcement

Fig. 2.12 Selection of bonding patterns for visual blockwork

Fig. 2.13 Reinforced blockwork

Typical changes of construction

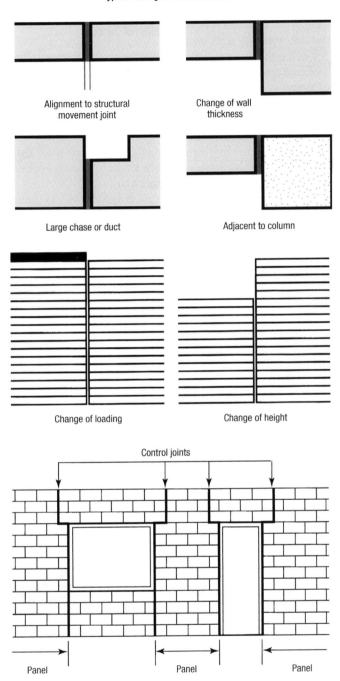

Alignment to structural
movement joint

Change of wall
thickness

Large chase or duct

Adjacent to column

Change of loading

Change of height

Control joints

Panel   Panel   Panel

Fig. 2.14 Blockwork movement joints

in thickness, height or loading of walls, above and below wall openings, and adjacent to movement joints in the adjoining structure (Fig. 2.14). External unreinforced non-load-bearing concrete masonry walls with a length to height ratio of 3 :1 or less must be separated into a series of panels with vertical movement-control joints at approximately 9 m centres or more frequently for masonry walls with a length to height ratio of more than 3 : 1 (NA to BS EN 1996–2: 2006).

Wall ties to PD 6697: 2010 should allow for differential movement between the leaves in cavity construction and should normally be spaced at 900 mm horizontally and 450 mm vertically for 50–300 mm cavities.

MORTARS

The mortar must always be weaker than the blocks to allow for movement. The usual mixes for standard 10 mm joints are by volume:

cement/lime/sand	1 : 1 : 5 to 1 : 1 : 6
cement/sand + plasticiser	1 : 5 to 1 : 6
masonry cement/sand	1 : 4 to 1 : 5

Below DPC level a stronger mix is required and sulphate-resisting cement may be necessary depending upon soil conditions:

cement/sand	1 : 4
cement/lime/sand	1 : 1½ : 4½

Where high-strength blockwork is required, stronger mortars may be necessary. Mortar joints should be slightly concave rather than flush. Bucket handle and weathered or struck joints are suitable for external use, but recessed joints should only be used internally. Coloured mortars should be ready mixed or carefully gauged to prevent colour variations. Contraction joints should be finished with a bond breaker of polythene tape and flexible sealant. For expansion joints, a flexible filler is required (e.g. bitumen-impregnated fibreboard with a polythene foam strip and flexible sealant). Where blockwork is to be rendered, the mortar should be raked back to a depth of 10 mm for additional key. Masonry should not normally be built when the temperature is at or below 3°C and falling or unless it is at least 1°C and rising. However, specially formulated thin-joint mortar is available for working between 0°C and 3°C. Aggregates for mortar are specified in BS EN 13139: 2013.

FINISHES

Internal finishes

Plaster should normally be applied in two coats to 13 mm. Blocks intended for plastering have a textured surface to give a good key. Dry lining may be fixed with battens or directly with adhesive dabs to the blockwork. Blockwork to be tiled should be first rendered with a cement/sand mix. Fairfaced blockwork may be left plain or painted. Where standard blocks are to be painted, the appropriate grade should be used.

Unfired-clay blocks should be finished with breathable materials, such as clay or lime plaster, clay boards, limewash or highly vapour-permeable paint.

External finishes

External boarding or hanging tiles should be fixed to battens, separated from the blockwork with a breather membrane. For external rendering a spatterdash coat should be applied initially on dense blockwork, followed by two coats of cement/lime/sand render. The first 10 mm coat should be the stronger mix (e.g. 1 : 1 : 6), and the 5 mm second coat must be weaker (e.g. 1 : 2 : 9). Cement/sand mixes are not recommended, as they are more susceptible to cracking and crazing than mixes incorporating lime. The render should terminate at damp-proof course level with a drip or similar weathering detail.

FOUNDATIONS

Foundation blocks laid flat offer an alternative to trench fill or cavity masonry. Portland cement blocks should not be used for foundations where sulphate-resisting cement mortar is specified, unless they are classified as suitable for the particular sulphate conditions. Sulphate and other chemically adverse ground conditions are classified in the BRE Special Digest 1 (2005) from DS1 (Design Sulphate Class 1) to the most aggressive, DS5. Foundation blocks can be of either dense or appropriate lightweight concrete, the latter providing enhanced floor-edge insulation. Interlocking foundation blocks, with a tongue and groove vertical joint, slot together with only bed-joint mortar being required. A handhold makes manipulating these blocks on site much easier than lifting standard rectangular blocks. Dimensions within the range 255–355 mm are available for various wall thicknesses.

Beam and block flooring

Beam and block flooring is the standard alternative to floating ground floors within domestic-scale construction (Fig. 2.15), and may also be used for upper storeys. Systems are described in BS EN 15037. Beams may be inverted T or I in form, or alternatively incorporate partially exposed lattice girder reinforcement to be incorporated within a concrete topping (BS EN 15037 Part 1: 2008).

The infill may be concrete blocks (BS EN 15037 Part 2: 2009), clay blocks (BS EN 15037 Part 3: 2009) or lightweight blocks of wood, plastic, GRP, metal, cardboard or polymer concrete (BS EN 15037 Part 5: 2013). Systems are overlaid with a structural floor topping incorporating the additional insulation required to achieve the appropriate U-value, typically 0.13 W/m² K.

BS EN 15037 Part 4: 2010 describes the use of cut or moulded expanded polystyrene blocks which may incorporate a tongue to give insulation under the concrete beams. A concrete structural topping, reinforced with steel mesh or polypropylene fibres, forms the floor finish. Typical U-values range from 0.20 to 0.07 W/m² K.

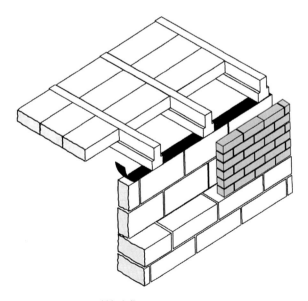

Fig. 2.15 Beam and block flooring

The following material combination achieves a U-value of 0.13 W/m² K, depending on the perimeter/area ratio:

structural floor topping
140 mm polyisocyanurate foam insulation
  (λ = 0.022 W/m K)
100 mm dense concrete block
  (λ = 1.13 W/m K)
dense concrete inverted T-beam
  (λ = 1.65 W/m K)
underfloor ventilated space.

Landscape blockwork

BLOCK PAVING

Concrete block paving units are manufactured to a wide range of designs as illustrated in Figure 2.16. Blocks may be of standard brick form (200 × 100 mm) to thicknesses of 60, 80 or 100 mm depending on the anticipated loading. Alternative designs include tumbled blocks, which emulate granite setts, and various interlocking forms giving designs based on polygonal and curvilinear forms. Colours range from red, brindle, buff, brown, charcoal and grey through to silver and white, with smooth, textured or simulated stone finishes. For most designs, a range of kerb blocks, drainage channels, edging and other accessory units are available. Concrete paving blocks are usually laid on a compacted sub-base with 50 mm of sharp sand. Blocks are frequently nibbed to create a narrow joint to be filled with kiln-dried sand. For the wider joints that occur between the simulated stone setts a coarser grit may be used to prevent loss by wind erosion.

Physical properties, including water absorption, freeze/thaw resistance, abrasion resistance and tolerances on size, are categorised in BS EN 1338: 2003. Guidance on the design and construction of pavements for a range of applications is given in BS 7533 Parts 1–13 inclusive. The types of tactile paving surfaces – blister, rib and groove – are defined and coded in the standard DD CEN/TS 15209: 2008.

Sustainable drainage systems (SuDS) are designed to reduce the environmental impact of impermeable hard landscaping surfaces which create rapid rainwater run-off. With permeable surfaces, including nibbed blocks, the rainwater permeates through the spacing and is dispersed by natural drainage into the underlying soil, or may be collected through rainwater harvesting systems for further use.

Where the appearance of grass is required, but with the traffic-bearing properties of a concrete block

Fig. 2.16 Selection of concrete pavers and hard landscape to the Gateshead Millennium Bridge. *Photographs*: Courtesy of Marshalls plc

pavement, a selection of porous (hollow) blocks is available which can be filled with soil and seeded to give the required effect. Different block depths and sub-bases may be specified according to the anticipated traffic loading. Sulphate-resisting blocks are available if dictated by the soil conditions.

EARTH-RETAINING BLOCKWORK

A range of precast-cellular concrete-interlocking blocks is manufactured for the construction of dry-bed retaining walls. Soil is placed in the pockets of each successive course to allow for planting. The rear is backfilled with granular material to allow for drainage. The size of the block determines the maximum construction height, but over 20 m can be achieved with very deep units. A face angle of 15°–22° is typical to ensure stability, but other gradients are possible with the appropriate block systems. Limited wall curvature is possible without cutting the standard blocks. The systems are used both for earth retention and to form acoustic barriers.

References

FURTHER READING

Architects' Journal, 11.2010: *Masonry AJ Specification.* London: AJ.

British Cement Association. 2005: *BCA guide to materials for masonry mortar.* Camberley: BCA.

Concrete Block Association. 2006: *Aggregate concrete blocks. Part L. Thermal insulation from April 2006. Guidance for designers and users.* Leicester: CBA.

Concrete Block Association. 2007: *Aggregate concrete blocks. Aggregate block sustainability.* Data Sheet 16. Leicester: CBA.

Concrete Society. 2007: *External in-situ concrete paving.* Technical Report No. 66. Camberley: The Concrete Society.

Durkin, J. 2012: *Brickwork and blockwork.* Oxford: Wiley-Blackwell.

Hugues, T., Greilich, K. and Peter, C. 2004: *Building with large clay blocks. Details, products, built examples.* Basel: Birkhäuser.

Robust Details. 2011: *Robust details handbook*, 3rd edn (2013 update). Milton Keynes: Robust Details Ltd.

STANDARDS

BS 743: 1970
    Materials for damp-proof courses.
BS 5977
    Lintels:
        Part 1: 1981   Method for assessment of load.
BS 6073
    Precast concrete masonry units:
        Part 2: 2008   Guide for specifying precast concrete masonry units.
BS 6100
    Glossary of building and civil engineering terms:
        Part 0: 2010   Introduction and index.
        Part 1: 2004   General terms.
        Part 6: 2008   Construction parts.
BS 6398: 1983
    Specification for bitumen damp-proof courses for masonry.
BS 6515: 1984
    Specification for polyethylene damp-proof courses for masonry.
BS 7533
    Pavements constructed with clay, natural stone or concrete pavers:
        Part 1: 2001   Guide for the structural design of heavy duty pavements constructed of clay pavers or precast concrete paving blocks.
        Part 2: 2001   Guide for the structural design of lightly trafficked pavements constructed of clay pavers or precast concrete paving blocks.
        Part 3: 2005   Code of practice for laying precast concrete paving blocks and clay pavers for flexible pavements.
        Part 4: 2006   Code of practice for the construction of pavements of precast concrete flags or natural stone slabs.
        Part 6: 1999   Code of practice for laying natural stone, precast concrete and clay kerb units.
        Part 8: 2003   Guide for the structural design of lightly trafficked pavements of concrete flags and natural stone flags.
        Part 9: 2010   Code of practice for the construction of rigid pavements of clay pavers.

Part 10: 2010　Guide for the structural design of trafficked pavements constructed of natural stone setts and bound construction with concrete paving blocks.

Part 12: 2006　Guide to the structural design of trafficked pavements constructed on a bound base using concrete paving flags and natural stone slabs.

Part 13: 2009　Guide for the design of permeable pavements constructed with concrete paving blocks and flags, natural stone slabs and setts and clay pavers.

BS 8000
Workmanship on building sites:
Part 3: 2001　Code of practice for masonry.

BS 8103
Structural design of low rise buildings:
Part 2: 2013　Code of practice for masonry walls for housing.

BS 8215: 1991
Code of practice for design and installation of damp-proof courses in masonry construction.

BS EN 413–1: 2011
Masonry cement. Composition, specifications and conformity criteria.

BS EN 771
Specification for masonry units:
Part 1: 2011　Clay masonry units.
Part 2: 2011　Calcium silicate masonry units.
Part 3: 2011　Aggregate concrete masonry units.
Part 4: 2011　Autoclaved aerated concrete masonry units.
Part 5: 2011　Manufactured stone masonry units.

BS EN 772
Methods of test for masonry units:
Part 1: 2011　Determination of compressive strength.

BS EN 845
Specification for ancillary components for masonry:
Part 1: 2013　Wall ties, tension straps, hangers and brackets.
Part 2: 2013　Lintels.
Part 3: 2013　Bed joint reinforcement of steel meshwork.

BS EN 934
Admixtures for concrete, mortar and grout:
Part 1: 2008　Common requirements.
Part 2: 2009　Concrete admixtures. Definitions, requirements, conformity, marking and labelling.
Part 3: 2009　Admixtures for masonry mortar. Definitions, requirements, conformity, marking and labelling.
Part 4: 2009　Admixtures for grout for prestressing tendons.

BS EN 998–2: 2010
Specification for mortar for masonry. Masonry mortar.

BS EN 1338: 2003
Concrete paving blocks. Requirements and test methods.

BS EN 1365–1: 2012
Fire resistance tests for loadbearing elements. Walls.

BS EN 1745: 2012
Masonry and masonry products. Methods for determining thermal properties.

BS EN 1806: 2006
Chimneys. Clay/ceramic flue blocks for single wall chimneys.

BS EN 1858: 2008
Chimneys. Components. Concrete flue blocks.

BS EN 1996
Eurocode 6: Design of masonry structures:
Part 1.1: 2005　General rules for reinforced and unreinforced masonry.
Part 1.2: 2005　Structural fire design.
Part 2: 2006　Design considerations, selection of materials and execution of masonry.
Part 3: 2006　Simplified calculation methods for unreinforced masonry structures.

BS EN 12859: 2011
Gypsum blocks. Definitions, requirements and test methods.

BS EN 12860: 2001
Gypsum based adhesives for gypsum blocks. Definitions.

BS EN 13139: 2013
Aggregates for mortar.

BS EN 13501

Fire classification of construction products and building elements:

Part 1: 2007   Classification using test data from reaction to fire tests.

Part 2: 2007   Classification using test data from fire resistance tests.

BS EN 13967: 2012

Flexible sheets for waterproofing. Plastic and rubber damp proof sheets including plastic and rubber basement tanking sheets. Definitions and characteristics.

BS EN ISO 14683: 2007

Thermal bridges in building construction. Linear thermal transmittance. Simplified methods and default values.

BS EN 14909: 2012

Flexible sheets for waterproofing. Plastic and rubber damp proof courses. Definitions and characteristics.

BS EN 15037

Precast concrete products. Beam-and-block floor systems:

Part 1: 2008   Beams.

Part 2: 2009   Concrete blocks.

Part 3: 2009   Clay blocks.

Part 4: 2010   Expanded polystyrene blocks.

Part 5: 2013   Lightweight blocks for simple formwork.

BS EN 15080–12: 2011

Extended application of results from fire resistance tests. Loadbearing masonry walls.

DD CEN/TS 15209: 2008

Tactile paving surface indicators produced from concrete, clay and stone.

BS EN 15254

Extended application of results from fire resistance tests. Nonloadbearing walls:

Part 2: 2009   Masonry and gypsum blocks.

BS EN 15318: 2007

Design and application of gypsum blocks.

BS EN 15435: 2008

Precast concrete products. Normal weight and lightweight concrete shuttering blocks. Product properties and performance.

PD CEN/TR 15728: 2008

Design and use of inserts for lifting and handling of precast concrete elements.

PD 6678: 2005

Guide to the specification of masonry mortar.

PD 6682–1: 2003

Aggregates for mortar. Guidance on the use of BS EN 13139.

PD 6697: 2010

Recommendations for the design of masonry structures to BS EN 1996–1-1 and 1996–2.

DD 140–2: 1987

Wall ties. Recommendations for design of wall ties.

BUILDING RESEARCH ESTABLISHMENT PUBLICATIONS

BRE Special digests

BRE SD1: 2005

Concrete in aggressive ground.

BRE SD4: 2007

Masonry walls and beam and block floors. U-values and building regulations.

BRE Digests

BRE Digest 460: 2001

Bricks, blocks and masonry made from aggregate concrete (Parts 1 and 2).

BRE Digest 461: 2001

Corrosion of metal components in walls.

BRE Digest 468: 2002

AAC 'aircrete' blocks and masonry.

BRE Digest 487: 2004

Structural fire engineering design. Part 4 Materials behaviour: Masonry.

BRE Good building guides

BRE GBG 44: 2000

Insulating masonry cavity walls (Parts 1 and 2).

BRE GBG 50: 2002

Insulating solid masonry walls.

BRE GBG 54: 2003

Construction site communication. Part 2 Masonry.

BRE GBG 58: 2003

Thin layer masonry mortar.

BRE GBG 62: 2004

Retro-installation of bed joint reinforcement in masonry.

BRE GBG 66: 2005

Building masonry with lime-based bedding mortars.

BRE GBG 67: 2006   Achieving airtightness (Parts 1, 2 and 3).

BRE GBG 68: 2006   Installing thermal insulation (Parts 1 and 2).

BRE Information papers

BRE IP 7/05
Aircrete tongue and grooved block masonry.

BRE IP 1/06
Assessing the effects of thermal bridging at junctions and around openings.

BRE IP 8/08
Determining the minimal thermal resistance of cavity closers.

BRE IP 16/11
Unfired clay masonry.

BRE Reports

BR 443: 2006
Conventions for U-value calculations.

BR 497: 2007
Conventions for calculating linear thermal transmittance and temperature factors.

EP 99: 2009
Sustainable masonry construction.

ADVISORY ORGANISATIONS

Aircrete Products Association, 4th floor, 60 Charles Street, Leicester LE1 1FB (0116 253 6161).

British Concrete Masonry Association, Grove Crescent House, 18 Grove Place, Bedford MK40 3JJ (01234 353745).

British Precast, 60 Charles Street, Leicester LE1 1FB (0116 253 6161).

Concrete Society, Riverside House, 4 Meadows Business Park, Station Approach, Blackwater, Camberley, Surrey GU14 9AB (01276 607140).

Mortar Industry Association, Gillingham House, 38–44 Gillingham Street, London SW1V 1HU (0207 963 8000).

3

LIME, CEMENT AND CONCRETE

Introduction

In its broadest sense, the term *cement* refers to materials which act as adhesives. However, in this context, its use is restricted to that of a binding agent for sand, stone and other aggregates within the manufacture of mortar and concrete. Hydraulic cements and limes set and harden by internal chemical reactions when mixed with water. Non-hydraulic materials will only harden slowly by absorption of carbon dioxide from the air.

Lime was used as a binding agent for brick and stone by the ancient civilisations throughout the world. The concept was brought to Britain in the first century AD by the Romans, who used the material to produce lime mortar. Outside Britain, the Romans frequently mixed lime with volcanic ashes, such as pozzolana from Pozzuoli in Italy, to convert a non-hydraulic lime into hydraulic cement suitable for use in constructing aqueducts, baths and other buildings. However, in Britain, lime was usually mixed with artificial pozzolanas, for example, crushed burnt-clay products, such as pottery, brick and tile. In the eighteenth century, so-called *Roman cement* was manufactured by burning of the *cement stone* (argillaceous or clayey limestone), collected from the coast around Sheppey and Essex.

In 1824, Joseph Aspdin was granted his famous patent for the manufacture of *Portland cement* from limestone and clay. Limestone powder and clay were mixed into a water slurry which was then evaporated by heat in *slip pans*. The dry mixture was broken into small lumps, calcined in a kiln to drive off the carbon dioxide, burnt to clinker and finally ground into a fine powder for use. The name *Portland* was used to enhance the prestige of the new concrete material by relating it to Portland stone which, to some degree, it resembled. Early manufacture of Portland cement was by intermittent processes within bottle, and later chamber, kilns. The introduction of the rotating furnace in 1877 offered a continuous burning process with consequent reductions in fuel and labour costs. The early rotating kilns formed the basis for the development of the various production systems that now exist. The year 1989 was the peak of production

when 18 million tonnes of cement were manufactured within the UK. About half of this was required by the ready-mixed concrete industry; the remainder was divided roughly equally between concrete product factories and bagged cement for general use.

The Cambridge University Sainsbury Laboratory in the University Botanic Garden (Fig. 3.1) illustrates high-quality concrete construction in the context of a Grade II listed garden setting. The building won the RIBA Stirling Prize for architecture in 2012.

Lime

MANUFACTURE OF LIME

Lime is manufactured by calcining natural calcium carbonate, typically hard-rock carboniferous limestone. The mineral is quarried, crushed, ground, washed and screened to the required size range. The limestone is burnt at approximately 950°C in either horizontal rotary kilns or vertical shaft kilns which drive off the carbon dioxide to produce the lime products. Limestone and chalk produce non-hydraulic or air

Fig. 3.1 Cambridge Botanic Garden Sainsbury Laboratory. *Architects*: Stanton Williams. *Photograph*: Arthur Lyons

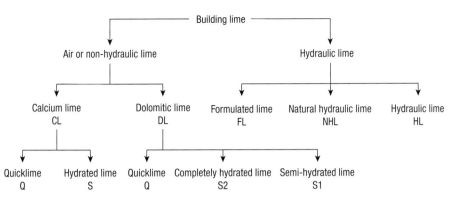

Fig. 3.2 Types of building limes

Table 3.1 Types of building lime (BS EN 459–1: 2010)

Designation	Notation		Designation	Notation
Air or non-hydraulic lime			Hydraulic lime	
Calcium lime 90	CL 90		Natural hydraulic lime 2	NHL 2
Calcium lime 80	CL 80		Natural hydraulic lime 3.5	NHL 3.5
Calcium lime 70	CL 70		Natural hydraulic lime 5	NHL 5
Dolomitic lime 90–30	DL 90–30		Formulated lime A 2	FL A 2
Dolomitic lime 90–5	DL 90–5		Formulated lime A 3.5	FL A 3.5
Dolomitic lime 85–30	DL 85–30		Formulated lime A 5	FL A 5
Dolomitic lime 80–5	DL 80–5		Formulated lime B 2	FL B 2
			Formulated lime B 3.5	FL B 3.5
			Formulated lime B 5	FL B 5
			Formulated lime C 2	FL C 2
			Formulated lime C 3.5	FL C 3.5
			Formulated lime C 5	FL C 5
			Hydraulic lime 2	HL 2
			Hydraulic lime 3.5	HL 3.5
			Hydraulic lime 5	HL 5

Notes:
– Air or non-hydraulic lime is classified according to its form – quicklime (Q) or hydrated lime (S).
– Calcium air lime is available as quicklime (Q), hydrated as powder (S), lime putty (S PL) or milk of lime (S ML).
– Dolomitic air lime is available as quicklime (Q), hydrated lime (S) or semi-hydrated dolomitic lime (S1).
– The notation numbers for calcium lime refer to the total (CaO + MgO) content.
– The notation numbers for dolomitic lime refer to the total (CaO + MgO) and (MgO) contents respectively.

Notes:
– Hydraulic lime is classified according to its compressive strength. The three categories are natural hydraulic lime (NHL), formulated lime (FL) and hydraulic lime (HL).
– The three grades of formulated lime refer to different levels of lime content as $Ca(OH)_2$. (FL A 40–80%, FL B 25–50% and FL C 15–40% lime respectively.)

limes, while limestone/clay mixtures produce hydraulic limes. Subsequent treatments produce the product range shown in Figure 3.2. The standard BS EN 459–1: 2010 describes the full range of building limes. Broadly, air or non-hydraulic lime is classified according to its calcium and magnesium oxide content, while hydraulic lime is classified according to its compressive strength (Table 3.1).

NON-HYDRAULIC OR AIR LIME

Quicklime

When burnt, limestone and chalk produce quicklime powder or lump lime (calcium oxide). Quicklimes include calcium lime (CL) and dolomitic lime (DL), depending on the composition of the starting mineral.

Dolomitic lime contains significant quantities of magnesium oxide (Table 3.1).

$$CaCO_3 \xrightarrow{950°C} CaO + CO_2$$
calcium carbonate　　　　　lime　　carbon dioxide

Slaking of lime

Slaking – the addition of water to quicklime – is a highly exothermic reaction. The controlled addition of water to quicklime produces hydrated lime (S) (mainly calcium hydroxide) as a dry powder:

$$CaO + H_2O \longrightarrow Ca(OH)_2$$
lime　　　water　　　　calcium hydroxide

It is suitable for use within mortars or in the manufacture of certain aerated concrete blocks. Generally, the addition of lime to cement mortar, render or plaster increases its water-retention properties, thus retaining workability, particularly when the material is applied to absorbent substrates such as porous brick. Lime also increases the cohesion of mortar mixes, allowing them to spread more easily. Hydrated lime absorbs moisture and carbon dioxide from the air, and should therefore be stored in a cool, draught-free building and used while still fresh.

Lime putty

Lime putty is produced by slaking quicklime with an excess of water for a period of several weeks until a creamy texture is produced. Alternatively, it can be made by stirring hydrated lime into water, followed by conditioning for at least 24 hours. However, the traditional direct slaking of quicklime produces finer particle sizes in the slurry, and the best lime putty is produced by maturing for at least six months. Lime putty may be blended with Portland cement in mortars where its water-retention properties are greater than those afforded by hydrated lime. In addition, lime putty, often mixed with sand to form *coarse stuff*, is used directly as a pure lime mortar, particularly in restoration and conservation work. It sets, not by reaction with sand and water, but only by carbonation, and is therefore described as *non-hydraulic*. Lime wash, as a traditional surface coating, is made by the addition of sufficient water to lime putty to produce a thin, creamy slurry.

Carbonation

Lime hardens by the absorption of carbon dioxide from the air, which gradually reconverts the calcium oxide back to calcium carbonate:

$$CaO + CO_2 \xrightarrow{slow} CaCO_3$$
lime　　carbon dioxide　　　calcium carbonate

The carbonation process is slow, as it is controlled by the diffusion of carbon dioxide into the bulk of the material. When sand or stone dust aggregate is added to the lime putty to form a mortar or render, the increased porosity allows greater access of carbon dioxide and a speedier carbonation process. The maximum size of aggregate mixed into lime mortars should not exceed half the mortar-joint width.

HYDRAULIC LIMES

Hydraulic limes are manufactured from chalk or limestone containing various proportions of clay impurities. The materials produced have some of the properties of Portland cement, and partially harden through hydration processes, rather than solely through carbonation as happens with non-hydraulic pure calcium oxide lime. Hydraulic limes rich in clay impurities are more hydraulic and set more rapidly than those with only a low silica and alumina content. Natural hydraulic limes (NHL) are traditionally categorised as *feebly, moderately* or *eminently hydraulic* depending on their clay content, which is in the ranges 0–8%, 8–18% and 18–25%, respectively. These traditional grades equate approximately to the 28-day compressive strengths of 2, 3.5 and 5 MPa, respectively, for NHL2, NHL3.5 and NHL5. Eminently hydraulic lime mortar is used for masonry in exposed situations, moderately hydraulic lime mortar for most normal masonry applications, and feebly hydraulic lime mortar is appropriate for conservation work and solid wall construction. Grey semi-hydraulic lime is still produced within the UK in small quantities from chalk containing a proportion of clay. It is used with very soft bricks and for conservation work.

Hydraulic lime (HL) is produced by blending lime with other constituents such as cement, blast furnace slag or fly ash in appropriate proportions. It is usually imported from France and is mainly used for the restoration of historic buildings, where the use of modern materials would be inappropriate. It is gauged with sand only, giving a mix which develops an initial set within a few hours, but which hardens over an extended period of time. The workable render or mortar mixes adhere well and, because the material is flexible, the risks of cracking and poor adhesion are reduced. The dried mortar is off-white in colour and contains very little alkali, which in Portland cement mortars can cause staining, particularly on limestone. Hydraulic lime may be used for interior lime washes, and also for fixing glass bricks where a flexible binding agent with minimum shrinkage is required. Unlike hydrated lime, hydraulic lime is little affected by exposure to dry air during storage.

Formulated lime (FL) is a blend of air lime (CL) and/or natural hydraulic lime (NHL) with additional hydraulic or pozzolanic material. It sets and hardens when mixed with water and by carbonation in the air.

LIME MORTAR

Various advantages of lime-based mortars over Portland cement mortars are reported. The production of building lime consumes less energy, thus reducing greenhouse emissions compared to the equivalent manufacture of Portland cement. The subsequent car-bonation process removes CO_2 from the atmosphere. Lime-based mortars remain sufficiently flexible to allow thermal and moisture movement, but, in addition, due to the presence of uncarbonated lime, any minor cracks are subsequently healed by the action of rainwater. The recycling of bricks and blocks is easier due to the lower adherence of the mortar. Lime mortar construction is more breathable than Portland cement masonry and lime mortars are more resistant to sulphate attack than standard Portland cement mixes due to their lower tricalcium aluminate content.

Typical lime mortars are within the range 1 : 2 and 1 : 3, lime : aggregate ratio. A 1 : 2 lime : sand mix made with NHL3.5 lime equates approximately to the NA to BS EN 1996–1-1: 2005 designation (iii) class M4 mortar, and a 1 : 3 mix equates approximately to a designation (iv) class M2 Portland cement mortar. A well-graded sharp sand should be used. Because of the slow carbonation process, masonry lifts are limited, and the mortar must be allowed some setting time to prevent its expulsion from the joints. Little hardening occurs at temperatures below 5°C.

HEMP LIME

Hemp is grown, particularly in France, for its fibre, which is used in the manufacture of certain grades of paper. The remaining 75% of the hemp stalks, known as hemp hurd or shiv, is a lightweight absorbent material which has the appearance of fine wood chips. When mixed with hydraulic lime it produces a cement mixture which sets within a few hours and gradually 'petrifies' to a lightweight solid due to the high silica content of hemp hurd. The material mix can be poured and tamped or sprayed as required and formwork may be removed after 24 hours or less. The set material, sometimes referred to as 'hempcrete', which has good thermal insulation properties and a texture similar to cork, has been used for the construction of floors, non-load-bearing walls using plywood or plastic formwork, as well as blocks and large panels. The material is also used as a solid infill for timber frame construction. In this case the combination of the moisture-absorbing properties of the hemp with the nature of lime affords some protection to the timber framing which it encloses. Hemp lime should not be used below ground level. The moisture-absorbing properties of hemp lime give rise to higher thermal efficiencies than are calcu-lated for the material based on conventional thermal transmittance data. To build a conventional house of hemp lime would require approximately 40 m^3 of the material containing 7–10 tonnes of hemp, produced from about 1 ha of land.

EXTERNAL LIME RENDERING

External lime rendering is usually applied in a two- or three-coat system, to give an overall thickness of up to 30 mm. In exposed situations, hydraulic lime is used and the thicker initial coat may be reinforced with horsehair. The final coat may be trowelled to receive a painted finish; alternatively, pebble dash or rough cast may be applied.

Cement

MANUFACTURE OF PORTLAND CEMENT

Portland cement is manufactured from calcium carbonate in the form of crushed limestone or chalk and an argillaceous material such as clay, marl or shale. Currently, approximately two-thirds of UK Portland cement production is based on limestone, with the remainder from chalk in the south and east of the country. Minor constituents such as iron oxide or sand may be added depending upon the composition of the raw materials and the exact product required.

In principle, the process involves the decarbonisation of calcium carbonate (chalk or limestone) by expulsion of the carbon dioxide, and sintering, at the point of incipient fusion, the resulting calcium oxide (lime) with the clay and iron oxide. Depending on the raw materials used and their water content at extraction, three key variations in the manufacturing process are currently in use. These are the *semi-wet*, *semi-dry* and *dry* processes. The UK produces over 10 million tonnes of Portland cement each year.

Semi-wet process

In the semi-wet process, chalk is broken down in water and blended into a marl clay slurry. The 40% water content within the slurry is reduced to 19% in a filter press; the resulting *filter-cake* is nodularised by extrusion onto a travelling preheater grate or reduced to pellets in a crusher/dryer. Heating to between 900°C and 1100°C precalcines the chalk; the mix is then transferred to a short kiln at 1450°C for the clinkering process.

Semi-dry process

In the semi-dry process, dry shale and limestone powders are blended. About 12% water is added to nodularise the blend, which is then precalcined and clinkered as in the semi-wet process.

Dry process

In the dry process (Fig. 3.3) limestone, shale and sand (typically 80%, 17% and 3%, respectively) are milled to fine powders, then blended to produce the *dry meal*,

which is stored in silos. The meal is passed through a series of cyclones, initially using recovered kiln gases to preheat it to 750°C, then with added fuel to precalcine at 900°C, prior to passage into a fast-rotating 60 m kiln for clinkering at 1450°C. In all processes an intimately mixed feedstock to the kiln is essential for maintaining quality control of the product. Most plants operate primarily with powdered coal, but in addition, other fuels including petroleum coke, waste tyre chips, smokeless fuel plant residues or reclaimed spoil heap coal are used when available. Oil and natural gas have also been used when economically viable. The grey/black clinker manufactured by all processes is cooled with full heat recovery and ground up with 5% added gypsum (calcium sulphate) retarder to prevent excessively rapid *flash* setting of the cement.

The older cement grinding mills are *open circuit* allowing one pass of the clinker, which produces a wide range of particle size. This product is typically used for the production of precast concrete products. The newer cement mills are closed circuit, with air separators to extract fine materials and with recycling of the oversized particles for regrinding. This product is frequently used in the ready-mixed market, as it can be controlled to produce cement with higher later strength. To reduce grinding costs, manufacturers accept load shedding and use off-peak electrical supplies where possible. The Portland cement is stored in silos prior to transportation in bulk, by road or rail, or in palletised packs. The standard bag is 25 kg for reasons of health and safety.

With the dry processing and additional increases in energy efficiency, a tonne of pulverised coal can now produce in excess of six tonnes of cement clinker. Because the cement industry's output is so large, the combined output of carbon dioxide to the atmosphere,

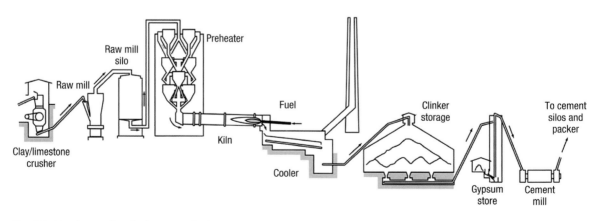

Fig. 3.3 *Manufacture of Portland cement – the dry process*

from fuel and the necessary decarbonation of the limestone or chalk, represents about 2% of the carbon dioxide emissions in Britain and up to 7.5% worldwide. Emissions of oxides of sulphur from the fuel are low, as these gases are trapped into the cement clinker; however, the escape of oxides of nitrogen and dust, largely trapped by electrostatic precipitators, can only be controlled by constantly improving process technologies. On the basis of the final production of concrete, cement manufacture releases considerably less carbon dioxide per tonne than does primary steel manufacture; however, the relative masses for equivalent construction and the recycling potential of each should also be considered.

COMPOSITION OF PORTLAND CEMENT

The starting materials for Portland cement are chalk or limestone and clay, which consist mainly of lime, silica, alumina and iron oxide. Table 3.2 illustrates a typical composition.

Minor constituents, including magnesium oxide, sulphur trioxide, sodium and potassium oxides, amount to approximately 2%. (The presence of the alkali oxides in small proportions may be the cause of the *alkali-silica* reaction, which leads to cracking of concrete when certain silica-containing aggregates are used.) During the clinkering process, these compounds react together to produce the four key components of Portland cement (Table 3.3).

The relative proportions of these major components significantly affect the ultimate properties of the cements and are therefore adjusted in the manufacturing process to produce the required product range. Typical compositions of Portland cements are shown in Table 3.4.

A small reduction in the lime content within the initial mix will greatly reduce the proportion of the tricalcium silicate and produce an equivalent large increase in the dicalcium silicate component of the product. The cement produced will harden more slowly, with a slower evolution of heat. As tricalcium aluminate is vulnerable to attack by soluble sulphates, it is the proportion of this component that is reduced in the manufacture of sulphate-resisting Portland cement.

Under the British Standard BS EN 197–1: 2011, except in the case of sulphate-resisting Portland

Table 3.2 Typical composition of starting materials for Portland cement manufacture

Component	Percentage
Lime	68
Silica	22
Alumina	5
Iron oxide	3
Other oxides	2

Table 3.3 Major constituents of Portland cement and their specific properties

Compound	Chemical formula	Cement notation	Properties
Tricalcium silicate	$3CaO.SiO_2$	C_3S	Rapid hardening giving early strength and fast evolution of heat
Dicalcium silicate	$2CaO.SiO_2$	C_2S	Slow hardening giving slow development of strength and slow evolution of heat
Tricalcium aluminate	$3CaO.Al_2O_3$	C_3A	Quick setting which is retarded by gypsum. Rapid hardening and fast evolution of heat but lower final strength; vulnerable to sulphate attack
Tetracalcium aluminoferrite	$4CaO.Al_2O_3.Fe_2O_3$	C_4AF	Slow hardening; causes grey colour in cement

Table 3.4 Typical compositions of Portland cements

Cement Type	Class	Composition %C_3S	%C_2S	%C_3A	%C_4AF	Fineness (m²/kg)
Portland cement	42.5	55	20	10	8	340
	52.5	55	20	10	8	440
White Portland cement	62.5	65	20	5	2	400
Sulphate-resisting Portland cement	42.5	60	15	2	15	380

cement, up to 5% of minor additional constituents may be added to cement. These constituents must be inorganic materials that do not increase the water requirements of the cement, reduce the durability of the mortar or concrete produced, or cause increased corrosion to any steel reinforcement. In the UK, they are typically limestone powder and dry meal or partially calcined material from the cement manufacturing processes.

SETTING AND HARDENING OF PORTLAND CEMENTS

Portland cement is hydraulic; when mixed with water it forms a paste, which sets and hardens as a result of various chemical reactions between the cementitious compounds and water. The setting and hardening processes are enhanced when excess moisture is present; indeed, Portland cement will harden under water, whereas premature drying out is detrimental to properties. Only a small proportion of the added water is actually required for the chemical hydration of the cementitious constituents to hydrated calcium silicates. The additional water is needed to ensure the *workability* of the mix when aggregates are added, so that concrete, for example, can be successfully placed within formwork containing steel reinforcement. Water in excess of that required for hydration will ultimately evaporate, leaving capillary pores in the concrete and mortar products. Typically, an increase in void space of 1% reduces crushing strength by 6%. It is therefore necessary to control carefully the water content of the mix by reference to the water/cement ratio. A minimum water/cement ratio of 0.23 is required to hydrate all the cement, although, as the cement powder is hydrated, it expands, and thus a ratio of 0.36 represents the point at which cement gel fills all the water space. However, a water/cement ratio of 0.42 more realistically represents the minimum water content to achieve full hydration without the necessity for further water to be absorbed during the curing process.

The setting and hardening processes should be distinguished. Setting is the stiffening of the cement paste which commences immediately the cement is mixed with water. Because the major cementitious constituents set at different rates it is convenient to refer to *initial set* and *final set*. Typically, initial set, or the formation of a plastic gel, occurs after one hour and final set, or the formation of a rigid gel, within 10 hours. The setting process is controlled by the quantity of gypsum added to the cement in the final stages of production. Hardening is the gradual gain in strength of the set cement paste. It is a process which continues, albeit at a decreasing rate, over periods of days, months and years. The rate of hardening is governed partially by the particle-size distribution of the cement powder. Finely ground cement hydrates more rapidly, and therefore begins to set and harden more quickly. Furthermore, the relative proportions of tricalcium silicate and dicalcium silicate have a significant effect on the rate of hardening, as indicated in Table 3.3.

During hydration, any sodium and potassium salts within the Portland cement are released into the pore water of the concrete, giving rise to a highly alkaline matrix. This effectively inhibits corrosion of any reinforcing steel embedded within the concrete, but if active silica is present in any of the aggregates, it may react to form an alkali-silica gel which absorbs water, swells and causes cracking of the concrete. This alkali-silica reaction can however be effectively prevented by limiting the total alkali content in the concrete to less than 3.5 kg/m^3. (Cement manufacturers normally specify alkali content in terms of equivalent percentage of sodium oxide.)

CLASSIFICATION OF CEMENTS

Cements are classified primarily on the type of the main constituents such as Portland cement or blast furnace cement. (In addition, there may be minor constituents up to 5% and also additives up to 1% by weight.)

The standard BS EN 197–1: 2011 lists five main types of cement:

CEM I	Portland cement
CEM II	Portland-composite cement
CEM III	Blast furnace cement
CEM IV	Pozzolanic cement
CEM V	Composite cement

Within these five main types of cement, a wide range of constituents is permitted, including silica fume, natural or industrial pozzolanas, calcareous or siliceous fly ash and burnt shale, may be incorporated. The full range of products is listed in Table 3.5, but not all are commercially available within the UK. In addition to Portland cement, CEM I, current UK factory production includes Portland-limestone cement, Portland-fly ash cement and low early strength blast furnace

Table 3.5 Cements to European Standard EN 197–1: 2011

Cement	Main type	Notation	Portland cement clinker content K (%)	Additional main constituent (%)	
Portland cement	CEM I	CEM I	95–100	0	
Portland slag cement	CEM II	CEM II/A-S	80–94	6–20	S
		CEM II/B-S	65–79	21–35 S	
Portland silica fume cement	CEM II	CEM II/A-D	90–94	6–10	D
Portland pozzolana cement	CEM II	CEM II/A-P	80–94	6–20	P
		CEM II/B-P	65–79	21–35	P
		CEM II/A-Q	80–94	6–20	Q
		CEM II/B-Q	65–79	21–35	Q
Portland fly ash cement	CEM II	CEM II/A-V	80–94	6–20	V
		CEM II/B-V	65–79	21–35	V
		CEM II/A-W	80–94	6–20	W
		CEM II/B-W	65–79	21–35	W
Portland burnt shale cement	CEM II	CEM II/A-T	80–94	6–20	T
		CEM II/B-T	65–79	21–35	T
Portland limestone cement	CEM II	CEM II/A-L	80–94	6–20	L
		CEM II/B-L	65–79	21–35	L
		CEM II/A-LL	80–94	6–20	LL
		CEM II/B-LL	65–79	21–35	LL
Portland composite cement	CEM II	CEM II/A-M	80–88	12–20	S,D,P,Q,V,W,T,L,LL
		CEM II/B-M	65–79	21–35	S,D,P,Q,V,W,T,L,LL
Blast furnace cement	CEM III	CEM III/A	35–64	36–65	S
		CEM III/B	20–34	66–80	S
		CEM III/C	5–19	81–95	S
Pozzolanic cement	CEM IV	CEM IV/A	65–89	11–35	D,P,Q,V,W
		CEM IV/B	45–64	36–55	D,P,Q,V,W
Composite cement	CEM V	CEM V/A	40–64	18–30	S + 18–30 P,Q,V
		CEM V/B	20–38	31–49	S + 31–49 P,Q,V

Notes:

The code letters used in the European Standard are:

D	silica fume	K	Portland cement clinker	L/LL	limestone
M	mixed	P	natural pozzolana	Q	natural calcined pozzolana
S	granulated blast furnace slag	T	burnt shale	V	siliceous fly ash
W	calcareous fly ash				

— Where additional main constituents are incorporated in composite and pozzolanic cements they must be declared by the designation.

— Minor additional constituents are within the range 0–5% mass.

— Limestone LL has a total organic carbon content limit of 0.2%, limestone L has a total organic carbon content limit of 0.5%.

Table 3.6 Strength classes of cements to European Standard BS EN 197–1: 2011

Strength class	Compressive strength (MPa)			
	Early strength		Standard strength	
	2-day minimum	7-day minimum	28-day minimum	28-day maximum
32.5L		12.0	32.5	52.5
32.5N		16.0	32.5	52.5
32.5R	10.0		32.5	52.5
42.5L		16.0	42.5	62.5
42.5N	10.0		42.5	62.5
42.5R	20.0		42.5	62.5
52.5L	10.0		52.5	
52.5N	20.0		52.5	
52.5R	30.0		52.5	

Notes:

The strength class code letters in the standards are: (N) ordinary early strength, (R) high early strength and (L) low early strength development. Strength class (L) is only applicable to the specific low early strength CEM III blast furnace cements.

cement, and these composite cements now account for about 25% of the total supply (i.e. bulk plus bagged/packed) in the UK market. Calcium aluminate cement (also known as high alumina cement (HAC)) has a totally different formulation compared with the range of Portland cements based on calcium silicates.

STRENGTH CLASSES OF CEMENT

The standard strength classes of cement are based on the 28-day compressive strength of mortar prisms, made and tested to the requirements of BS EN 196–1: 2005. The test uses specimens which are 40 × 40 × 160 mm, cast from a mix of 3 parts of CEN (European Committee for Standardisation) standard sand, 1 part of cement and 0.5 part of water. The sample is vibrated and cured for the appropriate time, then broken into halves and compression tested across the 40 mm face. Three specimens are used to determine a mean value from the six pieces.

Each standard strength class (32.5, 42.5 and 52.5) has sub-classes associated with the high early (R) and the ordinary (N) development of early strength (Table 3.6). The strength classes and sub-classes give production standards for cements, but do not specify how a particular mix of cement, aggregate and admixtures will perform as a concrete; this needs to be determined by separate testing.

The most commonly used cement within the UK (formerly ordinary Portland cement or OPC) is currently designated to the standard BS EN 197–1: 2011 as:

CEM I	42.5	N	EN 197–1-CEM I 42,5N
type of cement	strength class	ordinary early strength development	

High early strength Portland cement is designated as:

CEM I	42.5	R	EN 197–1-CEM I 42,5R
type of cement	strength class	high early strength development	

Ordinary early strength, low heat, sulphate-resisting blast furnace cement with a granulated blast furnace slag content of between 66 and 80% and a strength class of 32.5 is designated as:

CEM III/B 32.5		N – LH	EN 197–1-CEM III/B 32,5N-LH/SR
type of cement/ proportion of blast furnace slag	strength class	normal early strength development and low heat	

Portland limestone cement with between 6 and 20% limestone of 0.5% total organic content, a strength class of 32.5 and normal early strength is designated as:

CEM II/A	L	32.5	N	EN 197–1-CEM II/A-L 32,5N
type of cement/ proportion of cement clinker	sub-type limestone	strength class	ordinary early strength development	

PORTLAND CEMENTS

Portland cements – strength classes 32.5, 42.5 and 52.5

The Portland cement classes 32.5, 42.5 and 52.5 correspond numerically to their lower characteristic strengths at 28 days. The 32.5 and 42.5 classes have upper characteristic strengths which are 20 MPa greater than the lower characteristic strengths, as designated by the class number. Class 52.5 has no upper strength limit. Statistically, the tested strengths must fall, with no more than 5% of the tests below the lower limit or 10% of the tests above the upper limit. Thus class 42.5 Portland cement has a strength within the range 42.5–62.5 MPa, with a maximum of 5% of test results being below 42.5 MPa and a maximum of 10% of the test results being above 62.5 MPa.

Each class also has lower characteristic strength values at two days, except for class 32.5, which has a lower characteristic strength at seven days. Where high early strength is required, for example, to allow the early removal of formwork in the manufacture of precast concrete units, class 52.5 or class 42.5R is used. These Portland cements are more finely ground than class 42.5 to enable a faster hydration of the cement in the early stages. Class 32.5 cements, for general-purpose and DIY use, frequently contain up to 1% additives to improve workability and frost resistance, together with up to 5% minor additional constituents such as limestone fines, fly ash or granulated blast furnace slag. Portland cement of strength class 42.5 accounts for approximately 90% of the total cement production within the UK.

White Portland cement
White Portland cement is manufactured from materials virtually free of iron oxide and other impurities, which impart the grey colour to Portland cement. Generally, china clay and limestone are used and the kiln is fired with natural gas or oil rather than pulverised coal. Iron-free mills are used for the grinding process to prevent colour contamination. Because of the specialist manufacturing processes, it is approximately twice

Table 3.7 Sulphate resisting cements to European Standard BS EN 197–1: 2011

Cement type	Notation	Portland cement clinker K (% mass)	Other main constituent (% mass)	Additional minor constituents (% mass)
Sulphate resisting Portland cement CEM I	CEM I-SR 0	95–100	0	0–5
	CEM I-SR 3	95–100	0	0–5
	CEM I-SR 5	95–100	0	0–5
Sulphate resisting blast furnace cement CEM III	CEM III/B-SR	20–34	66–80 S	0–5
	CEM III/C-SR	5–19	81–95 S	0–5
Sulphate resisting pozzolanic cement CEM IV	CEM IV/A-SR	65–97	21–35 P,V	0–5
	CEM IV/B-SR	45–64	36–55 P,V	0–5

Notes:
The code letters used in the European Standard are:

K Portland cement clinker	P natural pozzolana	S granulated blast furnace slag	V siliceous fly ash

Where additional main constituents are incorporated in sulphate resisting pozzolanic cements they must be declared by the designation.

the price of the equivalent grey product. To further enhance the whiteness, up to 5% of white titanium oxide pigment may be added. The standard product is to strength class 52.5N. Typical applications include renderings, cast stone, precast and in situ structural concrete and pointing.

Sulphate-resisting Portland cement

Sulphate-resisting Portland cement (BS EN 197–1: 2011) is suitable for concrete and mortar in contact with soils and groundwater containing soluble sulphates up to the maximum levels (measured as sulphur trioxide) of 2% in soil or 0.5% in groundwater. In normal Portland cements the hydrated tricalcium aluminate component is vulnerable to attack by soluble sulphates, but in sulphate-resisting Portland cement this component is restricted. For maximum durability a high-quality, dense, non-permeable concrete is required.

Sulphate-resisting Portland cement is produced to three specifications: CEM I-SR 0, CEM I-SR 3 and CEM I-SR 5, with maximum levels of tricalcium aluminate within the clinker of 0, 3 and 5%, respectively.

Very low-heat special cements

Low-heat Portland cement (BS 1370: 1979) is appropriate for use in mass concrete, where the rapid internal evolution of heat could cause cracking. It contains a higher proportion of dicalcium silicate which hardens and evolves heat more slowly. The range of very low heat special cements, listed in Table 3.8, includes products based on blast furnace, pozzolanic and composite cements (BS EN 14216: 2004). Very low heat special cements are appropriate for use only in massive constructions such as dams, but not for bridges or buildings.

Factory-made composite cements

Composite Portland cements include not only masonry cement, with its specific end use, but also the wide range of additional materials classified within the European Standard EN 197–1: 2011.

Masonry cements

Portland cement mortar is unnecessarily strong and concentrates any differential movement within brickwork or blockwork into a few large cracks, which are unsightly and may increase the risk of rain penetration. Masonry cement produces a weaker mortar, which accommodates some differential movement, and ensures a distribution of hairline cracks within joints, thus preserving the integrity of the bricks and blocks. Masonry cements contain water-retaining mineral constituents, usually ground limestone or hydrated lime, and air-entraining agents to give a higher workability than unblended Portland cement. They should not normally be blended with further admixtures but mixed with building sand in ratios between 1 : 2.5 and 1 : 6.5, depending on the type of mineral constituent and on the degree of exposure of the brick or blockwork. The air entrained during mixing increases the durability and frost resistance of the hardened mortar. Masonry cement is also appropriate for use in renderings but not for floor screeds or concreting. It is therefore generally used as an alternative to Portland cement plus hydrated lime or plasticiser. Inorganic pigments, except for those containing carbon black, may be incorporated for visual effect. The strength

Table 3.8 Very low heat special cements to BS EN 14216: 2004

Cement	Type	Notation	Portland cement clinker content (%)	Additional main constituent (%)
Blast furnace cement	III	VLH III/B	20–34	66–80
		VLH III/C	5–19	81–95
Pozzolanic cement	IV	VLH IV/A	65–89	11–35
		VLH IV/B	45–64	36–55
Composite cement	V	VLH V/A	40–64	18–30 blast furnace slag
				18–30 pozzolana and fly ash
		VLH/VB	20–38	31–50 blast furnace slag
				31–50 pozzolana and fly ash

Table 3.9 Strength classes of masonry cements to European Standard BS EN 413–1: 2011

Strength classes	Compressive strength (MPa)	
	7-day early strength	28-day standard strength
MC 5	–	5 –15
MC 12.5	⩾7	12.5–32.5
MC 12.5X	⩾7	12.5–32.5
MC 22.5	⩾10	22.5–42.5
MC 22.5X	⩾10	22.5–42.5

Note:
Masonry cement is designated by MC. The X refers to cements which do not incorporate an air-entraining agent.

Table 3.10 Typical compositions of granulated blast furnace slag and Portland cement

	Granulated blast furnace slag (%)	Portland cement (%)
Lime	41	68
Silica	35	22
Alumina	11	5
Iron oxide	1	3
Other	12	2

classes for masonry cements conforming to BS EN 413–1: 2011 are listed in Table 3.9.

Portland slag and blast furnace cements
Granulated blast furnace slag (GBS) is a cementitious material, which in combination with Portland cement and appropriate aggregates makes a durable concrete. The material is a by-product of the iron-making process within the steel industry. Iron ore, limestone and coke are fed continuously into blast furnaces, where at 1500°C they melt into two layers. The molten iron sinks, leaving the blast furnace slag floating on the surface, from where it is tapped off at intervals. The molten blast furnace slag is rapidly cooled by quenching in a granulator or pelletiser to produce a glassy product. After drying, the blast furnace slag granules or pellets are ground to the fine off-white powder, ground granulated blast furnace slag. The composition of the material is broadly similar to that of Portland cement, as illustrated in Table 3.10.

Granulated blast furnace slag (i.e. slag which has not been prior ground) may be intimately ground with Portland cement clinker in the cement mill to form a factory-made Portland slag or blast furnace cement. Conversely, ground granulated blast furnace slag is generally mixed with Portland cement on site to produce a mixer combination. The standard BS EN 15167–1: 2006 gives the constituent specification for use with Portland cement at the mixer. In the case of factory-made cements the standard BS EN 197–1: 2011 refers to three types of blast furnace cements: CEM III/A, B and C with slag contents of 36–67%, 66–80% and 81–95%, respectively (Table 3.5). By using 50% of blast furnace slag as an alternative to Portland cement

in a typical mix, emissions of carbon dioxide in the overall concrete production process are reduced by approximately 40%. Concrete which incorporates blast furnace slag in sufficient proportion has a lower permeability than Portland cement concrete alone; this enhances resistance to attack from sulphates as well as to the ingress of chlorides which can cause rapid corrosion of steel reinforcement, for example, in marine environments and near roads subjected to de-icing salts. Sulphate attack is also reduced by the decrease in tricalcium aluminate content. Sulphate-resisting blast furnace cements are listed in Table 3.7 to BS EN 197–1: 2011.

The more gradual hydration of cements containing granulated blast furnace slag evolves less heat and more slowly than Portland cement alone; thus a 70% granulated blast furnace slag content may be used for mass concrete, where otherwise a significant temperature rise could cause cracking. The slower evolution of heat is associated with a more gradual development of strength over the first 28-day period. However, the ultimate strength of the mature concrete is comparable to that of the equivalent Portland cement. The initial set with blast furnace cements or Portland-slag cements is slower than for Portland cement alone, and the fresh concrete mixes are more plastic, giving better flow for placing and full compaction. The risk of alkali–silica reaction caused by reactive silica aggregates may be reduced by the use of blast furnace slag (either as part of a factory-made cement or a combination at the mixer) to reduce the active alkali content of the concrete mix to below the critical 3.5 kg/m³ level. The types of low early strength blast furnace cements are listed in Table 3.8.

Table 3.11 Composition of supersulphated cement to BS EN 15743: 2010

Type	Notation	Composition			
		Granulated blast furnace slag (%)	Calcium sulphate (%)	Portland cement clinker (%)	Other additional constituents (%)
SSC	Supersulphated cement	⩾ 75	5–20	0–5	0–5

Notes:

Percentages are by mass.

The percentage of Portland cement clinker and other minor additional constituents must not exceed 10% by mass.

Supersulphated cement

Supersulphated cement (BS EN 15743: 2010) consists of granulated blast furnace slag and calcium sulphate with only a small proportion of Portland cement clinker (Table 3.11). The hardening of the blast furnace slag component is activated by the calcium sulphate. The material has a lower heat of hydration and lower early strength than other common cements (Table 3.12), but it has high resistance to chemically aggressive environments including sulphates.

Portland-fly ash and pozzolanic cements

Pozzolanic materials are natural or manufactured materials containing silica, which react with the calcium hydroxide produced in the hydration of Portland cement to produce further cementitious products. Within the UK, natural volcanic pozzolanas are little used, but fly ash, the waste product from coal-fired electricity-generating stations, is used either factory mixed with Portland cement or combined on site. Portland-fly ash cement cures and evolves heat more slowly than Portland cement; it is therefore appropriate for use in mass concrete to reduce the risk of thermal cracking. Up to 30% of fly ash is often used; the

concrete produced is darker than with Portland cement alone. Concrete made with between 21 and 55% by weight of fly ash in the cement/combination has good sulphate-resisting properties (Table 3.7). However, in the presence of groundwater with high magnesium concentrations, sulphate-resisting Portland cement should be used. Fly ash concretes also have enhanced resistance to chloride ingress, which is frequently the cause of corrosion to steel reinforcement.

The fly ash produced in the UK by burning pulverised bituminous coal is siliceous, containing predominantly reactive silica and alumina. In addition to siliceous fly ash, the European Standard EN 197–1: 2011 allows for the use of calcareous fly ash, which also contains active lime, giving some self-setting properties. The range of fly ash suitable for concrete is defined in the standard BS EN 450–1: 2012, and includes the waste products from the co-combustion of coal with solid biofuels, sludge and animal waste. Natural pozzolana of volcanic origin (P) and natural calcined pozzolana activated by thermal treatment (Q) when blended with Portland cement are categorised under EN 197–1: 2011 as pozzolanic cements.

Table 3.12 Strength classes of supersulphated cement to BS EN 15743: 2010

Strength classes	Compressive strength (MPa)		
	2-day early strength	7-day early strength	28-day standard strength
32.5 L	—	⩾12	32.5–52.5
32.5 N	—	⩾12	32.5–52.5
42.5 L	—	⩾16	42.5–62.5
42.5 N	⩾10		42.5–62.5
52.5 L	⩾10	—	⩾52.5
52.5 N	⩾20	—	⩾52.5

Note:

L refers to low early strength and N to normal early strength.

Portland-limestone cement

The addition of up to 5% of limestone fines as a minor additional constituent to Portland cement has little effect on its properties. The addition of up to 25% of limestone gives a performance similar to that of Portland cement with a proportionally lower cementitious content; thus, if equivalent durability to Portland cement is required, then cement contents must be increased in the end-use product. The two categories of limestone for Portland-limestone cement are defined by their total organic carbon (TOC) content; LL refers to a maximum of 0.20% and L to a maximum of 0.50% by mass.

Silica fume

Silica fume or microsilica, a by-product from the manufacture of silicon and ferro-silicon, consists of ultrafine spheres of silica. The material, because of its high surface area, when used in a cement/combination increases the rate of hydration, giving the concrete a high early strength and also a reduced permeability. This in turn produces greater resistance to chemical attack and abrasion. Silica fume may be added up to 5% as a minor additional constituent, or in Portland silica fume cement to between 6% and 10%. The requirements for silica fume are described in BS EN 13263: 2005.

Burnt shale

Burnt shale is produced by heating oil shale to 800°C in a kiln. It is similar in chemical composition to blast furnace slag, containing mainly calcium silicate and calcium aluminate, but also silica, lime and calcium sulphate. It is weakly cementitious. The European Standard EN 197–1: 2011 allows for the use of burnt shale as a minor additional constituent to 5%, or between 6% and 35% in Portland-burnt shale cement.

Minor additional constituents

Minor additional constituents (MAC) of up to 5% by weight of the cement content may be added to cements to the standard EN 197–1: 2011. They should be materials which do not increase the water requirements of the cement. Minor additional constituents may be any of the permitted alternative main constituents (e.g. granulated blast furnace slag, pozzolanas, fly ash, burnt shale, silica fume, kiln dusts or limestone), or other inorganic materials, provided that they are not already present as one of the main constituents. The most common materials are limestone and either raw meal or partially calcined material from the cement-making process.

CEMENT ADMIXTURES

Admixtures may be defined as materials that are added in small quantities either to factory-made cements, or, more commonly, to mortars or concretes during mixing, in order to modify one or more of their properties or performance characteristics in the fresh wet and/or hardened state. Concrete admixtures are listed in BS EN 934–2: 2009 with their associated performance requirements.

Plasticisers

Plasticisers, or water-reducing admixtures, are added to increase the workability of a mix, thus enabling easier placing and compaction. Where increased workability is not required, water reducers may be used to lower the water/cement ratio, giving typically a 15% increase in strength and better durability. The plasticisers, which are usually lignosulphonates or hydroxylated polymers, act by dispersing the cement grains. Some air entrainment may occur with the lignosulphonates, causing a 6% reduction in crushing strength for every 1% of air entrained.

Superplasticisers

Superplasticisers, such as sulphonated naphthalene, sulphonated melamine formaldehyde or the newer polycarboxylate ether copolymers, when added to a normal 50 mm slump concrete produce a flowing, self-levelling or self-compacting concrete (SCC) which can be placed, even within congested reinforcement, without vibration. Alternatively, significantly reduced water contents may be used to produce early and ultimately higher strength concretes. As the effect of superplasticisers lasts for under an hour, the admixture is usually added to ready-mixed concrete on site prior to discharge and placing. Standard concrete additives, fillers and steel or polypropylene fibres may be incorporated into self-compacting concrete which can be pumped or delivered by skip or chute. Good-quality off-the-form surface finishes can be achieved, especially with timber formwork. Self-levelling mixes for screeds between 3 and 20 mm thick can be adjusted to take light foot traffic after 3 to 24 hours. Renovation mixes, usually incorporating fibre-mat reinforcement, may be used over a range of existing floor surfaces to thicknesses usually in the range of 4–30 mm.

Accelerators

Accelerators increase the rate of reaction between cement and water, thus increasing the rate of set and development of strength. This can be advantageous in precasting where early removal of the formwork is required, and in cold weather when the heat generated speeds up the hardening processes and reduces the risk of frost damage. Only chloride-free accelerators, such as calcium formate, should be used in concrete, mortar or grout where metal will be embedded, because calcium chloride accelerators can cause extensive metallic corrosion. Accelerators producing a rapid set are not normally used within structural concrete.

Retarders

Retarders, typically phosphates or hydroxycarboxylic acids, decrease the rate of set, thus extending the time between initial mixing and final compaction, but they do not adversely affect 28-day strength. Retarders may be applied to formwork, to retard the surface concrete where an exposed aggregate finish is required by washing after the formwork is struck. Retarders are also frequently used in ready-mixed mortars to extend their workable life by up to 36 hours. The mortars are usually delivered on site in date-marked containers of 0.3 m³ capacity.

Air-entraining admixtures

Air-entraining admixtures, typically wood resins or synthetic surfactants, stabilise the tiny air bubbles which become incorporated into concrete or mortar as it is mixed. The bubbles, which are between 0.05 and 0.5 mm in diameter, do not escape during transportation or vibration, improve the workability of the mix, reduce the risk of segregation and greatly enhance frost resistance. However, the incorporation of void space within concrete decreases its crushing strength by 6% for every 1% of air entrained; thus, for a typical 3% addition of entrained air, a reduction of 18% in crushing strength is produced. This is partially offset by the increase in plasticity, which generally produces a higher quality surface and allows a lower water content to be used. The increased cohesion of air-entrained concrete may trap air against moulded vertical formwork, reducing the quality of the surface.

Water-resisting admixtures

The water penetration through concrete can be reduced by the incorporation of hydrophobic materials, such as stearates and oleates, which coat the surface of pores and, by surface-tension effects, discourage the penetration of damp. The use of water-reducing admixtures also reduces water penetration by reducing the water/cement ratio, thus decreasing pore size within the concrete. To seal construction joints an encapsulated swellable waterstop strip of butyl rubber may be incorporated. For mortars and renders a styrene-butadiene latex emulsion admixture may be used to reduce permeability.

Foaming agents

Foamed concrete or mortar contains up to 80% by volume of void space, with densities as low as 300 kg/m³ and 28-day strengths of between 0.2 and 20 MPa. It is typically produced by blending cement, fine sand or fly ash and water into a preformed foam or by mechanically foaming the appropriate mix using a foaming surfactant. For the low densities below 600 kg/m³, no fillers are included but for densities up to 1600 kg/m³, limestone dust and concreting sand may be incorporated. Foamed concrete is free flowing, may be pumped and requires no compaction. When set, the material offers good frost resistance and thermal insulation. It is therefore used for trench reinstatement, filling cellars or to provide insulation under floors or in flat roofs.

Pumping agents

Not all concrete mixes are suitable for pumping. Mixes low in cement or with some lightweight aggregates tend to segregate and require thickening with a pumping agent. Conversely, high-cement content mixes require plasticising to make them pumpable. A range of pumping agents is therefore produced to suit the requirements of various concrete mixes. Lightweight aggregate concrete is often pumped into place for floor slabs.

Pigments

A wide range of coloured pigments is available for incorporation into concrete and mortars (BS EN 12878: 2014). Titanium oxide may be added to enhance the whiteness of white cement. Carbon black is used with grey Portland cement, although the black loses intensity with weathering. The most common colours are the browns, reds and yellows produced with synthetic iron, chromium and manganese oxides, and also with complexes of cobalt, aluminium, nickel and antimony. In addition, ultramarine and phthalocyanine extend the range of the blues and greens. The depth and shade of colour depends on the dose rate (between 1% and 10%), as well as on the colour of the sand and any other aggregates. To produce pastel shades, pigments may be added to white Portland cement.

CALCIUM ALUMINATE CEMENT

Calcium aluminate cement (CAC), also known as high alumina cement, is manufactured from limestone and bauxite (aluminium oxide). The ores, in roughly equal proportions, are charged together into a vertical furnace, which is heated to approximately 1600°C (Fig. 3.4). The mixture melts and is continuously run off into trays, where it cools to produce the clinker, which is then milled, producing calcium aluminate cement to BS EN 14647: 2005. The dark-grey cement composition differs from that of Portland cement as it is based on calcium aluminates rather than calcium silicates. Although calcium aluminate cement may be produced over a wide range of compositions, the standard product has 40% alumina content.

Calcium aluminate cement should not be used for foundations or structural purposes, but only for specific heat-resisting applications and where deterioration rates can be predicted. However, it is useful where rapid strength gain is required, allowing the fast removal of formwork within 6 to 24 hours. The fast evolution of heat allows concreting to take place in low temperatures. The material also has good heat-resistant properties, so it may be used to produce refractory concrete. Good-quality calcium aluminate cement is generally resistant to chemical attack by dilute acids, chlorides and oils, but not alkalis. When mixed with Portland cement it produces a rapid-setting concrete, suitable for non-structural repairs and sealing leaks.

Setting guide for calcium aluminate cement (CAC):

1 part CAC: 3 part Portland cement	Rapid set
1 part CAC: 2 part Portland cement	Very rapid set
1 part CAC: 1½ part Portland cement	Instant set

Some structural failures associated with calcium aluminate cement have been caused by *conversion* of the concrete, in which changes in the crystal structure, accelerated by high temperatures and humidity, have caused serious loss of strength, increased porosity and subsequent chemical attack. Depending upon the degree of conversion, calcium aluminate cement becomes friable and a deeper brown in colour; the

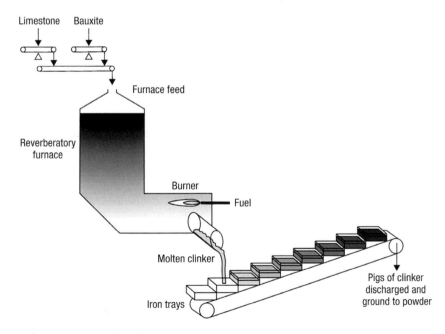

Fig. 3.4 Manufacture of calcium aluminate (high alumina) cement

exact degree of conversion can only be determined by chemical analysis of a core sample. It is now recognised that such failures can be prevented by using a minimum cement content of 400 kg/m^3, limiting the water/cement ratio to a maximum of 0.4, and by ensuring controlled curing during the 6- to 24-hour initial hardening stage. The concrete should be covered or sprayed to prevent excessive water loss, particularly where substantial increases in temperature may occur.

In addition, in order to prevent alkaline hydrolysis of the concrete, aggregates containing soluble alkalis should not be used; hard limestone is generally considered to be the best aggregate. Coloured calcium aluminate cement concrete has the advantage that it is free from calcium hydroxide which causes efflorescence in Portland cements. The BRE Special Digest SD3 (2002) gives methods for assessing existing calcium aluminate cement concrete (high alumina cement concrete [HACC]) constructions and suggests appropriate remedial actions. In some cases where the depth of conversion of HACC structural members is significant, with time there is an increasing risk of reinforcement corrosion.

LOW ENVIRONMENTAL IMPACT CEMENTS

Current greenhouse emissions, converted to carbon dioxide equivalents from the manufacture of Portland cement in the UK, are estimated at 913 kg of CO_2e per tonne of CEM I cement. At the current rate of world production this relates to approximately 7.5% of global CO_2 emissions. While the Portland cement industry has significantly increased production efficiency by using the dry manufacturing process and increasing the blended content, ultimately limestone has to be calcined at high temperatures to drive off the carbon dioxide. The overall carbon footprint of Portland cement may be reduced by the addition of granulated blast furnace slag (e.g. CEM III/B at 80% ggbs – 277 kg CO_2e/tonne), or fly ash (CEM II/B-V at 35% fly ash – 615 kg CO_2e/tonne). In Brazil and Cuba, the ash from the combustion of sugar cane after the sugar has been extracted is added to Portland cement. The resulting concrete with 20% ash is stronger than ordinary Portland cement concrete after one year, although the setting process is slower. Some new alternatives to Portland cement are being developed but the impact on the industry is modest to date. For example, *liquid granite*®, a silica-activated material containing 50% ground granulated blast furnace slag and pulverised fuel ash with only 4% Portland cement,

has enhanced fire resistance and lower water absorption than standard Portland cement concrete.

Lime-pozzolan cement

Lime-pozzolan cement is made by combining natural hydraulic lime (NHL5) with pozzolanic waste products such as alumina-siliceous waste ashes which govern the set and ultimate strength of the material. Concrete compressive strengths similar to that produced by Portland cement can be achieved. The embodied energy of lime-pozzolan concrete is approximately 10% less than that of equivalent Portland cement (CEM I) concrete.

Belite cement

The main components of Portland cement are alite (tricalcium silicate) and belite (dicalcium silicate). These are formed at 1450°C and 1200°C, respectively.

Belite cements may be manufactured from finely ground raw materials at kiln temperatures of 1350°C, and as the CaO content of belite is lower than that of alite, less CO_2 per tonne is driven off during the calcination process. The total CO_2 emissions for making belite cement are estimated at 0.69 tonne/tonne compared to 0.88 tonne/tonne for alite cement. This equates to an approximately 16% energy saving for belite cement compared to Portland cement. Belite cement has good long-term properties but develops its strength very slowly. Therefore its potential is enhanced by blending with a more active component.

Calcium sulphoaluminate cement

Calcium sulphoaluminate (CSA) is produced from a mixture of bauxite (aluminium oxide), anhydrite (calcium sulphate) and limestone at kiln temperatures between 1000°C and 1300°C.

$$3CaCO_3 + 3Al_2O_3 + CaSO_4 =$$
$$4CaO.3Al_2O_3.SO_3 + 3CO_2$$
calcium sulphoaluminate

Considerably less CO_2 is produced per tonne of calcium sulphoaluminate cement, compared to the production of alite or belite cement. Calcium sulphoaluminate cements develop early strength, so the blending of belite and calcium sulphoaluminate to produce belite-sulphoaluminate cement offers some potential as an alternative to Portland cement.

The addition of cheap Fe_2O_3 (iron oxide) in the raw meal leads to the formation of ferrite (calcium ferroaluminate – C_4AF), by the partial substitution of the Al_2O_3 (alumina) component in calcium sulphoaluminate. Belite-sulphoaluminate-ferrite cement therefore has the potential for reduced manufacturing costs. Typical compositions of belite-sulphoaluminate-ferrite cement currently used in China are 35–70% CSA, belite < 30% and ferrite 10–30%. Compared to manufacturing Portland cement CEM I, the reduction in CO_2 emissions is estimated at around 20%.

Magnesium oxide cements

Two types of magnesium oxide-based cements are currently under development, based on magnesium carbonate (magnesite) and magnesium silicate, respectively.

Magnesium oxide is produced from magnesium carbonate by heating to 650°C. After the bound carbon dioxide has been evolved, the magnesium oxide is hydrated to magnesium hydroxide (brucite), which acts as the cement binder. This material is rapidly carbonated in use by reabsorbing most of the CO_2 liberated in its manufacture.

Magnesium oxide may also be produced from magnesium silicates by heating the mineral to between 650°C and 700°C. In this case no carbon dioxide is evolved in the manufacture, but CO_2 is subsequently absorbed in its use as a cement. Manufacturing 1 tonne of this cement would produce 0.4 tonnes of CO_2, but its use would absorb 1.1 tonnes, leaving a reduction of 0.7 tonnes from the atmosphere. This product therefore has the potential to be 'zero carbon' or better. Magnesium silicate minerals are abundant worldwide, but not in large quantities within the UK.

Additional novel cements

Additional novel cements, which have lower carbon footprints than Portland cement, are currently under development in Australia and the USA, and these include alkali-activated cements in which sodium silicate or sodium hydroxide activates aluminate material derived from fly ash, slag or municipal solid waste incinerator ash (MSWIA). MSWIA may also be used in a Japanese process to produce traditional Portland cement, in which half the raw materials are MSWIA or sewage sludge and the heat is provided by burning waste oils and non-recyclable plastics.

Concrete

Concrete is a mixture of cement, aggregates and water, together with any admixtures which may be added to modify the placing and curing processes or the ultimate physical properties. Initially, when mixed, concrete is a *plastic* material, which takes the shape of the mould or formwork. When hardened it may be a dense, load-bearing material or a lightweight, thermally insulating material, depending largely on the aggregates used. It may be reinforced or prestressed by the incorporation of steel.

Most concrete is crushed and recycled at the end of its useful life, frequently as hard core for new construction work. However, a growth in the use of recycled aggregates for new concrete may be anticipated, as this will have a significant environmental gain in reducing the demand for new aggregate extraction.

AGGREGATES FOR CONCRETE

Aggregates form a major component of concretes, typically approximately 80% by weight in cured mass concrete. Aggregate properties, including crushing strength, size, grading and shape, have significant effects on the physical properties of the concrete mixes and hardened concrete. In addition, the appearance of visual concrete may be influenced by aggregate colour and surface treatments. The standard BS EN 12620: 2013 specifies the appropriate properties including materials, size, grading and shape.

Aggregates for concrete are normally classified as lightweight, dense or high density. Standard dense aggregates are classified by size as fine (e.g. sand) or coarse (e.g. gravel). In addition, steel or polypropylene fibres or air/gas bubbles may be incorporated into the mix for specialist purposes.

Dense aggregates

Source and shape
Dense aggregates are quarried from pits and from the seabed. In the south-east of England, most land-based sources are gravels, typically flint, whereas further north and west, both gravels and a variety of crushed quarried rocks are available. Marine aggregates which account for 18% of production in England and Wales may require washing to remove deleterious matter such as salts, silt and organic debris. The total chloride content should be monitored to ensure that it is within the

limits to BS 8500: 2006 for reinforced or unreinforced concrete as appropriate. This may be achieved by using well-drained unwashed marine sand in conjunction with land-based coarse aggregates.

The shape of aggregates can significantly affect the properties of the mix and cured concrete. Generally, rounded aggregates require a lower water content to achieve a given mix workability, compared to the equivalent mix using angular aggregates. However, cement paste ultimately bonds more strongly to angular aggregates with rough surfaces than to the smoother gravels, so a higher crushing strength can be achieved with crushed rocks as aggregate. Excessive proportions of long and flaky coarse aggregate should be avoided, as they can reduce the durability of concrete.

Aggregate size

For most purposes the maximum size of aggregate should be as large as possible consistent with ease of placement within formwork and around any steel reinforcement. Typically, 20 mm aggregate is used for most construction work, although 40 mm aggregate is appropriate for mass concrete, and a maximum of 10 mm for thin sections. The use of the largest possible aggregate reduces the quantity of sand and therefore cement required in the mix, thus controlling shrinkage and minimising cost. Large aggregates have a low surface area/volume ratio, and therefore produce mixes with greater workability for a given water/cement ratio, or allow water/cement ratios to be reduced for the same workability, thus producing a higher crushing-strength concrete.

Grading

To obtain consistent quality in concrete production, it is necessary to ensure that both coarse and fine aggregates are well graded. A typical *continuously graded* coarse aggregate will contain a good distribution of sizes, such that the voids between the largest stones are filled by successively smaller particles down to the size of the sand. Similarly, a well-graded sand will have a range of particle sizes, but with a limit on the proportion of fine clay or silt, because too high a content of *fines* (of size less than 0.063 mm) would increase the water and cement requirement for the mix. Usually a maximum of 3% fines is considered non-harmful. This overall grading of aggregates ensures that all void spaces are filled with the minimum proportion of fine material and expensive cement powder. In certain circumstances, coarse aggregate may be graded as *single-sized* or *gap graded*. The former is

Fig. 3.5 Riffle box

used for controlled blending in *designed mixes* while the latter is used particularly for exposed aggregate finishes on visual concrete. Sands are classified into three categories according to the proportion passing through a 0.500 mm sieve: coarse C (5–45%), medium M (30–70%) and fine F (55–100%). Only the coarse and medium categories of sands should be used for heavy-duty concrete floor finishes.

Sampling and sieve analysis

To determine the grading of a sample of coarse or fine aggregate, a representative sample has to be subjected to a sieve analysis. Normally at least ten samples would be taken from various parts of the stockpile, and these would be reduced down to a representative sample using a *riffle box*, which successively divides the sample by two until the required test volume is obtained (Fig. 3.5).

Aggregate gradings are determined by passing the representative sample through a set of standard sieves (BS EN 12620: 2013). Aggregate size is specified by the lower (d) and upper (D) sieve sizes and designated by the ratio d/D. For coarse aggregates the sieve sizes are 63, 31.5, 16, 8, 4, 2 and 1 mm, and for fine aggregates 4, 2, 1, 0.250 and 0.063 mm. Coarse aggregates are usually defined as having a minimum size (d) of 1 mm, while fine aggregates often have a maximum size (D) of 4 mm. Gradings for coarse, fine and all-in aggregates are the particle size distributions, expressed as percentages by mass, passing a specified set of standard sieves.

Coarse aggregate gradings are categorised as G_C and fine aggregate gradings are G_F. G_G refers to grit and G_A to all-in aggregates.

Coarse aggregate G_C85/20 (BS EN 12620: 2013)

The grading requirements for a category G_C85/20 coarse aggregate with an upper size limit (D) of 63 mm and a minimum size limit (d) of 4 mm are that 85–99% must pass through a 63 mm sieve and not more than 20% passes through a 4 mm sieve.

Aggregates for concreting are normally *batched* from stockpiles of 20 mm coarse aggregate and concreting sand in the required proportions to ensure consistency, although *all-in aggregate*, which contains both fine and coarse aggregates, is also available as a less well-controlled, cheaper alternative, where a lower grade of concrete is acceptable. Where exceptionally high control on the mix is required, single-size aggregates may be batched to the customer's specification. The batching of aggregates should normally be done by weight, since free surface moisture, particularly in sand, can cause *bulking*, which is an increase in volume by up to 40% (Fig. 3.6). Accurate batching must take into account the water content in the aggregates in the calculations of both the required weight of aggregates and the quantity of water to be added to the mix.

Impurities within aggregates

Where a high-quality exposed concrete finish is required, the aggregate should be free of iron pyrites, which causes spalling and rust staining of the surface. Alkali–silica reaction (ASR) may occur when active silica, present in certain aggregates, reacts with the alkalis within Portland cement, potentially causing cracking.

Recycled aggregates

The proportion and nature of constituent materials in recycled aggregate for concrete must be determined according to BS EN 933–11: 2009 and declared according to BS EN 12620: 2013 as percentage limits by mass.

Constituent categories of recycled coarse aggregate:

Rc	Concrete, concrete products, mortar, concrete masonry units
Ru	Unbound aggregate, natural stone, hydraulically bound aggregate
Rb	Clay masonry units (bricks and tiles), calcium silicate masonry units, aerated non-floating concrete
Ra	Bituminous materials
Rg	Glass
X	Other (clay, soil, metals (ferrous and non-ferrous), non-floating wood, plastic, rubber, gypsum plaster).

Quantities of deleterious material within recycled aggregates must be declared and carefully controlled, to prevent adverse effects on the quality of the concrete.

The by-products from the extraction of china clay in Devon and Cornwall are 'stent' which is a waste rock and 'tip sand' which is predominantly quartz with some mica. Recycled china clay waste was used extensively as aggregate for the foundations of the 2012 London Olympic Aquatics Centre and Stadium.

High-density aggregates

Where radiation shielding is required, high-density aggregates such as barytes (barium sulphate), magnetite (iron ore), lead or steel shot are used. Hardened concrete densities between 3000 and 5000 kg/m³, double that for normal concrete, can be achieved.

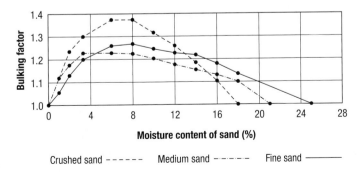

Fig. 3.6 Bulking of sands in relation to moisture content

Lightweight aggregates

Natural stone aggregate concretes typically have densities within the range 2200–2500 kg/m³, but where densities below 2000 kg/m³ are required an appropriate lightweight concrete must be used.

Lightweight concretes in construction exhibit the following properties in comparison with dense concrete:

- enhanced thermal insulation but reduced compressive strength;
- increased high-frequency sound absorption but reduced sound insulation;
- enhanced fire resistance over most dense aggregate concretes (e.g. granite spalls);
- easier to cut, chase, nail, plaster and render than dense concrete;
- reduced self-weight of the structure offers economies of construction;
- lower formwork pressures enable the casting of higher lifts.

The three general categories of lightweight concrete are lightweight aggregate concrete, aerated concrete and no-fines concrete (Fig. 3.7).

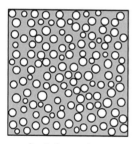

Aerated concrete –
voids in a cement matrix

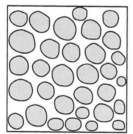

Lightweight aggregate
concrete – voids in the
aggregate pellets

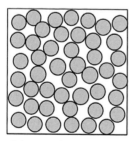

Polystyrene-bead aggregate
cement – very light aggregate

No-fines concrete –
dense aggregate with
voids between

Fig. 3.7 Lightweight concretes

Many of the lightweight aggregate materials are produced from by-products of other industrial processes or directly from naturally occurring minerals. The key exception is expanded polystyrene, which has the highest insulation properties, but is expensive due to its manufacture from petrochemical products.

Pulverised fly ash
Fly ash is the residue from coal-fired electricity-generating stations. The fine fly ash powder is moistened, pelleted and sintered or cold bonded to produce a uniform lightweight aggregate, which may be used in load-bearing applications.

Foamed blast furnace slag
Blast furnace slag is a by-product from the steel industry. Molten slag is subjected to jets of water, steam and compressed air to produce a pumice-like material. The foamed slag is crushed and graded to produce aggregate, which may be used in load-bearing applications. Where rounded pelletised expanded slag is required the material is further processed within a rotating drum.

Expanded clay and shale
Certain naturally occurring clay materials are pelletised, then heated in a furnace. This causes the evolution of gases which expands and aerates the interior, leaving a hardened surface crust. These lightweight aggregates may be used for load-bearing applications.

Expanded perlite
Perlite is a naturally occurring glassy volcanic rock, which, when heated almost to its melting point, evolves steam to produce a cellular material of low density. Concrete made with expanded perlite has good thermal insulation properties but low compressive strength and high drying shrinkage.

Exfoliated vermiculite
Vermiculite is a naturally occurring mineral, composed of thin layers like mica. When heated rapidly the layers separate, expanding the material by up to 30 times, producing a very lightweight aggregate. Exfoliated vermiculite concrete has excellent thermal insulation properties but low compressive strength and very high drying shrinkage.

Expanded polystyrene
Expanded polystyrene beads offer the highest level of thermal insulation, but with little compressive strength.

Polystyrene bead aggregate cement (PBAC) is frequently used as the core insulating material within precast concrete units. The standard BS EN 16025: 2013 describes the use of pre-mixed expanded polystyrene mortar/plaster for thermal and/or impact sound insulation on walls, ceilings, roofs and floors.

Aerated concrete

Aerated concrete (aircrete) is manufactured using foaming agents or aluminium powder, as previously outlined in the section on foaming agents. Densities in the range 400–1600 kg/m^2 give compressive strengths ranging from 0.5 to 20 MPa. Drying shrinkages for the lowest-density materials are high (0.3%), but thermal conductivity may be as low as 0.1 W/m K, offering excellent thermal insulation properties. Factory-autoclaved aerated concrete blocks have greatly reduced drying shrinkages and enhanced compressive strength over site-cured concrete. Aerated concrete is generally frost resistant but should be rendered externally to prevent excessive water absorption. The material is easily worked on site, as it can be cut and nailed.

No-fines concrete

No-fines concrete is manufactured from single-sized aggregate (usually between 10 and 20 mm) and cement paste. Either dense or lightweight aggregates may be used, but care has to be taken in placing the mix to ensure that the aggregate remains coated with the cement paste. The material should not be vibrated. Drying shrinkage is low, as essentially the aggregate is stacked up within the formwork, leaving void spaces; these increase the thermal insulation properties of the material in comparison with the equivalent dense concrete. The rough surface of the cured concrete forms an excellent key for rendering or plastering which is necessary to prevent rain, air or sound penetration. Dense aggregate no-fines concrete may be used for load-bearing applications.

Fibres

Either steel or polypropylene fibres may be incorporated into concrete as an alternative to secondary reinforcement, particularly in heavily trafficked floor slabs. The fibres reduce the shrinkage and potential cracking that may occur during the initial setting, and give good abrasion and spalling resistance to the cured concrete. The low-modulus polypropylene fibres, which do not pose a corrosion risk after carbonation of the concrete, enhance the energy-absorbing characteristics of the concrete, giving better impact resistance. Steel fibres increase flexural strength as well as impact resistance but are more expensive. Alternatively, stainless steel fibres may be used where rust spots on the surface would be unacceptable. Typically, polypropylene fibres are added at the rate of 0.2% by weight (0.5% by volume) and steel at the rate of 3–4% by weight. Both polypropylene and steel fibre concretes may be pumped. Steel fibres to BS EN 14889–1: 2006 may be straight or deformed cold-drawn wire, or alternatively sheet fibres. Polymer fibres to BS EN 14889–2: 2006 may be thick or thin monofilaments or fibrillated. (Glassfibre-reinforced cement is described in Chapter 11.)

Ultra-high-performance concrete

Ultra-high-performance concrete (UHPC) has six to eight times the compressive strength of traditional concrete. It is produced from a mixture of Portland cement, crushed quartz, sand, silica fume, superplasticiser, fibres and water with no aggregates larger than one millimetre. Wollastonite (calcium silicate) filler may also be included in the mix. The fibres most frequently used are either high-strength steel for maximum strength or polypropylene (PP) of approximately 12 mm in length for lower load applications. A low water/cement ratio between 0.2 and 0.26 is used. The concrete may be cast into traditional moulds by gravity or pumped or even injection cast under pressure. When cast into traditional moulds, the material is self-levelling, so only slight external vibration of the formwork may be required to ensure complete filling. The material is designed for use without steel reinforcement bars.

Structural components in ultra-high-performance concrete may, after setting, be subjected to steam treatment for 48 hours at 90°C. This enhances durability and mechanical properties, eliminates shrinkage and reduces creep. The material does not spall under fire test conditions.

The enhanced compressive and flexural strengths of ductile fibre-reinforced ultra-high-performance concrete enable lighter and thinner sections to be used for structural components such as shell roofs and bridges, creating an enhanced sleek aesthetic. A high-quality durable surface is produced from appropriate moulds (e.g. steel) coated with proprietary release agent.

Light transmitting concrete

By embedding 4% fibre-optic threads into fine concrete, the material is made translucent without any appreciable loss of compressive strength. Light transmitting concrete can be manufactured as blocks or panels provided that the fibres run transversely from face to face. If one face is illuminated, any shadow cast onto the bright side is clearly visible on the other face, while the colour of the transmitted light is unchanged. Blocks up to 1200 × 400 mm with thicknesses from 25 to 500 mm are standard and can be manufactured in a variety of colours including white, grey or black. The material has many potential applications including walls, floor surfaces and illuminated pavements.

Concrete canvas

Concrete canvas is a flexible cement-impregnated fabric which when hydrated forms a thin waterproof durable layer of concrete. The material is available in rolls of 1.0 and 1.1 m wide in thicknesses of 5, 8 and 13 mm. The material, which can be laid in any weather, sets in one to two hours and hardens 24 hours after hydration by spraying. Concrete canvas has a Euroclass B fire rating to BS EN 13501–1: 2007 + A1: 2009.

The material has many applications including drainage, structural fire protection and soil stabilisation.

Insulating concrete formwork (ICF)

Polystyrene
Blocks or panels of expanded or extruded polystyrene fit together to create permanent insulating formwork, which is then filled with in situ concrete to produce a monolithic structure. A range of systems is available giving a central core of 100 to 300 mm concrete and insulation thicknesses between 100 and 300 mm according to the structural and thermal requirements. The two faces of the insulation are connected by a matrix of polystyrene links which become embedded in the concrete. The units interlock to ensure correct location, and horizontal steel reinforcement may be incorporated if required for additional structural strength. Special blocks are available for lintels, wall ends and curved walls. A pumpable grade of concrete (high slump) will fill the void space by gravity flow without the need for mechanical vibration. Some temporary support for the formwork is required during construction to ensure accurate alignment. Internal and external finishes may be applied directly to the polystyrene which is keyed for plaster or lightweight render. Alternatively, masonry, timber or other claddings may be used externally and dry linings (e.g. plasterboard) may be attached to the inner leaf with appropriate adhesives.

Lightweight concrete
Concrete shuttering blocks are manufactured in normal, lightweight or wood chip concrete to standards BS EN 15435: 2008 and BS EN 15498: 2008, respectively. Systems are available with and without additional thermal insulation for use as internal, external and partition walls when filled with concrete. Some systems have lateral interlocking (e.g. by tongue and groove) and may be laid with or without mortar according to manufacturers' specifications. Hollow insulated blocks, 365 mm wide, manufactured from 80% wood chip, when filled with a 180 mm concrete core can produce a wall U-value of 0.19 W/m^2 K.

Polymer concrete

The incorporation of pre-polymers into concrete mixes, the pre-polymers then polymerising as the concrete sets and hardens, can reduce the penetration of water and carbon dioxide into cured concrete. Typical polymers include styrene-butadiene rubber and polyester-styrene. Epoxy resin and acrylic-latex-modified mortars are used for repairing damaged and spalled concrete due to their enhanced adhesive properties. Similarly, polymer-modified mortars are used for the cosmetic filling of blow-holes and blemishes in visual concrete. Resin-bound concrete construction products include street furniture, decorative elements and window-sills. The materials are covered by the standard BS EN 15564: 2008.

WATER FOR CONCRETE

The general rule is that if water is of a quality suitable for drinking, then it is satisfactory for making concrete. The standard BS EN 1008: 2002 gives the limits on impurities including sulphates.

CONCRETE MIXES

Concrete mixes are designed to produce concrete with the specified properties at the most economical price. The most important properties are usually strength and durability, although thermal and acoustic insulation, the effect of fire and appearance in visual concrete may also be critical.

In determining the composition of a concrete mix, consideration is given to the *workability* or ease of placement and compaction of the fluid mix, and to the properties required in the hardened concrete. The key factor which affects both of these properties is the free-water content of the mix after any water is absorbed into the aggregates. This quantity is defined by the water/cement ratio.

Water/cement ratio

$$\text{water/cement ratio} = \frac{\text{weight of free water}}{\text{weight of cement}}$$

The free water in a mix is the quantity remaining after the aggregates have absorbed water to the *saturated surface-dry* condition. The free water is used to hydrate the cement and make the mix workable. With low water/cement ratios below 0.4, some of the cement is not fully hydrated. At a water/cement ratio of 0.4, the hydrated cement just fills the space previously occupied by the water, giving a dense concrete. As the water/cement ratio is increased above 0.4, the mix becomes increasingly workable but the resulting cured concrete is more porous owing to the evaporation of the excess water leaving void spaces. Figure 3.8 shows the typical relationship between water/cement ratio and concrete crushing strength.

Workability

Workability describes the ability of the concrete mix to be placed within the formwork, around any

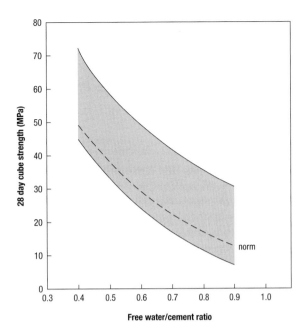

Fig. 3.8 Typical variation of crushing strengths about the published norm for the range of water/cement ratios

reinforcement, and to be successfully compacted by hand or mechanical means to remove trapped air pockets. Mixes should be cohesive, so that they do not segregate during transportation or placing. Workability is affected by not only the water/cement ratio but also the aggregate content, size, grading and shape, and the addition of admixtures. It is measured on site with the slump test (Fig. 3.9). Table 3.13 shows the relationship between water/cement ratio and workability for crushed and uncrushed aggregates at different cement contents.

Table 3.13 Typical relationship between water/cement ratio, workability and Portland cement 42.5 content for uncrushed and crushed aggregates

| Water/cement ratio | Type of aggregate (20 mm maximum) | Workability | | |
		Low slump 10–30 mm Cement content (kg/m³)	Medium slump 25–75 mm Cement content (kg/m³)	High slump 65–135 mm Cement content (kg/m³)
0.7	Uncrushed	230	260	285
	Crushed	270	300	330
0.6	Uncrushed	265	300	330
	Crushed	315	350	380
0.5	Uncrushed	320	360	400
	Crushed	380	420	460
0.4	Uncrushed	400	450	500
	Crushed	475	525	575

Free water

The workability of concrete is highly dependent on the free water within the mix. An increase in free-water content causes a significant increase in workability, which would result in a greater slump measured in a slump test.

Aggregate shape

Rounded aggregates make a mix more workable than if crushed angular aggregates are used with the same water/cement ratio. However, because the bonding between cured cement and crushed aggregate is stronger than that to rounded aggregates, when other parameters are comparable, crushed aggregates produce a stronger concrete.

Aggregate size

The size of aggregate also affects the workability of the mix. The maximum practical size of coarse aggregate, compatible with placement around reinforcement and within the concrete section size, should be used to minimise the water content necessary for adequate workability. With fine aggregates, excessive quantities of the fine material (passing through a 0.063 mm test sieve) would increase considerably the water requirement of a particular mix to maintain workability. This is because the smaller particles have a larger surface area/volume ratio and therefore require more water to wet their surfaces. As additional water in the mix will decrease the cured concrete strength, for good-quality dense concrete, well-graded coarser sands are preferable.

Aggregate/cement ratio

For a particular water/cement ratio, decreasing the aggregate/cement ratio, which therefore increases proportionally both the cement and water content, increases workability. However, as cement is the most expensive component in concrete, cement-rich mixes are more costly than the lean mixes.

Air-entraining

Workability may be increased by air-entraining, although 1% voids in the cured concrete produce a decrease in compressive strength of approximately 6%. Thus, in air-entraining, there is a balance between the increased workability and resultant improved compaction, versus the void space produced with its associated reduced crushing strength.

Slump test

The slump test is used for determining the workability of a mix on site. It gives a good indication of consistency from one batch to the next, but it is not effective for very dry or very wet mixes. The slump test is carried out as shown in Figure 3.9. The base plate is placed on level ground and the cone filled with the concrete mix in three equal layers, each layer being tamped down 25 times with the 16 mm diameter tamping rod. The final excess of the third layer is struck off and the cone lifted off from the plate to allow the concrete to slump. The drop in level (mm) is the recorded slump, which may be a *true slump*, a *shear slump* or a *collapse slump*. In the case of a shear slump the material is retested. In the case of a collapse slump the mixture is too wet for most purposes. Typical slump values would be 0–25 mm for very dry mixes, frequently used in road making; 10–40 mm (low workability) for use in foundations with light reinforcement; 50–90 mm (medium workability) for normal reinforced concrete placed with vibration and over 100 mm for high-workability concrete. Typically, slump values between 10 and 175 mm may be measured, although accuracy and repeatability are reduced at both extremes of the workability range. The slump test is not appropriate for aerated, no-fines or gap-graded concretes. The European Standard BS EN 206: 2013 classifies consistency classes of concrete mixes by results from the standard tests of slump (Table 3.14), Vebe consistency (a form of mechanised slump test), compaction and flow.

Table 3.14 Slump test classes to European Standard BS EN 206: 2013

Slump class	Slump (mm)
S1	10–40
S2	50–90
S3	100–150
S4	160–210
S5	⩾220

Compaction

After being placed within the formwork, concrete requires compaction to remove air voids trapped in the mix before it begins to stiffen. Air voids weaken the concrete, increase its permeability and therefore reduce durability. In reinforced concrete, lack of compaction reduces the bond to the steel, and on exposed visual

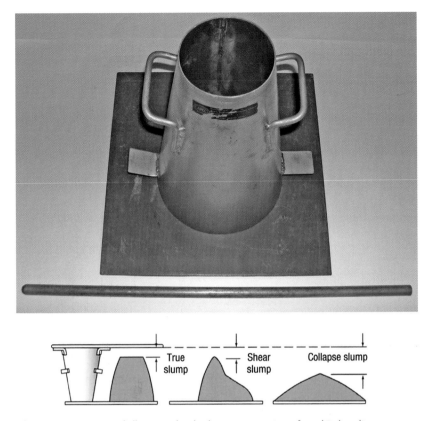

Fig. 3.9 Slump test (after Everett A. 1994: *Mitchell's Materials.* 5th edition. Longman Scientific and Technical)

concrete, blemishes such as blow-holes and honey-combing on the surface are aesthetically unacceptable and difficult to make good successfully. Vibration, to assist compaction, may be manual by rodding or tamping for small works, but normally poker vibrators and beam vibrators are used for mass and slab concrete, respectively. Vibrators which clamp onto the formwork are sometimes used when the reinforcement is too congested to allow access for poker vibrators.

The degree of compaction achieved by a standard quantity of work may be measured by the compacting factor test. In this test a fresh concrete sample is allowed to fall from one hopper into another. The weight of concrete contained in the lower hopper, when struck off flush, compared with a fully compacted sample gives the compacting factor. The compacting factor for a medium workability-concrete is usually about 0.9.

Concrete cube and cylinder tests

To maintain quality control of concrete, representative test samples should be taken, cured under controlled conditions and tested for compressive strength after the appropriate 3-, 7- or 28-day period. Steel cylinder and cube moulds (Fig. 3.10) are filled in layers with either hand or mechanical vibration. For hand tamping, a 100 mm cube would be filled in two equal layers, each tamped 25 times with a 25 mm square-end standard compacting bar; mechanical vibration would normally be with a vibrating table or pneumatic vibrator. The mix is then trowelled off level with the mould. Test samples are cured under controlled moisture and temperature conditions for 24 hours, then stripped and cured under water at 18–20°C until required for testing.

The European Standard BS EN 12390–1: 2012 lists dimensions of cubes, cylinders and prisms ranging in critical dimension from 100 to 300 mm. The critical mould dimension must be at least 3.5 times the size of the largest aggregate. The appropriate method of filling test samples is described in BS EN 12390–2: 2009. Generally, in Europe, cylinders 150 mm in diameter and 300 mm high rather than cubes are used, as they tend to give more uniform results for nominally similar concrete specimens. For a particular concrete, the

Fig. 3.10 Cylinder and cube test

characteristic compressive strength as determined by the cylinder test is lower than that obtained from the equivalent cube test. The compressive strength classes (Table 3.15, page 94), therefore, have a two-number notation (e.g. C 20/25). The first number, which is used in the European structural design codes, refers to the characteristic cylinder compressive strength, and the second number is the characteristic 150 mm cube compressive strength.

DURABILITY OF CONCRETE

While good-quality, well-compacted concrete with an adequate cement content and a low water/cement ratio is generally durable, concrete may be subjected to external agencies which cause deterioration, or, in certain circumstances, such as alkali–silica reaction, it may be subject to internal degradation. The standard BS EN 206: 2013 specifies requirements for the specification, constituents, composition, production and properties of concrete.

Sulphate attack

Sulphates are frequently present in soils, but the rate of sulphate attack on concrete is dependent on the soluble sulphate content of the groundwater. Thus, the presence of sodium or magnesium sulphate in solution is more critical than that of calcium sulphate, which is relatively insoluble. Soluble sulphates react with the tricalcium aluminate (C_3A) component of the hardened cement paste, producing calcium sulpho-aluminate (ettringite). This material occupies a greater volume than the original tricalcium aluminate; therefore, expansion causes cracking, loss of strength and increased vulnerability to further sulphate attack. The continuing attack by sulphates depends on the movement of sulphate-bearing groundwater and, in some cases, delayed ettringite formation may not be apparent for 20 years. Delayed ettringite formation is sometimes observed in precast concrete which had been steam-cured, or when the temperature within the in situ mass concrete had risen excessively during the curing process. With magnesium sulphates, deterioration may be more serious, as the calcium silicates within the cured concrete are also attacked. The use of sulphate-resisting Portland cement or combinations of Portland cement and fly ash or granulated blast furnace slag reduces the risk of sulphate attack in well-compacted concrete. In the presence of high soluble sulphate concentrations, concrete requires surface protection. The criteria which increase the resistance of the cement matrix to sulphate attack are described in the document PD CEN/TR 15697: 2008.

The BRE Special Digest 1: 2005 also describes provision for combating sulphate deterioration, including the more rapid form of attack in which the mineral thaumasite is formed. Thaumasite sulphate attack has seriously affected concrete foundations and substructures including some bridges on the UK M5 motorway. This type of sulphate attack is most active at temperatures below 15°C. It occurs under damp conditions in the presence of groundwater sulphates and limestone within the aggregate, casing decomposition of the concrete into a soft friable material.

Frost resistance

Weak, permeable concrete is particularly vulnerable to the absorption of water into capillary pores and cracks. On freezing, the ice formed will expand, causing frost damage. The use of air-entraining agents, which produce discontinuous pores within concrete, reduces the risk of surface frost damage. Concrete is particularly vulnerable to frost damage during the first two days of early hardening. Where new concrete is at risk, frost precautions are necessary to ensure that the mix temperature does not fall below 5°C until a strength of 2 MPa is achieved. Eurocode 2 (BS EN 1992–1-1: 2004) refers to four levels of exposure class (XF1 to XF4) with respect to freeze/thaw deterioration (Table 3.17, page 99).

Fire resistance

Up to 250°C, concrete shows no significant loss of strength, but by 450°C, depending on the duration of heating, the strength may be reduced to half and by 600°C little strength remains. However, as concrete is a good insulator, it may take four hours within a building fire for the temperature 50 mm below the surface of the concrete to rise to 650°C (Fig. 3.11).

The effect of heat on the concrete causes colour changes to pink at 300°C, grey at 600°C and buff by 1000°C. The aggregates used within concrete have a significant effect on fire resistance. For fire protection, limestone aggregates perform slightly better than granites and other crushed rocks, which spall owing to differential expansion. The inclusion of synthetic fibres also reduces spalling. Where the concrete cover over reinforced steel is greater than 40 mm, secondary reinforcement with expanded metal gives added protection to the structural reinforcement. Lightweight-aggregate concretes, owing to their enhanced thermal properties, perform significantly better in fires with respect to both insulation and spalling.

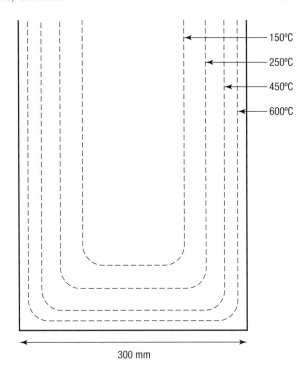

Fig. 3.11 Temperature profile within dense concrete exposed to fire for 60 minutes

Typically, a 100 mm-thick concrete load-bearing wall will give 120 minutes of fire resistance. Concrete manufactured without organic materials is Class A1 with respect to reaction to fire. If more than 1% of organic materials are incorporated into the mix, the material will require testing to the standard BS EN 13501–1: 2007.

Chemical attack and aggressive ground

The resistance of cured concrete to acid attack is largely dependent upon the quality of the concrete, although the addition of granulated blast furnace slag or fly ash increases the resistance to acids. Limestone-aggregate concrete is more vulnerable to acid attack than concretes with other aggregates. The resistance of cured concrete to chemical attack is defined by the design chemical class number, ranging from DC1 (low resistance) to DC4 (high resistance). The required design chemical class (DC Class) of the concrete is calculated by combining the effects of the sulphate content of the ground, the nature of the groundwater and the anticipated working life of the construction (BRE Special Digest 1: 2005).

Determining the design chemical class required for concrete in a particular ground environment is a three-stage process. The first stage is to determine the design sulphate class (DS) of the site. This is a five-level classification based primarily on the sulphate content of the soil and/or groundwater. It takes into account the concentrations of calcium sulphate as well as the more soluble magnesium and sodium sulphates and the presence of chlorides and nitrates if the pH is less than 5.5 (acid).

Design sulphate class	Limits of sulphate (mg/l)
DS1	< 500
DS2	500–1500
DS3	1600–3000
DS4	3100–6000
DS5	> 6000

The next stage is to determine the aggressive chemical environment for concrete (ACEC) classification. Adverse ground conditions such as acidity (low pH), often found in *brownfield* sites, and/or mobile groundwater lead to a more severe ACEC classification. Static water is more benign and leads to a less severe ACEC classification. The aggressive chemical environments for concrete classes range from AC1 (the least aggressive) to AC5 (the most aggressive), and are based on a combination of the design sulphate class, groundwater mobility and pH.

The design chemical class (DC1 to DC4) defines the qualities of the concrete required to resist chemical attack. It is determined from the ACEC class of the ground together with factors relating to the concrete, such as section size and intended working life (e.g. 100 years). As there are only four design chemical classes against five ACEC classes, for the severest grade of ACEC (i.e. AC5) there are additional protective measures (APMs) which can be specified to combat the more adverse conditions. Usually APM3 (surface protection to the concrete) is appropriate for AC5 environments, but for increasing the intended working life from 50 to 100 years under the less aggressive AC3 or AC4 conditions, any one APM may be applied.

Additional protective measures (APMs) for buried concrete:

APM1	enhance the concrete quality;
APM2	use controlled permeability formwork;
APM3	provide surface protection to the concrete;
APM4	increase the thickness of the concrete as a sacrificial layer;
APM5	reduce groundwater by drainage of the site.

Careful consideration of all these additional factors is required to ensure that a suitably durable concrete, appropriate to the job, is delivered on site for use in aggressive ground and chemical environments (BRE Special Digest 1: 2005).

Crystallisation of salts

The crystallisation of salts, particularly from sea water, within the pores of porous concrete can cause sufficient internal pressure to disrupt the concrete.

Alkali–silica reaction

Alkali–silica reaction (ASR) may occur between cements containing sodium or potassium alkalis and any active silica within the aggregate. In severe cases, expansion of the gel produced by the chemical reaction causes *map* cracking of the concrete, which is characterised by a random network of very fine cracks bounded by a few larger ones. Aggregates are defined as having low, normal or high reactivity. The risk of alkali–silica reaction when using normal reactivity aggregates can be controlled by restricting the alkali content of the Portland cement to a maximum of 0.5% (low alkali cement) or the soluble alkali content of the concrete to 3 kg Na_2O equivalent/m^3. Additions of controlled quantities of silica fume, ground granulated blast furnace slag or fly ash may be used with low or normal-reactivity aggregates to reduce the risk of alkali–silica reaction. Alternative methods of minimising the risk of alkali–silica reaction include the addition of lithium salts or metakaolin to the concrete mix. Alkali–silica reaction can only occur in the presence of some moisture.

Carbonation

Carbon dioxide from the atmosphere is slowly absorbed into moist concrete and reacts with the calcium hydroxide content to form calcium carbonate. The process occurs mainly at the surface and only penetrates very slowly into the bulk material. The rate of penetration is dependent upon the porosity of the concrete, the temperature and humidity; generally it becomes problematic only when the concrete surrounding steel reinforcement is affected. Carbonation turns strongly alkaline hydrated cement (pH 12.5) into an almost neutral medium (pH 8.3) in which steel reinforcement will corrode rapidly if subjected to moisture.

$$Ca(OH)_2 + CO_2 \longrightarrow CaCO_3$$

hydrated cement	carbon dioxide	calcium carbonate

Good-quality dense concrete may only show carbonation to a depth of 5–10 mm after 50 years, whereas a low-strength permeable concrete may carbonate to a depth of 25 mm within 10 years. If reinforcement is not correctly located with sufficient cover it corrodes, causing expansion, spalling and rust staining. The depth of carbonation can be determined by testing a core sample for alkalinity using phenolphthalein chemical indicator, which turns pink in contact with the uncarbonated alkaline concrete. Where steel reinforcement has become exposed due to carbonation and rusting, it may be coated with a rust-inhibiting cement and the cover restored with polymer-modified mortar which may contain fibre reinforcement. Additional protection against further attack can be achieved by the final application of an anti-carbonation coating which acts as a barrier to carbon dioxide. Thermosetting polymers such as polyurethane and chlorinated rubber, as well as certain acrylic-based polymers, give some protection against carbonation.

Self-healing concrete

Current research is investigating the use of limestone-producing bacteria packed in micro-capsules to generate limestone within concrete micro-cracks. The spores remain dormant until the ingress of water into the cracks activates the bacteria which then generate limestone from the nutrients within their capsules, thus sealing the cracks. The micro-capsules require a robust coating to survive the initial concrete mixing process.

PHYSICAL PROPERTIES OF CONCRETE

Thermal movement

The coefficient of thermal expansion of concrete varies between $(7 \text{ and } 14) \times 10^{-6}$ °C, according to the type of aggregate used, the mix proportions and curing conditions.

Moisture movement

During the curing process, concrete exhibits some irreversible shrinkage which must be accommodated within the construction joints. The extent of the shrinkage is dependent on the restraining effect of the aggregate and is generally larger when smaller or lightweight aggregates are used. High-aggregate content mixes with low workability tend to have small drying shrinkages.

The reversible moisture movement for cured concrete is typically $(2–6) \times 10^{-4}$ °C, depending upon the aggregate.

Creep

Creep is the long-term deformation of concrete under sustained loads (Fig. 3.12). The extent of creep is largely dependent upon the modulus of elasticity of the aggregate. Thus an aggregate with a high modulus of elasticity offers a high restraint to creep. The extent of creep may be several times that of the initial elastic deformation of the concrete under the same applied load. Where rigid cladding is applied to a concrete-frame building, compression joints at each storey must be sufficiently wide to take up any deformation due to creep in addition to normal cyclical movements.

CONCRETE STRENGTH CLASSES

Concrete should be specified, placed and cured according to BS EN 206: 2013. The preferred strength classes of concrete are shown in Table 3.15, in which the numbers refer to the test sample crushing strengths of a 150×300 mm cylinder and a 150 mm cube, respectively.

SPECIFICATION OF CONCRETE MIXES

There are five methods for specifying concrete described in BS 8500–1: 2006. All should conform to the standards BS 8500–1: 2006 and BS EN 206: 2013.

The five methods are:

- designated concrete
- designed concrete
- prescribed concrete
- standardised prescribed concrete
- proprietary concrete.

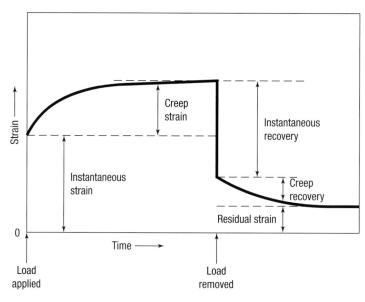

Fig. 3.12 Creep and creep recovery in concrete

If the application may be considered to be routine, then designated concrete is usually appropriate. If, however, the purchaser requires specific performance criteria and accepts the higher level of responsibility in the specification, then designed or prescribed concretes may be used. For housing and similar applications, standardised prescribed mixes should give the required performance, provided that there is sufficient control over the production and quality of materials used.

Designated concrete

Designated concretes are appropriate for most concrete construction including general-purpose work, foundations, reinforced concrete and air-entrained pavement concrete. The purchaser is responsible for correctly specifying the proposed use and the concrete mix designation. In addition, the purchaser must specify whether the concrete is to be reinforced, the exposure (or soil) conditions, the nominal aggregate size if other than 20 mm and the consistence class (slump). The producer must ensure that the mix fulfils all the performance criteria. Thus, normally for foundations in design chemical class soil conditions DC3, the designated mix FND 3 would be required. This mix may be supplied with sulphate-resisting Portland cement at 340 kg/m^3 and a maximum water/cement ratio of 0.5, or as Portland cement with 25% fly ash or 75% granulated blast furnace slag. Any of these mixes will perform to the required criteria for the specified purpose. For routine work, designated mixes produced by quality-assured plants offer

Table 3.15 Compressive strength classes for dense and lightweight concrete

Compressive strength classes for dense concrete								
C8/10	C12/15	C16/20	C20/25	C25/30	C30/37	C35/45	C40/50	
C45/55	C50/60	C55/67	C60/75	C70/85	C80/95	C90/105	C100/115	

Compressive strength classes for lightweight concrete								
LC8/9	LC12/13	LC16/18	LC20/22	LC25/28	LC30/33	LC35/38	LC40/44	LC45/50
LC50/55	LC55/60	LC60/66	LC70/77	LC80/88				

Note:
Within each compressive strength class the numbers indicate the 28-day crushing strength in MPa (N/mm^2) as determined by the 150 mm diameter by 300 mm cylinder and 150 mm cube test, respectively.

Table 3.16 Designated and standardised prescribed concrete for housing and other applications (BS 8500–1: 2006)

Typical application	Designated concrete	Standardised prescribed concrete	Consistence class
Foundations (Design Chemical Class1):			
Blinding and mass concrete fill	GEN 1	ST2	S3
Strip footings	GEN 1	ST2	S3
Mass concrete foundations	GEN 1	ST2	S3
Trench fill foundations	GEN 1	ST2	S4
Drainage works – immediate support	GEN 1	ST2	S1
Other drainage works	GEN 1	ST2	S3
Oversite below suspended slabs	GEN 1	ST2	S3
Foundations (Design Chemical Class 2 to 4):			
Foundations DC 2	FND 2		S3
Foundations DC 3	FND 3		S3
Foundations DC 4	FND 4		S3
Foundations DC 4m	FND 4M		S3
General applications:			
Kerb bedding and backing	GEN 0	ST1	S1
Floors:			
House floors with no embedded metal for screeding	GEN 1	ST2	S2
House floors with no embedded metal – no finish	GEN 2	ST3	S2
Garage floors with no embedded metal	GEN 3	ST4	S2
Wearing surface – light foot and trolley traffic	RC 25/30	ST4	S2
Wearing surface – general industrial	RC 32/40		S2
Wearing surface – heavy industrial	RC 40/50		S2
Paving:			
House drives, domestic parking and external paving	PAV 1		S2
Heavy-duty external paving for rubber tyre vehicles	PAV 2		S3

Note:
m refers to resistance to the higher magnesium levels in the various sulphate classes.

the specifier the least risk of wrong specification. Table 3.16 illustrates typical housing applications for designated mixes.

Designed concrete

The producer is responsible for selecting a designed concrete which will meet the performance criteria listed by the specifier. The specifier must clearly indicate the required use, curing conditions, exposure conditions, surface finish, maximum aggregate size and any excluded materials. In addition, the compressive strength class, the maximum water/cement ratio, the minimum cement content, the consistence (slump) and permitted cement types should be quoted. Within these constraints, the producer is responsible for producing a concrete which conforms to the required properties and any additional stated characteristics. Designed concrete would be used when the user specifications are outside those covered by designated concrete. Specialist requirements include low heat of hydration, exposure to chlorides or lightweight concrete.

Prescribed concrete

The purchaser fully specifies all materials by weight (kg/m^3), including admixtures but not the concrete strength. The purchaser is therefore responsible for the performance characteristics of the concrete. Prescribed concretes are used particularly for specialist finishes such as exposed aggregate visual concrete.

Standardised prescribed concrete

Standardised prescribed concretes are a set of five standard mixes, which may be mixed on site or delivered by a non-third-party certified producer. Standard mixes ST1 to ST5 may be made to S1, S2, S3 or S4 slump classes, giving low, medium, high or very high workability. The specification must record a maximum aggregate size and whether the concrete is to be reinforced or not. Table 3.16 illustrates typical housing applications for standardised prescribed concrete.

Strength classes for standardised prescribed concrete:

ST1	C 6/8
ST2	C 8/10
ST3	C 12/15
ST4	C 16/20
ST5	C 20/25

Proprietary concrete

Proprietary concrete must conform to the standards BS 8500–2: 2006 and BS EN 206: 2013 and be properly identified. This category allows for a concrete supplier to produce a concrete mix with an appropriate performance but without indicating its composition.

IN SITU CONCRETE TESTING

The compressive strength of hardened concrete may be estimated in situ by mechanical or ultrasonic measurements. The Schmidt hammer or sclerometer measures the surface hardness of concrete by determining the rebound of a steel plunger fired at the surface. In the *pull-out* test, the force required to extract a previously cast-in standard steel cone gives a measure of concrete strength. Ultrasonic devices determine the velocity of ultrasound pulses through concrete. Since pulse velocity increases with concrete density, the technique may be used to determine variations within similar concretes. The test gives a broad classification of the quality of concrete, but not absolute data for concretes of different materials in unknown proportions.

Reinforced concrete

Concrete is strong in compression, with crushing strengths typically in the range 20–40 MPa, and up to 100 MPa for high-strength concretes. However, the tensile strength of concrete is usually only 10% of the compressive strength. Steel is the universally accepted reinforcing material, as it is strong in tension, forms a good bond and has a similar coefficient of thermal expansion to concrete. The location of the steel within reinforced concrete is critical, as shown in Figure 3.13, to ensure that the tensile and shear forces are transferred to the steel. The longitudinal bars carry the tensile forces while the links or stirrups combat the shear forces and also locate the steel during the casting of the concrete. Links are therefore more concentrated around locations of high shear, although inclined bars may also be used to resist the shear forces. Fewer or thinner steel bars may be incorporated into reinforced concrete to take a proportion of the compressive loads in order to minimise the beam dimensions.

Steel reinforcement for concrete is manufactured, largely from recycled scrap, into round, ribbed, indented or ribbed and twisted bars (Fig. 3.14). Mild steel is frequently used for the plain bars to form bent links. Hot rolled, high-yield steel is used for ribbed and indented bars. British Standard BS 4482: 2005 refers to 250 MPa yield strength steel for plain bars and to the higher grade 500 MPa steel for plain, ribbed and indented reinforcement of diameters between 2.5 and 12 mm. British Standard BS 4449: 2005 specifies high-yield steel (grade 500 MPa) with three levels of ductility A, B and C (highest) for ribbed bars of 6–50 mm diameters. Welded steel mesh reinforcement to BS 4483: 2005 is used for slabs, roads and within sprayed concrete.

Austenitic stainless steels may be used for concrete reinforcement where failure due to corrosion is a potential risk. Grade 1.4301 (18% chromium, 10% nickel) stainless steel is used for most applications, but the higher grade 1.4436 (17% chromium, 12% nickel, 2.5% molybdenum) is used in more corrosive environments. Where long-term performance is required in highly corrosive environments, the duplex grades of stainless steel may be used. The initial cost of stainless steel reinforcement is approximately eight times that of standard steel reinforcement, but in situations where maintenance costs could be high, for example, due to chloride attack from sea water or road salts, the overall life cycle costs may be reduced by its use. In addition, stainless steels have higher strengths than the standard carbon steels. Suitable stainless steels for the reinforcement of concrete are specified in BS 6744: 2001 + A2: 2009.

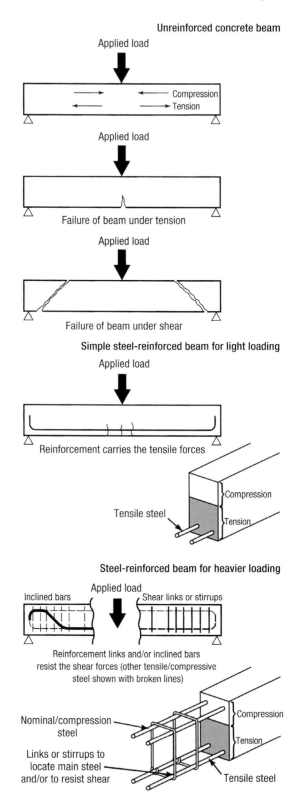

Fig. 3.13 Reinforced concrete

Bond between steel and concrete

For reinforced concrete to act efficiently as a composite material the bond between the concrete and steel must be secure. This ensures that any tensile forces within the concrete are transferred to the steel reinforcement. The shape and surface condition of the steel and the quality of the concrete all affect the bond strength.

To obtain the most efficient mechanical bond with concrete, the surface of the steel should be free of flaky rust, loose scale and grease, but the thin layer of rust, typically produced by short-term storage on site, should not be removed before use. The use of hooked ends in round bars reduces the risk of the steel being pulled out under load, but high bond strength is achieved with ribbed or indented bars which ensure a good bond along the full length of the steel. Steel *rebars* are usually either supplied in stock lengths, or cut and bent ready for making up into cages. Sometimes the reinforcement may be supplied as prefabricated cages, which may be welded rather than fixed with iron wire as on site. Steel reinforcement, although weldable, is rarely welded on site. Rebar joints can easily be made with proprietary fixings, such as steel sleeves fastened by shear bolts. Spacers are used to ensure the correct separation between reinforcement and formwork.

Good-quality dense concrete gives the strongest bond to the steel. Concrete should be well compacted around the reinforcement; thus, the maximum aggregate size must not bridge the minimum reinforcement spacing.

Corrosion of steel within reinforced concrete

Steel is protected from corrosion provided that it has adequate cover of a good-quality, well-compacted and cured concrete. The strongly alkaline environment of the hydrated cement renders the steel passive. However, insufficient cover caused by the incorrect fixing of the steel reinforcement or the formwork can allow the steel to corrode. Rust expansion causes surface spalling; then exposure of the steel allows corrosion, followed by rust staining of the concrete surface (Fig. 3.15). Calcium chloride accelerators should not normally be used in reinforced concrete, as the residual chlorides cause accelerated corrosion of the steel reinforcement.

Additional protection from corrosion can be achieved by the use of galvanised, epoxy-coated or stainless steel reinforcement. The protective alkalinity of the concrete is reduced at the surface by *carbonation*. The depth of carbonation depends upon the

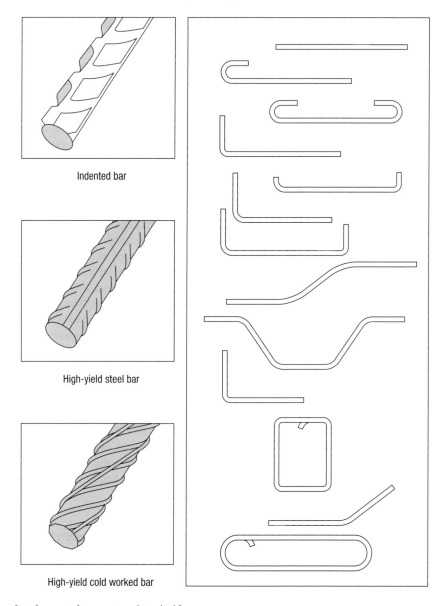

Indented bar

High-yield steel bar

High-yield cold worked bar

Fig. 3.14 Types of reinforcement for concrete and standard forms

permeability of the concrete, the moisture content and any surface cracking. The nominal cover for concrete reinforcement is therefore calculated from the anticipated degree of exposure (Table 3.17) and the concrete strength class as in Table 3.18. The recommended cover specified relates to all reinforcement, including any wire ties and secondary reinforcement. Some reduction in carbonation rate can be achieved by protective coatings to the concrete surface. It should be noted that the choice of an adequately durable concrete for the protection of the concrete itself against attack and

for the prevention of reinforcement corrosion may result in a higher compressive strength concrete being required than is necessary for the structural design (Table 3.19).

Where the depth of concrete cover over reinforcement is in doubt it can be measured with a *covermeter*. If reinforcement is corroding, cathodic protection by the application of a continuous direct current to the steel reinforcement may prevent further deterioration and lead to realkalisation of the carbonated concrete.

Fig. 3.15 Corrosion of steel reinforcement

Fibre-composite reinforced concrete

In most situations steel is used for reinforcing or pre-stressing concrete. However, for structures in highly aggressive environments high-modulus continuous fibres embedded in resin offer an alternative. The fibres, either glass, carbon or aramid, are encased in a thermosetting resin and drawn through a die by pultrusion to produce the required cross-section. The extruded material is then overwound with further fibres to improve its bond with concrete. The fibre-composite

Table 3.17 Concrete exposure classes to BS 8500: 2006 + A1: 2012 and BS EN 1992–1-1: 2004

Exposure classes		Typical environmental conditions
No risk of corrosion or attack		
X0	Concrete without reinforcement.	Interior unreinforced concrete
	Very dry concrete with reinforcement	Unreinforced concrete in non-aggressive water or soil
		Reinforced concrete in very dry conditions
Corrosion induced by carbonation		
XC1	Dry or permanently wet	Interior reinforced and prestressed concrete or permanently wet concrete in non-aggressive water
XC2	Wet, rarely dry	Reinforced concrete in non-aggressive soil
XC3 and XC4	Moderate humidity or cyclic wet and dry	External concrete and moderate or high humidity interiors, concrete exposed to alternate wetting and drying, concrete beneath waterproofing
Corrosion induced by chlorides other than from sea water		
XD1	Moderate humidity	Components exposed to airborne chlorides
XD2	Wet, rarely dry	Concrete immersed in water containing chlorides
XD3	Cyclic wet and dry	Exposed external concrete surfaces within 10 m of a carriageway, pavements, car parks
Corrosion induced by chlorides from sea water		
XS1	Exposure to sea air	External coastal structures
XS2	Submerged under sea water	Submerged marine structures
XS3	Tidal and sea spray zone	Parts of marine structures within the tidal zones
Freeze-thaw attack		
XF1	Moderate saturation without de-icing	Vertical and unsaturated non-vertical surfaces agent exposed to rain and freezing
XF2	Moderate saturation with de-icing agent	Vertical surfaces exposed to rain, freezing and de-icing
XF3	High saturation without de-icing agent	Horizontal surfaces exposed to rain and freezing, surfaces exposed to frequent splashing and freezing
XF4	High water saturation with de-icing agent	Surfaces exposed to rain, spray and de-icing agents or sea water
Chemical attack		
XA1	Slightly aggressive agencies	Soil and groundwater
XA2	Moderately aggressive agencies	Soil and groundwater
XA3	Highly aggressive agencies	Soil and groundwater

Table 3.18 Minimum cover required to ensure durability of steel reinforcement in structural concrete for exposure classes to Eurocode 2 (BS EN 1992–1-1: 2004)

Exposure class	X0	XC1	XC2/XC3	XC4	XD1/XS1	XD2/XS2	XD3/XS3
Recommended cover (mm)	10	15	25	30	35	40	45
Minimum cover (mm)	10	10	10	15	20	25	30
Strength class	$\geqslant$ C30/37	$\geqslant$ C30/37	$\geqslant$ C35/45	$\geqslant$ C40/50	$\geqslant$ C40/50	$\geqslant$ C40/50	$\geqslant$ C45/55

Notes:
- The recommended cover relates to standard production with a design working life of 50 years.
- Increased cover is required for a design working life of 100 years.
- The minimum cover relates to very specific conditions combining high-quality control for positioning of the reinforcement and the concrete production, as well as the use of 4% (minimum) air entrainment.
- The standard BS 8500–1: 2006 gives a more detailed set of recommendations relating strength classes to nominal cover including the option to add additional cover (Δc) for workmanship deviation.

Table 3.19 Indicative minimum strength classes for durability of concrete to Eurocode 2 (BS 8500–1: 2006 + A1: 2012)

Corrosion risk	XC1	XC2	XC3 and XC4	XD1 and XD2	XD3	XS1	XS2	XS3
Indicative strength class	C20/25	C25/30	C30/37	C40/50	C45/55	C45/55	C40/50	C45/55
Damage to concrete	XF1	XF2	XF3	XF4				
Indicative strength class	C25/30	C25/30	C25/30	C28/35				

Notes:
- The standard BS 8500–1: 2006 + A1: 2012 details a fully comprehensive relationship between Minimum Strength Class and Exposure Class for durability in relation to different maximum aggregate sizes and minimum cement contents.
- The Exposure Class coding letters are defined in Table 3.17.

rods are used as reinforcement or as pre-stressing tendons within standard concrete construction.

Bendable concrete

Bendable concrete is manufactured with 2% of short fibres in a fine aggregate mix. It may be continuously extruded into various sections to produce sheets, cylinders or tubes. The product has a 300% greater strain capacity and significantly higher impact strength than ordinary concrete. Bendable concrete may be drilled, cut and nailed without damage. It is lighter than ordinary concrete and with its good fire resistance may be used as an alternative to other wall boards. The hairline cracks produced in bending will heal by absorption of carbon dioxide in the presence of moisture. The material currently costs approximately three times that of standard concrete.

Fibre-reinforced aerated concrete

Polypropylene fibre-reinforced aerated concrete is used for making lightweight blocks, floor, wall and roofing panels, offering a combination of strength and insulation properties. The material, like standard aerated concrete, can be cut and worked with standard hand tools. Where additional strength is required, steel fibre-reinforced aerated concrete may be used for cast in situ or factory-produced units. The fibre-reinforced material has a greater resilience than standard aerated concrete. Roofing membranes and battens for tiling can be directly nailed to roofing panels, while floor panels accept all the standard floor finishes.

Fire resistance of reinforced concrete

Concrete manufactured without organic materials is Class A1 with respect to reaction to fire. If more than 1% of organic materials are incorporated into the mix, then the material will require testing to the standard BS EN 13501–1: 2007.

The depth of concrete cover over the steel reinforcement, to ensure various periods of fire resistance, is listed in Table 3.20. Where cover exceeds 40 mm,

Table 3.20 Typical cover to concrete reinforcement for fire resistance to Eurocode 2 (BS EN 1992–1-2: 2004)

Fire resistance (minutes)		Typical cover to reinforcement (mm)	
Beams	Width (mm)	Simply supported	Continuous beams
R 30	80	25	15
R 60	120	40	25
R 90	150	55	35
R 120	200	65	45
R 180	240	80	60
R 240	280	90	75
Columns	Minimum dimensions (mm)	One face exposed	
R 30	155	25	
R 60	155	25	
R 90	155	25	
R 120	175	35	
R 180	230	55	
R 240	295	70	
Walls	Minimum dimensions (mm)	One face exposed	
REI 30	100	10	
REI 60	110	10	
REI 90	120	20	
REI 120	150	25	
REI 180	180	40	
REI 240	230	55	
Slabs	Slab thickness (mm)	One-way slabs	Two-way slabs
REI 30	60	10	10
REI 60	80	20	10–15
REI 90	100	30	15–20
REI 120	120	40	20–25
REI 180	150	55	30–40
REI 240	175	65	40–50

Notes:
- Fire-resistance Class: R, load-bearing criterion, E, integrity criterion and I, insulation criterion in standard fire exposure.
- All reinforcement cover requirements are also dependent on the dimensions and geometry of the concrete components and the degree of fire exposure (BS EN 1992–1-2: 2004).
- Where low cover thicknesses are required for fire protection, a higher depth of cover may be required for corrosion protection (BS EN 1992–1-1: 2004).

additional reinforcement will be required to prevent surface spalling of the concrete. The cover should prevent the temperature of the steel reinforcement from exceeding 550°C (or 450°C for pre-stressing steel).

PRESTRESSED CONCRETE

Concrete has a high compressive strength but is weak in tension. Pre-stressing with steel wires or tendons ensures that the concrete component of the composite material always remains in compression when subjected to flexing up to the maximum working load. The tensile forces within the steel tendons act on the concrete putting it into compression, such that only under excessive loads would the concrete go into tension and crack. Two distinct systems are employed: in pre-tensioning the tendons are tensioned before the concrete is cured, and in post-tensioning the tendons are tensioned after the concrete is hardened (Fig. 3.16).

Pre-tensioning

Large numbers of precast concrete units, including flooring systems, are manufactured by the pre-tensioning process. Tendons are fed through a series of beam moulds and the appropriate tension is applied. The concrete is placed, vibrated and cured. The tendons are cut at the ends of the beams, putting the concrete into compression. As with precast reinforced concrete it is vital that pre-stressed beams are installed the correct way up according to the anticipated loads.

Post-tensioning

In the post-tensioning system the tendons are located in the formwork within sheaths or ducts. The concrete is placed, and when sufficiently strong, the tendons are stressed against the concrete and locked off with special anchor grips incorporated into the ends of the concrete. Usually reinforcement is incorporated into post-tensioned concrete, especially near the anchorages, which are subject to very high localised forces. In the bonded system, after tensioning the free space within the ducts is grouted up, which then limits the reliance on the anchorage fixing; however, in the unbonded system the tendons remain free to move independently of the concrete. Tendon ducts are typically manufactured from galvanised steel strip or high-density polythene.

Post-tensioning has the advantage over pre-tensioning that the tendons can be curved to follow the

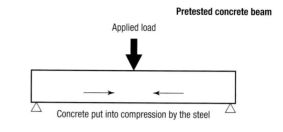

Pretested concrete beam

Applied load

Concrete put into compression by the steel

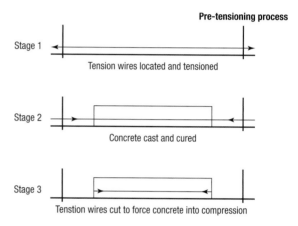

Pre-tensioning process

Stage 1
Tension wires located and tensioned

Stage 2
Concrete cast and cured

Stage 3
Tenstion wires cut to force concrete into compression

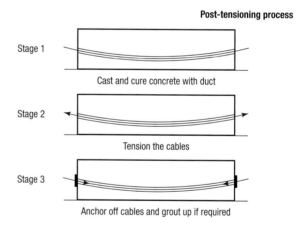

Post-tensioning process

Stage 1
Cast and cure concrete with duct

Stage 2
Tension the cables

Stage 3
Anchor off cables and grout up if required

Fig. 3.16 Prestressed concrete

most efficient pre-stress lines. In turn, this enables long spans of minimum thickness to be constructed. During demolition or structural alteration work, unbonded post-tensioned structures should be de-tensioned, although experience has shown that if demolished under tension, structures do not fail explosively. In alteration work, remaining severed tendons may subsequently require re-tensioning and re-anchoring to recover the structural performance. However, the use of post-tensioning does not preclude subsequent structural modifications.

Visual concrete

The production of visual concrete, whether precast or in situ, requires not only a high standard of quality control in manufacture, but also careful consideration to the correct specification and detailing of the material to ensure a quality finish which weathers appropriately. The exposed concrete at St John's College, Oxford (Fig. 3.17) illustrates the visual qualities of the material when designed, detailed and executed under optimum conditions.

The Angel Building in London (Fig. 3.18) illustrates high-quality interior concrete with appropriate detailing in a recently refurbished office complex.

The appearance of visual concrete is affected by four key factors:

- the composition of the concrete mix;
- the formwork used;
- any surface treatment after casting;
- the quality of workmanship.

DESIGN CONSIDERATIONS

The satisfactory production of large areas of smooth concrete is difficult due to variations in colour and the inevitability of some surface blemishes, which can be improved, but not eradicated, by remedial work. Externally smooth concrete weathers unevenly due to the buildup of dirt deposits and the flow of rainwater. Therefore, if concrete is to be used externally as a visual material, early design considerations must be given to the use of textured or profiled surfaces to control the flow of rainwater. Generally, the range of finishes and quality control offered by precasting techniques are wider than those available for in situ work, but frequently construction may involve both techniques. The use of external renderings offers an alternative range of finishes for concrete and other substrates. Figure 3.19 illustrates the range of processes available in the production of visual concrete.

Fig. 3.17 High-quality exterior concrete – St John's College, Oxford. *Architects*: MacCormac Jamieson Prichard. *Photograph*: Courtesy of Peter Cook

Fig. 3.18 High-quality interior concrete – Angel Building, Islington, London. *Photographs*: Arthur Lyons – Courtesy of Derwent London plc.

PRECAST CONCRETE

Precast concrete units may be cast vertically or horizontally, although most factory operations use the latter, either face-up or face-down, as better quality control can be achieved by this method. Moulds are usually manufactured from plywood or steel. While steel moulds are more durable for repeated use, plywood moulds are used for the more complex forms; they can also be more readily modified for non-standard units. Moulds are designed to be dismantled for the removal of the cast unit and must be manufactured to tight tolerances to ensure quality control on the finished product. As high costs are involved in the initial production of the moulds, economies of construction can be achieved by limiting the number of variations. This can have significant effects on the overall building aesthetic. Fixing and lifting systems for transportation must be incorporated into precast units, usually in conjunction with the steel reinforcement. In addition to visual concrete panels, units faced with natural stone, brickwork or tiles extend the range of precast architectural claddings (Figs 3.20 and 3.21). The document PD CEN/TR 15739: 2008 categorises the range of precast concrete finishes according to flatness (P), texture (T) and colour (C).

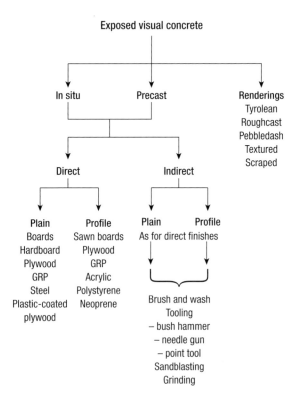

Fig. 3.19 Types of visual concrete according to formwork and surface treatment

Fig. 3.20 Slate surfaced precast concrete cladding – Swansea Museum. *Photograph*: Courtesy of Trent Concrete Ltd

Fig. 3.21 Reconstructed stone cladding – Experian Data Centre, Nottingham. *Photograph*: Courtesy of Trent Concrete Ltd

Precast aircrete panels

Precast aircrete panels, 100 mm thick and to standard storey heights, are suitable for the inner leaf of standard cavity construction and internal walls. Tolerances are close, similar to those required by the equivalent thin-joint masonry system. Maximum dimensions are 600 × 3000 mm with a standard thermal conductivity of 0.11 W/m K. The thin mortar jointing system with 2 mm joints is used for fixing the panels. Larger units, 200 mm thick, are manufactured for commercial projects.

IN SITU CONCRETE

The quality of in situ visual concrete is heavily dependent upon the formwork, as any defects will be mirrored in the concrete surface. The formwork must be strong enough to withstand, without distortion, the pressure of the fresh concrete, and the joints must be tight enough to prevent leakage, which may cause honeycombing of the surface. A wide range of timber products, metals and plastics is used as formwork, depending on the surface finish required.

The Millau Viaduct in France (Fig. 3.22), completed in 2004, is an elegant cable-stayed bridge supported by seven slender piers 270 m over the Tarn Valley. The main columns of 1.5 m-thick concrete were cast in situ

Fig. 3.22 Concrete columns – Millau Viaduct above the Tarn Valley, France. *Architect:* Foster + Partners. *Photograph:* Arthur Lyons

with self-climbing steel formwork externally and by crane hoist internally. The concrete piers are surmounted by 90 m steel pylons, which support the steel box section road platform. Each pier is supported by four reinforced concrete piles splayed at the foot to spread the loading. During construction work of the 2.46 km viaduct, rapid-hardening concrete was being placed at the rate of 80 m³ per hour and ultimately a total mass of 205,000 tonnes was used.

CONCRETE FINISHES

Smooth finishes

In direct as-cast concrete, the surface texture and water absorbancy of the formwork or any formwork lining directly determine the final exposed fairfaced finish. A high level of quality control is therefore required to ensure a visually acceptable finish. Hard, shiny, non-absorbent formwork materials, such as steel, glassfibre-reinforced polyester (GRP) or plastic-coated plywood,

can give surfaces which suffer from *map crazing* due to differential shrinkage between the surface and underlying bulk material. In addition, *blow-holes* caused by air bubbles trapped against the form face may spoil the surface if the concrete has not been sufficiently vibrated. Where the absorbency of the formwork varies, because of the mixing of new and reused formwork, or variations within the softwood timbers, or because of differing applications of release agent to the formwork, permanent colour variations may be visible on the concrete surface. Release agents prevent bonding between the concrete and the formwork, which may cause damage to the concrete on striking the formwork. Cream emulsions and oils with surfactant are typically used as release agents for timber and steel, respectively. Formwork linings with controlled porosity can improve the quality of *off-the-form* finishes, by substantially reducing the number of blow-holes. The linings allow the escape of air and excess moisture, but not cement, during vibration. A good-quality direct-cast concrete should exhibit only a few small blow-holes and modest colour variation.

A black and white photographic image can be imparted onto smooth concrete from a plastic liner. The picture is represented on the liner by pixel dots and these are treated with retarder. For light colours the dots of retarder are small, and proportionately larger for the darker colours. Using an appropriate concrete mix, the sheet liner produces varying degrees of retardation, which are then exposed when the formwork sheet is removed revealing the transferred image.

The application of paint to *off-the-form* concrete will emphasise the surface blemishes such as blow-holes. These become particularly noticeable if a light colour gloss paint is used. Surface defects must therefore be made good with filler before priming and subsequent painting of the concrete.

Textured finishes

A variety of textured finishes can be achieved by the use of rough-sawn boards as formwork. The grain effect can be enhanced by abrasive blasting, and a three-dimensional effect can be achieved by using variations in board thickness. Plastic materials, such as glassfibre-reinforced polyester (GRP), vacuum-formed thermoplastic sheeting, neoprene rubber and polystyrene, may be used as formwork linings to give different pattern effects. Colour variations are reduced by the use of matt finishes, which retain the mould release agent during compaction of the concrete. The number of blow-holes is reduced by the use of the slightly absorbent materials such as timber and polystyrene. Concrete panels cast face-up can be textured by rolling or tamping the concrete while it is still plastic.

Unformed concrete finishes

If a textured finish is required on an unformed surface such as a floor slab it is usually applied immediately after placing the concrete. Standard finishes to freshly placed concrete are tamping and brushing to create the required texture. Alternatively, a pattern such as herringbone brickwork or cobbles can be rolled onto the surface. Manual or machine floating produces a smooth surface, although usually with some float marks.

Ribbed and profiled finishes

Ribbed concrete is typically cast in situ against vertical timber battens fixed to a plywood backing. In order to remove the formwork without damage to the cured concrete, the battens must be splayed and smooth. A softer ribbed appearance is achieved by hammering off the projecting concrete to a striated riven finish. Profiled steel formwork and rope on plywood produce alternative finishes. Where deep profiles are required, expanded polystyrene and polyurethane foam can be carved out to produce highly sculptural designs.

Abraded, acid-etched and polished finishes

Light abrasion with sandpaper may be applied to in situ or precast concrete. Acid etching is normally limited to precast concrete due to the hazards associated with the use of acids on site. Both techniques remove the surface laitance (cement-rich surface layer) to create a more stone-like finish with some exposure of the aggregate. Polishing with carborundum abrasives to remove 2–3 mm of the surface produces a hard, shiny finish, imparting full colour brightness to the aggregate. It is, however, a slow and therefore expensive process.

Exposed aggregate finishes

The exposure of the coarse aggregate in concrete, by removal of the surface smooth layer formed in contact with the formwork, produces a concrete with a more durable finish and better weathering characteristics, which is frequently aesthetically more pleasing. Smooth, profiled and deeply moulded concrete can all be treated, with the visual effects being largely dependent upon the form and colour of the coarse aggregates used. While gap-graded coarse aggregates may be used both in precast and in situ exposed aggregate finishes, precasting gives additional opportunities for the uniform placement of the aggregate. In face-down casting, flat stones can be laid on the lower face of the mould, which can be pre-treated with retardant to slow the hardening of the surface cement. In face-up casting, individual stones can be pressed into the surface either randomly or to prescribed patterns without the use of retardants. Alternatively, a special facing mix may be used on the fairfaced side of the panel, with the bulk material made up with a cheaper standard mix. The aggregate has to be exposed by washing and brushing when the concrete has cured sufficiently to be self-supporting. The use of a retarder applied to the formwork face enables the timing of this process to be less critical. The surface should be removed to a depth of no more than one-third of the thickness of the aggregate to eliminate the risk of it becoming detached. An alternative method of exposing the aggregate in both precast and in situ concrete involves the use of abrasive blasting. Depending on the size of grit used and the hardness of the concrete, a range of finishes including sculptural designs can be obtained.

Tooled concrete finishes

A range of textures can be obtained by tooling hardened concrete either by hand or mechanically. Generally, a high-quality surface must be tooled, as blemishes can be accentuated rather than eliminated by tooling. Only deep tooling removes minor imperfections such as blow-holes and the effects of slight formwork mis-alignment. Hand tooling is suitable for a light finish on plain concrete and club hammering may be used on a ribbed finish. Where deep tooling is anticipated, allowance must be made for the loss of cover to the steel reinforcement. The exposed aggregate colour in tooled concrete is less intense than that produced by wash-and-brush exposure due to the effect of the hammering on the aggregate. Standard mechanical tools are the needle-gun, the bush hammer and the point-tool (Fig. 3.23). A range of visual concrete finishes is illustrated in Figure 3.24.

Coloured concrete finishes

Colour may be applied to concrete either as an integral pigment in the mix, a dry-shake topping or as a surface applied stain. Dry-shake applications are made on freshly placed concrete whereas stains are used on hardened concrete. Dry-shake finishes are combinations of cement, pigments, surface hardener and fine aggregate which are trowelled into the surface as soon as the excess bleed water has evaporated. Acid-etch stains contain metallic compounds which react with the surface lime content to product permanent colouring. Surface colourisers deposit the stain in the pores of the concrete.

Weathering of concrete finishes

The weathering of exposed visual concrete is affected by the local microclimate, the concrete finish itself and the detailing used to control the flow of rainwater over the surface. It is virtually impossible to ensure that all sides of a building are equally exposed, as inevitably there will be a prevailing wind and rain direction which determines the weathering pattern. It is therefore likely that weathering effects will differ on the various elevations of any building. Some elevations will be washed regularly, while others may suffer from an accumulation of dirt which is rarely washed. However, this broad effect is less likely to cause unsightly weathering than the pattern streaking on individual façades.

The choice of concrete finish can have a significant effect upon the weathering characteristics. Good-quality dense uniform concrete is essential if patchy weathering is to be avoided, and generally a rougher

Point tool   Bush hammer   Needle gun

Fig. 3.23 Tools for indirect visual concrete finishes

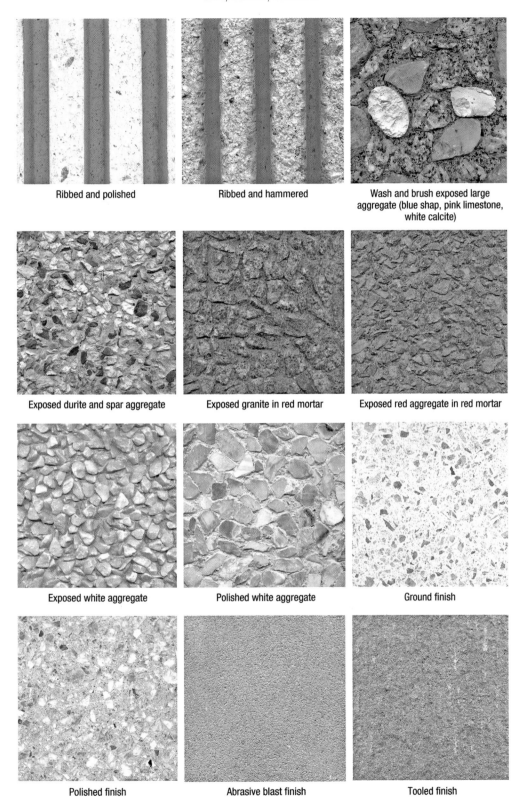

Fig. 3.24 Selection of visual concrete finishes

finish is likely to perform better than a smooth as-cast finish. Profiling and the use of exposed aggregates have the advantage of dictating the flow of rainwater, rather than letting it run in a random manner, but dirt becomes embedded in the hollows. Dark aggregates and bold modelling minimise the change in appearance on weathering, but generally, exposed non-absorbent aggregates are likely to give the best weathering performance. Horizontal surfaces may be subject to organic growths and this effect is increased by greater surface permeability.

Careful detailing is necessary to ensure a dispersed and controlled flow of water over the washed areas. The water should then be collected or shed clear by bold details to prevent pattern staining below. Water collected onto horizontal surfaces should not be allowed to run down façades below, so copings, sills and string courses should all be provided with drips to throw the water off the building face; alternatively, water should be removed by gutters. Multi-storey façades should be articulated with horizontal features to throw the water off, at least at each storey height. Only on seriously exposed façades, where strong winds are likely to cause rain to be driven upwards, should small horizontal drip projections be avoided. Where concrete is modelled, due consideration should be given to the direction of flow and the quantity of rainwater anticipated.

EXTERNAL RENDERING

Renders are used to provide a durable and visually acceptable skin to sound but unattractive construction.

Renders can reduce rain penetration and maintain the thermal insulation of walls. The finishes illustrated in Figure 3.25 are all appropriate for external use. In each case it is essential to ensure good adhesion to the background. Where a good mechanical key, such as raked-out brickwork joints, is not present, an initial stipple coat of sand, cement, water and appropriate bonding agent (e.g. styrene-butadiene-rubber) is required to create a key. Bonding is also affected by the suction or absorbency of the background; where suction is very high, walls may be lightly wetted before the rendering is applied. Metal lathing may be used over timber, steel or friable masonry to give a sound background. Two or three coats of rendering are normally applied; in either case the successive coats are weaker by a reduction in thickness or strength of the mix. Smooth renders require careful workmanship for external work, as they may craze if finished off with a steel rather than a wooden float.

Generally, permeable renders are more durable than dense impermeable renders, as the latter may suffer cracking and subsequent localised water penetration. Sands for external renderings should be sharp rather than soft. The design detailing of rendering is important to ensure durability. The top edges of rendering should be protected from the ingress of water by flashings, copings or eaves details. Rendering should stop above damp-proof course level and be formed into a drip with an appropriate edging bead. Rainwater run-off from sills and opening heads should be shed away from the rendering to prevent excessive water absorption at these points, which would lead to deterioration and detachment of the rendering. Figure 3.26 illustrates the striking visual effect of the rendered blockwork student

Textured finish

Dry-dash

Tyrolean finish

Fig. 3.25 Typical render finishes

Fig. 3.26 Rendered blockwork – University of East London, Docklands Campus. *Architects*: Edward Cullinan Architects. *Photograph*: Arthur Lyons

halls of residence at the University of East London adjacent to the Royal Albert Dock.

Roughcast render

Roughcast consists of a wet mix of cement (1 part), lime (0.5 part), sand (3 parts) and 5–15 mm shingle or crushed stone (1.5 parts) which is applied to walls by throwing from a hand scoop.

Dry-dash render

A 10 mm coat of cement (1 part), lime (1 part) and sand (5 parts) is applied to the wall, and while it is still wet, calcined flint, spar or shingle is thrown onto the surface and tamped in with a wooden float.

Scraped finish

A final coat of cement (1 part), lime (2 parts) and sand (9 parts) is applied and allowed to set for a few hours, prior to scraping with a rough edge (e.g. saw blade) to remove the surface material. After it has been scraped, the surface is lightly brushed over to remove loose material.

Textured finishes

A variety of finishes may be obtained by working the final rendering coat with a float, brush, comb or other tool to produce a range of standard textured patterns. Pargeting, in which more sophisticated patterns are produced, has its cultural roots in Suffolk and Essex.

Tyrolean finish

For a Tyrolean finish, cement mortar is spattered onto the wall surface from a hand-operated machine. Coloured mixes may be used.

Painted rendered finishes

Most renderings do not necessarily need painting; however, smooth renderings are frequently painted with masonry paint to reduce moisture absorption and give colour. Once painted, walls will need repainting at regular intervals.

Concrete components

In addition to the use of concrete for the production of large in situ and precast units, concrete bricks (Chapter 1) and concrete blocks (Chapter 2), the material is widely used in the manufacture of small components, particularly concrete tiles, slates and paving slabs.

CONCRETE ROOFING TILES AND SLATES

Concrete plain and interlocking slates and tiles form a group of highly competitive pitched roofing materials, with concrete interlocking tiles remaining the cheapest visually acceptable unit pitched-roof product. Plain and feature double-lap and interlocking tiles are manufactured to a range of designs, many of which emulate the traditional clay tile forms (Fig. 3.27). Concrete plain tiles may be used on pitches down to 35°, while the ornamental tiles are appropriate for vertical hanging and pitches down to 70°. The ranges of colours usually include both granular and through-colour finishes. Standard ranges of concrete interlocking tiles may be used in certain cases down to roof pitches of 15°, and for some shallow-pitched roofs the concrete tiles are laid to broken bond. One interlocking product emulates the appearance of plain tiles, but may be used down to a minimum pitch of 22.5°. Colours include brown, red, rustic and grey in granular and smooth finish. A limited range of concrete interlocking tiles may be used at roof rafter pitches as low as 10° for single-storey construction in sheltered positions, provided that all tiles are clipped to prevent wind lift.

Concrete interlocking slates are manufactured with either a deep, flat profile, giving a stone/slate appearance, or with a thin square or chamfered leading edge to simulate natural slate. Surfaces can be simulated riven or smooth in a range of colours including grey, blue, brown, buff and red. Matching accessories for either traditional mortar bedding or dry-fixing for ridges, hips and verges are available, together with appropriate ventilation units. The minimum pitch is normally 15° which may also be achieved with a limited range of fibre cement slates.

CONCRETE PAVING SLABS AND TILES

Grey concrete paving slabs are manufactured from Portland cement mixes with pigments added to produce the standard buff, pink and red colours. Standard sizes include 900 × 600 mm, 750 × 600 mm and 600 × 600 mm × 50 mm, but a wide range of smaller and thinner units is available for the home improvement market including 600 × 600 mm, 600 × 450 mm, 450 × 450 mm and 400 × 400 mm by 30–40 mm. Thicker units (65 and 70 mm) are manufactured to withstand light traffic. Plain pressed slabs may have slightly textured surfaces, while cast slabs are available with smooth, simulated riven stone, terrazzo or textured finishes. Tooled textured-finished slabs and

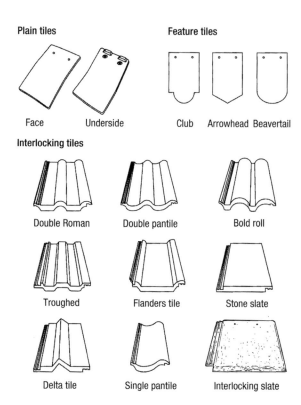

Fig. 3.27 Concrete roofing tiles and slates

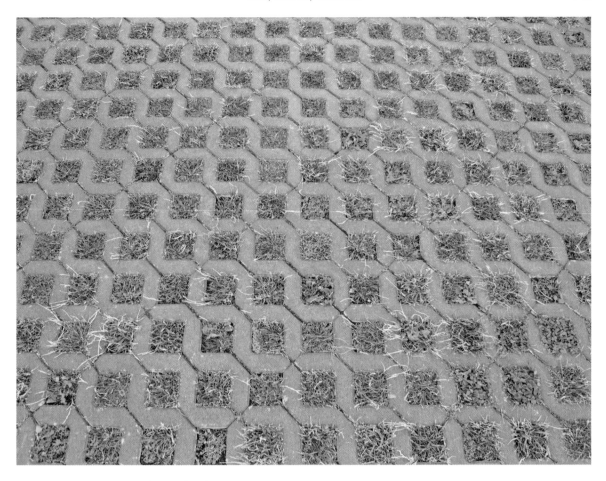

Fig. 3.28 Grass block system. *Photograph*: Arthur Lyons

associated products are available for use in visually sensitive locations. In addition to the standard square and rectangular units, a wide range of decorative designs including hexagonal, simulated bricks and edging units is generally available. Grass block systems (Fig. 3.28) offer stabilised surfaces for light pedestrian and vehicular traffic while allowing rainwater to infiltrate into the ground. Attenuated water run-off, preventing flash flooding, is an important component of Sustainable Drainage Systems (SuDS) schemes.

Paving tile units for roof terraces, balconies and external pedestrian areas in frost-resistant Portland cement concrete are manufactured to square and hexagonal designs in a range of standard red, brown and buff colours. They are suitable for laying on asphalt, bitumen membrane roofing, inverted roofs and sand/cement screed. Typical sizes are 305 × 305 mm and 457 × 457 mm with thicknesses ranging from 25 to 50 mm.

Nitrogen oxide-absorbing concrete tiles and pavers

Roof tiles incorporating titanium dioxide in the granular surface absorb oxides of nitrogen and convert them in the presence of sunlight to nitrates which then react with the excess calcium in the concrete to form soluble calcium nitrate.

Titanium oxide-coated paving stones similarly absorb and convert nitrogen oxides produced by road traffic, thus reducing the harmful pollution within trafficated zones. Nitrogen oxide levels can be reduced in urban areas by between 10% and 20%. Furthermore, the titanium-coated paving slabs are easier to clean than standard concrete pavers. Cement incorporating nano-crystalline titanium oxide is commercially available for construction use where pollution control is particularly required.

References

FURTHER READING

Aitcin, P-C. and Mindess, S. 2011: *Sustainability of concrete (Modern concrete technology)*. Abingdon: Spon Press.

Allbury, K., Franklin, E. and Anderson, J. 2013: *Environmental impact of brick stone and concrete*. Garston: IHS BRE.

Allen, G. 2003: *Hydraulic lime mortar for stone, brick and block masonry*. Abingdon: Routledge.

Beall, C. 2004: *Masonry design and detailing for architects and contractors*, 5th edn. Maidenhead: McGraw-Hill.

Bennett, D. 2005: *The art of precast concrete. Colour texture expression*. Basel: Birkhäuser.

Bennett, D. 2007: *Architectural in situ concrete*. London: RIBA Publishing.

Bennett, D. 2010: *Sustainable concrete architecture*. London: RIBA Publishing.

Cohen, J-L. and Mueller, G. (eds) 2006: *Liquid stone. New architecture in concrete*. Basel: Birkhäuser.

Concrete Centre. 2010: *Concrete credentials. Sustainability. A quick reference guide to the sustainability and performance benefits of concrete*. Camberley: Concrete Centre.

Croft, C. 2005: *Concrete architecture*. London: Laurence King Publishing.

Dhir, R.K. 2005: *Cement combinations for durable concrete*. London: Thomas Telford.

Dyer, T. 2013: *Concrete durability*. Abingdon: CRC Press.

Fröhlich, B. (ed.) 2002: *Concrete architecture. Design and construction*. Basel: Birkhäuser.

Gaventa, S. 2006: *Concrete design*. London: Mitchell Beazley.

Glass, J. 2000: *Future for precast concrete in low rise housing*. Leicester: British Precast Concrete Federation.

Hewlett, P.C. 2003: *Lea's chemistry of cement and concrete*, 4th edn. London: Butterworth Heinemann.

Holmes, S. and Wingate, M. 2002: *Building with lime: A practical introduction*. London: Intermediate Technology Publications.

Jahren, P. and Tongbo, S. 2013: *Concrete and sustainability*. Abingdon: CRC Press.

Johnson, R.D. and Anderson, D. 2004: *Design of composite of steel and concrete structures*. London: Thomas Telford.

King, B. 2007: *Making better concrete*. California: Green Building Press.

Koren, L. and Hall, W. 2012: *Concrete*. London: Phaidon Press Ltd.

Links. 2012: *Architecture and construction in concrete*. Barcelona S.A.: Links International.

Matthews, S. 2014: *Design of durable concrete structures*. Garston IHS BRE.

Meyhôefer, D. 2007: *Concrete creations. Contemporary buildings and interiors*. Hamburg: Braun.

Mosley, W.H., Hulse, R. and Bungey, J.H. 2012: *Reinforced concrete design to Eurocode 2*, 7th edn. Basingstoke: Palgrave Macmillan.

NHBC. 2013: *NHBC Standards 2013 Part 2 Materials*. Milton Keynes: NHBC.

Peck, M. (ed.). 2006: *Design practice. Concrete design construction examples*. Basel: Birkhäuser.

Ramezanianpour, A.A. 2013: *Cement replacement materials. Properties, durability, sustainability*. London: Springer.

Raupach, M., Elsener, B., Polder, R. and Mietz, J. 2006: *Corrosion of reinforcement in concrete*. Cambridge: CRC Press – Woodhead Publishing.

Sakai, K. and Noguchi, T. 2012: *Sustainable use of concrete*. Abingdon: CRC Press.

Slessor, C. 2000: *Concrete regionalism*. London: Thames and Hudson.

Stacey, M. 2010: *Concrete. A studio guide*. London: RIBA Publishing.

Taschen 2008: *Architectural materials. Concrete*. Cologne: Taschen GmbH.

Thomas, M. 2013: *Supplementary cementing materials in concrete*. Abingdon: CRC Press.

True, G. 2012: *Decorative and innovative use of concrete*. Caithness: Whittles Publishing.

Winter, N.B. 2012: *Understanding cement*. St Ives: WHD Microanalysis Consultants Ltd.

Yates, T. and Ferguson, A. 2008: *The use of lime-based mortars in new build*. Publication NF12. Amersham: NHBC Foundation.

STANDARDS

BS 410
    Test sieves:
        Part 1: 2000   Test sieves of metal wire cloth.
        Part 2: 2000   Test sieves of perforated metal.
BS 812
    Testing aggregates:
        Parts 2, 103, 104, 109–112, 114, 121, 123–124.
BS 1370: 1979
    Specification for low heat Portland cement.

BS 1881
  Testing concrete:
      Parts 112, 113, 119, 121, 122, 124,125, 129–131,
      201, 204, 206–210.
BS 4449: 2005
  Steel for the reinforcement of concrete. Weldable
  reinforcing steel. Bar, coil and decoiled product.
  Specification.
BS 4482: 2005
  Steel wire for the reinforcement of concrete
  products. Specification.
BS 4483: 2005
  Steel fabric for the reinforcement of concrete.
  Specification.
BS 4486: 1980
  Specification for hot rolled and hot rolled and
  processed high tensile alloy steel bars for the
  prestressing of concrete.
BS 4550
  Methods of testing cement:
      Parts 0, 3.1, 3.4, 3.8 and 6.
BS 4551: 2005
  Mortar. Methods of test for mortar. Chemical
  analysis and physical testing.
BS 5642
  Sills and copings:
      Part 1: 1978   Specification for window sills
                      of precast concrete, cast stone,
                      clayware, slate and natural
                      stone.
      Part 2: 1983   Specification for coping of pre-
                      cast concrete, cast stone, clay-
                      ware, slate and natural stone.
BS 5838
  Specification for dry packaged cementitous
  mixes:
      Part 1: 1980   Prepacked concrete mixes.
BS 5896: 2012
  High tensile steel wire and strand for the
  prestressing of concrete. Specification.
BS 5977
  Lintels:
      Part 1: 1981   Method for assessment of load.
BS 6073
  Precast concrete masonry units:
      Part 2: 2008   Method for specifying precast
                      concrete masonry units.
BS 6089: 2010
  Assessment of in-situ compressive and precast
  concrete components.
BS 6100
  Building and civil engineering vocabulary:

      Part 0: 2010   Introduction.
      Part 1: 2004   General terms.
      Part 9: 2007   Work with concrete and plaster.
BS 6463
  Quicklime, hydrated lime and natural calcium
  carbonate:
      Part 101: 1996 Methods of preparing samples
                      for testing.
      Part 102: 2001 Methods for chemical analysis.
      Part 103: 1999 Methods for physical testing.
BS 6744: 2001
  Stainless steel bars for the reinforcement of and
  use in concrete.
pr BS ISO 6935–2: 2014
  Steel for the reinforcement of concrete. Ribbed
  bars.
BS 7542: 1992
  Method of test for curing compounds for
  concrete.
BS 7979: 2001
  Specification of limestone fines for use with
  Portland cement.
BS 8000
  Workmanship on building sites:
      Part 2: 1990   Code of practice for concrete
                      work.
      Part 9: 2003   Cementitious levelling screeds
                      and wearing screeds.
BS 8204
  Screeds, bases and in-situ floorings:
      Part 1: 2003   Concrete bases and cement sand
                      levelling screeds to receive
                      floorings.
      Part 2: 2003   Concrete wearing surfaces.
      Part 3: 2004   Polymer modified cementitious
                      levelling screeds and wearing
                      surfaces.
      Part 4: 2004   Cementitious terrazzo wearing
                      surfaces.
      Part 7: 2003   Pumpable self-smoothing
                      screeds.
BS 8297: 2000
  Code of practice for design and installation of
  non-loadbearing precast concrete cladding.
BS 8443: 2005
  Specification for assessing the suitability of special
  purpose concrete admixtures.
BS 8500
  Concrete. Complementary British Standard to
  BS EN 206–1:
      Part 1: 2006   Method of specifying and
                      guidance to the specifier.

Part 2: 2006 Specification for constituent materials and concrete.

BS ISO 13270: 2013
 Steel fibres for concrete. Definitions and specifications.

BS ISO 14656: 1999
 Epoxy powder and sealing material for the coating of steel for the reinforcement of concrete.

BS ISO 16020: 2005
 Steel for the reinforcement and prestressing of concrete. Vocabulary.

BS EN 196
 Methods of testing cement:
  Part 1: 2005 Determination of strength.
  Part 2: 2013 Chemical analysis of cement.
  Part 3: 2005 Determination of setting time and soundness.
  PD CEN/TR Part 4: 2007 Quantitative determination of constituents.
  Part 5: 2011 Pozzolanicity test for pozzolanic cements.
  Part 6: 2010 Determination of fineness.
  Part 7: 2007 Methods of taking and preparing samples of cement.

BS EN 197
 Cement:
  Part 1: 2011 Composition specifications and conformity criteria for common cements.
  Part 2: 2014 Conformity evaluation.

BS EN 206: 2013
 Concrete. Specification, performance, production and conformity.

BS EN 413
 Masonry cement:
  Part 1: 2011 Composition, specifications and conformity criteria.
  Part 2: 2005 Test methods.

BS EN 445: 2007
 Grout for prestressing tendons. Test methods.

BS EN 446: 2007
 Grout for prestressing tendons. Grouting procedures.

BS EN 447: 2007
 Grout for prestressing tendons. Basic requirements.

BS EN 450
 Fly ash for concrete:
  Part 1: 2012 Definitions, specification and conformity criteria.
  Part 2: 2005 Conformity evaluation.

BS EN 451
 Method of testing fly ash:

Part 1: 2003 Determination of free calcium oxide content.
Part 2: 1995 Determination of fineness by wet sieving.

BS EN 459
 Building lime:
  Part 1: 2010 Definitions, specifications and conformity criteria.
  Part 2: 2010 Test methods.
  Part 3: 2011 Conformity evaluation.

BS EN 480
 Admixtures for concrete, mortar and grout:
  Part 1: 2006 Test methods. Reference concrete and reference mortar for testing.
  Part 2: 2006 Test methods. Determination of setting time.

BS EN 490: 2011
 Concrete roofing tiles and fittings for roof covering and wall cladding. Product specifications.

BS EN 491: 2011
 Concrete roofing tiles and fittings for roof covering and wall cladding. Test methods.

BS EN 845
 Specification for ancillary components for masonry:
  Part 2: 2013 Lintels.

BS EN 933
 Tests for geometrical properties of aggregates:
  Part 1: 2012 Determination of particle size distribution. Sieving method.
  Part 11: 2009 Classification test for the constituents of coarse recycled aggregate.

BS EN 934
 Admixtures for concrete, mortar and grout:
  Part 1: 2008 Common requirements.
  Part 2: 2009 Concrete admixtures. Definitions, requirements, conformity.
  Part 3: 2009 Admixtures for masonry mortar. Definitions, requirements, conformity.
  Part 4: 2009 Admixtures for grout for prestressing tendons.
  Part 5: 2007 Admixtures for sprayed concrete.
  Part 6: 2001 Sampling, conformity control and evaluation of conformity.

BS EN 998
 Specification of mortar for masonry:

Part 1: 2010   Rendering and plastering mortar.

Part 2: 2010   Masonry mortar.

BS EN 1008: 2002
Mixing water for concrete. Specification for sampling, testing and assessing suitability.

BS EN 1015
Methods of test of mortar for masonry.

BS EN 1168: 2005
Precast concrete products. Hollow core slabs.

BS EN 1504
Products and systems for the protection and repair of concrete structures.

BS EN 1771: 2004
Products and systems for the repair of concrete structures.

BS EN 1992
Eurocode 2: Design of concrete structures:

Part 1.1: 2004   General rules and rules for buildings.

Part 1.2: 2004   General rules. Structural fire design.

NA Part 1.1: 2004   UK National Annex to Eurocode 2. Design of concrete structures. General rules and rules for building.

NA Part 1.2: 2004   UK National Annex to Eurocode 2. Design of concrete structures. General rules. Structural fire design.

DD CEN/TS 1992
Design of fastenings for use in concrete:

Part 4–1: 2009   General.

Part 4–2: 2009   Headed fasteners.

Part 4–3: 2009   Anchor channels.

Part 4–4: 2009   Post-installed fasteners. Mechanical systems.

Part 4–5: 2009   Post-installed fasteners. Chemical systems.

BS EN 1994
Eurocode 4: Design of composite steel and concrete structures:

Part 1.1: 2004   General rules and rules for buildings.

Part 1.2: 2005   General rules. Structural fire design.

NA Part 1.1: 2004   UK National Annex to Eurocode 4. Design of composite steel and concrete structures. General rules and rules for building.

NA Part 1.2: 2005   UK National Annex to Eurocode 4. Design of composite steel and concrete structures. General rules. Structural fire design.

BS EN 10080: 2005
Steel for the reinforcement of concrete. Weldable reinforcing steel. General.

BS EN 12350
Testing fresh concrete:

Part 1: 2009   Sampling.

Part 2: 2009   Slump test.

Part 3: 2009   Vebe test.

Part 4: 2009   Degree of compactability.

Part 5: 2009   Flow table test.

Part 6: 2009   Density.

Part 7: 2009   Pressure methods.

Part 8: 2010   Self-compacting concrete. Slump-flow test.

Part 9: 2010   Self-compacting concrete. V-funnel test.

Part 10: 2010   Self-compacting concrete. L-box test.

Part 11: 2010   Self-compacting concrete. Sieve segregation test.

Part 12: 2010   Self-compacting concrete. J-ring test.

BS EN 12390
Testing hardened concrete:

Part 1: 2012   Shape, dimensions for specimens and moulds.

Part 2: 2009   Making and curing specimens for strength tests.

Part 3: 2009   Compressive strength of test specimens.

Part 4: 2000   Specification for testing machines.

Part 5: 2009   Flexural strength of test specimens.

Part 6: 2009   Tensile splitting strength of test specimens.

Part 7: 2009   Density of hardened concrete.

Part 8: 2009   Depth of penetration of water under pressure.

DD CEN/TS   Part 9: 2006 Freeze thaw resistance.

DD CEN/TS Part 10: 2007 Determination of the relative carbonation resistance of concrete.

DD CEN/TS Part 11: 2010 Determination of chloride resistance of concrete.

DD CEN/TS Part 13: 2012 Determination of secant modulus of elasticity in compression.

BS EN 12504

  Testing concrete in structures:

    Part 1: 2009 Cored specimens. Taking, examining and testing under compression.

    Part 2: 2012 Non-destructive testing. Determination of rebound number.

    Part 3: 2005 Determination of pull-out force.

    Part 4: 2004 Determination of ultrasonic pulse velocity.

BS EN 12602: 2008

  Prefabricated reinforced components of autoclaved aerated concrete.

BS EN 12620: 2013

  Aggregates for concrete.

BS EN ISO 12696: 2012

  Cathodic protection of steel in concrete.

BS EN 12794: 2005

  Precast concrete products. Foundation piles.

BS EN 12878: 2014

  Pigments for the colouring of building materials based on cement and/or lime.

pr EN 13055: 2012

  Lightweight aggregates for concrete, mortar, grout, bituminous mixtures, surface treatments and for bound and unbound applications.

BS EN 13139: 2013

  Aggregates for mortar.

BS EN 13263

  Silica fume for concrete:

    Part 1: 2005 Definitions, requirements and conformity.

    Part 2: 2005 Conformity evaluation.

BS EN 13369: 2013

  Common rules for precast concrete products.

BS EN 13501

  Fire classification of construction products and building elements:

    Part 1: 2007 Classification using test data from reaction to fire tests.

    Part 2: 2007 Classification using data from fire resistance tests.

BS EN 13747: 2005

  Precast concrete products. Floor plates for floor systems.

BS EN 13791: 2007

  Assessment of in-situ compressive strength in structures and precast concrete components.

BS EN 13813: 2002

  Screed materials and floor screeds. Screed material. Properties and requirements.

BS EN 13888: 2009

  Grout for tiles. Requirements, evaluation of conformity, classification and designation.

pr EN 13914–1: 2013

  Design, preparation and application of external rendering and internal plastering. External rendering.

BS EN 14216: 2004

  Cement. Composition, specifications and conformity criteria for very low heat special cements.

BS EN 14474: 2004

  Precast concrete products. Concrete with wood chips as aggregate. Requirements and test methods.

BS EN 14487

  Sprayed concrete:

    Part 1: 2005 Definitions, specifications and conformity.

    Part 2: 2006 Execution.

BS EN 14488: 2005

  Testing sprayed concrete.

BS EN 14647: 2005

  Calcium aluminate cement. Composition, specification and conformity criteria.

BS EN 14650: 2005

  Precast concrete. General rules for factory production control of metallic fibred concrete.

BS EN 14721: 2005

  Test method for metallic fibre concrete.

BS EN 14843: 2007

  Precast concrete products. Stairs.

BS EN 14845

  Test methods for fibres in concrete:

    Part 1: 2007 Reference concretes.

    Part 2: 2006 Effect on concrete.

BS EN 14889

  Fibres for concrete:

Part 1: 2006   Steel fibres. Definitions,
                specifications and conformity.
Part 2: 2006   Polymer fibres. Definitions,
                specifications and conformity.

BS EN 14991: 2007
Precast concrete elements. Foundation elements.
BS EN 14992: 2007
Precast concrete elements. Wall elements.
BS EN 15037
Precast concrete products. Beam and block floor
systems:

Part 1: 2008   Beams.
Part 2: 2009   Concrete blocks.
Part 3: 2009   Clay blocks.
Part 4: 2010   Expanded polystyrene blocks.
Part 5: 2013   Lightweight blocks for simple
                formwork.

BS EN 15167
Ground granulated blast furnace slag for use in
concrete, mortar and grout:

Part 1: 2006   Definitions, specifications and
                conformity criteria.
Part 2: 2006   Conformity evaluation.

BS EN 15183: 2006
Products and systems for the protection and
repair of concrete structures. Test methods.
BS EN 15191: 2009
Precast concrete products. Classification of glass
fibre reinforced concrete performance.
BS EN 15422: 2008
Precast concrete products. Specification of glass
fibres for reinforcement of mortars and concretes.
BS EN 15435: 2008
Precast concrete products. Normal and
lightweight concrete shuttering blocks. Product
properties and performance.
BS EN 15498: 2008
Precast concrete products. Woodchip concrete
shuttering blocks. Product properties and
performance.
BS EN 15564: 2008
Precast concrete products. Resin bound concrete.
Requirements and test methods.
BS EN ISO 15630
Steel for the reinforcement and prestressing of
concrete. Test methods:

Part 1: 2010   Reinforcing bars, wire rod and
                wire.
Part 2: 2010   Welded fabric.
Part 3: 2010   Prestressing steel.

PD CEN/TR 15697: 2008
Cement. Performance testing for sulfate
resistance. State of the art report.
PD CEN/TR 15739: 2008
Precast concrete products. Concrete finishes.
Identification.
BS EN 15743: 2010
Supersulfated cement. Composition, specifications
and conformity criteria.
BS EN 16025: 2013
Thermal and/or sound insulating products in
building construction. Bound EPS ballasting:

Part 1: 2013   Requirements for factory
                premixed EPS dry plaster.
Part 2: 2013   Processing of the factory
                premixed EPS dry plaster.

PD 6678: 2005
Guide to the specification of masonry mortar.
PD 6682
Aggregates:

Part 1: 2009   Aggregates for concrete.
                Guidance on the use of BS EN
                12620.
Part 3: 2003   Aggregates for mortar.
                Guidance on the use of BS EN
                13139.
Part 4: 2003   Lightweight aggregates for
                concrete, mortar and grout.
                Guidance on the use of
                BS EN 13055–1.
Part 9: 2003   Guidance on the use of
                European test method
                standards.

PD 6687–1: 2010
Background paper to the National Annexes to
BS EN 1992–1 and BS EN 1992–3.
PD 6687–2: 2008
Recommendations for the design of structures to
BS EN 1992–2: 2005.

BUILDING RESEARCH ESTABLISHMENT PUBLICATIONS

BRE Special digests

BRE SD1: 2005
Concrete in aggressive ground.
BRE SD3: 2002
HAC concrete in the UK: Assessment, durability
management, maintenance and refurbishment.

BRE Digests

BRE Digest 330: 2004
Alkali-silica reaction in concrete (Parts 1–4).
BRE Digest 444: 2000
Corrosion of steel in concrete (Parts 1, 2 and 3).
BRE Digest 451: 2000
Tension tests for concrete.
BRE Digest 455: 2001
Corrosion of steel in concrete: Service life design and prediction.
BRE Digest 487: 2004
Structural fire engineering design: Materials behaviour. Part 1 Concrete.
BRE Digest 491: 2004
Corrosion of steel in concrete.
BRE Digest 507: 2008
Marine aggregates in concrete.
BRE Digest 527: 2012
Effects of chemical, physical and mechanical processes on concrete.
BRE Digest 530: 2013
Applications, performance characteristics and environmental benefits of alkali-activated binder concretes.

BRE Good building guides

BRE GBG 39: 2001
Simple foundations for low-rise housing: Rule of thumb design.
BRE GBG 64 Part 2: 2005
Tiling and slating pitched roofs: Plain and profiled clay and concrete tiles.
BRE GBG 66: 2005
Building masonry with lime-based mortars.

BRE Information papers

BRE IP 9/01
Porous aggregates in concrete: Jurassic limestones.
BRE IP 11/01
Delayed ettringite formation: In-situ concrete.
BRE IP 18/01
Blastfurnace slag and steel slag: Their use as aggregates.
BRE IP 1/02
Minimising the risk of alkali–silica reaction: Alternative methods.
BRE IP 7/02
Reinforced autoclaved aerated concrete panels.

BRE IP 15/02
Volumetric strain of concrete under uniaxial compression with reference to sustained loading and high grade concrete.
BRE IP 4/03
Deterioration of cement-based building materials: Lessons learnt.
BRE IP 16/03
Proprietary renders.
BRE IP 3/04
Self-compacting concrete.
BRE IP 6/04
Porous aggregates in concrete.
BRE IP 12/04
Concrete with minimal or no primary aggregate content.
BRE IP 11/05
Innovation in concrete frame construction.
BRE IP 17/05
Concretes with high ggbs contents for use in hard/firm secant piling.
BRE IP 3/06
Reinforced concrete service life design (Parts 1, 2 and 3).
BRE IP 9/07
Performance-based intervention for durable concrete repairs.
BRE IP 7/08
Cements with lower environmental impact.
BRE IP 3/09
Lessons learnt from the Barratt green house. Delivering a zero carbon home on innovative concrete systems.
BRE IP 5/09
Silica fume in concrete.
BRE IP 9/10
Alkaline ash binders.
BRE IP 14/11
Hemp lime. An introduction to low-impact building materials.
BRE IP 26/12
Asset management and service life of concrete structures and components.
BRE IP 5/13
Durability of alkali activated binder concretes. Early age performance data.

BRE Reports

BR 421: 2001
Low energy cements.

BR 429: 2001
High alumina cement and concrete.
BR 451: 2002
High alumina cement. BRAC rules – revised 2002.
BR 468: 2004
Fire safety of concrete structures.
BR 496: 2007
Calcium sulfoaluminate cements.
EP 85: 2008
Hemp lime construction.
EP 99: 2009
Sustainable masonry construction.

BRITISH CEMENT ASSOCIATION

BCA Fact Sheet 7: 2012
Alternative fuels and raw materials in cement kilns.
BCA Fact Sheet 13: 2007
Specifying factory-made CEM II cements for use in masonry mortars.
BCA Fact Sheet 17: 2012
Cement and cement clinker.
BCA Fact Sheet 18: 2012
Embodied CO_2e of UK cement, additions and cementitious material.

MINERAL PRODUCTS ASSOCIATION

MPA Fact Sheet 1: 2013
Fire resistance of concrete.
MPA Fact Sheet 2: 2013
Thaumasite form of sulphate attack.
MPA Fact Sheet 3: 2013
Delayed ettringite formation.
MPA Fact Sheet 4: 2013
Alkali silica reaction.
MPA Fact Sheet 5: 2013
Self-compacting concrete.
MPA Fact Sheet 6: 2013
Use of recycled aggregate in concrete.
MPA Fact Sheet 8: 2011
Factory made Portland limestone cements.
MPA Fact Sheet 12: 2013
Novel cements. Low energy, low carbon cements.
MPA Fact Sheet 14a: 2013
Modern cements (packed).
MPA Fact Sheet 14b: 2013
Modern cements (bulk).

CONCRETE CENTRE

MPA Concrete Centre: 2008
Concrete and fire safety.
MPA Concrete Centre: 2009
Concrete and the Green Guide.
MPA Concrete Centre: 2009
Concrete and the Code for Sustainable Homes.
MPA Concrete Centre: 2010
Concrete credentials – sustainability.
MPA Concrete Centre: 2010
How to achieve acoustic performance in masonry homes.
MPA Concrete Centre: 2011
Fabric for the future.
MPA Concrete Centre: 2011
Thermal performance Part L1A.
MPA Concrete Centre: 2011
Specifying sustainable concrete.
MPA Concrete Centre: 2011
Performance of concrete structures in fire.
MPA Concrete Centre: 2011
Zero carbon performance – cost effective concrete and masonry homes.
MPA Concrete Centre: 2011
How to achieve good levels of airtightness in masonry homes.
MPA Concrete Centre: 2012
Thermal mass explained (2012 update).
MPA Concrete Centre: 2012
Insulating concrete formwork (ICF) Guide.
Report TCC/05/20: 2010
Concrete credentials. Sustainability. A quick reference guide to the sustainability and performance of concrete.

CONCRETE SOCIETY REPORTS

Technical Report 18: 2002
A guide to the selection of admixtures for concrete.
Technical Report 46: 1997
Calcium aluminate cements in construction – a reassessment.
Technical Report 51: 1998
Guidance on the use of stainless steel reinforcement.
Technical Report 52: 1999
Plain formed concrete finishes.
Technical Report 55: 2000
Design guidance for strengthening concrete structures using fibre composite materials.

Technical Report 61: 2004
Enhancing reinforced concrete durability.
Technical Report 62: 2005
Self-compacting concrete – a review.
Technical Report 63: 2007
Guidance for the design of steel-fibre-reinforced concrete.
Technical Report 65: 2007
Guidance on the use of macro-synthetic fibre reinforced concrete.
Technical Report 66: 2007
External in-situ concrete paving.
Technical Report 72: 2010
Durable post-tensioned concrete structures.
Technical Report 73: 2011
Cathodic protection of steel in concrete.
Technical Report 74: 2011
Cementitious materials.

Concrete Advice 07: 2003
Galvanised steel reinforcement.
Concrete Advice 14: 2003
Concrete surfaces for painting.
Concrete Advice 16: 2003
Assessing as struck in situ concrete surface finishes.
Concrete Advice 43: 2012
Spacers and visual concrete.

CS 023: 2003
The new concrete standards – getting started.
CS 030: 2012
Formwork – a guide to good practice, 3rd edn.
CS 152: 2004
National structural concrete specification for building construction.
CS 170: 2013
Visual concrete – finishes.
CS 171: 2013
Visual concrete – planning and assessment.
CS 172: 2013
Visual concrete – control of blemishes.
CS 173: 2013
Visual concrete – weathering stains and efflorescence.

CS 174: 2013
Influence of integral water-resisting admixtures on the durability of concrete.

Good Concrete Guide 7: 2009
Foamed concrete – application and specification.
Good Concrete Guide 9: 2009
Designed and detailed – Eurocode 2.

ADVISORY ORGANISATIONS

Architectural Cladding Association, 60 Charles Street, Leicester LE1 1FB (0116 253 6161).

British Precast, 60 Charles Street, Leicester LE1 1FB (0116 253 6161).

Cement Admixtures Association, 38a Tilehouse Green Lane, Knowle, West Midlands B93 9EY (01564 776362).

Cementitious Slag Makers Association, The Coach House, West Hill, Oxted, Surrey RH8 9JB (01708 682439).

Concrete Society Advisory Service, Riverside House, 4 Meadows Business Park, Station Approach, Blackwater, Camberley, Surrey GU17 9AB (01276 607140).

Construct Concrete Structures Group Ltd, Riverside House, 4 Meadows Business Park, Station Approach, Blackwater, Camberley, Surrey GU17 9AB (01276 38444).

Lime Centre, Long Barn, Morestead, Winchester, Hampshire SO21 1LZ (01962 713636).

MPA (Mineral Products Association), The Concrete Centre, Gillingham House, 38–44 Gillingham Street, London SW1V 1HU (0207 963 8000).

Prestressed Concrete Association, 60 Charles Street, Leicester LE1 1FB (0116 253 6161).

Quarry Products Association, 156 Buckingham Palace Road, London SW1W 9TR (0207 730 8194).

Sprayed Concrete Association, Kingsley House, Ganders Business Park, Kingsley, Bordon, Hampshire GU35 9LU (01420 471622).

Structural Precast Association, 60 Charles Street, Leicester LE1 1FB (0116 253 6161).

4

TIMBER AND TIMBER PRODUCTS

CONTENTS

Introduction

Timber, arguably the original building material, retains its prime importance within the construction industry because of its versatility, diversity and aesthetic properties. About 20% of the earth's land mass is covered by forests, divided between approximately two-thirds as hardwoods in temperate and tropical climates and one-third as softwoods within temperate and colder regions. Approximately one-third of the annual worldwide timber harvest is used in construction, and the rest is consumed for paper production, as a fuel, or wasted during the logging process.

Environmental issues, raised by the need to meet the current and future demands for timber, can only be resolved by sustainable forest developments. In temperate climate forests, clear cutting, in which an area is totally stripped, followed by replanting, is the most economical, but the shelterwood method, involving a staggered harvest over several years, ensures that replacement by young trees becomes established as the mature ones are felled. The managed forests of North America and Scandinavia are beginning to increase in area due to additional planting for future use. The deforestation of certain tropical regions has allowed wind and rain to erode the thin topsoil, leaving inhospitable or desert conditions; furthermore, the overall reduction in world rain forest areas is contributing significantly to the *greenhouse effect* by reducing the rate of extraction of carbon dioxide from the atmosphere.

Compared to many other major construction materials including aluminium, plastics and steel, timber has a relatively low embodied energy of production. Trees require little energy for their conversion into usable timber, and young replacement trees are particularly efficient at absorbing carbon dioxide and releasing oxygen into the atmosphere. Temperate and tropical hardwoods, when suitably managed, can be brought to maturity within a human lifespan, whereas softwoods take half that period to mature. Timber products manufactured from reconstituted and waste wood add to the efficient use of forestry. Timber frame, one of the modern methods of construction (MMC), now accounts for approximately 22% of new-build housing within the UK.

Great emphasis is now placed on timber certification schemes, which track the material as a chain of custody from source to user, to ensure the accuracy of environmental claims being made. In the UK, the two international schemes – the Forest Stewardship Council (FSC) and the Programme for the Endorsement of Forest Certification (PEFC) – ensure that timber with their labels has been harvested from properly managed sustainable sources. Products purchased under the labels of the Canadian Standards Association (CSA) or the American Sustainable Forestry Initiative (SFI) are also certified by PEFC. One scheme accredited by PEFC for tropical timber is the Malaysian Timber Certification Scheme (MTCS). In addition, TRADA, as an accreditation body, has developed its own chain of custody and certifies other routes under its Forest Products scheme. Materials available under these certified schemes include structural timber, joinery timber and the timber products plywood, particleboard, flooring, MDF and OSB. The European Union Timber Regulation (2013) requires companies using EU-produced timber or importing timber into the EU to have a documented Due Diligence Process in place to reduce the risk that illegal timber is involved. The FSC and PEFC schemes provide the necessary Chain of Custody.

Timber

The Study Centre at Darwin College, Cambridge (Fig. 4.1), which occupies a narrow site overlooking the River Cam, is designed to accommodate both books and computers. It is a load-bearing masonry and timber building which features the extensive use of English oak, including massive paired columns to the first-floor reading room which is partly cantilevered over the river. The columns in green oak have characteristic shakes and splits giving an impression of great age, and these contrast with the refined oak and oak veneer of the floors, window frames and furniture. Joints in the green oak are held by stainless steel fixings, which can be tightened as the timber dries and shrinks. The use of oak throughout gives unity to the building, which sits comfortably within its highly sensitive location.

METABOLISM OF THE TREE

The tree, a complex living organism, may be considered in three main sections: the branches with their leaves, the trunk (or bole) and the roots (Fig. 4.2). The roots anchor the tree to the ground and absorb water with dissolved minerals from the soil. The leaves absorb carbon dioxide from the air and in the presence of sunlight, together with chlorophyll as a catalyst, combine carbon dioxide with water to produce sugars. Oxygen, a by-product of the process, diffuses out of the leaves. The sugars in aqueous solution are transported down the branches and trunk to be subsequently converted, where required for growth, into the cellulose of the tree. The trunk gives structural strength to the tree and acts as a store for minerals and food such as starch and also as a two-way transport medium.

The tree is protected from extremes of temperature and mechanical damage by the bark, inside which is the bast layer which transports the sugars synthesised in the leaves downward. Radial rays then move the food into the sapwood cells for storage. Inside the bast is the thin and delicate cambium, which is the growing layer for the bark and sapwood. Growth only takes place when the cambium layer is active, which in temperate climates is during the spring and summer seasons.

A transverse section through the bole shows the growth rings. These are sometimes referred to as annual rings, but unusual growth patterns can lead to multiple rings within one year, and in tropical climates, where seasonal changes are less pronounced, growth rings

Fig. 4.1 Green Oak construction – Darwin College Study Centre, Cambridge. *Architects*: Dixon Jones. *Photographs*: Courtesy of Dennis Gilbert

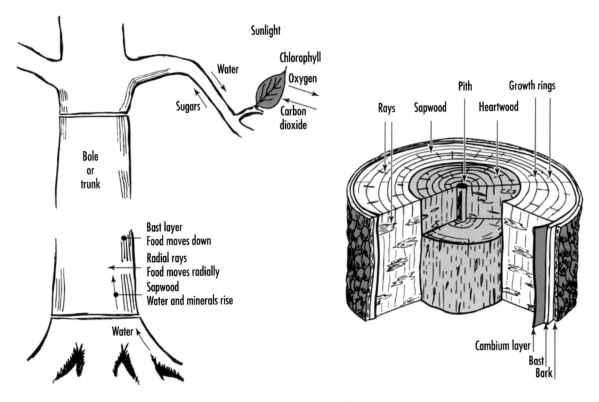

Fig. 4.2 Metabolism of the tree (after Everett A. 1994: *Mitchell's Materials*. 5th edition. Longman Scientific & Technical)

Photosynthesis

$$6CO_2 + 6H_2O \xrightarrow{\text{sunlight/chlorophyll}} C_6H_{12}O_6 + 6O_2$$

carbon   water   glucose   oxygen
dioxide

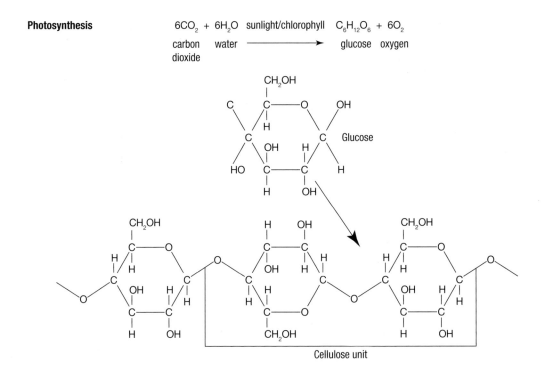

Fig. 4.3 Structure of cellulose

may be indistinct and not annual. The growth rings are apparent because the *early wood* produced at the start of the growing season tends to be made from larger cells of thinner walls and is thus softer and more porous than the *late wood* produced towards the end of the growing season. Each year as the tree matures with the production of an additional growth ring, the cells of an inner ring are strengthened by a process of *secondary thickening*. This is followed by lignification in which the cell dies. These cells are no longer able to act as food stores, but now give increased structural strength to the tree. The physical changes are often associated with a darkening of the timber due to the incorporation into the cell walls of so-called *extractives*, such as resins in softwoods or tannins in oak. These are natural wood preservatives which make heartwood more durable than sapwood.

CONSTITUENTS OF TIMBER

The main constituents of timber are cellulose, hemicellulose and lignin, which are natural polymers. Cellulose, the main constituent of the cell walls, is a polymer made from glucose, a direct product of photosynthesis within the leaves of the tree. Glucose molecules join together to form cellulose chains containing typically 10,000 sugar units (Fig. 4.3). Alternate cellulose chains, running in opposite directions to each other, form a predominantly well-ordered crystalline material. It is this crystalline chain structure which gives cellulose its fibrous properties, and accounts for approximately 45% of the dry weight of the wood.

Hemicelluloses, which account for approximately 25% of the weight of wood, have more complex partially crystalline structures composed of a variety of other sugars. The molecular chains are shorter than those in cellulose, producing a more gelatinous material. Lignin (approximately 25% by weight of the timber) is an insoluble non-crystalline polymeric material. Its main constituents are derivatives of benzene combined to form a complex branched-chain structure.

The three major components are combined to form *microfibrils* which are in turn the building blocks for the cell walls. Crystalline cellulose chains are surrounded by semi-crystalline hemicellulose, then a layer of non-crystalline cellulose, and are finally cemented together with lignin (Fig. 4.4). Millions of these microfibrils are built up in layers to form the individual cell walls. It is this composite structure which gives timber its physical strength, with the cellulose contributing mainly to the

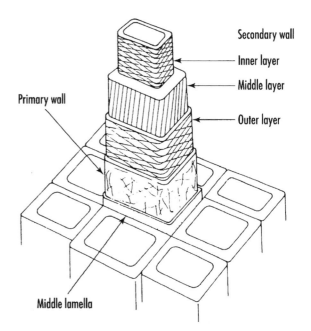

Fig. 4.4 Cell structure of timber (after Desch H.E. 1981: *Timber: its structure properties and utilisation*, 6th edition. Macmillan education – Crown Copyright)

tensile properties and the hemicellulose and lignin to the compressive strength and elasticity.

In addition to the three major constituents and significant quantities of water, timbers contain many minor constituents; some, such as resins, gums and tannins, are associated with the conversion of sapwood to heartwood. Starch present in sapwood is attractive to fungi, and inorganic granules such as silica make working certain tropical hardwoods, such as teak, difficult. The various colours present in different timbers arise from these minor constituents, as the various celluloses and lignin are virtually colourless. Some colours are fixed to the polymeric chains, but others are light-sensitive natural dyes which fade upon prolonged exposure to sunlight, unless the timber is coated with an ultraviolet-absorbing finish.

HARDWOODS AND SOFTWOODS

Commercial timbers are defined as hardwoods or softwoods according to their botanical classification rather than their physical strength. Hardwoods (angiosperms) are from broad-leafed trees, which in temperate climates are deciduous, losing their leaves in autumn, although in tropical climates, where there is little seasonal variation, old leaves are constantly being replaced by new. Softwoods (gymnosperms) are from conifers, characteristically with needle-shaped leaves, and growing predominantly in the northern temperate zone. They are largely evergreen, with the notable exception of the European larch (*Larix decidua*), and they include the Californian redwood (*Sequoia sempervirens*), the world's largest tree with a height of over 100 metres.

Although the terms *hardwood* and *softwood* arose from the physical strength of the timbers, paradoxically, balsa (*Ochroma lagopus*) used for model-making is

botanically a hardwood, while yew (*Taxus baccata*), a strong and durable material, is defined botanically as a softwood. Under microscopic investigation, softwoods show only one type of cell which varies in size between the rapid growth of spring and early summer (early wood) and the slow growth of late summer and autumn (late wood). These cells, or tracheids, perform the food- and water-conducting functions and give strength to the tree. Hardwoods, however, have a more complex cell structure with large cells or vessels for the conducting functions and smaller cells or wood fibres

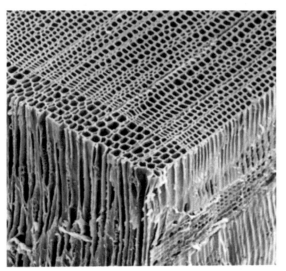

Softwood – Scots Pine (*Pinus sylvestris*)

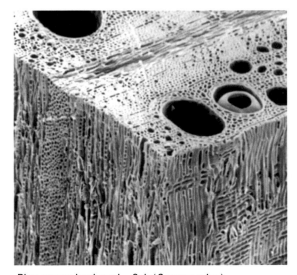

Ring-porous hardwood – Oak (*Quercus robur*)

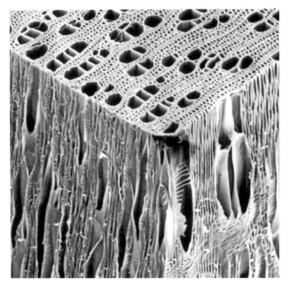

Diffuse-porous hardwood – Birch (*Betula pendula*)

Fig. 4.5 Cell structures of hardwoods and softwoods

Fig. 4.6 Limed Oak – Jerwood Library, Trinity Hall, Cambridge. *Architects*: Freeland Rees Roberts. *Photograph*: Arthur Lyons

which provide the mechanical support. According to the size and distribution of the vessels, hardwoods are divided into two distinct groups. Diffuse-porous hardwoods, which include beech (*Fagus sylvatica*), birch (*Betula pendula*) and most tropical hardwoods, have vessels of a similar diameter distributed approximately evenly throughout the timber. Ring-porous hardwoods, however, including oak (*Quercus robur*), ash (*Fraxinus excelsior*) and elm (*Ulmus procera*), have large vessels concentrated in the earlywood, with only small vessels in the latewood (Fig. 4.5). The Jerwood Library of Trinity Hall, Cambridge (Fig. 4.6) illustrates the visual quality of limed oak as an architectural feature within the context of a sensitive built environment.

TIMBER SPECIES

Any specific timber can be defined through the correct use of its classification into family, genus and species. Thus oak and beech are members of the Fagaceae family; beech is one genus (*Fagus*) and oak (*Quercus*) another. The oak genus is subdivided into several

species, including the most common, the pedunculate oak (*Quercus robur*) and the similar but less common sessile oak (*Quercus petraea*). Such exact timber nomenclature is, however, considerably confused by the use of lax terminology within the building industry; for example, both Malaysian meranti and Philippine lauan are frequently referred to as Philippine mahogany, and yet they are from a quite different family and genus to the true mahogany (*Swietenia*) from the West Indies or Central America. This imprecision can cause the erroneous specification or supply of timber, with serious consequences. Where there is the risk of confusion, users should specify the correct genus and species.

The standard BS EN 13556: 2003 lists both hardwoods (dicotyledons) and softwoods (gymnosperms) used within Europe with a four-letter code. The first two letters are a distinctive combination referring to the genus (e.g. oak – *Quercus* – QC). The third and fourth letters refer to the particular species; thus, European oak – *Quercus petraea* is QCXE and American red oak – *Quercus rubra* is QCXR. Typical softwoods are western red cedar – *Thuja plicata* – THPL and Scots pine – *Pinus sylvestris* – PNSY. The standard pr EN 14081–1: 2013 lists the coding for a range of species combinations; for example, British pine which includes *pinus nigra* and *pinus sylvestris* is coded commercially as WPNN, British spruce (*picea abies* and *picea sitchensis*) is WPCS, and Larch which includes *Larix decidua*, *Larix X eurolepis* and *Larix kaempferi* is coded WLAD.

Softwood accounts for approximately 80% of the timber used in the UK construction industry. Pine (European redwood) and spruce (European whitewood) are imported from Northern and Central Europe, while western hemlock, spruce, pine and fir are imported in quantity from North America. Forest management in these areas ensures that supplies will continue to be available. Smaller quantities of western red cedar, as a durable lightweight cladding material, are imported from North America, together with American redwood from California, pitch pine from Central America and parana pine from Brazil. Increasingly, New Zealand, South Africa and Chile are becoming significant exporters of renewable timber. However, over 40% of the total of sawn timber used in the UK is from British forests. Most British-grown softwood (pine and spruce) is machine graded to strength class C16.

Over 100 different hardwoods are used in the UK, although together beech, oak, sweet chestnut, meranti, lauan, American mahogany and ramin account for over half of the requirements. Approximately half of the hardwoods used in the UK come from temperate forests in North America and Europe including Britain, but the remainder, including the durable timbers such as iroko, mahogany, sapele and teak, are imported from the tropical rain forests. Predominant UK production includes oak, sweet chestnut, ash, beech and sycamore. The Great Oak Hall at Westonbirt Arboretum, Gloucestershire (Fig. 4.7) illustrates the use of 'green oak construction' within a modern building using traditional dowel and wedge carpentry joints. Chestnut timber is also used for 'green oak construction'. Allowances must be made for drying shrinkage and if metal connectors are employed they should be corrosion-resistant due to the acidic tannins in the unseasoned timber.

As measured by satellite photography, since 1970 18% of the Amazon forest in Brazil has been lost, but much of this deforestation has been for agricultural purposes, with more than three-quarters of the timber felled used as a local fuel rather than exported as timber. With the growing understanding of the environmental effects of widespread deforestation, some producer governments are now applying stricter controls to prevent clear felling and encourage sustainable harvesting through controlled logging. Other imported naturally durable hardwoods, available in long lengths, include ekki, greenheart and opepe, while UK-produced sweet chestnut is durable and an appropriate structural timber. Some timbers not previously used within the UK, such as jatoba (*Hymenaea courbaril*), are now being imported from South America.

CONVERSION

Conversion is the process of cutting boles or logs into sections prior to seasoning. Subsequent further cutting into usable sizes is called manufacture. Finishing operations involving planing and sanding produce a visually smooth surface but reduce the absorption of penetrating wood stains. Timber for solid sections is sawn, whereas thin layers for plywood are peeled and veneers are usually sliced across the face of the log to maximise the visual effect of colour and figure, which is the pattern effect seen on the longitudinal surface of cut wood.

Types of cut

The two main types of cut, *plain sawn* and *quarter sawn*, refer to the angle between the timber face and the

Fig. 4.7 Traditional Oak construction – Great Oak Hall, Westonbirt Arboretum, Gloucestershire. *Architects*: Roderick James Architects. *Photograph*: Arthur Lyons

growth rings. This is best observed from the end of the timber, as shown in Figure 4.8. If the cut is such that the growth rings meet the surface at less than 45° then the timber is plain sawn. Timber with this type of cut tends to have a more decorative appearance but a greater tendency to distort by *cupping*. Timber cut with the growth rings meeting the surface at not less than 45° is quarter sawn. Such timber is harder wearing, weather-resistant and less likely to flake. If a log is cut *through and through*, which is most economical, then a mixture of plain and quarter-sawn timber is produced. Quarter sawing is more expensive, as the log requires resetting for each cut and more waste is produced; however, the larger sections will be more dimensionally stable. The centre of the tree, the pith, is frequently soft and may be weakened by splits or shakes. In this case, the centre is removed as a *boxed heart*.

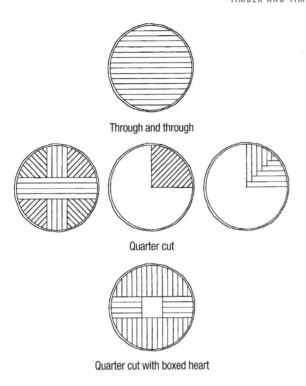

Through and through

Quarter cut

Quarter cut with boxed heart

Fig. 4.8 Conversion of timber

Sizes

BS EN 1313–1: 2010 defines the standard sizes of sawn softwood timbers at 20% moisture content (Table 4.1). Widths over 225 mm and lengths over 5 m are scarce and expensive, but finger jointing (BS EN 14080: 2013 and BS EN 15497: 2014), which can be as strong as the continuous timber, does allow longer lengths to be specified. Regularising, which ensures uniformity of width of a sawn timber, reduces the nominal section by 3 mm (5 mm over 150 mm) and planing on all faces or 'processed all round' (PAR) reduces, for example, a 47 × 100 mm section to 44 × 97 mm. General tolerances for acceptable deviations from target sizes for softwoods are given in BS EN 1313–1: 2010 and for structural softwood in BS EN 336: 2013 (Table 4.2). The latter defines two tolerance levels for sawn surface dimensions (tolerance classes T1 and T2) with T2 specifying the smaller tolerance limits.

Hardwood sizes are more variable due to the diversity of hardwood species, but preferred sizes to BS EN 1313–2: 1999 are specified in Table 4.1. Hardwoods are usually imported in random widths and lengths; certain structural hardwoods such as Iroko (*Chlorophora excelsa*) are available in long lengths (6–8 m) and

large sections. Preferred lengths for hardwood are given in Table 4.1. Maximum reductions by planing of sawn hardwoods are given in Table 4.3.

MOISTURE CONTENT AND SEASONING

As a tree is a living organism, the weight of water within it is frequently greater than the dry weight of wood itself. The water content of a tree is equal in winter and in summer, but one advantage of winter felling is that there is a reduced level of insect and fungal activity. After felling, the wood will lose the water held within the cell cavities without shrinkage, until the *fibre saturation point* is reached when the cells are empty. Subsequently, water will be removed from the cell walls, and it is during this process that the timber becomes harder and shrinkage occurs. As cellulose is a hygroscopic material, the timber will eventually equilibrate at a moisture content dependent on the atmospheric conditions. Subsequent reversible changes in dimension are called *movement*. The controlled loss of moisture from green timber to the appropriate moisture content for use is called *seasoning*.

$$\text{Moisture content} = \frac{\text{weight of wet specimen} - \text{dry weight of specimen}}{\text{dry weight of specimen}} \times 100\%$$

The primary aim of seasoning is to stabilise the timber to a moisture content that is compatible with the equilibrium conditions under which it is to be used, so that subsequent movement will be negligible. At the same time, the reduction in water content to below 20% will arrest any incipient fungal decay, which can only commence above this critical level. Drying occurs with evaporation of water from the surface, followed by movement of moisture from the centre of the timber outwards due to the creation of a vapour–pressure gradient. The art of successful seasoning is to control the moisture loss to an appropriate rate. If the moisture loss is too rapid then the outer layers shrink while the centre is still wet and the surface sets in a distended state (case hardening) or opens up in a series of cracks or checks. In extreme cases, as the centre subsequently dries out and shrinks it may honeycomb.

Air seasoning

Timber, protected both from the ground and from rain, is stacked in layers separated by strips of wood called stickers which, depending on their thickness, control the passage of air (Fig. 4.9). The air, warmed

Table 4.1 Common sizes of softwoods and hardwoods

Softwood sizes of 38 mm and over

Common target sizes of sawn softwood based on imperial measure to BS EN 1313–1: 2010.

Thickness (mm)	Width (mm)											
	75	100	115	125	138	150	175	200	225	250	275	300
38	✗	✓	✗	✓	✗	✓	✗	✗	✗	✗	✗	✗
47	✗	✗		✗		✗	✗	✗	✗	✗		✗
50	✗	✓		✓		✓	✓	✓	✓	✗		✗
63		✓		✓		✓	✓	✗	✗			
75		✗		✗		✓	✓	✓	✓	✗	✗	✗
100						✗		✓	✗	✗	✗	✗
150						✗		✗				✗
250										✗		
300												✗

Sizes marked with a tick ✓ indicate the common EU target sizes.
Sizes marked with a cross ✗ are the complementary UK preferred sizes.

Common target sizes of sawn softwood based on metric measure to BS EN 1313–1: 2010.

Thickness (mm)	Width (mm)							
	80	100	120	140	160	180	200	220
50	✓	✓	✓	✓	✓	✓	✓	✓
60		✓	✓	✓	✓	✓		
80	✓	✓	✓	✓	✓	✓	✓	
100	✓	✓	✓	✓	✓		✓	
120	✓	✓	✓		✓	✓	✓	
140	✓	✓		✓				
160	✓	✓	✓		✓	✓	✓	

Sizes marked with a tick ✓ indicate the common EU target sizes.

Hardwood sizes

Common sizes of sawn hardwood (20% moisture content) to BS EN1313–2: 1999.

Preferred thicknesses										
EU	20	27	32	40	50	60	65	70	80	100 mm

Complementary thicknesses								
UK	19	26		38	52		63	75 mm

Preferred widths
EU   10 mm intervals for widths between 50 mm and 90 mm,
            20 mm intervals for widths of 100 mm or more.

Preferred lengths
EU   100 mm intervals for lengths between 2.0 m and 6.0 m,
            50 mm intervals for lengths less than 1.0 m.

Table 4.2 Permitted cross-sectional deviations on structural timber sizes to BS EN 336: 2013

Maximum deviations from target sizes	Tolerance Class T1	Tolerance Class T2
Thicknesses and widths ≤ 100 mm	− 1 to + 3 mm*	− 1 to + 1 mm
Thicknesses and widths > 100 mm and ≤ 300 mm	− 2 to + 4 mm*	− 1.5 to + 1.5 mm
Thicknesses and widths > 300 mm	− 3 to + 5 mm	− 2 to + 2 mm

Notes:
Negative deviations on length are not permitted.
* The deviations permitted to BS EN 1313–1: 2010 for general sawn softwood correspond only to Tolerance Class T1.

Table 4.3 Maximum permitted reduction from target sawn sizes of hardwoods by planing two opposed faces (BS EN 1313–2: 1999).

Typical application	Reduction from basic size (mm)				
	15–25	26–50	51–100	101–150	151–300
Flooring, matchings, interlocked boarding and planed all round	5	6	7	7	7
Trim	6	7	8	9	10
Joinery and cabinet work	7	9	10	12	14

by the sun and circulated by the wind, removes moisture from the surface of the timbers. The timber ends are protected by waterproof coatings (bituminous paint) to prevent rapid moisture loss, which would cause splitting. Within the UK a moisture content of between 17 and 23% may be achieved within a few months for softwoods, or over a period of years for hardwoods.

Kiln drying

Kiln drying or seasoning is effected by heating within a closed chamber, which may be programmed to a precise schedule of temperature and humidity. Thus, drying to any desired moisture content can be achieved without significant degradation of the timber, although some early examples of kiln-dried timber showed serious damage through the use of inappropriate drying schedules. For economic reasons, timber is frequently air seasoned to fibre saturation point, followed by kiln drying to the required moisture content. This roughly halves the necessary kiln time and fuel costs. A typical softwood load would be dried from fibre saturation point within a few days and hardwood within two to three weeks.

Seasoned timber, if exposed to rain on site, will reabsorb moisture. Good site management is therefore necessary to protect timber both from physical damage and wetting prior to its use. The heating up of new

Fig. 4.9 Air seasoning of Jarrah wood in Western Australia

buildings by central heating systems can cause rapid changes in the moisture content of joinery timber and lead to shrinkage, cracking and splitting.

MOISTURE MOVEMENT

Wood is an anisotropic material, with differing moisture movements along the three principal axes; tangential, radial and longitudinal (Fig. 4.10). The highest moisture movement is tangential to the grain, the next being radial with the least along the grain.

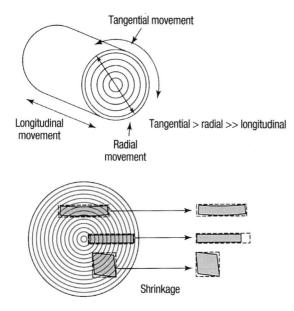

Fig. 4.10 Moisture movement and initial drying shrinkage

Typical figures are given in Table 4.4. The larger the ratio between tangential and radial movement, the greater the distortion. Moisture movements are conventionally quoted for a change in relative humidity from 90% to 60% at 25°C. The BRE classifies woods into three categories according to the sum of radial and tangential movement effected by this standard change in relative humidity. Small movement is defined as less than 3%, medium between 3% and 4.5%, and large over 4.5%. Large-movement timbers are not recommended for use as cladding.

TIMBER DEFECTS

Timber, as a natural product, is rarely free from blemishes or defects, although in some instances, such as knotty pine, waney-edge fencing timber or burr veneers, the presence of the imperfections enhances the visual quality of the material. Timber imperfections may be divided into three main categories: natural,

Table 4.4 Moisture movement of some hardwoods and softwoods normally available in the UK

Hardwoods

Small moisture movement (less than 3.0%)	Medium moisture movement (3.0–4.5%)	Large moisture movement (over 4.5%)
Afzelia	Ash	Beech, European
Agba	Cherry	Birch
Iroko	Elm, European	Keruing
Jelutong	Jarrah	Ramin
Lauan	Maple	
Mahogany, African	Oak, American	
Mahogany, American	Oak, European	
Meranti	Sapele	
Merbau	Utile	
Obeche	Walnut, European	
Teak		

Softwoods

Small moisture movement (less than 3.0%)	Medium moisture movement (3.0–4.5%)	Large moisture movement (over 4.5%)
Corsican pine	European redwood	
Douglas fir	European whitewood	
Sitka spruce	Parana pine	
Western hemlock	Radiata pine	
Western red cedar	Scots pine	

Note:
Moisture movement is assessed on the sum of the radial and tangential movements for a change in environmental conditions from 60 to 90% relative humidity.

conversion and seasoning defects, according to whether they were present in the living tree, or arose during subsequent processing. In addition, timber may be subject to deterioration by weathering, fungal and insect attack and fire. These latter effects are discussed later in the chapter.

Natural defects

Knots

Knots are formed where branches of the tree join the trunk (Fig. 4.11). Where the wood fibres of the branch are continuous with the trunk, then a live knot is produced. If, however, the branch is dead, or bark becomes incorporated into the trunk, a dead knot is produced. This is liable to be loose, lead to incipient decay and causes structural weakness.

Knots are described as face, edge, splay, margin or arris, depending on how they appear on the faces of converted timber. In addition, knots may appear as clusters, and range in size from insignificant to many millimetres across. Frequently they are hard to work, and in softwoods contain quantities of resin, which will continue to seep out unless the wood is sealed before painting.

Natural inclusions

Many minor defects occur to varying degrees in different varieties of timber. Bark pockets occur where pieces of bark have been enclosed within the timber as a result of earlier damage to the cambium or growth layer. Pitch pockets and resin streaks, containing fluid resin, are frequently seen along the grain of softwoods; their extent in usable timber is limited by BS EN 942: 2007.

Compression and tension wood

Trees leaning owing to sloping ground, or subject to strong prevailing winds, produce reaction wood to counteract these forces. In softwoods, compression wood is produced, which is darker in colour due to an increased lignin content. In hardwoods, tension wood is produced, which is lighter in colour owing to the presence of an extra cellulose layer in the cell walls. Both types of reaction wood have an abnormally high longitudinal shrinkage, causing distortion on seasoning; furthermore, tension wood tends to produce a rough surface when it is machined.

Abnormal growth rings

The width of the growth rings is an indicator of the growth rate and timber strength, with the optimum ranged around five rings per centimetre for softwoods and three rings per centimetre for hardwoods depending on the species. Excessively fast or slow growth rates give rise to weaker timber owing to a reduction in the proportion of the stronger late wood or its production with thinner walled fibres.

Conversion defects

Sloping grain

For maximum strength, timber should be approximately straight grained, as with increasing slope of the grain (Fig. 4.12) there is a proportionate reduction in bending strength, ranging typically from 4% at 1 in 25 to 19% at 1 in 10. The British Standard BS 4978: 2007 + A1: 2011 limits the slope of grain in visual strength graded softwoods to 1 in 6 for the general structural grade (GS) and 1 in 10 for the special structural grade (SS). The slope of grain in timber for internal and external windows, doors and door frames is limited to 1 in 10.

The British Standard BS 5756: 2007 + A1: 2011 limits the slope of grain in visual strength graded structural tropical hardwood (HS) to 1 in 11. For temperate hardwoods the visual strength grading limits for general structural timber are TH1 – 1 in 10 and TH2 – 1 in 4, and for heavy structural timber THA – 1 in 12 and THB – 1 in 4. (TH1 and THA are the higher grades with fewer natural defects.)

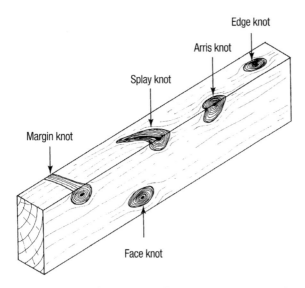

Edge knot

Arris knot

Splay knot

Margin knot

Face knot

Fig. 4.11 Knots (after Porter, B. and Rose, R. 1996: *Carpentry and joinery: Bench and site skills.* Arnold)

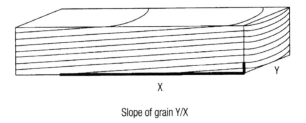

Slope of grain Y/X

Fig. 4.12 Sloping grain

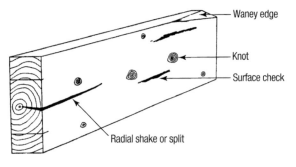

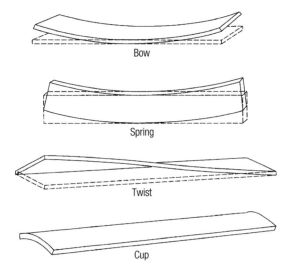

Fig. 4.13 Warping, splits and checks

Wane

Wane is the loss of the square edge of the cut timber due to the incorporation of the bark or the curved surface of the trunk. A degree of wane is acceptable in structural and floor timbers (BS 4978: 2007 and BS 1297: 1987), and is a special feature in waney-edge fencing.

Seasoning defects

Some of the commonest defects in timber are associated with the effects of seasoning. During the seasoning process, the contraction of the timber is different in the three major directions; furthermore, as described in the section on 'Moisture movement' (page 133), the outside of the timber tends to dry out more rapidly than the interior. These combined effects cause distortion of the timber including warping and the risk of rupture of the timber to produce surface checks and splits (Fig. 4.13).

Shakes

Major splits within timber are termed *shakes*, and these may result from the release of internal stresses within the living tree upon felling and seasoning; however, some fissures may be present within the growing timber. Commonly, shakes are radial from the exterior of the trunk, but star shakes which originate at the centre or pith may be associated with incipient decay. Ring shakes follow round a particular growth ring and are frequently caused by the freezing of the sap in severe winters.

Appearance grading

The standard BS EN 975 Part 1: 2009 and Part 2: 2004 sets the criteria for appearance grading of hardwoods, specifically for oak and beech, and European poplars, respectively. Grades relate to the number and size of the natural, seasoning and conversion defects visible on the timber face. Generally, four quality categories are listed for different forms such as boules, individual boards, edged boards and finished timber. The standard BS EN 1611–1: 2000 specifies five grades of appearance grading for the softwoods European spruces, firs, pines, Douglas fir and larches according to the number and size of observable defects.

SPECIFICATION OF TIMBER

The building industry uses timber for a wide range of purposes from rough-sawn structural members to claddings, trim and highly machined joinery. The specification of timber for each use may involve defining the particular hardwood or softwood, where particular visual properties are required. However, for the majority of general purposes, where strength and durability are the key factors, timber is specified by either a strength class, or a combination of timber species and strength grade.

In addition to strength class or grade, the specification of structural timber should include lengths and cross-section sizes, surface finish or tolerance class, moisture content and any preservative or special treatments (BRE Digest 416: 1996).

Strength grading

Strength grading is the measurement or estimation of the strength of individual timbers, which allows each piece to be used to its maximum efficiency. It may be done visually, a slow and skilled process, or within a grading machine which tests flexural rigidity. The European standard pr EN 14081–1: 2013 broadly outlines the requirements for visual and machine grading of timber, including the requirements for marking. The standards for the visual grading of hardwood and softwood within the UK are BS 5756: 2007 + A1: 2011 and BS 4978: 2007 + A1: 2011, respectively.

Visual strength grading
Each piece of timber is inspected for distortions, growth ring size and slope of grain, then checked against the set permissible limits for the number and severity of the natural defects, such as knots, waney edge and fissures. The timber is then assigned to a grade and stamped accordingly. Softwood timber is assessed as special structural grade (SS) or general structural grade (GS). The one visual strength grade for tropical hardwood is STH (structural tropical hardwood). The standard pr EN 16737: 2014 lists the full range of tropical hardwoods. Temperate hardwoods are graded to TH1 or TH2 (general structural temperate hardwood) for thicknesses of less than 100 mm and to THA or THB (heavy structural temperate hardwood) for larger sections. (The higher grades TH1 and THA in each category have fewer natural defects such as knots or sloping grain.)

Machine strength grading
Each piece of timber is quickly inspected for any distortions which may cause it to be rejected manually, or any serious defects within 500 mm of either end, at which points machine testing is ineffective. It is then tested, usually by one of two systems of contact or bending-type grading machines. In both techniques, the timber is moved through a series of rollers, and either the machine measures the load required to produce a fixed deflection, or it measures the deflection produced by a standard load. Either technique is measuring stiffness which is then related to timber strength and therefore a grading standard.

The three grading machines used within the UK are shown in Figure 4.14. In the constant deflection machine, the timber is then moved through a series of rollers which press it firmly against a curved metal plate. The force required to bend the timber to this standard deflection is determined by a series of transducers and from this data the timber strength is computed. However, a second pass through this machine is required to eliminate the effects of bow. The constant load system applies a defined lateral load, depending on the sample thickness, and the resulting deflection, with automatic adjustment for bow, indicates the timber grade. A more sophisticated system measures the forces necessary to bend the timber into an S-shape with two fixed deflections, thus neutralising the effects of any natural bow in the material.

As the timber leaves the machine it is stamped with the standard number, the strength class, M (machine graded), together with information on its species, wet/dry graded state, the producer and the certification body. Timber single species are coded with four letters to BS EN 13556: 2003 (e.g. Scots pine – *Pinus sylvestris* – PNSY). Species combinations have four letters commencing with W. (e.g. British spruce – *Picea sitchensis* and *Picea abies* – WPCS).

Other non-contact techniques for strength grading include X-ray and stress wave systems. X-ray machines assess the density of the timber, which is then related to strength. Stress wave techniques measure either the speed through the timber or the natural frequency of a stress wave produced by a small impact, and relate this to strength. Both techniques offer the potential for faster throughput than conventional contact strength grading systems, although they are sensitive to timber moisture content. Some grading machines combine physical bending techniques with the use of X-ray or microwave systems for the detection of natural defects such as knots or sloping grain, respectively.

Strength classes

Strength classes to pr EN 338: 2013 (Table 4.5) are defined as C14 to C50 and D18 to D80, where the prefix C refers to softwoods (coniferous) and D to hardwoods (deciduous). The number refers to the characteristic bending strength in MPa. The full specification of the strength classes gives characteristic values for density and a wide range of strength and stiffness properties, all based on sample test values. The data do not take into account any safety factors to be included in the design process. For trussed rafters, the grades TR20 and TR26 defined in BRE Digest 445: 2000 are applicable.

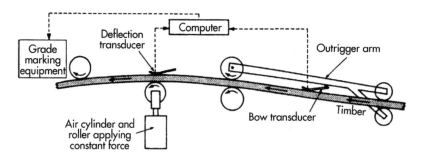

The MPC Computermatic grading machine

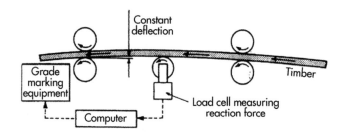

The MPC Cook-Bolinder grading machine

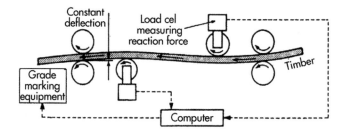

The Raute Timgrader grading machine

Typical timber certification marks

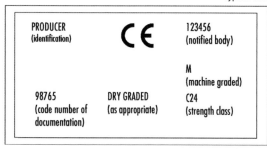

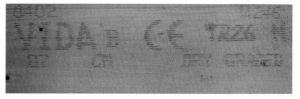

Fig. 4.14 Strength-grading machines and timber certification marks. *Diagram:* Reproduced by permission of IHS

Table 4.5 Relationship between strength classes and physical properties (pr EN 338: 2013)

Softwood species – Strength classes – Characteristic values

	C14	C16	C18	C20	C22	C24	C27	C30	C35	C40	C45	C50
Strength properties MPa												
Bending	14	16	18	20	22	24	27	30	35	40	45	50
Stiffness properties GPa												
Mean modulus of elasticity parallel to the grain	7	8	9	9.5	10	11	11.5	12	13	14	15	16
Mean density kg/m³	350	370	380	400	410	420	430	460	470	480	490	520

Hardwood species – Strength classes – Characteristic values

	D18	D24	D27	D30	D35	D40	D45	D50	D55	D60	D65	D70	D75	D80
Strength properties MPa														
Bending	18	24	27	30	35	40	45	50	55	60	65	70	75	80
Stiffness properties GPa														
Mean modulus of elasticity parallel to the grain	9.5	10	10.5	11	12	13	13.5	14	15.5	17	18.5	20	22	24
Mean density kg/m³	570	580	610	640	650	660	700	740	790	840	960	1080	1080	1080

Notes:
- Tabulated data refers to timber with a moisture content consistent with a temperature of 20°C and relative humidity of 65% which corresponds to a moisture content of 12% for most species.
- C refers to coniferous softwoods and D refers to deciduous hardwoods.

Table 4.6 shows generally available softwood species and their visually graded strength classes.

Service class

The service class defines the conditions in which the timber will be used and thus the anticipated moisture content. There are three categories defined within Eurocode 5: Design of timber structures: Part 1–1: (BS EN 1995–1-1: 2004). Timber to be used in service classes 1 and 2 must be adequately protected from the weather when on site.

Service classes of wood in use:

Service class 1   Timber with a moisture content corresponding to an ambient temperature of 20°C and a relative humidity of the surrounding air only exceeding 65% for a few weeks each year.
Average moisture content not exceeding 12% (e.g. internal walls, internal floors except ground floor, warm roofs).

Service class 2   Timber with a moisture content corresponding to an ambient temperature of 20°C and a relative humidity of the surrounding air only exceeding 85% for a few weeks each year.
Average moisture content not exceeding 20% (e.g. ground floors, inner leaf of cavity walls, single-leaf walls with external cladding).

Service class 3   Timber exposed to conditions leading to higher moisture contents than in service class 2.
Average moisture content 20% and above (e.g. exposed parts of buildings and marine structures).

Limit state design

Eurocode 5 (BS EN 1995–1-1: 2004) is based on *limit state design*, rather than permissible stress. There are generally two limit states to be considered; first, the *ultimate limit state* beyond which parts of the structure may fail or collapse, and second, the *serviceability limit state* beyond which excessive deformation, deflection or

Table 4.6 Softwood visual grades to BS 4978: 2007 + A1: 2011 allocated to strength classes

Species	Origin	European Standard BS EN 338: 2009 Strength classes						
		C14	C16	C18	C22	C24	C27	C30
British-grown softwoods								
Pine	UK	GS			SS			
Spruce	UK	GS		SS				
Douglas fir	UK	GS		SS				
Larch	UK		GS			SS		
Imported softwoods								
Douglas fir-larch	Canada/USA		GS			SS		
Hem-fir	Canada/USA		GS			SS		
Parana pine	Imported		GS			SS		
Pitch pine	Caribbean			GS			SS	
Redwood	Imported		GS			SS		
Sitka spruce	Canada	GS		SS				
Spruce-fir-pine	Canada/USA		GS			SS		
Southern pine	USA			GS		SS		
Western red cedar	Imported	GS		SS				
Western white woods	USA	GS		SS				
Whitewood	Imported		GS			SS		

Notes:
- GS and SS are General Structural and Special Structural visual grades respectively.
- Machine-graded timber is allocated directly to the appropriate strength class.

vibration would render the structure unfit for its purpose. The ultimate limit states are determined from the characteristic values of the timber properties, to which safety factors are applied. The characteristic values of various grades of timber, which are quoted in pr EN 338: 2013, are derived from laboratory tests without reductions for long-term loading or safety factors which become the responsibility of the designer.

MODIFIED TIMBER

The physical properties of timber can be significantly changed by either heat treatment to produce thermal modified timber (TMT), or reagent treatment producing chemical modifications or impregnation.

Thermal modification

Thermal modification involves heating the timber, usually pine or spruce but also hardwoods such as beech, maple, birch, oak and ash, to between 180°C and 240°C within an inert atmosphere to prevent combustion. Industrial processes variously use vacuum, an atmosphere of steam or nitrogen or a bath of heated oil, with process times of around 36 hours. The Finnish process at 190°C produces, from pine or spruce, *Thermowood* S to durability Class 3, or at 212°C *Thermowood* D to durability Class 2 (equivalent to European oak). The product is significantly more durable and more resistant to insect attack than untreated timber. It is less hygroscopic producing less moisture movement, has improved thermal insulation and the resins are removed from the knots. The timber is darker in colour and has a reduced strength in relation to splitting. In an alternative commercial process using temperatures between 150°C and 180°C, with aqueous, drying and heating cycles over several days, highly durable but brown softwoods and hardwoods are produced. Where thermal modified timber is to be used for structural purposes, allowances must be made for the reduced strength of the material. Frake or Limba (*Terminalia superba*), a non-durable West African timber, is currently being imported and thermally treated to produce a durable hardwood.

Thermal modified timber (TMT) is described in the British Standards document DD CEN/TS 15679: 2007. The John Foster Hall, University of Leicester (Fig. 4.15) is clad with thermally modified timber.

Thermally modified timber classes:

Softwoods	*Typical uses*
Class D (212°C)	Garden structures, external cladding, exterior doors, window frames, decking, internal floors and internal decoration
Class S (190°C)	Construction materials, structural components, garden furniture, interior floors, exterior doors and window frames

Hardwoods	*Typical uses*
Class D (212°C)	Garden furniture, patio floors, interior floors and interior decoration
Class S (190°C)	Constructional materials, mouldings, furniture, interior floors and interior decoration

Chemical modification

Chemical modification of timber involves the impregnation of permeable species of wood with chemical reagents which interact with the free hydroxyl groups within the cellulose macromolecules. Processes involve pressure treatment with either acetic anhydride for acetylation (*Accoya®*) or furfuryl alcohol for furfurylation (*Kebony®*). After impregnation, the timber is heated to activate the reaction between the water-binding sites in the wood cell walls and the modification chemical.

The modified timber with a reduced number of free hydroxyl groups has significantly reduced water absorption and thus greater dimensional stability. For a timber such as radiata pine, a maximum moisture content of approximately 6% is achieved by acetylation with no colour change, producing a timber durable to Class 1. The chemical processes do not significantly affect the physical properties of the timber, but increase resistance to biological deterioration and insect attack. Furfurylation significantly darkens the colour of the wood.

Impregnation modification

These processes involve the pressure impregnation of chemicals into the void spaces within dry timber. The chemicals then polymerise filling the void spaces, increasing the density and hardness of the wood. Dyes may be incorporated within the reagents to modify the colour of the wood throughout its section.

Fig. 4.15 Thermally modified timber cladding – John Foster Hall, University of Leicester. *Architects:* Goddard Manton. *Photographs:* Arthur Lyons

SPECIALIST TIMBER CONSTRUCTION

Timber-frame construction

The two standard forms of timber-frame construction are balloon frame and platform frame. The latter modern method of construction (MMC) is more common within the UK. For two-storey housing the balloon frame is constructed with double-storey height panels, with the intermediate floor supported from the framework. In platform frame, the panels are single-storey height, which are easier to manoeuvre, and the intermediate floor is supported directly on the lower-storey panels. Panels are normally constructed of 100 × 50 mm softwood, sheathed in plywood or particleboard, filled with insulation between the studs and with an internal finish of plasterboard. The subsequent external weatherproofing may be of brickwork or blockwork and rendering as required.

Gridshell construction

The Weald and Downland Open Air Museum in Sussex (Fig. 4.16) illustrates a different approach to timber construction combining modern computing technologies with traditional craftsmanship. The large barn-like construction (50 m long × 12 m wide × 10 m high) is formed from a double-skin grid of 35 × 50 mm × 40 m-long green oak laths at 1 m centres, which generate an undulating envelope of curved walls blending into three domes. Continuous curvature of the walls and roof is necessary for structural integrity.

Freshly sawn green oak was used, as it is supple and easily formed. Initially, it was bolted into a flat grid with stainless steel bolts. The supporting scaffolding was then gradually removed, allowing the construction to settle into its design form which was finally fixed around the perimeter. Once the correct form is established, the geometry is locked to ensure stability against wind and snow loading. The construction requires no interior supports, which would have inhibited the free use of the internal space. The structure was glazed with polycarbonate clerestory panels and clad in western red cedar vertical boarding. Gridshell construction has been used previously in Germany and Japan, but the Weald and Downland Museum by Edward Cullinan Architects was the first of its type within the UK. A more recent gridshell building is the Savill Building at Windsor Great Park by Glenn Howells Architects (Fig. 4.17), constructed from locally sourced larch and green oak cladding.

Fig. 4.16 Gridshell construction – Weald and Downland Open Air Museum. *Architects*: Edward Cullinan Architects. *Photographs*: Courtesy of Edward Cullinan Architects

Timber piles and foundations

Timber piled foundations have been used for many centuries and have a good record of durability. The city of Venice is largely built on timber piles and their use as an alternative construction system is current within North America for the foundations of bridges and other significant structures. The use of timber pile foundations, in appropriate ground conditions, offers an economical alternative to concrete, with the environmental advantage of creating carbon dumps to reduce global warming.

Historically, a range of softwoods and hardwoods has been used for timber piles, but in the UK, Douglas

Fig. 4.17 Gridshell construction – The Savill Building, Windsor Great Park. *Architects*: Glen Howells. *Photograph*: Warwick Sweeney / The Royal Landscape

Fig. 4.18 Timber pole construction – Hooke Park, Dorset – Lodge (Edward Cullinan Architects and Buro Happold), and workshop interior (Frei Otto, Ahrends Burton & Koralek and Buro Happold). *Photographs*: Courtesy of Allan Glennie

fir, up to 500 mm square and 12–15 m long, is a standard material. Other suitable timbers are treated Scots pine or larch, oak in non-salt-water soils, elm, beech and sycamore. Untreated timber below the water-table is virtually immune to decay, but it is at risk from biological degradation above this level. It is therefore appropriate to treat timber with preservatives if it is to be used above the water-table. Preservative-treated timber piles, cut off below ground level and capped with concrete, should have a service life of 100 years.

ROUND TIMBER

Forest thinnings, which are too small to be converted into rectangular standard sections for construction, have the potential to be used directly for certain low-technology forms of building. Currently, much of this material is used for paper and particleboard production or burnt as firewood.

The advantage of this material is that it is a renewable resource with a relatively short production cycle and rapid carbon dioxide sequestration. Timber poles are naturally tapered, but the effects on mechanical properties of defects such as knots and sloping grain, which are significant in converted timber, are virtually eliminated. In addition, as little machining is required, energy and labour costs are low.

Forest thinnings up to 200 mm in diameter are generally available, but typical small-scale construction usually requires poles in the 50–150 mm range, with lengths of between 3 and 15 m. Figure 4.18 illustrates an experimental building at Hooke Park using Norway spruce timber pole construction.

Dimensional classes for round timber

The standard BS EN 1315: 2010 describes dimensional classes for both hardwood and softwood round timber, measured at the mid-diameter for either under bark (D) or over bark (R), with diameters ranging from < 10 to ⩾ 80 cm in steps of 10 mm (Table 4.7).

Qualitative classification of round timber

Round timbers may be classified within four quality classes. Classification depends on the extent of natural defects and any deterioration by rot or insect attack. The criteria are listed BS EN 1927 Parts 1–3: 2008 for spruces and firs, pines, larches and Douglas fir, respectively. The qualitative classifications for oak and beech, poplar, and ash, maple and sycamore are described in BS EN 1316–1: 2012, BS EN 1316–2: 1997 and BS EN 1316–3: 1998, respectively. The classification

Table 4.7 Dimensional classification of round timber

	Mid diameter (cm)											
	< 10	10–14	15–19	20–24	25–29	30–34	35–39	40–49	50–59	60–69	70–79	⩾ 80
Class over bark	R 0	R 1a	R 1b	R 2a	R 2b	R 3a	R 3b	R 4	R 5	R 6	R 7	R 8
Class under bark	D 0	D 1a	D 1b	D 2a	D 2b	D 3a	D 3b	D 4	D 5	D 6	D 7	D 8

Note:
Mid diameter is measured to BS EN 1309–2: 2006.

is written as the Latin name or abbreviation of the species followed by the class letter. For example, QA is the highest grade of oak (*Quercus*) and FrB is the average quality of Ash (*Fraxinus*).

Quality class A	First-quality timber, generally corresponding to a butt log with clear timber or with only minor features which do not restrict use.
Quality class B	Timber of average to first quality with no specific requirements for clear wood. Knots are permitted to such an extent as is considered to be average for each species.
Quality class C	Timber of average to low quality, allowing all quality features which do not seriously reduce the natural features of the wood.
Quality class D	Timber which can be sawn into usable wood, which, because of its features, falls into none of the quality classes A, B or C.

SPECIALIST USES OF TIMBER

Softwood cladding

Western red cedar has long been the preferred timber for external timber cladding because of its durability and warm colour. The double-curved roof of the London Olympic Velodrome designed by Hopkins Architects (Fig. 4.19) is clad with 5000 m² of sustainably resourced cedar wood cladding. The cladding is perforated to allow natural ventilation of the building which requires no air conditioning.

However, recently the popularity of larch and Douglas fir as softwood cladding has increased as greater emphasis is placed on the use of renewable resources from sustainable forests. Large quantities of these materials, which are classified as moderately durable to decay, will become available as plantation-grown stocks reach maturity. Both larch and Douglas fir are more resistant to impact damage than western red cedar, and for cladding purposes they should not need additional preservative treatment. The timbers

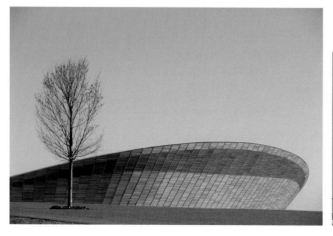

Fig. 4.19 Cedar cladding – London Olympic Velodrome. *Architects*: Hopkins Architects.
Photographs: Courtesy of London Legacy Development Corporation

have been used successfully for school and health centre buildings as both vertical and horizontal cladding. As the timbers are acidic, all fixings must be in corrosion-resistant materials; also, some resin bleed can be expected which will penetrate any applied surface finishes.

Hardwood flooring

Hardwood flooring has a proven track record for durability and aesthetic impact. Both solid timber and plywood laminates with a 4 mm hardwood wearing layer are commercially available. The standard timbers are the European oak, beech, birch, ash, chestnut, walnut and maple, but some imported hardwoods with darker grain colours are also available and interesting effects are produced with bamboo. The timbers are frequently offered with minimal knots and uniform graining, or as *rustic* with knots and a larger variation of colour. Laminates are usually pre-finished but solid timber may be sealed with oil or lacquer after installation on site.

Joinery timber

The term *joinery* applies to the assembly of worked timber and timber panel products, using timber which has been planed to a smooth finish. By contrast, carpentry refers to the assembly of the structural carcase of a building usually with rough-sawn timbers. Joinery work, including the production of windows, doors, staircases, fitted furniture, panelling and mouldings, requires timber that is dimensionally stable, appropriately durable with acceptable gluing properties, and can be machined well to a good finish. Joinery grade timber is categorised into seven quality classes (Table 4.8) according to the number and size of natural defects, particularly knots. These classes are subdivided

into two surface categories, visible and concealed, according to whether the timber is to be visible in use. (BRE Digest 407: 1995 and BS EN 942: 2007 list some softwoods and hardwoods suitable for joinery.) Softwood flooring, cladding and profiled boards should not be specified as joinery.

DETERIORATION OF TIMBER

The major agencies causing the deterioration of timber in construction are weathering, fungi, insects and fire. The natural durability of timber is defined into five categories in relation to the resistance of the heartwood to wood-decaying fungi (BS EN 350–1: 1994).

Natural durability of timber:

Class 1	Very durable
Class 2	Durable
Class 3	Moderately durable
Class 4	Slightly durable
Class 5	Not durable (perishable)

Weathering

On prolonged exposure to sunlight, wind and rain, external timbers gradually lose their natural colours and turn grey. Sunlight and oxygen break down some of the cellulose and lignin into water-soluble materials which are then leached out of the surface, leaving it grey and denatured. Moisture movements, associated with repeated wetting and drying cycles, raise the surface grain, open up surface checks and cracks, and increase the risk of subsequent fungal decay. Provided that the weathering is superficial, the original appearance of the timber can be recovered by removing the denatured surface.

Table 4.8 Classes of timber for joinery use and maximum knot sizes (BS EN 942: 2007)

Class	Visible faces							Concealed faces
	J2	J5	J10	J20	J30	J40	J50	
Maximum knot size (mm)	2	5	10	20	30	40	50	These knots are all permitted
Maximum percentage of width of finished piece (%)	10	20	30	30	30	40	50	

Note:
The standard also refers to spiral and slope of grain, fissures, shakes, resin pockets, bark, discoloured sapwood, exposed pith and Ambrosia beetle damage.

Fungal attack

Fungi are simple plants which, unlike green plants, cannot synthesise chlorophyll, and therefore must obtain their nutrients by metabolising organic material, breaking it down into soluble forms for absorption into their own system. For growth they need oxygen and a supply of food and water, a minimum moisture content of 20% being necessary for growth in timber. The optimum temperature for growth is different for the various species of fungi, but is usually within the range 20–30°C. Little growth takes place below 5°C and fungi will be killed by prolonged heating to 40°C. Some timbers, particularly the heartwoods of certain hardwoods, are resistant to attack because their minor constituents or *extractives* are poisonous to fungi.

All the timber-destroying fungi have a similar life cycle (Fig. 4.20), commencing with the microscopic spores which are always present in quantity in the air. Under favourable conditions, spores within the surface cracks of timber will germinate and produce fine filaments or hyphae, which feed on the cellulose of the timber. The hyphae branch and grow through the timber cells, feeding on both the walls and their contents. With increasing colonisation of the timber, the fine hyphae combine to produce a white matrix or mycelium, which is then visible to the eye. After a period of growth, the mycelium at the surface produces fruiting bodies which generate many thousands of spores to continue the life cycle. The spores, which are less than 10 μm in size, are readily distributed by air movement.

Certain living tree species are also at risk from fungal attack. The elm bark beetle (*Scolytus scolytus*) was responsible for the spread of *Dutch elm disease* in the 1970s. The larvae tunnelled under the bark, within the bast and cambium layers, preventing growth and spreading the destructive fungus which eventually killed large numbers of trees across the UK. Of current concern is the spread in Europe of ash dieback disease caused by the fungus *Chalara fraxinea*, which leads to leaf loss, crown dieback and potentially tree death.

Moulds and stains
Moulds and stains are fungi that metabolise only the starch and sugar food reserves stored within the timber

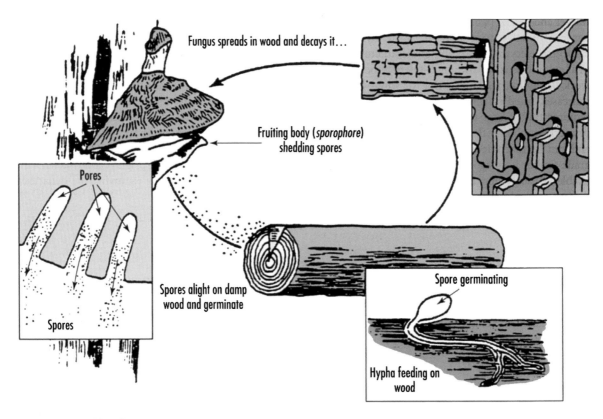

Fig. 4.20 Fungus life cycle

cells; therefore, sapwoods are generally more vulnerable than heartwoods, since during the conversion of sapwood to heartwood the stored food is removed. Generally, there is little loss of strength associated by such an attack, although one variety, *blue-stain*, aesthetically degrades large quantities of timber and its presence may indicate conditions for incipient wood-rotting fungal attack. It is best prevented by kiln drying to quickly reduce the surface moisture content unless infection has already occurred within the forest. Generally, softwoods are more susceptible to attack than hardwoods. However, the light-coloured hardwoods ramin, obeche and jelutong are sometimes affected.

Wet and dry rots

The name *dry rot*, attributed to one variety of fungus, is a misnomer, as all fungal growths require damp conditions before they become active. Destructive fungi may be categorised as soft, brown or white rots.

Soft rots, which belong to a group of micro-fungi, are restricted to very wet conditions such as timbers buried in the ground and are therefore not experienced within normal construction. They are usually found only within the timber surface, which becomes softened when wet and powdery when dry.

The brown rots preferentially consume the cellulose within timber, leaving more of the lignin, tannin and other coloured extractives; thus, the timber becomes progressively darker. In contrast, the white rots consume all the constituents of the cells; thus, the timber becomes lighter in colour as the attack proceeds.

A major cause of deterioration of timber within buildings is *Serpula lacrymans*, the so-called *dry rot*. Under damp conditions, above 20% moisture content, the mycelium forms cotton-wool-like masses over the surface of the timber, which becomes wet and slimy. The mycelium strands, up to 20 mm in diameter, can grow through brickwork and past inert materials to infect otherwise dry timber. Under drier conditions the mycelium forms a grey-white layer over the timber, with patches of bright yellow and occasionally lilac. The fruiting bodies, or fructifications, are plate-like forms, which disperse the rust-red spores. In some circumstances the fruit bodies may be the first signs of attack by dry rot. After an attack by dry rot the timber breaks up both along and across the grain into cube-shaped pieces, becoming dry and friable; hence the name *dry rot*.

Wet rot or cellar fungus (*Coniophora puteana*) is the most common cause of timber decay within buildings

Fig. 4.21 Deterioration of sapwood timber illustrating the relative durability of heartwood over sapwood

in the UK. It requires a higher moisture content than dry rot (40–50%), and is therefore frequently associated with water ingress due to leaks or condensation. The decayed timber is darkened and tends to crack mainly along the grain. The thin individual strands or hyphae are brown or black, and the fruit bodies, rarely seen, are olive green in colour. Frequently, the decay is internal without significantly affecting the exposed faces of the timber.

Phellinus contiguus (*Poria contigua*) causes decay to external softwood joinery, particularly window frames, causing the timber to decompose into fibrous lengths. Another variety, *Phellinus megaloporus*, is known to attack oak timbers, ultimately leaving a white mass.

Figure 4.21 illustrates the relative vulnerability of sapwood compared to the naturally more resistant darker heartwood, which has been partially protected from rot by secondary thickening and the inclusion of extractives.

Insect attack

Insect attack on wood within the UK is limited to a small number of species, and tends to be less serious than fungal attack on construction timbers. However, in hotter climates termites and other insects can cause catastrophic damage, and the unintentional importation and subsequent establishment of subterranean termites in North Devon shows that this species may pose a future threat to UK buildings. The main damage by insects within the UK comes from beetles, which during their larval stage bore through the timber, mainly within the sapwood, causing loss of mechanical strength. For other species, such as the pinhole borers (*Platypus cylindricus*), the adult beetle bores into the timber to introduce a fungus on which the larvae live.

The typical insect life cycle (Fig. 4.22) commences with eggs laid by the adult beetle in cracks or crevices of timber. The eggs develop into larvae, which tunnel through the timber leaving behind their powdery waste or frass. Depending on the species, the tunnelling process may continue for up to several years before the development of a pupa close to the surface of the timber, prior to the emergence of the fully developed adult beetle, which eats its way out leaving the characteristic flight hole. The insects which attack well-seasoned timber within the UK are the common furniture beetle, death-watch beetle, house longhorn beetle and powder post beetle. Wood-boring weevils only attack timber that has been previously affected by fungal decay (Fig. 4.23).

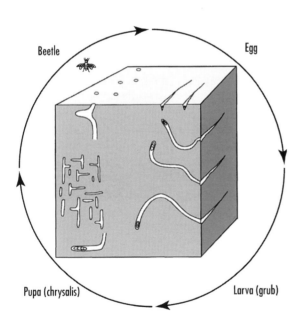

Fig. 4.22 Life cycle of wood-boring beetles

Common furniture beetle

The common furniture beetle (*Anobium punctatum*) attacks mainly the sapwood of both hardwoods and softwoods. It can be responsible for structural damage in the cases of severe attack, and is thought to be present in up to 20% of all buildings within the UK. The brown beetle is 3–5 mm long and leaves flight holes of approximately 2 mm in diameter. Both waterborne and organic-solvent insecticides offer effective treatments.

Death-watch beetle

The death-watch beetle (*Xestobium rufovillosum*) characteristically attacks old hardwoods, particularly oak, and is therefore responsible for considerable damage to historic buildings. Attack is normally on the sapwood, but heartwood softened by moisture and fungal decay will attract infestation; adjacent softwood may also be affected. The brown beetle is approximately 8 mm long and leaves a flight hole of 3 mm in diameter. Remedial measures should include the eradication of damp and the application of organic-solvent insecticides.

House longhorn beetle

The house longhorn beetle (*Hylotrupes bajulus*) is a serious pest in some parts of southern England, particularly Surrey. The Building Regulations 2010 Approved Document A Section 2B, specifies the key geographical areas at risk where additional precautions are required within roof spaces. Longhorn beetle can infest and cause serious structural damage to the sapwood of seasoned softwood roof timbers. With an average life cycle of six years and a larva that is up to 35 mm long, this beetle can cause serious damage before evidence of the infestation is observed. The affected timbers bulge where tunnelling occurs just below the surface, and the eventual flight holes of the black beetle are oval and up to 10 mm across. Where sufficient serviceable timber remains, remedial treatment with organic solvent or paste formulations is appropriate.

Powder post beetle

The powder post beetle (*Lyctus brunneus*) attacks the sapwood of certain hardwoods, particularly oak and ash. The sapwood of large-pored tropical hardwoods, such as ramin and obeche, may also be affected. Only timbers with sufficient starch content within the sapwood are vulnerable to attack as the larvae feed on starch rather than on the cell walls. The eggs are laid

Common furniture beetle (*Anobium punctatum*)

Actual size approx. 3 to 5 mm long

Flight holes

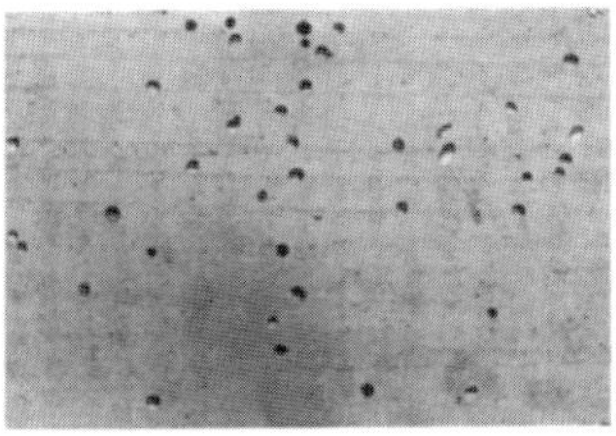

Bore dust

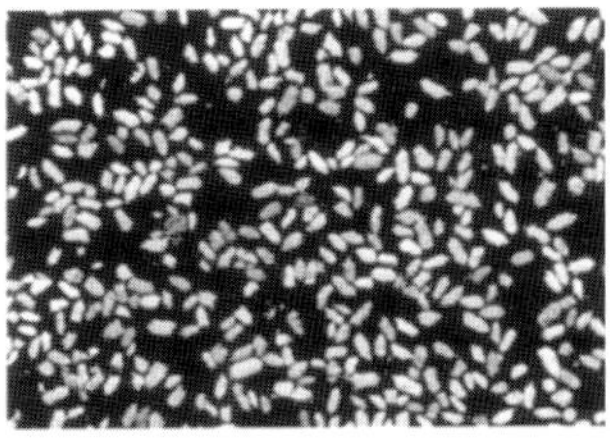

Death watch beetle (*Xestobium rufovillosum*)

Actual size approx. 6 mm long

Typical damage

Bore dust

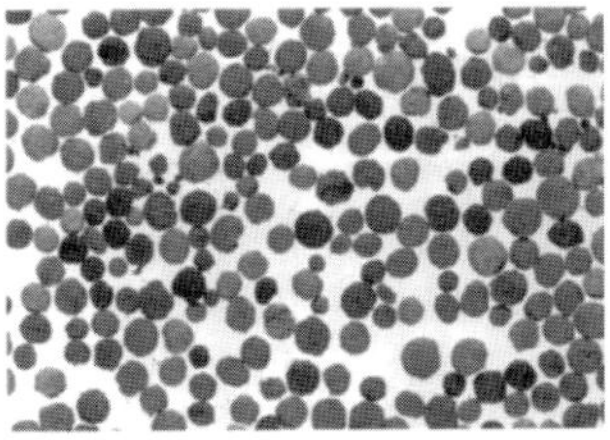

House longhorn beetle (*Hylotrupes bajulus*)

Actual size approx. 25 mm long

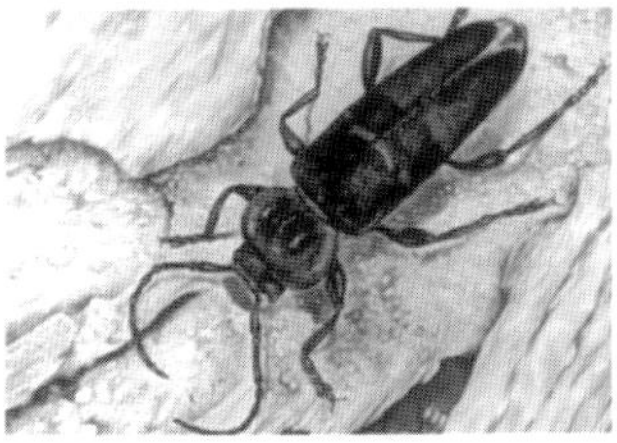

Typical damage to rafters

Bore dust

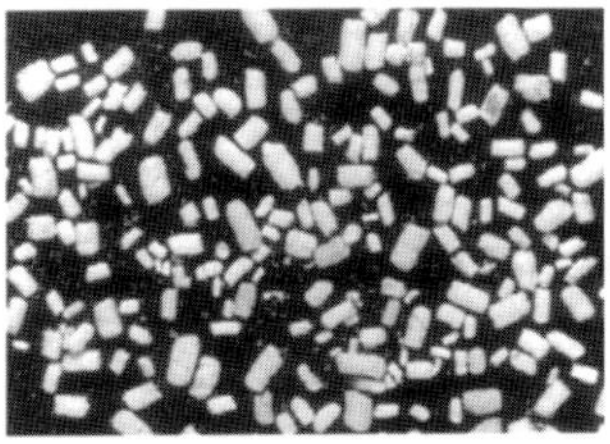

Powder post beetle (*Lyctus brunneus*)

Actual size approx. 5 to 6 mm long

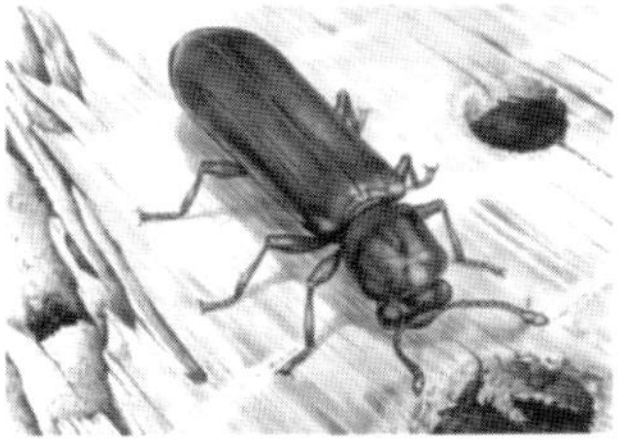

Severe internal damage and apparently
superficial external damage

Bore dust

Wood-boring weevil (*Pentarthrum huttoni*)

Actual size approx. 3 to 5 mm long

Typical internal damage

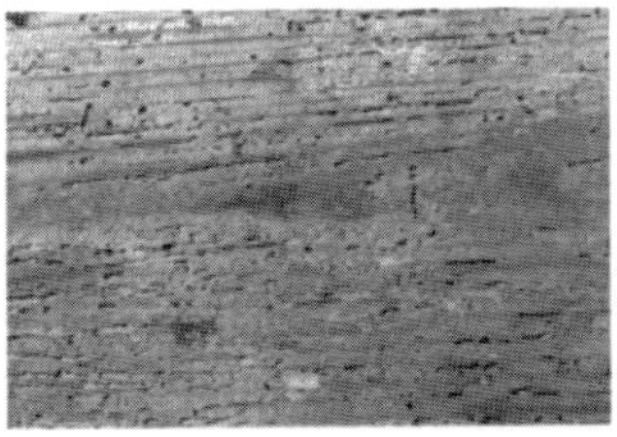

Bore dust

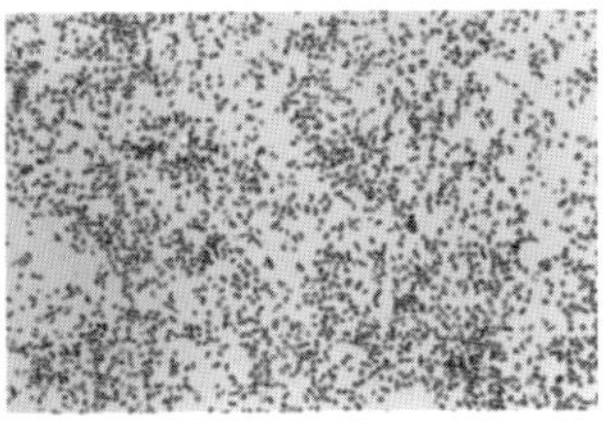

Fig. 4.23 Wood-boring beetles common within the UK

by the adult female beetle into the vessels which are the characteristically large cells within hardwoods. Timbers with low starch content or fine vessels are immune, and the extended soaking of vulnerable timbers in water can reduce the risk of attack, but owing to the long time scale involved, this is not commercially viable. The 4 mm reddish-brown beetle leaves a flight hole of approximately 1.5 mm in diameter. Timbers are attacked only until all the sapwood is consumed, so in older buildings damage is usually extinct. In new buildings, coatings of paint or varnish make treatment impractical, so replacement is the usual option.

Ambrosia beetle

A large number of ambrosia beetle species attack freshly felled hardwood and softwood logs, in both temperate and tropical regions. A high moisture content of over 35% is necessary for this beetle attack, which is therefore eliminated on seasoning. The circular pin-holes range from 0.5 to 3 mm depending on the particular beetle species and the tunnelling is across the grain of the timber.

Wood-boring weevils

Wood-boring weevils attack only timber previously softened by fungal decay. The most common weevil (*Pentarthrum huttoni*) produces damage similar in appearance to the common furniture beetle, but removal of decayed timber will eliminate the secondary infestation.

Termites

Termites are social insects, similar in size to ants (4–5 mm), which live in colonies containing millions of individuals. Most species of termites are beneficial to nature in breaking down organic matter, but a few varieties can cause catastrophic damage to buildings. The subterranean termite family *Rhinotermitidae* is the major cause of building damage, with the genus *Reticulitermes* being a significant threat in the UK. The species *R. santonensis* is already widely distributed in Europe and the species *R. lucifugus* was established at Saunton, North Devon. However, treatment with hexaflumuron, an insect growth regulator, within bait, appears to have reduced the infestation to a small localised pocket near the original infestation. The primary source of food for termites is the cellulose in wood, particularly from structural softwood timbers near to ground level or in partially decayed timber. Timber may be only slightly affected or it may be heavily excavated leaving only the surface and any protective coatings. In areas at risk of termites,

termiticide protection may be appropriate, but where attack has already occurred, specialist advice from the Forestry Commission is essential. Colony elimination by physical, chemical or biological techniques can be a slow process, taking months or even years to complete.

Preservation of timber

Wood preservatives contain pesticides in the form of insecticides and fungicides. Their use is therefore strictly controlled to limit unnecessary or accidental environmental damage. Preservative treatments should involve only materials registered in the EU (REACH 1907/2006), in current approval by the Biocidal Products Regulation (2013), and they should be used in accordance with the Control of Substances Hazardous to Health (COSHH) regulations, the current Code of Practice BS 8417: 2011, the manufacturers' instructions and by operatives wearing appropriate protective clothing. Timber treatments may be divided into the application of preservatives to new timber, and remedial treatments used to eradicate or reduce an existing problem.

Preservative treatments for new timber

A wide range of effective timber preservatives is commercially available for use under controlled industrial conditions. However, within the current climate of increasing health and environmental awareness, the drive towards more ecologically friendly products has led to considerable changes within industrial timber preservation processes. The use of creosote (BS 144: 1997) is now only permitted by EU legislation for external timber such as fencing posts and piles. The use of copper-chrome-arsenate (CCA) has been prohibited in the EU since 2006. New products are predominantly organic biocides.

The industrial timber preservation processes involve the use of vacuum and either high- or low-pressure impregnation, or alternatively dip-diffusion. Chemicals are either water-borne, solvent-based or water-based micro-emulsions.

The double-vacuum process, using organic solvent-borne preservatives, is suitable for low- and medium-risk timber such as external joinery. The timber, at less than 28% moisture content, is loaded into a low-pressure vessel which is evacuated to extract the air from within the timber. The vessel is flooded with preservative and a low positive pressure is applied for between several minutes and one hour depending on the permeability of the timber. The vessel is then drained and evacuated to remove excess preservative from the timber surface. Formulations consist of either fungicides or insecticides, or both, dissolved in volatile organic solvents. The solvents penetrate well into the timber but have a strong odour and are highly flammable. A water repellent may also be incorporated into the preservative formulation. Organic-solvent preservatives will eventually be limited in use to timbers where it is critical that the dimensions are not affected by the preservative treatment. To reduce the environmental effects of volatile organic compound (VOC) emissions, organic solvents are being replaced by micro-emulsions of biodegradable biocide formulations, with significantly reduced organic-solvent content. The treated timbers generally meet the requirements of Use Classes 1, 2 and 3.1 to BS EN 335: 2013.

The pressure/vacuum process is similar to the double-vacuum process, but uses water-borne preservatives and the application of low or high pressure within a pressure vessel to ensure the appropriate level of penetration. The standard preservatives are based on copper compounds and azole biocides. Some treatments also include boron derivatives. Timber treated with proprietary products such as *Tanalith* is coloured slightly green but may be directly painted. Low-pressure treatments are generally to Use Classes 1, 2 and 3.1 (Table 4.9). High-pressure treatment can give additional protection to Use Class 4, but causes raised grain and swelling of the timber.

Dip-diffusion treatments involve the immersion or spraying of the freshly sawn green timber using boron derivatives (disodium octaborate tetrahydrate). Two or three immersions are used to ensure complete coverage of all faces of the timber, and larger sections require a second treatment by spraying or immersion. After treatment the timber should be stored for an appropriate period to allow the diffusion of the preservative into the timber to the required depth of penetration. However, the product is water soluble and therefore not suitable for pressure impregnation of timber for external structures.

Timbers to be built into high-risk situations, such as industrial roofs, frames and floors, timbers embedded in masonry, sole plates, sarking boards, tile battens etc., should be treated with appropriate timber preservative.

Remedial treatment for timber

Remedial treatments to existing buildings should be limited to those strictly necessary to deal with the fungal or insect attack. The use of combined fungicides and insecticides is not advised when the attack is by one agent only. Within the UK, much timber decay is caused by building failures. As fungal decay can only occur in damp conditions, the first remedial measure must be to restore dry conditions. This should remove the necessity for frequent chemical applications.

The orthodox approach to the eradication of fungal and wood-boring beetle attack involves the removal of severely decayed or affected timber, followed by appropriate preservative treatment to the remaining timber. For fungal attack, 300 mm of apparently sound timber should be removed beyond the last visible sign of decay, and the adjoining timbers treated with fungicide. For wood-boring beetle, unless the infestation is widespread, preservative treatment should be applied only up to 300 mm beyond the visible holes. Organic-solvent fungicides and insecticides applied by brush or spray offer some protection from further attack, but applications of pastes or rods which deliver higher quantities of the active ingredients are usually more effective. Insecticidal smoke treatments need to be repeated annually, as they are only effective against emerging beetles, but they may be useful in situations where brushing or spraying is impracticable.

The environmental approach to the eradication of fungal decay relies heavily on the removal of the causes of damp. On the basis that fungal attack will cease when timber is at less than 20% moisture content, increased ventilation and the rectification of building defects should prevent further attack. Only seriously affected timbers need to be replaced, and affected masonry sterilised; however, continual monitoring is required, as dormant fungal decay will become active if the timber moisture content rises again above 20%. *Rothounds* (specially trained sniffer dogs), fibre optics and chemical detection systems offer non-destructive methods for locating active dry rot before it becomes visible.

Pesticides used professionally for remedial treatment include disodium octaborate or azoles as fungicides and flurox (an insect growth regulator) as insecticide. These pesticides are currently considered acceptable for treatment in areas inhabited by bats, which are a protected species under the Wildlife and Countryside Act 1981.

Guidance on timber treatments

Under the standards BS EN 351–1: 2007, timber preservative treatments against wood-destroying organisms are categorised by performance standards and not to the individual chemical preservative treatments. The standards define wood preservatives according to their effectiveness in a range of environmental use conditions. Table 4.9 (BS EN 335: 2013) relates the Use Classes to the occurrence of biological agents.

The standard BS EN 335: 2013 gives general guidance on the likely occurrence of biological agents under the various Use Class conditions. Additional guidance on preservation is given in BS EN 460: 1994 (Table 4.10) which relates the timber durability class to the use class, indicating where additional treatment may be necessary.

The level of treatment required to give the necessary performance is classified according to the depth of penetration into the timber and by retention or loading within the defined zone of the timber. The depth of penetration is defined in BS EN 351–1: 2007, by six classes (NP1–NP6) of increasing zones of preservative retention. As full preservative treatment is not appropriate in all cases, a range of service factors (A–D) define the level of safety and economic considerations appropriate to preservative treatment. These are listed in the British Standard BS 8417: 2011 (Table 4.11).

Table 4.9 Use classes and relevant attacking biological agents for wood and wood-based products (BS EN 335: 2013) and relation to Service classes (BS EN 1995–1-1: 2004 + A1: 2008)

Use Class	General use situation	Occurrence of biological agents				
		Discolouring fungi	Wood-destroying fungi	Wood-boring beetles	Termites	Marine borers
Use Class 1	Inside a construction and not exposed to weather or wetting			U	L	
Use Class 2	Under cover and not exposed to weather but where occasional but not persistent wetting may occur	U	U	U	L	
Use Class 3	Above ground and exposed to weather 3.1 Will not remain wet for long periods. Water will not accumulate 3.2 Will remain wet for long periods. Water may accumulate	U	U	U	L	
Use Class 4	Direct contact with ground and/or fresh water	U	U	U	L	
Use Class 5	Permanently or regularly submerged in salt water	U	U	U	L	U

Notes:
— U is ubiquitous in Europe.
— L is locally present in Europe, and currently confined to a local area within the UK.
— Use classes were previously described as Hazard Classes and this terminology remains in BS EN 460: 1994.
— Service classes as defined in BS EN 1995–1-1: 2004 + A1: 2008 are relevant to the strength values and the calculation of deformations of timber elements to be used in construction. They are determined by the wood moisture content corresponding to the humidity and temperature which will prevail in service. There is no direct linkage between Use Class and Service Class, but the table below (BS EN 335: 2013) gives guidance.

Service Class to BS EN 1995–1-1: 2004 (page 139)	Possible corresponding Use Class to BS EN 335: 2013
Service Class 1	Use Class 1
Service Class 2	Use Class 1 Use Class 2 if the component is in a situation where it could be subjected to occasional wetting
Service Class 3	Use Class 2 Use Class 3 or higher if used externally

Table 4.10 Relationship between environmental use class and timber durability class and requirement for preservative treatment against wood-destroying fungi (BS EN 460: 1994)

Use class	Durability class				
	1	2	3	4	5
1	O	O	O	O	O
2	O	O	O	A	A
3	O	O	A	B	B
4	O	A	C	X	X
5	O	C	C	X	X

Key:
O   natural durability sufficient.
A   preservative treatment may be advisable for certain end uses.
B   preservative treatment may be necessary for certain end uses.
C   preservative treatment is normally advised.
X   preservative treatment necessary.

Notes:
−   Sapwood of all wood species is regarded as durability class 5.
−   In the 1994 Standard the use class was described as hazard class.

Table 4.11 Service factors for preservative treatment of timber to BS 8417: 2011

Service factors	Need for preservation	Safety and economic considerations
A	Unnecessary	Negligible risk of failure
B	Optional	Where risk of failure is low and remedial action is not difficult, preservation is regarded as an insurance against potential cost of repairs
C	Desirable	Where risk of failure is high and where replacement or remedial action is difficult and expensive
D	Essential	Where risk of failure is very high and where failure would have serious consequences to the structure or persons

The British Standard BS 8417: 2011 recommends the durability classes of timber (BS EN 350–2: 1994) which may be used without preservative treatment in relation to their use and environmental factors; for example, dry roof timbers in an area not affected by house longhorn beetle:

- Building component – dry roof timbers in a non-*Hylotrupes* area
- Use Class of wood against biological attack – Class 1
- Service factor – safety and economic considerations (e.g. B for dry roof timbers)
- Desired service life – 60 years
- Durability Class 5 – non-durable timber is appropriate.

The British Standard BS 8417: 2011 also gives guidance on the types of preservative treatment appropriate for timbers to be used in more severe situations; for example, occasionally wet roof timbers:

- Building component – roof timbers (with risk of occasional wetting)
- Use Class of wood against biological attack – Class 2
- Timber species – e.g. permeable wood
- Preservative treatment required

- Desired service life – 60 years
- Penetrating treatment required to Class NP1 (preservative to BS EN 599–1: 2009)
- (The British Standard BS 8417: 2011 gives guidance on the required depth of penetration and retention of the preservative.)

Fire

Timber is an organic material and therefore combustible. As timber is heated it initially evolves any absorbed water as vapour. By about 230–250°C, decomposition takes place with the production of charcoal, and combustible gases such as carbon monoxide and methane are evolved, which cause the flaming. Finally, the charcoal smoulders to carbon dioxide and ash.

However, despite its combustibility, timber, particularly in larger sections, performs better in a fire than the equivalent sections of exposed steel or aluminium. Timber has a low thermal conductivity, which, combined with the protection afforded by the charred surface material, insulates the interior from rapid rises in temperature and loss of strength. The design charring rates for timber and timber products depend on the density of the timber and the extent of exposure to the fire (Table 4.12). It is therefore possible to predict the fire resistance of any timber component using the standard BS EN 1995–1-2: 2004. In addition, as all

Table 4.12 Design charring rates for each exposed face of solid timber and wood-based products

Timber or timber product		Exposed on one face (mm/min)	Exposed all round (mm/min)
Softwood and beech			
Solid timber	$\rho \geqslant 290$ kg/m³	0.65	0.80
Glulam	$\rho \geqslant 290$ kg/m³	0.65	0.70
Hardwood			
Solid or Glulam	$\rho \geqslant 290$ kg/m³	0.65	0.70
Solid or Glulam	$\rho \geqslant 450$ kg/m³	0.50	0.55
Laminated veneer lumber			
LVL	$\rho \geqslant 480$ kg/m³	0.65	0.70
Panels (20 mm)			
Wood panelling	$\rho \geqslant 450$ kg/m³	0.9	–
Plywood	$\rho \geqslant 450$ kg/m³	1.0	–
Wood-based panels other than plywood	$\rho \geqslant 450$ kg/m³	0.9	–

Note:
Rate of charring for each timber face is measured against standard fire exposure.

timbers have a low coefficient of expansion, timber beams will not push over masonry walls as sometimes occurs with steel beams and trusses during fires. Solid timber with a minimum density of 350 kg/m³ and a thickness equal to or greater than 10 mm may be assigned to Euroclass D with respect to fire without testing.

Flame retardants
Within a fire, volatile combustible components are evolved from the surface of the timber and these cause the flaming. The two alternative types of treatment which may be used on timber to reduce this effect are impregnation or the application of surface coatings.

Impregnation involves forcing into the timber, under high pressure and then vacuum, inorganic chemicals which on heating evolve non-combustible gases. Timbers should be machined to their final dimensions before treatment. For interior use, typical compositions include water-borne inorganic salts, such as ammonium sulphate or phosphate with sodium borate or zinc chloride. As these materials are hygroscopic, the timber should not be used in areas of high humidity. For exterior use, a leach-resistant flame-retardant

material based on an organo-phosphate is used, as this is heat fixed by polymerisation within the timber.

Surface treatments, which cause the evolution of non-combustible gases in fire, include antimony tri-oxide flame-retardant paints which are suitable for both interior and exterior use. However, correct application rates by brush or spray can only be reliably achieved under controlled factory conditions. Intumescent coatings, which swell up and char in fire, are suitable for most environments if overcoating is applied. However, the protection afforded by surface treatments may be negated by unsuitable covering or removal by redecoration.

Untreated timber, which is normally Class 3 Spread of Flame to BS 476–7: 1997, can be improved to Class 1 and/or Euroclass C by standard treatments. In some cases it is possible to achieve Euroclass B and/or Class 0 to BS 476 Parts 6 and 7 with heavy impregnation, a combination of impregnation and surface coating or certain very specific surface coating treatments.

While fire-retardant treatments can considerably improve the reaction of wood and wood-based products to fire, this positive effect may be adversely reduced by wet or humid conditions which can lead to migration of the fire-retardant chemicals or their loss by leaching. The publication PD CEN/TS 15912: 2012 lists the Durability of Reaction to Fire performance (DRF) appropriate to the intended use of the wood material.

Requirements for DRF Classes of fire retardant in relation to wood end use are:

DRF Class ST	Short-term use (e.g. under one year)
DRF Class INT1	Permanent interior use (Service Class 1, e.g. wall and ceiling products)
DRF Class INT2	Permanent interior use and certain protected exterior applications (Service Class 2, e.g. wall and ceiling products)
DRF Class EXT	Permanent exterior use (Service Class 3, e.g. façade claddings and exterior conditions)

Note: Service classes to BS EN 1995–1-1: 2004.

An alternative specification system uses the coding DI (dry interior), HR (humidity resistant) and LR (leach resistant), corresponding to the BS EN 1995–1-1: 2004 Service Classes 1, 2 and 3 respectively, for the product types.

European fire classification of construction materials

The European fire classification of construction products and building elements is defined in BS EN 13501–1: 2007. All construction products except floorings may be classified to one of the following seven classes: A1, A2, B, C, D, E or F based on performance. Class A1 represents products which do not contribute to the fire load even within a fully developed fire. Class A2 products do not significantly contribute to the fire load and growth, while the other classes reflect decreasingly stringent fire performance criteria down to Class E products which can resist a small flame for a short period of time without substantial flame spread. Class F products are outside the other classes or have no determined performance. In addition to fire load factors, the standard includes classifications relating to smoke production (s1, s2 and s3, where s1 is the most stringent criterion) and flaming droplets (d0, d1 and d2, where d0 indicates no flaming droplets or particles). For floorings the seven classes are $A1_{FL}$, $A2_{FL}$, B_{FL}, C_{FL}, D_{FL}, E_{FL} and F_{FL} with sub-classifications for smoke production (s1, s2 and s3). Correlation between the UK and European classes is not exact and materials require testing to the European Standard before assignment to a particular class to BS EN 13501–1: 2007.

The Euroclass fire performance rating under the conditions specified in BS EN 13986: 2004 for 12 mm untreated solid wood panels of minimum density 400 kg/m^3 is Class D-s2, d0 for non-floor use and Class D_{FL}-s1 for floorings. The Euroclass fire performance of structural timber with a minimum density of 350 kg/m^3 and a minimum thickness of 22 mm is Class D-s2, d0 to pr EN 14081–1: 2013.

TIMBER CONNECTORS

Timber connectors are metal components which fix together adjacent timbers in the correct orientation and permit the transfer of loads. Punched metal and nail plates are flat for two-dimensional joints. A variety of three-dimensional units includes joist hangers, roof truss clips and anchorages for steel tie bars. Trussed rafters, which account for a large market, are usually constructed with galvanised steel punched plates which are hydraulically pressed into both sides of the timbers to be connected at the butt joints (Fig. 4.24). Metal web joists (Fig. 4.25) are manufactured from strength graded timber flanges with pressed steel punched metal web units fixed to the sides of the softwood flanges. They have the advantage over solid timber joists of lightness and the open web spaces for services.

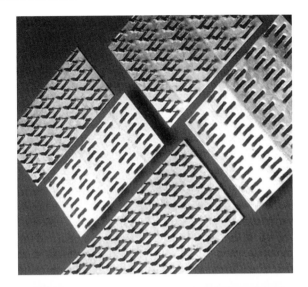

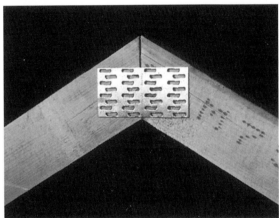

Fig. 4.24 Nail plates

Split-ring, shear-plate and toothed-plate connectors are fastened with a central bolt. Single-sided shear-plate and toothed connectors are used for fixing timber to steel. Plate and ring connectors are described in the standard BS EN 912: 2011. Joist hangers and tension

Fig. 4.25 Metal web floor joists

straps are described in BS EN 845–1: 2013. Structural timber adhesives are classified in BS EN 301: 2013.

Non-metallic timber connectors include bonded-in fibre-reinforced plastic rods and plates as well as dowels manufactured from modified wood products.

Timber products

A wide range of products is manufactured from wood material, ranging in size from small timber sections and thin laminates through chips and shavings down to wood fibres. The physical properties of the materials produced reflect a combination of the subdivision of the wood, the addition of any bonding material and the manufacturing process. The physical properties then determine the products' appropriate uses within the building industry. Many of the products are manufactured from small timber sections or timber by-products from the conversion of solid timber which would otherwise be wasted. Compressed straw slabs and thatch are additionally included in this section.

The product range includes:

- laminated timber;
- cross-laminated timber;
- structural insulated panels;
- laminated veneer lumber;
- plywood;
- blockboard and laminboard;
- particleboard;
- fibreboard;
- wood wool slabs;
- compressed straw slabs;

- thatch;
- shingles;
- '*Steko*' blocks;
- '*Brettstapel*';
- flexible veneers;
- decorative timbers.

Within the European Union, whenever wood-based panels are used in construction, compliance with the Construction Products Regulation (2013) must be demonstrated. This may be achieved by adherence to the European Harmonised Standard for wood-based panels BS EN 13986: 2004. This requires that products used in construction comply with its specifications and also to the additional performance-based criteria within the various standards listed for each specific material. European countries are now required to use the CE mark on boards and panels manufactured to this harmonised standard.

LAMINATED TIMBER

Manufacture

Large, solid timber sections are limited by the availability of appropriate lumber; in addition, their calculated strength must be based on the weakest part of the variable material. Laminated timber sections overcome both of these difficulties and offer additional opportunities to the designer. Laminated timber is manufactured by curing within a jig, layers of accurately cut smaller timber sections which are continuously glued together with a resin adhesive. Laminates may be vertically or horizontally orientated. The use of

Table 4.13 Strength classes for homogeneous and combined glulam (BS EN 14080: 2013)

Glulam strength classes	GL 20	GL 22	GL 24	GL 26	GL 28	GL 30	GL 32
Homogeneous glulam:							
Strength class of laminates	C16	C22	C24	C27	C30	C35	C40
Combined glulam:							
Strength class of outer laminates	C22	C22		C27	C35		C40
Strength class of inner laminates	C14	C14		C18	C18		C30
Proportion of inner zone laminates	34%	34%		34%	50%		66%

Notes:
- Glulam strength classes are h for homogeneous or c for combined (e.g. GL 20h or GL 20c).
- A range of other proportions of inner and outer laminates are permitted.
- Some glulam sections may have an intermediate zone of laminations.
- The C strength classes are to BS EN 338: 2009.

strength-graded timber and the staggering of individual scarf or finger joints ensure uniformity of strength within the product; although, under BS EN 14080: 2013, large finger joints through the whole section of a *glulam* member are permissible. The manufacturing process ensures greater dimensional stability and fewer visual defects than in comparable solid timber sections. Laminated timber may be homogeneous, with all laminates of the same strength class of timber, or combined, in which lower strength class laminates are used for the centre of the units. Table 4.13 gives the European strength classes for the two alternatives. The standard BS EN 14080: 2013 also refers to block glued glulam and glued solid timber. Block glued glulam is manufactured by fixing a series of glulam components together with gap-filling adhesive.

Forms

Sections may be manufactured up to any transportable size, typically 30 m, although spans over 50 m are possible. Standard-size straight sections (315 × 65 and 90 mm; 405 × 90 and 115 mm; 495 × 115 mm) are stock items, but common sizes range from 180 × 65 to 1035 × 215 mm. Sections may be manufactured to order, to any uniform or non-uniform linear or curved form. The majority of laminated timber structures are manufactured from softwoods such as European redwood or whitewood, although the rib members within the roof structure of the Thames Flood Barrier (Fig. 4.26) were manufactured from the West African hardwood, iroko. Figure 4.27 illustrates typical laminated timber arches, columns and portal frames as generators of structural forms. The aesthetic properties of laminated timber may be enhanced by the application of suitable interior or exterior timber finishes. Steel fixing devices and joints may be visually expressed (Fig. 4.27) or almost unseen (Fig. 4.28) by the use of concealed bolted steel plates.

Laminated timber of 22 mm minimum thickness ($\rho \geqslant 350$ kg/m^3) and manufactured to the standards in BS EN 14080: 2013 is classified (excluding floorings) as D-s2, d0 with respect to fire. When exposed to fire on all sides, the charring rate for softwood glulam ($\rho \geqslant 290$ kg/m^3) is typically 0.70 mm/min (Table 4.12). Preservative treatments are necessary when the material is to be used under conditions in which the moisture content is likely to exceed 20%. The three Service Classes of glulam structures relate to the environmental conditions.

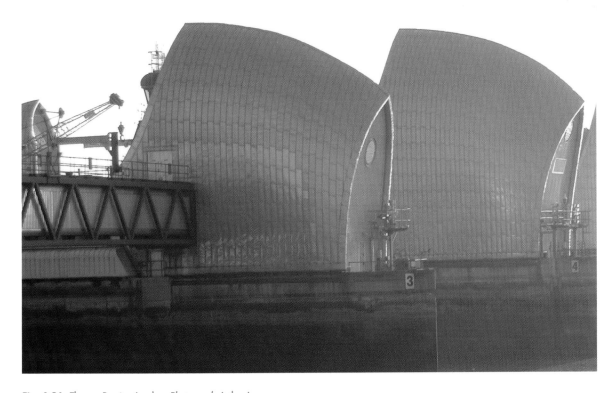

Fig. 4.26 Thames Barrier, London. *Photograph:* Arthur Lyons

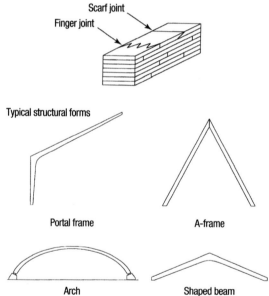

Fig. 4.27 Glued laminated-timber beams. *Photograph*: Arthur Lyons

Service Classes for glulam:

Service Class 1　Internal conditions with heating
　　　　　　　　and protection from damp
　　　　　　　　(typical moisture content < 12%)

Service Class 2　Protected, but unheated conditions
　　　　　　　　(typical moisture content < 20%)

Service Class 3　Exposed to the weather
　　　　　　　　(typical moisture content > 20%)

CROSS-LAMINATED TIMBER

Cross-laminated timber (CLT or X-LAM) is similar to conventional plywood, except that the laminates are thicker, and the panel thicknesses are generally between 50 and 300 mm, although 500 mm can be produced. The maximum panel size is governed by transportation limits which are normally 15 × 3 m, but the standard production limit is 24 m. The wide range of timber

should be factory formed. On site the material should preferably be kept dry and only used above damp-proof level.

The fire rating for cross-laminated timber is D-s2, d0 for non-flooring applications and D_{FL}-s1 for flooring. The thermal properties of CLT panels depend upon the timber used. Density is usually within the range 470–590 kg/m^3, and an average of 500 kg/m^3 has a thermal conductivity λ value 0.13 W/m K.

STRUCTURAL INSULATED PANELS

Structural insulated panels (SIPs) are prefabricated lightweight building components, used for load-bearing internal and external walls and roofs. Unlike cladding sandwich panels, structural insulated panels can support considerable vertical and horizontal loads without internal studding. They are manufactured from two high-density face layers separated by a light-weight insulating core. The three layers are strongly bonded together to ensure that the composite acts as a single structural unit. The outer layers of oriented strand board (OSB), cement-bonded particleboard or gypsum-based products are typically 8–15 mm thick. The core is composed of a rigid cellular foam, such as polyurethane (PUR), polyisocyanurate (PIR), phenolic foam (PF), expanded (EPS) or extruded (XPS) poly-styrene, giving an overall unit thickness of between 70 and 252 mm. The structural performance is pre-dominantly influenced by the thickness and physical properties of the outer layers, and the thermal performance is largely determined by the width and insulating characteristics of the core material. A U-value of 0.10 W/m^2 K can be achieved with 252 mm panels. Large panels (e.g. 6.0 m × 2.5 m × 180 mm, U-value 0.15 W/m^2 K) may be used for two-storey construction to the Code for Sustainable Homes Level 4.

Structural insulated panels (typically 1.2 m wide by 2.4 m high) offer a thermally efficient and airtight form

Fig. 4.28 Laminated timber beam metal joint detail – Wells Cathedral Restaurant. *Photograph:* Arthur Lyons

species and adhesives permitted for use are described in pr EN 16351: 2011. Most CLT is manufactured from either Grade C24 or C16 softwood. Large finger jointing across the full cross-sectional area is permitted only when factory produced. The structural panels, which may be used to form walls, roofs and floors, are easily clad with brickwork, tiling or rendering as appropriate. Using cross-laminated timber requires careful design coordination, as all openings for windows and doors

Table 4.14 Thermal performance data for structural insulated panels (SIPs)

Core material	Face material	Panel thickness (mm)	Thermal performance (W/m^2 K)
Polyurethane foam	cement bonded particleboard	86	0.28
Polyurethane foam	oriented strand board	100	0.23
Polyisocyanurate foam	oriented strand board	140	0.22

Note:
The thermal performance data (U-values) are typical for the listed SIPs when constructed with a brick outer leaf and 50 mm clear vented cavity.

of construction, which is rapidly erected on site. Jointing between panels is usually some form of tongue and groove system. Sound reduction for separating walls is typically 58 dB depending on construction details. External cladding may be brickwork, wooden panelling or rendering, with plasterboard as the standard internal finish. Table 4.14 gives typical thermal performance data for structural insulated panels. The use of SIPs is one of the Modern Methods of Construction (MMC) promoted by the UK government to reduce energy waste in new building. Currently, SIPs account for approximately 10% of new building methods construction.

LAMINATED VENEER LUMBER

Laminated veneer lumber (LVL) (Fig. 4.29), also known as *microlam*, is more economical than laminated timber as there is little waste in the production process. It is manufactured to three grades by laminating timber strands with polyurethane resin under heat and pressure. In one process, logs are cut into flat timber strands 300 mm long; these are then treated with resin, aligned and hot-pressed into billets of reconstituted wood. In the other processes, 3 mm-thick timber strands or sheets of veneer are coated with waterproof adhesive and bundled together with the grain parallel. The strands or veneers are pressed together and microwave cured to produce structural timber billets or sheets up to 26 m long. The versatile material is suitable for use in columns, beams, purlins and trusses,

and can be machined as solid timber (Fig. 4.30). I-section joists, with LVL flanges and OSB webs, are suitable for flat and pitched roofs and floor construction. Metal-web timber joists combine LVL flanges with metal strutting webs. Untreated LVL has a Class 3 surface spread of flame classification (BS 476–7: 1997). Three grades of laminated veneer lumber are classified by BS EN 14279: 2004 according to their serviceability in dry and exposed conditions. BS EN 14374: 2004 details the physical properties of laminated veneer lumber.

Grades of laminated veneer lumber:

Purpose loading	Environmental conditions	Type
Load-bearing	Dry (Service Class 1)	LVL/1
Load-bearing	Humid (Service Class 2)	LVL/2
Load-bearing – exterior conditions (subject to testing or appropriate finish)	Exterior (Service Class 3)	LVL/3

Notes:
Service Classes are according to BS EN 1995–1-1: 2004.
Laminated veneer lumber should be marked 'S' for structural application or 'NS' for non-structural application to BS EN 14279: 2004 + A1: 2009.

Monocoque structures

Interesting and innovative built forms can be created by using LVL (and other timber products) to produce flat or curved form monocoque structures. These work on the well-established principles from the motor industry, in which the hard body skin acts in concert with any stiffeners to form the structure. Using this technology, structurally efficient and elegant forms, which may be slender, tapered, flat or curved, can be produced. LVL is quickly becoming a significant material to complement the more established products such as plywood, OSB and glulam, particularly because of its availability in very large sections.

PLYWOOD

Manufacture

Plywood is manufactured by laminating a series of thin timber layers, or plies, to the required thickness. The timber log is softened by water or steam treatment and rotated against a full-length knife to peel off a veneer

Fig. 4.29 Laminated veneer lumber (LVL)

Fig. 4.30 Laminated veneer lumber (LVL) construction – Finnforest Office, Boston, Lincolnshire. *Architects:* Arosuo and Vapaavuori Oy. *Photograph:* Courtesy of Finnforest

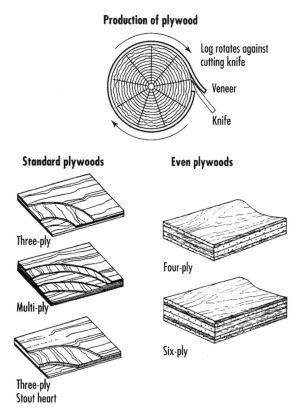

Fig. 4.31 Manufacture and standard types of plywood

or ply of constant thickness (Fig. 4.31). The ply is then cut to size, dried and coated with adhesive prior to laying up to the required number of layers. Not all the plies are of the same thickness; often, thicker plies of lower-grade material are used in the core. However, the sheets must be balanced at about the centre to prevent distortions caused by differential movement. Plies are normally built up with adjacent grain directions at right-angles to each other to give uniform strength and reduce overall moisture movement, although with even plywoods, the central pair of plies has parallel grains. The laminate of plies and glue is cured in a hot press, sanded and trimmed to standard dimensions for packaging. Decorative veneers of hardwood or plastic laminate may be applied to one or both faces. Most plywood imported into the UK is made from softwood (largely pine and spruce), from North America and Scandinavia. Smaller quantities of plywood produced from temperate hardwoods are imported from Finland (birch) and Germany (beech), while tropical hardwood products are imported from Indonesia, Malaysia, South America and Africa. Bamboo plywood is made from a core of tightly compressed fibres with bamboo veneers on either face. It is coated with a lacquer but should not be used externally without preservative pre-treatment, as bamboo is inherently not durable.

The standard sheet size is 2440 × 1220 mm, with some manufacturers producing sheet sizes of up to 3050 × 1525 mm or slightly larger. Sheet thicknesses range from 4 to 25 mm for normal construction use, although thinner sheets down to 1.5 mm are available for specialist purposes.

Under the European fire classification of construction materials (BS EN 13501–1: 2007), an untreated plywood panel would normally achieve a class D-s2, d0 rating, excluding its use as flooring when the rating is class D_{FL}-s1 (depending on a minimum thickness of 9 mm, a minimum density of 400 kg/m^3 and fixing to a non-combustible substrate [class A1 or A2] without an air gap. The secondary classifications 's' and 'd' relate to smoke production and flaming droplets).

Grades

Plywood is classified according to its general appearance and physical properties (BS EN 313–1: 1996). The key characteristics are the form of construction, durability and nature of the surface. The durability of plywood is largely determined by the bonding class of the adhesive used. This ranges from Class 1, to the most durable Class 3 (BS EN 314–2: 1993), which may be used externally without delamination, provided that the timber itself is durable or suitably protected against deterioration.

Bonding classes for plywood – corresponding to Service Classes in BS EN 1995–1-1: 2004:

Class 1   Dry conditions (suitable for interior use).
Class 2   Humid conditions (protected external applications, e.g. behind cladding or under roof coverings).
Class 3   Exterior conditions (exposed to weather over sustained periods).

Plywood should be marked 'S' for structural application or 'NS' for non-structural application to BS EN 636: 2012.

Phenol formaldehyde resins are the most frequently used for the most durable plywoods. Marine plywood (BS 1088–1: 2003) is a combination of a moderately durable timber bonded with phenolic or melamine-formaldehyde resin. The standard class of marine plywood is suitable for regular wetting or permanent exposure to salt or fresh water. The lower grades of plywood are bonded with melamine-urea formaldehyde or urea-formaldehyde resins. The relationship between the natural durability of the timber against wood-destroying fungi and the Use Class to which the plywood may be assigned is described in DD CEN/TS 1099: 2007.

The quality of plywood is also affected by the number of plies for a particular thickness and the surface condition of the outer plies which range from near perfect, through showing repaired blemishes to imperfect. Factory-applied treatments to improve timber durability and fire resistance are normally available.

The standard BS EN 635: 1995 describes five classes of allowable defects (E, and I to IV) according to decreasing quality of surface appearance; Class E is practically without surface defects. These are related to hardwood and softwood surfaced plywoods in BS EN 635: 1995 Parts 2 and 3, respectively. The publication DD CEN/TS 635–4: 2007 describes the range of 15 surface conditions (A to O) appropriate for the application of finishes such as paint, varnish or stain, according to the anticipated dry, humid or exterior end-use conditions.

The performance specifications for plywood to be used in dry, humid or exterior conditions depend upon a combination of bonding quality (BS EN 314–2: 1993) and biological durability (BS EN 335: 2013) as collated in the standard BS EN 636: 2012.

Requirements for plywood for use in dry conditions:

Bonding Class 1 and Biological Durability Class 1 (or 2 and 3)

Requirements for plywood for use in humid conditions:

Bonding Class 2 and Biological Durability Class 2 (or 3)

Requirements for plywood for use in exterior conditions:

Bonding Class 3 and Biological Durability Class 3.

Dry, humid and exterior conditions correspond to the Service Classes in BS EN 1995–1-1: 2004.

Board marking should include the intended use conditions (dry, humid, exterior), the formaldehyde release class and 'S' for structural or 'NS' for general-purpose non-structural application. The standard BS EN 636: 2012 also has a classification system based on the bending strength which may be included on the documentation. The characteristic values for structural design using plywood are described in BS EN 12369–2: 2011 which also lists the load duration classes.

Plywood load duration classes:

Load duration class	Typical loading period	Examples of loading
Permanent	More than 10 years	Self-weight
Long term	6 months to 10 years	Storage
Medium term	1 week to 6 months	Imposed load
Short term	Less than 1 week	Snow and wind
Instantaneous		Accidental load

Uses

Considerable quantities of plywood are used by the construction industry because of its strength, versatility and visual properties. The strength of plywood in shear is used in the manufacture of plywood box and I-section beams in which the plywood forms the web. Increased stiffness can be generated by forming the plywood into a sinusoidal web. Plywood box beams can be manufactured to create pitched and arched roof forms, as illustrated in Figure 4.32. Stiffened and stressed skin panels, in which plywood and softwood timbers are continuously bonded to act as T or I-beams, will span greater distances as floor structures than the same depths of traditional softwood joists with nailed boarding. Such structural units may also be used to form pitched roofs, or to form folded plate roof

Plywood beams

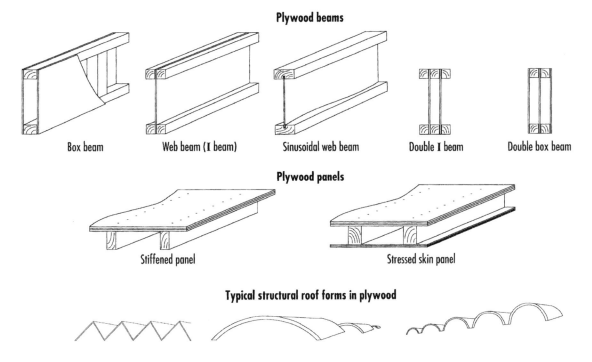

Box beam Web beam (**I** beam) Sinusoidal web beam Double **I** beam Double box beam

Plywood panels

Stiffened panel Stressed skin panel

Typical structural roof forms in plywood

Fig. 4.32 Structural uses of plywood

structures or barrel vaulting (Fig. 4.32). Plywood of 8–10 mm thickness is frequently used as the sheeting material in timber frame construction and for complex roof forms such as domes. The lower-grade material is extensively used as formwork for in situ concrete.

CORE PLYWOOD

The standard core plywood products are blockboard and laminboard. Both are manufactured with a core of usually softwood strips sandwiched between one or two plies (Fig. 4.33). In blockboard the core strips are between 7 and 30 mm wide, but in laminboard, the more expensive product, they are below 7 mm in width and continuously glued throughout. As with plywood, the grain directions are perpendicular from layer to layer. Most core plywoods are bonded with urea formaldehyde adhesives appropriate to interior applications only. The standard sheet size is 2440 × 1220 mm with a thickness range of 12–25 mm, although larger sheets up to 45 mm thick are available. Blockboard may be finished with a wide range of decorative wood, paper or plastic veneers for use in fitted furniture. Variants on the standard products include plywood with phenolic foam, polystyrene or a particleboard core.

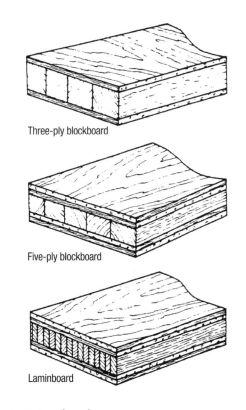

Three-ply blockboard

Five-ply blockboard

Laminboard

Fig. 4.33 Core plywoods

PARTICLEBOARDS

Particleboards are defined as panel materials produced under pressure and heat from particles of wood, flax, hemp or other similar lignocellulosic materials. The wood particles may be in the form of flakes, chips, shavings, sawdust, wafers or strands (BS EN 309: 2005). Boards may be uniform through their thickness or of a multi-layered structure. Wood particleboard and cement-bonded particleboard are made from wood chips with resin and cement binder, respectively. Oriented strand board is manufactured from large wood flakes and is classified in BS EN 300: 2006.

Wood particleboard

Manufacture

Wood particleboard (chipboard) is manufactured from wood waste or forest thinnings which are converted into wood chips, dried and graded according to size. The chips are coated with adhesive to approximately 8% by weight and then formed into boards (Fig. 4.34). The wood chips are either formed randomly into boards giving a uniform cross-section or distributed with the coarse material in the centre and the finer chips at the surface to produce a smoother product. The boards are then compressed and cured between

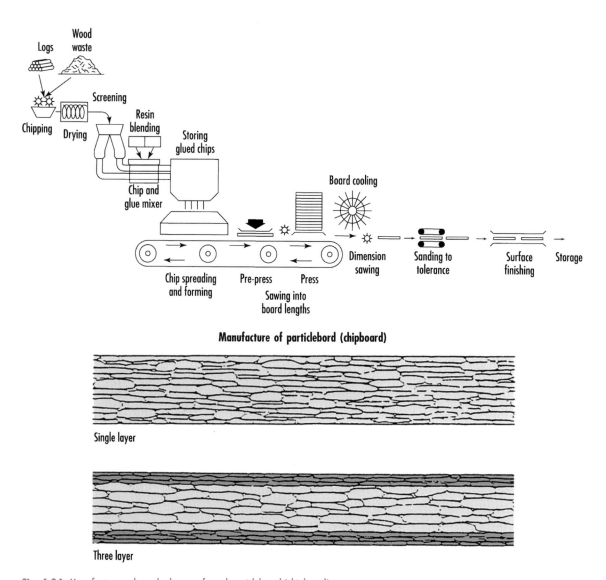

Manufacture of particlebord (chipboard)

Single layer

Three layer

Fig. 4.34 Manufacture and standard types of wood particleboard (chipboard)

the plates of a platen press at 200°C. Boards are finally trimmed, sanded and packed. In the *Mende* process a continuous ribbon of 3–6 mm particleboard is produced by calendering the mix around heated rollers. Over 70% of UK particleboard consumption is sourced from British-grown timber.

The standard sizes are 2440 × 1220 mm, 2750 × 1220 mm, 3050 × 1220 mm and 3660 × 1220 mm, with the most common thicknesses ranging from 12–38 mm, although much larger sheet sizes and thicknesses from 2.5 mm are available.

Extruded particleboard (BS EN 14755: 2005) is manufactured by extruding the mixture of wood chip and resin through a die into a continuous board; however, in this method the wood chips are predominantly orientated at right-angles to the board face, thus generating a weaker material. Extruded particleboard is specified within four grades according to its density and whether it is solid or has tube voids.

Types

The durability of particleboards is dependent upon the resin adhesive. Much UK production uses urea formaldehyde resin, although the moisture-resistant grades are manufactured with melamine-urea formaldehyde or phenol formaldehyde resins. Wood chipboards are categorised into seven types (P1–P7 with a voluntary coding system) to BS EN 312: 2010 according to the anticipated loading and environmental conditions. The standard specifies requirements for mechanical and swelling properties and also formaldehyde emissions. The first colour code defines the loading and the second colour the moisture conditions.

Grades of wood particleboard:

Purpose/loading	Environmental conditions	Optional colour codes		Type
General purpose	Dry	White, white	blue	P1
Interior fitments	Dry	White	blue	P2
Non-load-bearing	Humid	White	green	P3
Load-bearing	Dry	Yellow, yellow	blue	P4
Load-bearing	Humid	Yellow, yellow	green	P5
Heavy duty, load-bearing	Dry	Yellow	blue	P6
Heavy duty, load-bearing	Humid	Yellow	green	P7

Grades of extruded particleboard:

ES	Extruded Solid: Board with a minimum density of 550 kg/m³.	
ET	Extruded Tubes: Board with a minimum solid density of 550 kg/m³.	
ESL	Extruded Solid Light: Board with a density of less than 550 kg/m³.	
ETL	Extruded Tubes Light: Board with a solid density of less than 550 kg/m³.	

(Grade ET must have at least 5 mm of material over the void spaces.)

Lightweight particleboards with a density of less than 600 kg/m³ used mainly in furniture and other non-structural applications are described in PD CEN/TS 16368: 2014. Two grades, LP1 and LP2, are for use only in dry conditions. LP2 has the higher physical properties specification.

Standard particleboards are hygroscopic and respond to changes in humidity. A 10% change in humidity will typically increase the sheet length and breadth by 0.13% and the thickness by 3.5%. Dry grades should not be exposed to moisture even during construction. Humid-tolerant grades are resistant to occasional wetting and relative humidities over 85%. However, no particleboards should be exposed to prolonged wetting, as they are all susceptible to wet rot fungal attack.

All untreated wood particleboards have Class 3 spread of flame (BS 476–7: 1997). However, they may be treated to the requirements of Class 1 by chemical addition in manufacture, by impregnation or by the use of intumescent paints. Class 0 may also be achieved. For untreated particleboard with a minimum density of 600 kg/m³ and a minimum thickness of 9 mm, the Euroclass fire performance rating under the conditions specified in BS EN 13986: 2004 is Class D-s2, d0 for non-floor use and Class D_{FL}-s1 for floorings.

A wide range of wood veneer, primed/painted, paper and plastic (PVC, phenolic film or frequently melamine) finishes is available as standard products. Pre-cut sizes are available edged to match. Domestic flooring grade particleboard, usually 18 or 22 mm, may be square-edged or tongued and grooved. The industrial flooring grades are typically from 38 mm upward in thickness.

Uses

Significant quantities of wood particleboard (chipboard) are used in the furniture industry. Much flat-pack DIY furniture is manufactured from painted or veneered particleboard. Particleboard can be effectively jointed by the use of double-threaded particleboard wood screws and various specialist fittings. Where high humidity is anticipated the moisture-resistant grades should be used. The domestic housing market uses large quantities of flooring-grade particleboard as it is competitively priced compared to traditional tongued and grooved softwood. Joist centres should be at 450 and 610 mm centres maximum for 18/19 and 22 mm particleboard, respectively. Edges should be tongued and grooved or fully supported and the standard panel size is 2400 × 600 mm. For heavy-duty flooring, flat-roof decking and structural work, the moisture-resistant structural grade must be used. Phenolic film-coated particleboard offers a suitable alternative to plywood as formwork to concrete.

Cement-bonded particleboard

Manufacture
Cement-bonded particleboard is manufactured from a mixture of wood particles or filaments (usually softwood) and cement. The boards, which are light grey in colour, have a uniform cementitious surface. The material has up to 75% cement by weight, with the cement filling all the void spaces, producing a material with a density of 1000–1250 kg/m^3 (cf. 650–690 kg/m^3 for standard grade particleboard).

Types and uses
The material based on Portland cement has good resistance to fire, water, fungal attack and frost. The standard BS EN 634–2: 2007 specifies only one grade, which is suitable for use both internally and externally. It should be colour-coded white, white (non-load-bearing) and brown (suitable for dry, humid and exterior conditions) with 25 mm vertical stripes near one corner of the board. Within the one grade there are two Technical Classes 1 and 2 which relate only to modulus of elasticity in bending. The standard BS EN 633: 1994 refers to both Portland and magnesium-based cements. Magnesite-bonded particleboard is used as a lining board but it is not frost-resistant and is unsuitable for use in humid conditions.

Boards frequently have a core of coarse wood chips, sandwiched between finer material, producing a good finish, which may be further treated by sanding and priming. Because of its density, cement-bonded particleboard has good sound-insulation properties. Typically, 18 mm board will give sound reduction of 31–33 dB. The material is frequently used for soffits, external sheathing and roofing on both modular and timber frame buildings, particularly where racking resistance is required. The heavier grades, generally

tongued and grooved, are suitable for flooring, due to their resistance to moisture, fire, impact and airborne sound.

The material has a Class 0 Surface Spread of Flame to Building Regulations (Class 1 to BS 476: Part 7: 1997). The Euroclass fire performance rating under the conditions specified in BS EN 13986: 2004 for 10 mm cement-bonded particleboard is Class B-s1, d0 for non-floor use and Class B$_{FL}$-s1 for floorings.

Board sizes are typically 1200 × 2440, 2600 or 3050 mm, with standard thicknesses of 12 and 18 mm, although sheets up to 40 mm in thickness are made. However, because of the density of the material, a 1200 × 2440 × 12 mm board weighs approximately 45 kg and should not be lifted by one operative alone.

Gypsum-bonded particleboard
Gypsum-bonded particleboard, available in sheets of 6 mm thickness upward, is an alternative multipurpose building board. It is not included in the scope of BS EN 633: 1994.

Oriented strand board

Manufacture
Oriented strand board (OSB) is manufactured from 0.5 mm-thick timber flakes tangentially cut and measuring approximately 75 × 35 mm. These are dried and coated with wax and 2.5% of either phenol formaldehyde or melamine-urea formaldehyde resin. The mix is laid up in three (or occasionally five) layers with the strands running parallel to the sheet on the outer faces and across or randomly within the middle layer. The boards are then cured under heat and pressure, sanded and packaged (Fig. 4.35). Standard panel sizes are 2440 or 3660 × 1220 mm with densities usually in the range 600–680 kg/m^3.

Grades and uses
Orientated strand board is graded according to the anticipated loading and environmental conditions (BS EN 300: 2006). Large quantities are used as sheathing in timber-frame housing. The moisture-resistant grade is suitable for roof sarking, while the higher specification grade with enhanced strength properties is suitable for flat roof decking. Thicker panels are used for heavy-duty flooring; the web material in timber I-beams and the manufacture of Structural Insulated Panels (SIPs). Oriented strand board is produced to a thickness range of 6–40 mm. In Europe it is manufactured from Scots pine and spruce, but in North

Oriented strand board

Wood wool slab

Compressed straw slab

Fig. 4.35 Oriented strand board, wood wool and compressed straw slabs

America it is made from aspen and Southern pine. Within the UK and Ireland, OSB is manufactured from managed forest thinnings certified by the Forest Stewardship Council.

Grades of oriented strand board:

Grade	Purpose/loading	Environmental conditions	Colour codes	
OSB 1	General purpose, interior fitments	Dry conditions (Use Class 1)	White	blue
OSB 2	Load-bearing	Dry conditions (Use Class 1)	Yellow, yellow	blue
OSB 3	Load-bearing	Humid conditions (Use Class 2)	Yellow, yellow	green
OSB 4	Heavy duty, load-bearing	Humid conditions (Use Class 2)	Yellow	green

Note:
The Use Class environmental/biological conditions related to durability (BS EN 335: 2013) were previously called hazard class.

The Euroclass fire performance rating under the conditions specified in BS EN 13986: 2004 for 9 mm untreated oriented strand board with a minimum density of 600 kg/m^3 is Class D-s2, d0 for non-floor use and Class D_{FL}-s1 for floorings.

Flaxboard

Manufacture
Flaxboard is manufactured from a mixture of at least 70% flax shives (thin slices) and adhesive. Other raw materials such as wood flakes, chips and sawdust may be incorporated. The boards are formed under heat and pressure.

Grades and uses
The standard BS EN 15197: 2007 describes four types of flaxboard, according to their potential use and environmental conditions.

Grades of flaxboard:

Grade	Purpose/loading	Environmental conditions
FB1	General purpose/filling	Dry
FB2	Non-load-bearing/veneering	Dry
FB3	Non-load-bearing/interior fitment and furniture	Dry
FB4	Non-load-bearing	Humid

Scrimber

Scrimber is manufactured from small forest thinnings, typically radiata pine, aspen, birch and larch. The small section timber, in the range 70–120 mm in diameter, is debarked and crushed to a web of fibrous material, largely maintaining the orientation of the timber fibres. The material is dried and reconstituted with adhesive under heat and pressure into a solid product. The scrimber produced has many of the good properties of natural timber, including ease of nailing, and may be manufactured to any required dimensions, restricted only by the size of the press. Scrimber has also been manufactured from crushed bamboo.

FIBREBOARDS

Fibreboards are manufactured from wood or other plant fibres by the application of heat and/or pressure. They are bonded by the inherent adhesive properties and felting of the fibres or by the addition of a synthetic binder. In the *wet* process used for the manufacture of hardboard, mediumboard and softboard, no adhesive is added to the wood fibres. In the case of medium-density fibreboard (MDF), a resin-bonding agent is incorporated during the production process. Fibreboards to BS EN 316: 2009 are classified according to the production process, use conditions and typical applications.

Manufacture

Wet process
Forest thinnings and wood waste are chipped and then softened by steam heating. The chips are ground down into wood fibres and made into a slurry with water. The slurry is fed onto a moving wire-mesh conveyor, where the excess water is removed by suction and light rolling which causes the fibres to felt together. The *wet lap* is then cut to lengths and transferred to a wire mesh for further pressing and heat treatment to remove the remaining water and complete the bonding process. Boards are then conditioned to the correct moisture content and packaged. The range of products arises primarily from the differing degrees of compression applied during the manufacturing process (Fig. 4.36).

Dry process: medium-density fibreboard
The manufacture of medium-density fibreboard (MDF) involves the addition of adhesive, usually urea

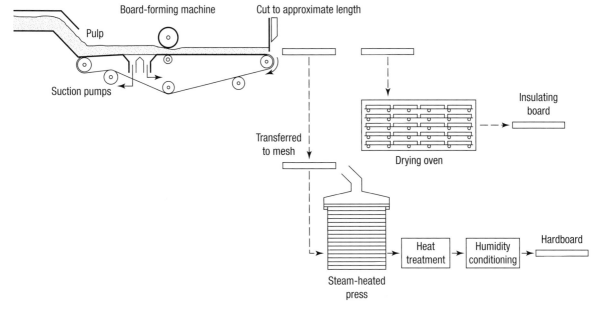

Fig. 4.36 Manufacture of fibreboards

formaldehyde, to the dry wood fibres, which are laid up to an appropriate thickness, slightly compressed to a density of at least 450 kg/m³ and cut to board lengths. The boards are cured under heat and pressure in a press, trimmed to size and sanded. MDF has the advantage of a high-quality, machinable finish, and is currently used for the production of various mouldings as well as boards. Decorative profiled sheets can be manufactured by laser cutting of MDF panels to individual client designs. Because of the uniformity of the material, solid sections may be routed to any form. It is therefore widely used for furniture panels as well as internal load-bearing applications.

Nine grades are described in BS EN 622–5: 2009 relating to their anticipated loading and environmental conditions. Most MDF is based on urea formaldehyde resin, but in order to ensure safety, the quantities of formaldehyde used are strictly controlled by appropriate standards (BS EN 622–1: 2003). Where improved moisture resistance is required, a melamine-urea formaldehyde resin is used, but this material is not suitable for exterior applications. MDF sheets and mouldings may be finished with a range of coatings including paints, lacquers, stains, plastic laminates, wood veneers and foils.

Moisture-resistant dense MDF (690–800 kg/m³) with all-through colour is available in a range of colours and thicknesses from 8–30 mm. The material is manu-

factured from organic-dyed, fade-resistant wood fibres and melamine resin with a low residual formaldehyde content. The material can be machined to decorative forms and patterns with a high-quality surface finish which requires only sealing to enhance the colours.

Grades of medium-density fibreboard (MDF):

Grade	Purpose/loading	Environmental conditions
MDF	General purpose	Dry
MDF.H	General purpose	Humid
MDF.LA	Load-bearing	Dry
MDF.HLS	Load-bearing	Humid
L-MDF	Non-load-bearing/general purpose	Dry
L-MDF.H	Non-load-bearing/general purpose	Humid
UL1-MDF	Non-load-bearing/general purpose	Dry
UL2-MDF	Non-load-bearing/general purpose	Dry
MDF.RWH	Rigid underlays in roofs and walls	

Notes:
– L-MDF refers to light MDF and UL-MDF to ultra-light MDF, respectively.
– UL2-MDF has a higher specification than UL1-MDF.
– Grades to BS EN 622–5: 2009.

The Euroclass fire performance rating under the conditions specified in BS EN 13986: 2004 for 9 mm untreated MDF with a minimum density of 600 kg/m³ is Class D-s2, d0 for non-floor use and Class D_{FL}-s1 for floorings.

Acylated MDF (*Medite Tricoya®*) has enhanced dimensional stability, durability and fungal resistance compared to standard MDF. It is therefore suitable for internal and external use as a non-structural material, with a durability Class 1 equivalent to teak.

Hardboard

Hardboards are the densest fibreboards, with a minimum density of 900 kg/m³. The boards range in colour from light to dark brown, usually with one smooth surface and a mesh-textured surface on the underside, although *duo-faced* hardboard – smooth on two faces – is available. Standard sheet sizes are 1220 × 2440 to 3600 mm and 1700 × 4880 mm; there are also door sizes. Standard thicknesses range from 3.2 to 6.4 mm, although a wider range is available.

The standard BS EN 622–2: 2004 specifies six grades of hardboard according to load-bearing properties and environmental conditions.

Grades of hardboard:

Grade	Purpose/loading	Environmental conditions	Colour codes	
HB	General purpose	Dry	White, white	blue
HB.H	General purpose	Humid	White, white	green
HB.E	General purpose	Exterior	White, white	brown
HB.LA	Load-bearing	Dry	Yellow, yellow	blue
HB.LA1	Load-bearing	Humid	Yellow, yellow	green
HB.LA2	Heavy duty, load-bearing	Humid	Yellow	green

Standard hardboard is suitable for internal use, typically panelling, wall and ceiling linings, floor underlays and furniture. A range of perforated, embossed and textured surfaces is available. Applied coatings include primed or painted and various printed woodgrain, PVC or melamine foils.

The Euroclass fire performance rating under the conditions specified in BS EN 13986: 2004 for 6 mm untreated hardboard with a minimum density of 900 kg/m³ is Class D-s2, d0 for non-floor use and Class D_{FL}-s1 for floorings.

Tempered hardboard
Tempered hardboards, impregnated with oils during manufacture, are denser and stronger than the standard hardboards, with enhanced water and abrasion resistance. Tempered hardboards are dark brown to black in colour and have a density usually exceeding 960 kg/m³. Tempered hardboards are suitable for structural and exterior applications. The high shear strength of the material is used within hardboard-web structural box and I beams. Typical exterior applications include claddings, fascias and soffits, where weather resistance is important. The moisture resistance of tempered hardboard makes it suitable for lining concrete formwork.

Mediumboard and softboard
Mediumboard and softboard are manufactured by the *wet* process. Mediumboard (high density and low density) and softboard exhibit a range of physical properties which reflects the degree of compression applied during the manufacturing process. High-density mediumboard (density 560–900 kg/m³) has a dark-brown shiny surface like hardboard. Low-density mediumboard (density 400–560 kg/m³) has a light-brown soft finish. Softboard (density 210–400 kg/m³) is light in colour with a fibrous, slightly textured finish. Softboard impregnated with bitumen offers an increased moisture resistance over the untreated material.

The general-purpose exterior grades (E) may be used for exterior cladding. The higher density grades (H) are used for wall linings, sheathing, partitioning, ceilings and floor underlays. Low-density mediumboard (L) is used for wall linings, panelling, ceilings and notice-boards. Softboard is used for its acoustic and thermal insulating properties. Bitumen impregnated softboard is suitable for use as a floor underlay to chipboard on concrete.

The standard BS EN 622–3: 2004 specifies ten grades of low- (L) and high- (H) density mediumboard according to load-bearing requirements and environmental conditions.

Grades of mediumboard:

Grade	Purpose/loading	Environmental conditions	Optional colour codes	
MBL	General purpose	Dry	White, white	blue
MBH	General purpose	Dry	White, white	blue
MBL.H	General purpose	Humid	White, white	green
MBH.H	General purpose	Humid	White, white	green
MBL.E	General purpose	Exterior	White, white	brown
MBH.E	General purpose	Exterior	White, white	brown
MBH.LA1	Load-bearing	Dry	Yellow, yellow	blue

MBH.LA2	Heavy duty, load-bearing	Dry	Yellow	blue
MBH.HLS1	Load-bearing	Humid	Yellow, yellow	green
MBH.HLS2	Heavy duty, load-bearing	Humid	Yellow	green

The Euroclass fire performance rating under the conditions specified in BS EN 13986: 2004 for 9 mm untreated high-density mediumboard of 600 kg/m³ is Class D-s2, d0 for non-floor use and Class D_{FL}-s1 for floorings. For untreated low-density mediumboard of 400 kg/m³ the equivalent rating is Class E, pass for non-floor use and Class E_{FL} for floorings.

The standard BS EN 622–4: 2009 specifies five grades of softboard according to load-bearing properties and environmental conditions.

Grades of softboard:

Grade	Purpose/loading	Environmental conditions	Optional colour codes	
SB	General purpose	Dry	White, white	blue
SB.H	General purpose	Humid	White, white	green
SB.E	General purpose	Exterior	White, white	brown
SB.LS	Load-bearing	Dry	Yellow, yellow	blue
SB.HLS	Load-bearing	Humid	Yellow, yellow	green

The Euroclass fire performance rating under the conditions specified in BS EN 13986: 2004 for 9 mm untreated softboard of 250 kg/m³ is Class E, pass for non-floor use and Class E_{FL}-s1 for floorings.

WOOD WOOL SLABS

Manufacture

Wood wool slabs are manufactured by compressing long strands of chemically stabilised wood fibres coated in Portland cement (Fig. 4.35, page 167). The grey product has an open texture which may be left exposed, spray painted or used as an effective substrate for plastering. It is also a suitable material for permanent shuttering for concrete. Slabs are available in a range of thicknesses from 25–150 mm, typically 500, 600 or 625 mm wide and up to 3 m in length. Standard sizes are listed in BS EN 13168: 2012.

Types and uses

Wood wool slabs are available either plain-edged or with interlocking galvanised steel channels to the longitudinal edges. Thicknesses in the range 15–50 mm are suitable for ceilings, partitions, wall linings and permanent concrete shuttering. The thicker grades from 50–150 mm may be used for roof decking, with spans up to 3 m according to the anticipated loading. Some composite wood wool slabs incorporate a core or a single layer of additional insulation to enhance thermal properties.

The material is rated as Class 0 with respect to Building Regulations and Class I (BS 476–7: 1997) in terms of surface spread of flame. Classification to the European Standard BS EN 13501–1: 2007 is subject to manufacturer's testing. The material is resistant to fungal attack and is unaffected by wetting. Wood wool slabs offer good sound-absorption properties due to their open-textured surface. This is largely unaffected by the application of sprayed emulsion paint. The material is therefore appropriate for partitions, internal walls and ceilings where sound absorption is critical. Acoustic insulation for a pre-screeded 50 mm slab is typically 30 dB. The relatively high proportion of void space affords the material good thermal insulating properties with a typical thermal conductivity of 0.077 W/m K at 8% moisture content.

The material is workable, being easily cut and nailed. Where a gypsum plaster or cement/lime/sand external rendering is to be applied, all joints should be reinforced with scrim. Wood wool slabs form a suitable substrate for flat roofs finished with reinforced bitumen membranes, mastic asphalt or metals.

COMPRESSED STRAW SLABS

Compressed straw slabs are manufactured by forming straw under heat and pressure, followed by encapsulation in a fibreglass mesh and plastering grade paper (Fig. 4.35, page 167). Typically used for internal partitioning, the panels are mounted onto a timber sole plate and butt-jointed with adhesive or dry-jointed with proprietary sheradised clip fixings. All joints are jute scrimmed and the whole partitioning finished with a 3 mm skim of board plaster. The slabs are 58 mm thick by 1200 mm wide in a range of standard lengths from 2270–2400 mm. Service holes may be incorporated into the panels at 300 mm centres for vertical electrical wiring. While normal domestic fixtures can be fitted directly to the panels, heavier loads require coach-bolt fixings through the panels. The product should not be used where it will be subjected to moisture. Compressed straw slabs, when skim plastered, have a 30-minute fire resistance rating, a Class 0 spread of flame, and a sound reduction of typically 35 dB over the range 100–3200 Hz. The standard BS 4046: 1991 describes four types of compressed straw slabs depending on boron-based insecticide treatment and the provision of continuous longitudinal voids for services. Alternative finishes are plain paper for direct decoration or showerproof paper.

Types of compressed straw slabs:

Type A	Untreated	Solid core
Type B	Insecticide treated	Solid core
Type C	Untreated	Continuous voids in core
Type D	Insecticide treated	Continuous voids in core

THATCH

Thatch was the roof covering for most buildings until the end of the Middle Ages, and remained the norm in rural areas until the mid-nineteenth century. For most of the twentieth century thatch was only used in conservation work; however, with the new resurgence of interest in the material, partially associated with the reconstruction of the Globe Theatre in London (Fig. 4.37), thatch has once again become a current construction material.

Materials

The three standard materials for thatching within the UK are water reed (*Phragmites australis*), long straw (usually wheat) and combed wheat reed (also known as Devon reed). Water reed (Fig. 4.37) is associated with the Norfolk broads, the fens, south Hampshire and the Tay estuary, but much is imported from Turkey, Poland, Romania and China. Long straw is the standard thatch in the Midlands and Home Counties, while combed wheat reed is more common in Devon and Cornwall. Water reed is the most durable, lasting typically 50–60 years, but long straw and combed wheat reed last approximately 20–30 years, respectively, depending on location and roof pitch. All thatched roofs will need reridging at 10–15-year intervals; in the case of water reed this is often done with saw sedge (*Cladium mariscus*) which is more flexible than the reed itself.

Both long straw and combed wheat reed are often grown and harvested specifically as thatching materials to ensure long, undamaged stems. Long straw is

threshed winter wheat, whereas combed wheat straw is wheat with any leaves and the grain head removed. An alternative to combed wheat reed is triticale (*Triticale hexaploide*), which is a cross between wheat and rye. Triticale produces a more reliable harvest than other forms of wheat straw and it is indistinguishable from combed wheat reed when used as a thatching material. Water reed for thatching is usually between 915 and 1830 mm in length. Typical lengths for long straw and combed wheat reed are 760 and 915 mm, respectively.

Appearance

Long straw roofs show the lengths of the individual straws down the roof surface and are also characterised by the use of split hazel rodding around the eaves and gables to secure the thatch. To prevent attack by birds they are frequently covered in netting. Combed wheat reed and water reed both have a closely packed finish with the straw ends forming the roof surface. A pitch of about 50° is usual for thatch with a minimum of 45°, the steeper pitches being more durable. The ridge, which may be a decorative feature, is produced by either wrapping wheat straw over the apex or butting up reeds from both sides of the roof. Traditionally, hazel twigs are used for fixings, although these can be replaced with stainless steel wires. The durability of thatch is significantly affected by the climate. All materials tend to have shorter service lives in warmer locations with high humidity, which encourage the development of fungi. Chemical treatment, consisting of an organic heavy metal compound, may be used, preferably on new thatch, to delay the biological decomposition. Thatch is usually laid to a thickness of between 220 and 400 mm.

Properties

Fire

The fire hazards associated with thatched roofs are evident; however, fire retardants may be used, although these may denature the material. In the case of the Globe Theatre in London, a sparge water-spray system has been installed. In other new installations, the granting of planning consent has been facilitated by the location of permanent water-drenching systems near the ridge and by the use of fire-resisting board and foil under the thatch to prevent internal fire spread. Electrical wiring and open-fire chimneys are the most common causes of thatch fires, although maintenance work on thatched roofs is also a risk if not carefully managed.

Globe Theatre, London

Norfolk reed thatching

Fig. 4.37 Thatched roof – Globe Theatre, London and Norfolk reed thatching *Photographs*: Arthur Lyons

Insulation

Thatch offers good insulation, keeping buildings cool in summer and warm in winter, a typical 300 mm of water reed achieving a U-value of 0.35 W/m² K.

SHINGLES

Western red cedar (*Thuja plicata*), as a naturally durable material, is frequently used as shingles or shakes for roofing or cladding. Shingles are cut to shape, whereas shakes are split to the required thickness, usually between 10 and 13 mm. Both shakes and shingles may be tapered or straight. Shingles are typically 400, 450 or 600 mm long and between 75 and 355 mm wide. Shingles and shakes should be pressure treated with suitable preservatives to enhance their durability; additionally, they should be treated with flame retardant to satisfy the AA fire rating of BS 476–3: 1975. Shingles and shakes should be fixed with corrosion-resistant nails, leaving a spacing of 6–9 mm and 9–14 mm respectively between adjacent shingles and shakes. A minimum pitch of 14° is necessary and three layers are normally required. While the standard laying pattern is straight coursing, staggered patterns and the use of profiled shingles on steeper pitches can create decorative effects. Shingles must be laid over a waterproof breather membrane for low pitches between 14° and 22°. The standard BS 5534: 2003 suggests a minimum side lap of 38 mm for shakes and shingles.

Figure 4.38 illustrates the typical detailing of cedar shingles as a lightweight cladding material. The natural colour of the material gives an overall warmth to the exterior envelope of the building. In exposed locations the red-brown cedar wood surface gradually weathers to a silver grey, while in very sheltered locations the shingles become green with lichen. Decorative shape shingles include fish-scale, arrow, hexagonal, half cove, diagonal and round.

SPECIALIST PRODUCTS

Steko blocks

Steko® is an innovative wall system, which uses very accurately engineered large hollow timber blocks which slot together. No fixings or glue is needed to form load-bearing wall construction. The blocks are manufactured from small offcuts of spruce timber, which can be readily obtained from renewable sources. The blocks (640 × 320 × 160 mm wide), weighing 6.5 kg, are manufactured from two 20 mm-thick panels glued to

Fig. 4.38 Western red cedar shingles. *Photograph*: Arthur Lyons

horizontal battens and separated by vertical studs (Fig. 4.39). Units fit snugly together with their tongued and grooved profiles and vertical dowels. Special units include quarter, half and three-quarter blocks as well as components for corners, lintels, wall closers, base plates, wall plates and solid blocks for point loads. Walls can be up to 20 m long without additional bracing, and 3 m high, but up to four or five storeys high, if the necessary horizontal bracing is provided by the intermediate floors and roof. Where two-storey unrestrained walls are required they can be post-tensioned with threaded steel rods. The blocks are untreated except when borax protection against house longhorn beetle is required. Electrical services can run in the voids, which are finally dry-injected with cellulose insulation. *Steko* construction must work from a 300 mm upstand to avoid contact with surface water, and should be protected externally with a breather paper to prevent rain damage during construction. Internal finishes may be left exposed or finished with 15 mm gypsum plasterboard, which gives 30 minutes' fire resistance. Additional insulation is required externally to achieve the current Building Regulation requirements. Typically, 100 mm of mineral wool (λ = 0.035 W/m K) and a 20 mm render externally on cellulose filled and internally plastered blocks will achieve a U-value of 0.18 W/m² K. The first UK house built with the *Steko* system, which originated in Switzerland, is located on a cliff-top site in Downderry, Cornwall.

Brettstapel

Brettstapel, also known as 'Dowellam', is a solid wood construction fabricated with softwood posts connected

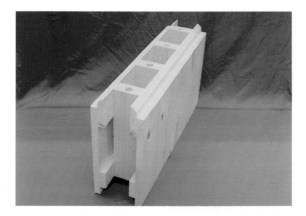

Fig. 4.39 'Steko' block. *Photograph:* Courtesy of Construction Resources

Fig. 4.40 Palm wood

only with hardwood dowels and not using nails or adhesive. The softwood posts of fir or spruce at 12–15% moisture content are fixed with perpendicular hardwood dowels (usually beech) at 8% moisture content. The moisture contents equalise, causing the hardwood to expand which locks the posts together. The potential for laminates to separate can be reduced by using diagonal dowels or cross-lamination of the timber layers. This form of construction is used particularly in Austria, Switzerland and Germany. Units up to 600 mm wide × 80–300 mm thick can span up to 8 m.

Flexible veneers

Flexible veneers are sheets of wood veneer which have been backed with paper or fibre reinforcement to allow the material to be handled without splitting. Flexible veneers can be moulded onto profiled components of MDF or plywood by a rolling process. Typical standard components are cornice and pelmet trims for kitchen furniture. The material can be rolled up for storage without damage, unlike traditional veneer. The veneer is fixed to the substrate with polyvinyl acetate (PVA) or urea formaldehyde (UF) adhesive under pressure or with contact adhesive. The material is increasingly being specified by designers for creating high-quality polished wood finishes to complex curved forms such as reception desks and wall panelling.

Decorative timbers

Palm and coconut wood as well as coconut shell are available as decorative materials for surface finishes including flooring. Palm wood is also being used to manufacture palm plywood, as illustrated in Figure 4.40. Bamboo is used in the interior of Madrid Barajas International Airport (Fig. 4.41) to create a relaxing atmosphere while harnessing a renewable resource. Potentially, bamboo may be used as a structural material as in South America, but durability and jointing systems need resolution.

Translucent timber panels

Translucent timber panels are manufactured from 5 mm laminates of wood interleaved with 0.25 mm glass fibres at 1.4 mm centres and bonded with high-strength adhesive. Panels of 3000 mm × 500, 1000 or 1250 mm dimensions range in thickness from 4 mm to 50 mm. Sunlight or artificial light can be transmitted, and graphics applied to the backs of the panels produce customised visual effects.

Recycling timber

According to the Timber Research and Development Association (TRADA), the UK generates approximately 4.1 million tonnes of timber waste annually of which less than 30% now enters landfill sites. This generates some of the greenhouse gas methane as it degrades. About half the total of timber waste arises from building construction and demolition. Poor handling and bad

Fig. 4.41 Undulating bamboo interior – Barajas Airport, Madrid. *Architects*: Rogers Stirk Harbour + Partners. *Photograph*: AENA / Manuel Renau

storage of new timber products on site leads to an average 10% wastage, but government initiatives such as Site Waste Management Plans have been developed to target this loss. Of the 2.8 million tonnes of timber waste which is recycled, nearly half is used to produce particleboard (chipboard) and a quarter is used as biomass fuel. The remainder is used largely for animal bedding and horticultural products.

Timber products such as particleboard with high quantities of polymer resin, and timber treated with some of the older hazardous wood preservatives such as creosote and chromated copper arsenate (CCA), cannot be recycled into new products. It is estimated that the quantity of timber waste contaminated with these hazardous preservatives will increase to the mid-twenty-first century as most of the wood will have a service life of 60 years or more. The pyrolysis of waste timber can produce biofuels, while anaerobic digestion will produce fuel for heat and electricity production. These technologies are currently under development.

References

FURTHER READING

BRE. 2000: *Handbook of hardwoods*, 2nd edn. Bracknell: IHS BRE Press.

Breyer, D., Fridley, K., Pollock, D. and Cobeen, K. 2007: *Design of wood structures*. New York: McGraw Hill Professional.

Brunskill, R.W. 2006: *Timber in building*. New Haven, CT: Yale University Press.

Dinwoodie, J.M. 2000: *Timber. Its nature and behaviour*, 2nd edn. London: E. & F.N. Spon.

English Heritage. 2012: *Timber (Practical building conservation)*. Farnham: Ashgate Publishing.

Fearn, J. 2004: *Thatch and thatching*. Princes Risborough: Shire Publications.

Hugues, T., Steiger, L. and Weber, J. 2004: *Detail practice. Timber construction: Details, products, case studies*. Basel: Birkhäuser.

Jayanetti, L. and Follett, P. 2000: *Timber pole construction*. London: ITDG Publishing.

Larsen, O.P., Tang, G. and Chilton, J. 2013: *Timber gridshell architecture*. Abingdon: Routledge.

Law, L.K.B. 2010: *Roundwood timber framing*. New York: Permanent.

Lefteri, C. 2005: *Materials for inspirational design*. East Sussex: Rotovision.

McKenzie, W.M.C. 2007: *Design of structural timber to EC5*. London: Palgrave.

Minke, G. 2012: *Building with bamboo. Design and technology of a sustainable architecture*. Basel: Birkhäuser.

Mueller, C. 2000: *Laminated timber construction*. Berlin: Birkhäuser.

Newman, R. 2005: *Oak-framed buildings*. Lewes: Guild of Master Craftsman Publications.

Porteous, J. and Kermani, A. 2013: *Structural timber design to Eurocode 5*, 2nd edn. Oxford: Wiley-Blackwell.

Porteous, J. and Ross, P. 2013: *Designers guide to Eurocode 5. Design of timber buildings*. London: ICE Publishing.

Pryce, W. 2005: *Architecture in wood, a world history*. London: Thames and Hudson.

Ruske, W. 2004: *Timber construction for trade, industry, administration*. Berlin: Birkhäuser.

Sanders, M. and Angold, R. 2012: *Thatches and thatching. A handbook for owners, thatchers and conservators*. Ramsbury: Crowood Press.

Scheer, J. 2004: *How to build with bamboo*. Utah: Gibbs Smith Publishers.

Seidel, F., Schleifer, S., Brooke, A. and Debard, C. 2008: *Architectural materials. Wood*. Cologne: Taschen GmbH.

Steurer, A. 2006: *Developments in timber engineering. The Swiss contribution*. Basel: Birkhäuser.

Thelandersson, S. and Larsen, H.J. 2003: *Timber engineering*. Chichester: John Wiley and Sons.

TRADA. 2007: *Green oak in construction*. High Wycombe: TRADA Technology Ltd.

TRADA. 2008: *Timber and the sustainable home*. High Wycombe: TRADA Technology Ltd.

TRADA. 2009: *Wood flooring*. High Wycombe: TRADA Technology Ltd.

TRADA. 2009: *Wood windows*. High Wycombe: TRADA Technology Ltd.

TRADA. 2009: *Timber in contemporary architecture. A designer's guide*. High Wycombe: TRADA Technology Ltd.

TRADA. 2010: *Off site and modern methods of timber construction*. High Wycombe: TRADA Technology Ltd.

TRADA. 2010: *Timber in refurbishment*. High Wycombe: TRADA Technology Ltd.

TRADA. 2011: *Timber frame construction*, 5th edn. High Wycombe: TRADA Technology Ltd.

TRADA. 2011: *Low energy timber frame buildings*. High Wycombe: TRADA Technology Ltd.

TRADA. 2012: *Concise illustrated guide to timber connections*. High Wycombe: TRADA Technology Ltd.

TRADA. 2012: *Innovative timber construction. New ways to achieve energy efficiency*. High Wycombe: TRADA Technology Ltd.

TRADA. 2012: *Getting started with Chain of Custody Certification for forest products*. High Wycombe: BM TRADA Group.

TRADA. 2013: *External timber cladding*, 3rd edn. High Wycombe: BM TRADA Group.

TRADA. 2013: *Getting started with Chain of Custody project certification*. High Wycombe: BM TRADA Group.

Villegas, M. 2003: *New bamboo architecture and design*. Villegas Asociados.

Wood Protection Association. 2010: *Use of CCA treated timber*. Publication W10. Castleford: The Wood Protection Association.

Wood Protection Association. 2010: *Use of creosote and creosote-treated timber*, 2nd edn. W11. Castleford: The Wood Protection Association.

Wood Protection Association. 2011: *Modified wood specification*. W24. Castleford: The Wood Protection Association.

Wood Protection Association. 2011: *Flame retardant specification manual*. Castleford: The Wood Protection Association.

Wood Protection Association. 2012: *Industrial wood preservation. Specification and practice*, 2nd edn. W16. Castleford: The Wood Protection Association.

Wood Protection Association. 2012: *Wood selection guide. A guide to selecting suitable wood and wood-based construction products*. Guidance Note 10/12. Castleford: The Wood Protection Association.

STANDARDS

BS 144: 1997
Specification for coal tar creosote for wood preservation.

BS 373: 1957
Methods for testing small clear specimens of timber.

BS 476
Fire tests on building materials and structures:
Parts 3, 4, 6, 7, 10–13, 15, 20–24, 31–33.

BS 644: 2012
Timber windows and doorsets. Fully finished factory-assembled windows and doorsets of various types. Specification.

BS 1088
Marine plywood:
Part 1: 2003   Requirements.
Part 2: 2003   Determination of bonding quality.

BS 1186
Timber for and workmanship in joinery:
Part 2: 1988   Specification for workmanship.
Part 3: 1990   Specification for wood trim and its fixing.

BS 1187: 1959
Wood blocks for floors.

BS 1203: 2001
Hot-setting phenolic and aminoplastic wood adhesives.

BS 1297: 1987
Specification for tongued and grooved softwood flooring.

BS 1336: 1971
Knotting.

BS 4046: 1991
Compressed straw building slabs.

BS 4050
  Specification for mosaic parquet panels:
    Part 1: 1977   General characteristics.
    Part 2: 1966   Classification and quality
                    requirements.
BS 4787
  Internal and external wood doorsets, door leaves
  and frames:
    Part 1: 1980   Specification for dimensional
                    requirements.
BS 4965: 1999
  Specification for decorative laminated plastics
  sheet veneered boards and panels.
BS 4978: 2007
  Visual strength grading of softwood. Specification.
BS 5277: 1976
  Doors. Measurement of defects of general flatness
  of door leaves.
BS 5278: 1976
  Doors. Measurement of dimensions and of defects
  of squareness of door leaves.
BS 5369: 1987
  Methods of testing doors; behaviour under
  humidity variations of door leaves placed in
  successive uniform climates.
BS 5395
  Stairs, ladders and walkways:
    Part 1: 2010   Code of practice for straight
                    stairs.
    Part 2: 1984   Code of practice for the design
                    of helical and spiral stairs.
    Part 3: 1985   Code of practice for the design
                    of industrial type stairs,
                    permanent ladders and
                    walkways.
    Part 4: 2011   Code of practice for the design
                    of stairs for limited access.
BS 5534: 2003
  Code of practice for slating and tiling (including
  shingles).
BS 5666
  Methods of analysis of wood preservatives and
  treated timber:
    Parts 2–7
BS 5707: 1997
  Specification for preparations of wood
  preservatives in organic solvents.
BS 5756: 2007
  Visual strength grading of hardwood.
  Specification.

BS 6100
  Building and civil engineering. Vocabulary:
    Part 0: 2010   Introduction and index.
    Part 8: 2007   Work with timber and wood-
                    based panels.
BS 6446: 1997
  Specification for manufacture of glued
  structural components of timber and wood
  based panels.
BS 8000–5: 1990
  Code of practice for carpentry, joinery and
  general fixings.
BS 8103
  Structural design of low-rise buildings:
    Part 3: 2009   Code of practice for timber
                    floors and roofs for housing.
BS 8201: 2011
  Code of practice for installation flooring of wood
  and wood-based panels.
BS 8417: 2011
  Preservation of timber. Code of practice.
pr BS ISO 12122
  Timber structures. Determination of characteristic
  values:
    Part 1: 2013   Basic requirements.
    Part 2: 2013   Sawn timber.
pr BS ISO 13910: 2013
  Timber structures. Strength graded timber. Test
  method for structural properties.
pr BS ISO 16598: 2014
  Timber structures. Structural classification for
  sawn timber.
pr BS ISO 16893: 2013
  Wood-based panels. Particleboard.
pr BS ISO 16895: 2013
  Wood-based panels. Dry-process fibreboard.
pr BS ISO 24294: 2012
  Round and sawn timber. Vocabulary.
BS EN 204: 2001
  Classification of thermoplastic wood adhesives
  for non-structural applications.
BS EN 300: 2006
  Oriented strand board (OSB). Definitions,
  classification and specifications.
BS EN 301: 2013
  Adhesives, phenolic and aminoplastic for
  loadbearing timber structures.
BS EN 309: 2005
  Wood particleboards. Definition and
  classification.

BS EN 311: 2002
Wood-based panels. Surface soundness.
Test method.

BS EN 312: 2010
Particleboards. Specifications.

BS EN 313
Plywood. Classification and terminology:
Part 1: 1996   Classification.
Part 2: 2000   Terminology.

BS EN 314
Plywood. Bonding quality:
Part 1: 2004   Test methods.
Part 2: 1993   Requirements.

BS EN 315: 2000
Plywood. Tolerances for dimensions.

BS EN 316: 2009
Wood fibreboards. Definition, classification and
symbols.

BS EN 317: 1993
Particleboards and fibreboards. Determination
of swelling in thickness after immersion in
water.

BS EN 318: 2002
Wood-based panels. Determination of
dimensional changes associated with changes in
relative humidity.

BS EN 319: 1993
Particleboards and fibreboards. Determination of
tensile strength perpendicular to the plane of the
board.

BS EN 320: 2011
Fibreboards. Determination of resistance to axial
withdrawal of screws.

BS EN 321: 2002
Wood-based panels. Determination of moisture
resistance.

BS EN 322: 1993
Wood-based panels. Determination of moisture
content.

BS EN 323: 1993
Wood-based panels. Determination of density.

BS EN 324
Wood-based panels. Determination of dimensions
of boards:
Part 1: 1993   Determination of thickness,
width and length.
Part 2: 1993   Determination of squareness
and edge straightness.

BS EN 325: 2012
Wood-based panels. Determination of dimensions
of test pieces.

BS EN 326
Wood-based panels. Sampling, cutting and
inspection:
Part 1: 1994   Sampling and cutting of test
pieces and expression of test
results.
Part 2: 2010   Initial type testing and factory
production control.
Part 3: 2003   Inspection of an isolated lot of
panels.

BS EN 330: 1993
Wood preservatives. Field test method for
determining the relative protective effectiveness of
a wood preservative for use under a coating.

BS EN 335: 2013
Durability of wood and wood-based products. Use
classes, definitions, application to solid wood and
wood-based products.

BS EN 336: 2013
Structural timber. Sizes, permissible deviations.

pr EN 338: 2013
Structural timber. Strength classes.

BS EN 350
Durability of wood and wood-based products.
Natural durability of solid wood:
Part 1: 1994   Guide to the principles of
testing and classification of the
natural durability of wood.
Part 2: 1994   Guide to the natural durability
and treatability of selected wood
species of importance in Europe.

BS EN 351
Durability of wood and wood-based products.
Preservative-treated solid wood:
Part 1: 2007   Classification of preservative
penetration and retention.
Part 2: 2007   Guidance on sampling for the
analysis of preservative-treated
wood.

BS EN 380: 1993
Timber structures. Test methods. General
principles for static load testing.

BS EN 382
Fibreboards. Determination of surface absorption:
Part 1: 1993   Test method for dry process
fibreboards.
Part 2: 1994   Test methods for hardboards.

BS EN 383: 2007
Timber structures. Test methods. Determination
of embedment strength and foundation values for
dowel type fasteners.

pr EN 384: 2013
Structural timber. Determination of characteristic values of mechanical properties and density.

BS EN 408: 2010
Timber structures. Structural timber and glued laminated timber. Determination of some physical and mechanical properties.

BS EN 409: 2009
Timber structures. Test methods. Determination of the yield moment of dowel type fasteners.

BS EN 460: 1994
Durability of wood and wood-based products. Natural durability of solid wood. Guide to the durability requirements for wood to be used in hazard classes.

BS EN 594: 2011
Timber structures. Test methods. Racking strength and stiffness of timber frame wall panels.

BS EN 595: 1995
Timber structures. Test methods. Test of trusses for the determination of strength and deformation behaviour.

BS EN 596: 1995
Timber structures. Test methods. Soft body impact test of timber framed walls.

BS EN 599
Durability of wood and wood-based products. Efficacy of wood preservatives as determined by biological tests:

Part 1: 2009	Specification according to use class.
Part 2: 1997	Classification and labelling.

BS EN 622
Fibreboards. Specifications:

Part 1: 2003	General requirements.
Part 2: 2004	Requirements for hardboards.
Part 3: 2004	Requirements for medium boards.
Part 4: 2009	Requirements for softboards.
Part 5: 2009	Requirements for dry process boards (MDF).

BS EN 633: 1994
Cement-bonded particleboards. Definition and classification.

BS EN 634
Cement-bonded particleboards. Specification:

Part 1: 1995	General requirements.
Part 2: 2007	Requirements for OPC bonded particleboards for use in dry, humid and exterior conditions.

BS EN 635
Plywood. Classification by surface appearance:

Part 1: 1995	General.
Part 2: 1995	Hardwood.
Part 3: 1995	Softwood.
DD CEN/TS	Part 4: 2007 Parameters of ability for finishing, guideline.
Part 5: 1999	Methods for measuring and expressing characteristics and defects.

BS EN 636: 2012
Plywood. Specifications.

BS EN 789: 2004
Timber structures. Test methods. Determination of mechanical properties of wood-based panels.

pr EN 839: 2013
Wood preservatives. Determination of the protective effectiveness against wood destroying basidiomycetes. Application by surface treatment.

BS EN 844
Round and sawn timber. Terminology:

Part 1: 1995	General terms common to round timber and sawn timber.
Part 2: 1997	General terms relating to round timber.
Part 3: 1995	General terms relating to sawn timber.
Part 4: 1997	General terms relating to moisture content.
Part 5: 1997	Terms relating to dimensions of round timber.
Part 6: 1997	Terms relating to dimensions of sawn timber.
Part 7: 1997	Terms relating to anatomical structure of timber.
Part 8: 1997	Terms relating to features of round timber.
Part 9: 1997	Terms relating to features of sawn timber.
Part 10: 1998	Terms relating to stain and fungal attack.
Part 11: 1998	Terms relating to degrade by insects.
Part 12: 2001	Additional terms and general.

BS EN 845
Specification for ancillary components for masonry:

Part 1: 2013	Ties, tension straps, hangers and brackets.

BS EN 912: 2011
Timber fasteners. Specifications for connectors for timber.

BS EN 942: 2007
Timber in joinery. General requirements.

BS EN 951: 1999
Door leaves. Method for measurement height, width, thickness and squareness.

BS EN 975
Sawn timber. Appearance grading of hardwoods:
Part 1: 2009   Oak and beech.
Part 2: 2004   Poplars.

BS EN 1001
Durability of wood and wood-based products. Terminology:
Part 1: 2005   List of equivalent terms.
Part 2: 2005   Vocabulary.

BS EN 1014
Wood preservatives. Creosote and creosoted timber:
Part 1: 2010   Procedure for sampling creosote.
Part 2: 2010   Procedure for obtaining a sample of creosote from creosoted timber.
Part 3: 2010   Determination of the benzo(a)pyrene content of creosote.
Part 4: 2010   Determination of the water-extractable phenols content of creosote.

BS EN 1026: 2000
Windows and doors. Air permeability.

BS EN 1027: 2000
Windows and doors. Watertightness.

BS EN 1058: 2009
Wood-based panels. Determination of characteristic 5-percentiles values and characteristic mean values.

BS EN 1087
Particleboards. Determination of moisture resistance:
Part 1: 1995   Boil test.

DD CEN/TS 1099: 2007
Plywood. Biological durability. Guidance for the assessment of plywood for use in different use classes.

BS EN 1128: 1996
Cement-bonded particleboards. Determination of hard body impact resistance.

BS EN ISO 1182: 2010
Reaction to fire tests for products. Non-combustibility test.

BS EN 1193: 1998
Timber structures. Structural timber and glued laminated timber. Determination of shear strength and mechanical properties.

BS EN 1194: 1999
Timber structures. Glued laminated timber. Strength classes and determination of characteristic values.

BS EN 1195: 1998
Timber structures. Performance of structural floor decking.

BS EN 1294: 2000
Door leaves. Determination of the behaviour under humidity variations.

BS EN 1309
Round and sawn timber. Method of measurement of dimensions:
Part 1: 1997   Sawn timber.
Part 2: 2006   Requirements for measurement and volume calculation rules.

BS EN 1310: 1997
Round and sawn timber. Method of measurement of features.

BS EN 1311: 1997
Round and sawn timber. Method of measurement of biological degrade.

BS EN 1312: 1997
Round and sawn timber. Determination of the batch volume of sawn timber.

BS EN 1313
Round and sawn timber. Permitted deviations and preferred sizes:
Part 1: 2010   Softwood sawn timber.
Part 2: 1999   Hardwood sawn timber.

BS EN 1315: 2010
Dimensional classification of round timber.

BS EN 1316
Hardwood round timber. Qualitative classification:
Part 1: 2012   Oak and beech.
Part 2: 2012   Poplar.
Part 3: 1998   Ash, maples and sycamore.

BS EN 1365–1: 2012
Fire resistance tests for loadbearing elements. Walls.

BS EN 1390: 2006
Wood preservatives. Determination of the eradicant action against Hylotrupes bajulus larvae.

BS EN 1611–1: 2000
  Sawn timber. Appearance grading of softwoods.
  European spruces, firs, pines and Douglas fir.
BS EN ISO 1716: 2010
  Reaction to fire tests for products. Determination
  of the gross heat of combustion.
BS EN 1912: 2012
  Structural timber. Strength classes. Assignment of
  visual grades and species.
BS EN 1927
  Qualitative classification of round timber:
    Part 1: 2008   Spruces and firs.
    Part 2: 2008   Pines.
    Part 3: 2008   Larches and Douglas fir.
BS EN 1995
  Eurocode 5: Design of timber structures:
    Part 1.1: 2004   Common rules and rules for
                      buildings.
    Part 1.2: 2004   Structural fire design.
BS EN ISO 11925–2: 2010
  Reaction to fire tests for building products.
  Ignitability of products subjected to direct
  impingement of flame. Single flame source test.
pr EN 12211: 2013
  Windows and doors. Resistance to wind load.
  Test method.
BS EN 12369
  Wood-based panels. Characteristic values for
  structural design:
    Part 1: 2001   OSB, particleboards and
                      fibreboards.
    Part 2: 2011   Plywood.
    Part 3: 2008   Solid-wood panels.
BS EN 12436: 2002
  Adhesives for load-bearing timber structures,
  Casein adhesives. Classification.
BS EN 12490: 2010
  Durability of wood and wood-based products.
  Preservative treated solid wood.
BS EN 12512: 2001
  Timber structures. Cyclic testing of joints made
  with mechanical fasteners.
BS EN 12871: 2013
  Wood-based panels. Determination of
  performance characteristics for load bearing
  panels for use in floors, walls and roofs.
DD CEN/TS 12872: 2007
  Wood-based panels. Guidance on the use of load-
  bearing boards in floors, walls and roofs.

BS EN 13017
  Solid wood panels. Classification by surface
  appearance:
    Part 1: 2001   Softwood.
    Part 2: 2001   Hardwood.
BS EN 13168: 2012
  Thermal insulation products for building. Factory
  made wood wool products. Specification.
BS EN 13183
  Moisture content of a piece of sawn timber:
    Part 1: 2002   Determination by oven dry
                      method.
    Part 2: 2002   Estimation by electrical
                      resistance method.
    Part 3: 2005   Estimation by capacitance
                      method.
BS EN 13226: 2009
  Wood flooring. Solid wood parquet floor elements
  with grooves and/or tongues.
BS EN 13228: 2011
  Wood flooring. Solid wood overlay flooring
  elements including blocks.
BS EN 13271: 2002
  Timber fasteners. Characteristic load-carrying
  capacities.
BS EN 13353: 2008
  Solid wood panels (SWP). Requirements.
BS EN 13501
  Fire classification of construction products and
  building elements:
    Part 1: 2007   Classification using test data
                      from reaction to fire tests.
    Part 2: 2007   Classification using data from
                      fire resistance tests.
BS EN 13556: 2003
  Round and sawn timber. Nomenclature of timbers
  used in Europe.
BS EN 13756: 2002
  Wood flooring. Terminology.
BS EN 13823: 2010
  Reaction to fire tests for building products.
  Building products excluding floorings exposed to
  the thermal attack by a single burning item.
BS EN 13986: 2004
  Wood-based panels for use in construction.
  Characteristics, evaluation of conformity and
  marking.
BS EN 13990: 2004
  Wood flooring. Solid softwood floorboards.

BS EN 13991: 2003
Derivatives from coal pyrolysis. Coal tar based oils. Creosotes. Specifications and test methods.

BS EN 14080: 2013
Timber structures. Glued laminated timber and glued solid timber. Requirements.

BS EN 14081
Timber structures. Strength graded timber with rectangular cross-section:

pr Part 1: 2013	General requirements.
Part 2: 2010	Machine grading. Additional requirements for initial type testing.
Part 3: 2012	Machine grading. Additional requirements for factory production control.
Part 4: 2009	Machine grading. Grading machine settings.

BS EN 14220: 2006
Timber and wood-based materials in external windows, external door leaves and external doorframes.

BS EN 14221: 2006
Timber and wood-based materials in internal windows, internal door leaves and internal doorframes.

BS EN 14250: 2010
Timber structures. Product requirements for prefabricated structural members assembled with punched metal plate fasteners.

BS EN 14272: 2011
Plywood. Calculation method for some mechanical properties.

BS EN 14279: 2004
Laminated veneer lumber (LVL). Definitions, classification and specification.

BS EN 14298: 2004
Sawn timber. Assessment of drying quality.

BS EN 14342: 2013
Wood flooring. Characteristics, evaluation of conformity and marking.

BS EN 14351–1: 2006
Windows and doors, product standard, performance characteristics.

BS EN 14374: 2004
Timber structures. Structural laminated veneer lumber. Requirements.

BS EN 14519: 2005
Solid softwood panelling and cladding.

BS EN 14545: 2008
Timber structures. Connectors. Requirements.

BS EN 14592: 2008
Timber structures. Dowel type fasteners.

pr EN 14732: 2011
Timber structures. Prefabricated wall, floor and roof elements. Requirements.

BS EN 14755: 2005
Extruded particleboard. Specifications.

BS EN 14761: 2006
Wood flooring. Solid wood parquet.

BS EN 14762: 2006
Wood flooring. Sampling procedures for evaluation of conformity.

BS EN 14915: 2013
Solid wood panelling and cladding. Characteristics, evaluation of conformity and marking.

BS EN 14951: 2006
Solid hardwood panelling and cladding. Machined profiles elements.

BS EN 14964: 2006
Rigid underlays for discontinuous roofing. Definitions and characteristics.

BS EN 15197: 2007
Wood-based panels. Flaxboards. Specifications.

BS EN 15228: 2009
Structural timber. Structural timber preservative treated against biological attack.

DD CEN/TS 15397: 2006
Wood preservatives. Method for natural preconditioning out of ground contact of treated wood specimens.

BS EN 15497: 2014
Structural finger jointed solid timber. Performance requirements and minimum production requirements.

BS EN 15534
Composites made from cellulose-based materials and thermoplastics:

Part 1: 2014	Test methods for characterisation of compounds and products.
DD CEN/TS	Part 2: 2007 Characterisation of WPC materials.
DD CEN/TS	Part 3: 2007 Characterisation of WPC products.
Part 4: 2014	Specifications for decking profiles and tiles.
Part 5: 2014	Specifications for cladding profiles and tiles.
pr Part 6: 2014	Specifications for fencing profiles and systems.

BS EN 15644: 2008
　Traditionally designed prefabricated stairs made of solid wood.
DD CEN/TS 15679: 2007
　Thermal modified timber. Definitions and characteristics.
DD CEN/TS 15680: 2007
　Prefabricated timber stairs. Mechanical test methods.
DD CEN/TS 15717: 2008
　Parquet flooring. General guidance for installation.
PD CEN/TS 15912: 2012
　Durability of reaction to fire performance. Classes of fire-retardant treated wood-based product.
pr EN 16351: 2011
　Timber structures. Cross laminated timber. Requirements.
PD CEN/TS 16368: 2014
　Lightweight particleboards. Specifications.
BS EN 16485: 2014
　Round and sawn timber. Environmental Product Declarations. Product category rules for wood and wood-based products for use in construction.
pr EN 16737: 2014
　Structural timber. Visual strength grading of tropical hardwood.
pr EN 16755: 2014
　Durability of reaction to fire performance. Classes of fire-retardant treated wood products in interior and exterior use applications.
BS EN 26891: 1991
　Timber structures. Joints made with mechanical fasteners.
BS EN 28970: 1991
　Timber structures. Testing of joints made with mechanical fasteners.
BIP 2198: 2012
　Concise Eurocode for the design of timber structures BS EN 1995.
PD 6693–1: 2012
　Recommendations for the design of timber structures to Eurocode 5.
PAS 111: 2012
　Specification for the requirements and test methods for processing waste wood.

REGULATIONS

Biocidal Products Regulation 2013.
Control of Substances Hazardous to Health 2002.

REACH – Registration, Evaluation, Authorisation & Restriction of Chemicals 2013.
Wildlife & Countryside Act 1981.

BUILDING RESEARCH ESTABLISHMENT PUBLICATIONS

BRE Special digests

BRE SD2: 2006
　Timber-frame dwellings. Conservation of fuel and power.
BRE SD6: 2008
　Timber-frame dwellings. Section six of the *Domestic Technical Handbook* (Scotland). Energy.
BRE SD7: 2008
　Insulation of timber-frame construction. U-values and regulations for the UK, Republic of Ireland and the Isle of Man.

BRE Digests

BRE Digest 416: 2014
　Specifying structural timber.
BRE Digest 445: 2000
　Advances in timber grading.
BRE Digest 470: 2002
　Life cycle impacts of timber.
BRE Digest 476: 2003
　Guide to machine strength grading of timber.
BRE Digest 477 Part 1: 2003
　Wood-based panels: Oriented strand board (OSB).
BRE Digest 477 Part 2: 2003
　Wood-based panels: Particleboard (chipboard).
BRE Digest 477 Part 3: 2003
　Wood-based panels: Cement-bonded particleboard.
BRE Digest 477 Part 4: 2004
　Wood-based panels: Plywood.
BRE Digest 477 Part 5: 2004
　Wood-based panels: Medium density fibreboard (MDF).
BRE Digest 477 Part 6: 2004
　Wood-based panels: Hardboard, medium board and softboard.
BRE Digest 477 Part 7: 2004
　Wood-based panels: Selection.
BRE Digest 479: 2003
　Timber piles and foundations.
BRE Digest 487 Part 4: 2004
　Structural fire engineering design: Materials behaviour. Timber.

BRE Digest 492: 2005
    Timber grading and scanning.
BRE Digest 494: 2005
    Using UK-grown Douglas fir and larch for
    external cladding.
BRE Digest 496: 2005
    Timber frame building.
BRE Digest 500: 2006
    Using UK-grown Sitka spruce for external
    cladding.
BRE Digest 503: 2007
    External timber structures. Preservative treatment
    and durability.
BRE Digest 504: 2007
    Modified wood. An introduction to products in
    UK construction.
BRE Digest 514: 2010
    Drying distortion of timber.
BRE Digest 517: 2010
    Assessment of timber structures.
BRE Digest 521: 2011
    Timber cladding.

BRE Good building guides

BRE GBG 32: 1999
    Ventilating thatched roofs.
BRE GBG 60: 2004
    Timber frame construction. An introduction.

BRE Information papers

BRE IP 2/01
    Evaluating joinery preservatives.
BRE IP 13/01
    Preservative-treated timber – the UK's code of
    best practice.
BRE IP 14/01
    Durability of timber in ground contact.
BRE IP 1/03
    European Standards for wood preservatives and
    treated wood.
BRE IP 9/03
    Best practice of timber waste management.
BRE IP 13/04
    An introduction to building with Structural
    Insulated Panels.
BRE IP 10/05
    Green gluing of timber. A feasibility study.
BRE IP 13/05
    Incising UK-grown Sitka spruce.

BRE IP 3/07
    Modern methods of construction (MMC) in
    housing (Parts 1–4).
BRE IP 17/11
    Cross-laminated timber. An introduction to low-
    impact building materials.
BRE IP 16/12
    Service life prediction for timber cladding.
BRE IP 17/12
    Innovation in timber supply at the London 2012
    Olympic and Paralympic Games.
BRE IP 3/14
    British grown Douglas fir. Growth rate and
    density relating to visual grading and strength
    class attribution.

BRE Report

BR 453: 2003
    Recognising wood rot and insect damage in
    buildings, 3rd edn.

TRADA PUBLICATIONS

Wood information sheets

WIS 0–3: 2012
    Introduction to timber framed construction.
WIS 0–5: 2013
    Timber frame building. Material specification.
WIS 1–6: 2012
    Glued laminated timber.
WIS 1–17: 2010
    Structural use of hardwoods.
WIS 1–29: 2013
    Trussed rafters.
WIS 1–37: 2012
    Introduction to Eurocode 5.
WIS 1–42: 2012
    Timber I-joists applications and design.
WIS 1–46: 2013
    Decorative timber flooring.
WIS 1–47:2011
    Timber external doors.
WIS 1–48: 2013
    Sole plates in timber frame construction.
WIS 1–49: 2014
    Cladding for timber frame buildings.
WIS 1–50: 2011
    Timber cladding for building refurbishment.

WIS 2/3–1: 2012
> Finishes for external timber.

WIS 2/3–3: 2013
> Flame-retardant treatments for timber products.

WIS 2/3–10: 2002
> Timbers – their properties and uses.

WIS 2/3–11: 2013
> Specification and use of wood-based panels in external situations.

WIS 2/3–16: 2012
> Preservative treatment for timber. A guide to specification.

WIS 2/3–23: 2012
> Introduction to wood-based panel products.

WIS 2/3–28: 2012
> Introducing wood.

WIS 2/3–31: 2014
> Adhesives for structural use.

WIS 2/3–32: 2011
> Fungi and insect pests in timber.

WIS 2/3–33: 2005
> Wood preservation – chemicals and processes.

WIS 2/3–36: 2013
> Design of structural timber connections.

WIS 2/3–37: 2005
> Softwood sizes.

WIS 2/3–51: 2011
> Timber engineering hardware and connectors.

WIS 2/3–52: 2012
> Fasteners for structural timber – nails, staples screws, dowels and bolts.

WIS 2/3 56: 2012
> CE marking. Implications for timber products.

WIS 2/3–57: 2013
> Specifying wood-based panels for structural use.

WIS 2/3–58: 2013
> Sustainable timber sourcing.

WIS 2/3–59: 2012
> Recovering and minimising wood waste.

WIS 2/3–60: 2011
> Specifying timber exposed to weathering.

WIS 2/3–61: 2011
> Cross laminated timber. Information for specifiers.

WIS 2/3–62: 2011
> Cross laminated timber. Structural principles.

WIS 2/3–63: 2013
> Modified wood products.

WIS 2/3–65: 2011
> Principles of green oak construction.

WIS 2/3–67: 2013
> Specifying British-grown timbers.

WIS 4–7: 2011
> Timber strength grading and strength classes.

WIS 4–11: 2014
> Wood-based panel products and timber in fire.

WIS 4–12: 2012
> Care of timber and wood-based products on building sites.

WIS 4–14: 2011
> Moisture in timber.

WIS 4–16: 2012
> Timber in joinery.

WIS 4–17: 2011
> Insect pests in houses.

WIS 4–22: 2013
> Adhesive-bonded repair systems for structural timber.

WIS 4–28: 2012
> Durability by design.

WIS 4–30: 2012
> Fire performance of timber framed houses.

WIS 4–31: 2011
> Life cycle costing.

WIS 4–32: 2013
> Acoustic performance in residential timber frame developments.

WIS 4–33: 2012
> Life cycle assessment.

ADVISORY ORGANISATIONS

British Woodworking Federation, The Building Centre, 26 Store Street, London WC1E 7BT (0844 209 2610).

Coed Cymru, The Old Sawmill, Tregynon, Newtown, Powys SY16 3PL (01686 650777).

Glued Laminated Timber Association, Chiltern House, Stocking Lane, Hughenden Valley, High Wycombe, Buckinghamshire HP14 4ND (01494 565180).

Thatching Advisory Services, 8–10 Queen Street, Seaton, Devon EX12 2NY (08455 204060).

Timber Research and Development Association, Stocking Lane, Hughenden Valley, High Wycombe, Buckinghamshire HP14 4ND (01494 569600).

Trussed Rafter Association, The Building Centre, 26 Store Street, London WC1E 7BT (0203 205 0032).

UK Timber Frame Association Ltd, The e-Centre, Cooperage Way, Business Village, Alloa FK10 3LP (01259 272140).

Wood for Good Ltd, 211 High Road, London N2 8AN (0208 365 2700).

Wood Panel Industries Federation, Autumn Park Business Centre, Dysart Road, Grantham, Lincolnshire NG31 7EU (01467 512381).

5

FERROUS AND NON-FERROUS METALS

CONTENTS

Introduction

A wide range of ferrous and non-ferrous metals and their alloys are used within the construction industry, but iron, steel, aluminium, copper, lead and zinc predominate. Over the past decades titanium has featured significantly in construction, having previously been used mainly in the chemical process industry and for military purposes. Recent trends have been towards the development of more durable alloys and the use of coatings to both protect and give visual diversity to the product ranges. Generally, the metals require a large energy input for their production from raw materials; however, this high embodied energy is partially offset by the long life and recycling of most metals. The recovery rate of steel from demolition sites in the UK is 99% for structural steelwork and 94% for all steel construction products, with 10% being reused and the remainder recycled. Approximately 60% of current steel production is from scrap, and steel can be recycled any number of times without any degradation of the material. Current steel production generates 2.15 tonnes of carbon dioxide per tonne of steel, but the commercial aim is to reduce CO_2 emissions by 20% from 1997 levels by 2020. A revolutionary cyclone converter which produces iron directly from coal and iron ore without their pre-treatment is under development and should reduce emissions in the production of iron by 20%.

According to the Construction Products Regulation (CPR) all fabricated structural steelwork delivered to site must be C€ marked from July 2014. This applies to all elements made from C€ marked materials which are covered by a harmonised standard or a European Technical Assessment. The mandatory C€ marking of all constituent construction products came into force in 2013.

Ferrous metals

Ferrous metals are defined as those in which the element iron predominates. The earliest use of the metal was for the manufacture of implements and weapons in the Iron Age commencing in Europe *circa* 1200 BC. Significant developments were the use by Wren in 1675 of a wrought iron chain in tension to restrain the outward thrust from the dome of St Paul's Cathedral, the use of cast iron in compression for the Ironbridge at Coalbrookdale in 1779, and by Paxton in the prefabricated sections of the Crystal Palace in 1851. Steel is a relatively recent material, only being available

Fig. 5.1 Structural steelwork – King's Cross Terminal, London. *Architects*: John McAslan + Partners. *Photographs*: Arthur Lyons

in quantity following the development of the Bessemer converter in the late nineteenth century. The first steel-frame high-rise building of 10 storeys was built in 1885 in Chicago by William le Baron Jenney.

The elegance of steel is illustrated by the redevelopment of the King's Cross rail terminal in London (Fig. 5.1). Designed by John McAslan and Partners, it creates a large open concourse with improved passenger facilities and retail outlets. The fan-shaped lattice of welded tubular steel forms a semicircular light and airy space focusing on the central cluster of columns.

MANUFACTURE OF STEEL

The production of steel involves a sequence of operations which are closely interrelated in order to

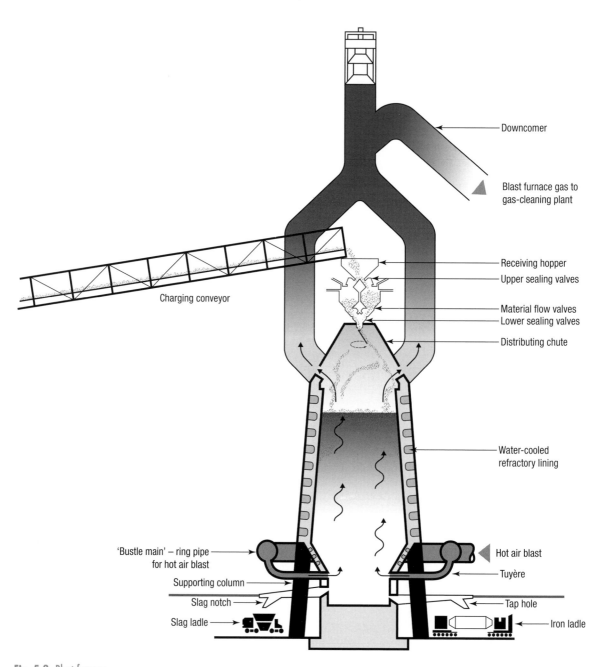

Fig. 5.2 Blast furnace

Fig. 5.3 Structural steelwork – London Olympic Stadium. *Architects*: Populous. *Photograph*: Courtesy of London Legacy Development Corporation

ensure maximum efficiency of a highly energy-intensive process. The key stages in the production process are the manufacture of pig iron, its conversion into steel, the casting of the molten steel and its formation into sections or strips. Finally, coils of steel strip are cold-rolled into thin sections and profiled sheet.

Manufacture of pig iron

The raw materials for the production of iron are iron ore, coke and limestone. Most iron ore is imported from America, Australia and Scandinavia, where the iron content of the ore is high. Other large world producers are China, Brazil, India and Russia. Coke is produced from coking coal, mainly imported from Europe, in batteries of coking ovens. Some of this coke is then sintered with iron ore prior to the iron-making process.

Iron ore, coke, sinter and limestone are charged into the top of the blast furnace (Fig. 5.2). A hot air blast, sometimes enriched with oxygen, is fed through the tuyères into the base of the furnace. This heats the furnace to white heat, converting the coke into carbon monoxide, which then reduces the iron oxide to iron. The molten metal collects at the bottom of the furnace. The limestone forms a liquid slag, floating on the surface of the molten iron. Purification occurs as impurities within the molten iron are preferentially absorbed into the slag layer.

$$2C \quad + \quad O_2 \quad \longrightarrow \quad 2CO$$
carbon (coke)   oxygen   carbon monoxide

$$Fe_2O_3 \quad + \quad 3CO \quad \longrightarrow \quad 2Fe \quad + \quad 3CO_2$$
iron ore   carbon   iron   carbon
(hematite)   monoxide   dioxide

The whole process is continuous, as relining the blast furnace with the special refractory bricks is expensive and time-consuming. From time to time, as the molten slag level rises, excess is tapped off for subsequent disposal as a by-product of the steel-making industry. When *hot metal* is required for the subsequent steel-making process it is tapped off into huge ladles for direct transportation to the steel converter. At this stage the iron is only 90–95% pure with sulphur, phosphorus, manganese and silicon as impurities and a carbon content of 4–5%. Waste gases from the blast furnace are cleaned and recycled as fuel within the plant. A blast furnace will typically operate non-stop for 10 years, producing 40,000 tonnes per week.

Steel making

There are two standard processes used within the UK for making steel. The basic oxygen process is used for the manufacture of bulk quantities of standard-grade steels, and the electric arc furnace process is used for the production of high-quality special steels and particularly stainless steel. The London 2012 Olympic Stadium (Fig. 5.3) used approximately 10,700 tonnes of structural steel. The 80,000-capacity athletics stadium, designed by Populous and Buro Happold, was built to be scaled down after the games to a 25,000-seat arena for football and other sports.

Basic oxygen process
Bulk quantities of steel are produced by the basic oxygen process in a refractory lined steel furnace which can be tilted for charging and tapping. A typical furnace (Fig. 5.4) will take a charge of 350 tonnes and convert it into steel within 30 minutes. Initially, scrap metal, accounting for a quarter of the charge, is loaded into the tilted furnace, followed by the remainder of the charge as hot metal direct from the blast furnace. A water-cooled lance is then lowered to blow high-pressure oxygen into the converter. This burns off impurities and reduces the excess carbon content while raising the temperature. Argon and a small quantity of nitrogen are introduced at the bottom of the furnace. Lime is added to form a floating slag to remove further impurities and alloying components are added to adjust the steel composition, prior to tapping. Finally, the furnace is inverted to run out any remaining slag, prior to the next cycle.

Electric arc process
The electric arc furnace (Fig. 5.5) consists of a refractory-lined hearth, covered by a removable roof, through which graphite electrodes can be raised and lowered. With the roof swung open, scrap metal is charged into the furnace, the roof is closed and the electrodes lowered to near the surface of the metal. A powerful electric arc is struck between the electrodes and the metal, which heats it up to melting point. Lime and fluorspar are added to form a slag, and oxygen is blown into the furnace to complete the purification process. When the temperature and chemical analysis are correct, the furnace is tilted to tap off the metal, to which appropriate alloying components may then be added. A typical furnace will produce 150 tonnes of high-grade or stainless steel within 90 minutes.

Casting

Traditionally, the molten steel was cast into ingots, prior to hot rolling into slabs and then sheets. However, most steel is now directly poured, or teemed, and cast into continuous billets or slabs, which are then cut to appropriate lengths for subsequent processing. Continuous casting (Fig. 5.6), which saves on reheating, is not only more energy efficient than processing through the ingot stage but also produces a better surface finish to the steel.

Hot-rolled steel

Sheet steel is produced by passing 25-tonne hot slabs at approximately 1250°C through a series of computer-controlled rollers which reduce the thickness to typically between 1.5 and 20 mm prior to water cooling

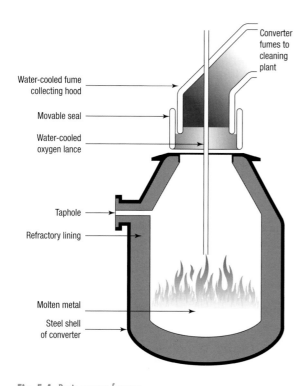

Converter fumes to cleaning plant

Water-cooled fume collecting hood

Movable seal

Water-cooled oxygen lance

Taphole

Refractory lining

Molten metal

Steel shell of converter

Fig. 5.4 Basic oxygen furnace

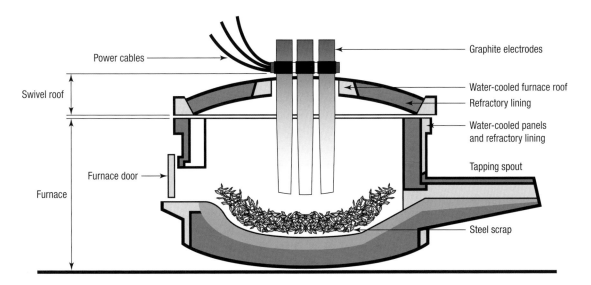

Fig. 5.5 Electric arc furnace

and coiling. A 25-tonne slab would produce 1 km coil of 2 mm sheet. Steel sections such as universal beams and columns, channels and angle (Fig. 5.7) are rolled from hot billets through a series of *stands* to the appropriate section.

Cold-rolled steel

Sheet steel may be further reduced by cold rolling, which gives a good surface finish and increases its tensile strength. Light round sections may be processed into steel for concrete reinforcement, while coiled sheet may be converted into profiled sheet or light steel sections (Fig. 5.7). Cold-reduced steel for construction is frequently factory finished with zinc, alloys including terne (lead and tin) or plastic coating. Cold-reduced sheet steel of structural quality is covered by pr BS ISO 4997: 2014.

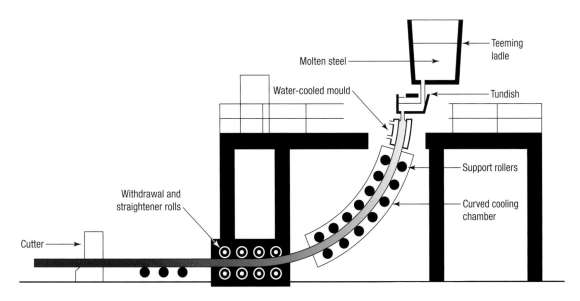

Fig. 5.6 Continuous casting

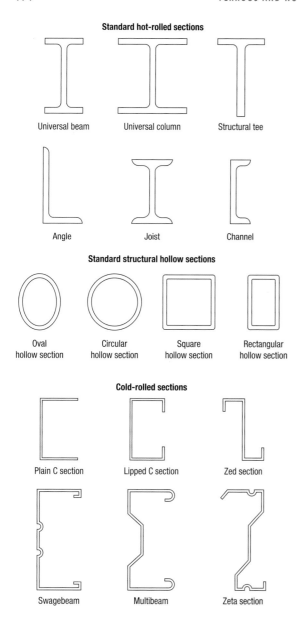

Fig. 5.7 Hot-rolled and cold-rolled sections (after Trebilcock, P.J. 1994: *Building design using cold formed steel sections: an architect's guide.* Steel Construction Institute)

CARBON CONTENT OF FERROUS METALS

The quantity of carbon alloyed with iron has a profound influence on the physical properties of the metal due to its significant effect on the microscopic crystal structure (Fig. 5.8). At ambient temperature a series of crystal forms (ferrite, pearlite and cementite) associated with different proportions of iron and carbon are stable. However, upon increasing the temperature, crystal forms that were stable under ambient conditions become unstable and are recrystallised into the high-temperature form (austenite). This latter crystal structure may be trapped at room temperature by the rapid quenching of red-hot steel, thus partially or completely preventing the natural recrystallisation processes which would otherwise occur on slow cooling. These effects are exploited within the various heat treatments that are applied to steels in order to widen the available range of physical properties.

Wrought iron

Wrought iron contains only about 0.02% of carbon. It was traditionally made by remelting and oxidising pig iron in a reverberatory furnace. The process was continued until virtually all the high carbon content of the pig iron had been burnt off to produce a pasty wrought iron, which was withdrawn from the furnace and then hammered out. Wrought iron is fibrous in character due to the incidental incorporation within the metal of slag residues and impurities such as magnesium sulphide, which are formed into long veins by the hammering process. Wrought iron has a high

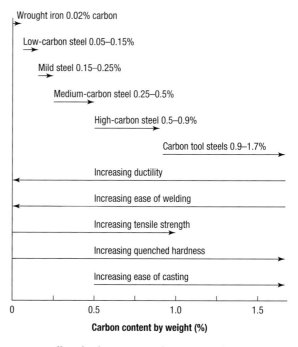

Fig. 5.8 Effect of carbon content on the properties of wrought iron and steels

melting point, approaching 1540°C, depending upon its purity. It was traditionally used for components in tension due to its tensile strength of about 350 MPa. It is ductile and easily worked or forged when red-hot, and thus eminently suitable for crafting into ornamental ironwork, an appropriate use owing to its greater resistance to corrosion than steel. Because of its high melting point, wrought iron cannot be welded or cast. Production ceased in the UK in 1974 and modern wrought iron is either recycled old material or, more frequently, low-carbon steel, with its attendant corrosion problems.

Cast iron

Cast iron contains in excess of 2% carbon in iron. It is manufactured by the carbonising of pig iron and scrap with coke in a furnace. The low melting point of around 1130°C and its high fluidity when molten give rise to its excellent casting properties but, unlike wrought iron, it cannot be hot worked and is generally a brittle material. The corrosion resistance of cast iron has been exploited in its use for boiler castings, street furniture and traditional rainwater goods. Modern foundries manufacture castings to new designs and as reproduction Victorian and Edwardian components.

Differing grades of cast iron are associated with different microscopic crystal structures. The common grey cast iron contains flakes of graphite, which cause the characteristic brittleness and impart the grey colour to fractured surfaces. White cast iron contains the carbon as crystals of cementite (iron carbide, Fe_3C) formed by rapid cooling of the melt. This material may be annealed to reduce its brittle character. A more ductile cast iron (spheroidal cast iron) is produced by the addition of magnesium and ferrosilicon and annealing which causes the carbon to crystallise into graphite nodules. This material has an increased tensile strength and significantly greater impact resistance. All cast irons are strong in compression. The designation system for cast irons is described in BS EN 1560: 2011.

Road iron goods, such as manhole covers, made from largely recycled grey cast iron, are heavy but brittle. Where increased impact resistance is required for public roads, lighter and stronger ductile iron components are used. Traditional sand-cast rainwater goods are usually manufactured from grey cast iron, while cast iron drainage systems are manufactured from both grey and ductile iron. Unlike steel, cast iron does not soften prematurely in a fire, but may crack if cooled too quickly with water from a fire hose. Cast iron

drainage systems in both grey and spheroidal cast iron are covered by the standard BS EN 877: 1999. Cast iron drainage systems are particularly appropriate in heritage and conservation areas.

Steels

A wide range of steels are commercially available, reflecting the differing properties associated with carbon content, the various heat treatments and the addition of alloying components.

Carbon contents of steels range typically between 0.05% and 1.7%, and this alone is reflected in a wide spectrum of physical properties. The low-carbon (0.05–0.15%) and mild steels (0.15–0.25%) are relatively soft and may be subjected to extensive cold working. Medium-carbon steels (0.25–0.5%), which are often heat treated, are hard-wearing. High-carbon steels (0.5–0.9%) and carbon tool steels (0.9–1.7%) exhibit increasing strength and wear resistance with increasing carbon content. In addition, extra-low-carbon steel (<0.02%) and ultra-low-carbon steel (<0.01%), with carbon contents similar to traditional wrought iron, are used for high formability and drawing applications.

HEAT TREATMENT OF STEELS

The physical properties of steels may be modified by various heat treatments which involve heating to a particular temperature followed by cooling under controlled conditions. The full range of heat treatments is described in BS EN 10052: 1994. Construction steels for the manufacture of component parts requiring quenching and tempering (+QT) or normalising (+N) are covered by BS EN 10343: 2009.

Hardening

Rapidly quenched steel, cooled quickly from a high temperature in oil or water, thus retaining the high-temperature crystalline form, is hard and brittle. This effect becomes more pronounced for the higher carbon content steels, which are mostly unsuitable for engineering purposes in this state. Quenching is often followed by tempering to reduce excessive hardness and brittleness.

Annealing and normalising

These processes involve softening of the hard steel by recrystallisation, which relieves internal stresses within the material and produces a more uniform grain structure. For annealing, the steel is reheated and soaked at a temperature of approximately 650°C, then cooled slowly at a controlled rate within a furnace or cooling pit. This produces the softest steel for a given composition. With normalising, the steel is reheated to 830–930°C for a shorter period and then allowed to cool more rapidly in air. This facilitates subsequent cold working and machining processes.

Tempering

Reheating the steel to a moderate temperature (400–600°C), followed by cooling in air, reduces the brittleness, by allowing some recrystallisation of the metal. The magnitude of the effect is directly related to the tempering temperature with ductility increasing, and tensile strength reducing, for the higher process temperatures.

Carburising

Components may be case hardened to produce a higher carbon content on the outer surface while leaving the core relatively soft, thus giving a hard-wearing surface without embrittlement and loss of impact resistance to the centre. Usually, this process involves heating the components surrounded by charcoal or other carbon-based material to approximately 900°C for several hours. The components are then heat treated to develop fully the surface hardness.

SPECIFICATION OF STEELS

Hot-rolled structural steels within the European Union are designated by a series of European Standards, pr EN 10025: 2011.

Hot-rolled structural steels:

pr EN 10025 -1: 2011	General technical data
pr EN 10025 -2: 2011	Non-alloy structural steels
pr EN 10025 -3: 2011	Normalised weldable fine-grained structural steels
pr EN 10025 -4: 2011	Thermomechanical-rolled weldable fine grained structural steels
pr EN 10025 -5: 2011	Steels with improved atmospheric corrosion resistance
pr EN 10025 -6: 2011	Flat products of high-yield strength structural steels in the quenched and tempered condition

In addition, BS EN 10210–1: 2006 and BS EN 10219–1: 2006 relate to hot- and cold-formed structural hollow sections, respectively. The standard grades and their associated characteristic strengths are illustrated in Tables 5.1–5.5. In the standards, S refers to structural steel and the subsequent coding numbers relate to the minimum yield strength. The sub-grade letters refer to impact resistance and other production conditions and compositions, such as W for weather-resistant steel. Steel numbers for each grade of steel are defined by pr BS EN 10027–2: 2013.

The following example illustrates the two coding systems for one standard grade of steel:

S275JR (BS EN 10027–1: 2005)
1.0044 (pr EN 10027–2: 2013)

S275JR	S refers to structural steel. The yield strength is 275 MPa. J is the lower impact strength at room temperature R.
1.0044	The first digit is the material group number with steel 1. The second pair of digits is the steel group number with 00 referring to a non-alloy base steel. The final digits refer to the particular grade of non-alloy steel.

STRUCTURAL STEELS

Weldable structural steels, as used in the Wembley Stadium, London (Fig. 5.9), have a carbon content within the range 0.16–0.25%. Structural steels are usually normalised by natural cooling in air after hot rolling. The considerable size effect, which causes the larger sections to cool more slowly than the thinner sections, gives rise to significant differences in physical properties; thus, an 80 mm section may typically have a 10% lower yield strength compared to a 16 mm section of the same steel. While grade S275 had previously been considered to be the standard-grade structural steel and is still used for most small beams, flats and angles, the higher grade S355 is increasingly being used for larger beams, columns and hollow sections.

Table 5.1 Designations for standard grade steels to pr EN 10025–2: 2011 (hot rolled products of non-alloy structural steels)

Designation		Properties	
BS EN 10027–1: 2005 and pr EN 10027–2 : 2013		pr EN 10025–2: 2011 limits	
Grade	Number	Tensile strength (MPa)	Minimum yield strength (MPa)
S235JR	1.0038	360–510	235
S235J0	1.0114	360–510	235
S235J2	1.0117	360–510	235
S275JR	1.0044	410–560	275
S275J0	1.0143	410–560	275
S275J2	1.0145	410–560	275
S355JR	1.0045	470–630	355
S355J0	1.0553	470–630	355
S355J2	1.0577	470–630	355
S355K2	1.0596	470–630	355
S460JR		550–720	460
S460J0		550–720	460
S460J2		550–720	460
S460K2		550–720	460
S500J0		580–760	500

Notes:
- Sub-grades JR, J0, J2 and K2 indicate increasing impact resistance as measured by the Charpy V-notch test.
- K has a higher impact energy than J, the symbols R, 0 and 2 refer to the impact test at room temperature, 0°C and −20°C respectively.
- Minimum yield strength data is for thicknesses of 16 mm or less.
- Tensile strength data is for thicknesses between 3 and 100 mm.

Table 5.2 Designations for higher grade structural steels to pr EN 10025–3: 2011 (normalised rolled weldable fine grain structural steels)

Designation		Properties	
BS EN 10027–1: 2005 and pr EN 10027–2 : 2013		pr EN 10025–3 : 2011 limits	
Grade	Number	Tensile strength (MPa)	Minimum yield (strength MPa)
S275N	1.0490	370–510	275
S275NL	1.0491	370–510	275
S355N	1.0545	470–630	355
S355NL	1.0546	470–630	355
S420N	1.8902	520–680	420
S420NL	1.8912	520–680	420
S460N	1.8901	550–720	460
S460NL	1.8903	550–720	460

Notes:
- Sub-grade N (normalised or normalised rolled) relates to the physical state of the steel and L (low temperature impact) to high impact resistance.
- Minimum yield strength data is for thicknesses of 16 mm or less.
- Tensile strength data is for thicknesses of 100 mm or less.

Table 5.3 Designations for higher grade steels to pr EN 10025—4: 2011 (thermomechanical rolled weldable fine grain structural steels)

Designation		Properties	
BS EN 10027—1: 2005 and pr BS EN 10027—2: 2013		pr EN 10025—4 : 2011 limits	
Grade	Number	Tensile strength (MPa)	Minimum yield (strength MPa)
S275M	1.8818	370—530	275
S275ML	1.8819	370—530	275
S355M	1.8823	470—630	355
S355ML	1.8834	470—630	355
S420M	1.8825	520—680	420
S420ML	1.8836	520—680	420
S460M	1.8827	540—720	460
S460ML	1.8838	540—720	460
S500M		580—760	500

Notes:

- Sub-grade M (thermomechanical rolled) relates to the physical state of the steel and L (low temperature impact) to high-impact resistance.
- Minimum yield strength data is for thicknesses of 16 mm or less.
- Tensile strength data is for thicknesses of 40 mm or less.

Table 5.4 Designations for structural steels with improved atmospheric corrosion resistance grades to pr EN 10025—5: 2011

Designation		Properties	
BS EN 10027—1: 2005 and pr EN 10027—2: 2013		pr EN 10025—5: 2011 limits	
Grade	Number	Tensile strength (MPa)	Minimum yield strength (MPa)
S235J0W	1.8958	360—510	235
S235J2W	1.8961	360—510	235
S355J0WP	1.8945	470—630	355
S355J2WP	1.8946	470—630	355
S355J0W	1.8959	470—630	355
S355J2W	1.8965	470—630	355
S355K2W	1.8967	470—630	355
S355J4W		470—630	355
S420J0W		500—660	420
S420J2W		500—660	420
S420J4W		500—660	420
S460J0W		530—710	460
S460J2W		530—710	460
S460J4W		530—710	460

Notes:

- Sub-grades J0, J2, J4 and K2 indicate increasing impact resistance, respectively.
- The symbols 0, 2 and 4 refer to the impact test at 0°C, −20°C and −40°C respectively.
- Sub-grade W refers to weather resistant steel.
- P indicates a high phosphorus grade.
- Minimum yield strength data is for thicknesses of 16 mm or less.
- Tensile strength data is for thicknesses between 3 and 100 mm.

Table 5.5 Designations for high-yield strength quenched and tempered structural steels to pr EN 10025–6: 2011

Designation		Properties	
BS EN 10027–1: 2005 and pr EN 10027–2: 2013		pr EN 10025–6: 2011 limits	
Grade	Number	Tensile strength (MPa)	Minimum yield strength (MPa)
S460Q	1.8908	550–720	460
S460QL	1.8906	550–720	460
S460QL1	1.8916	550–720	460
S500Q	1.8924	590–770	500
S500QL	1.8909	590–770	500
S500QL1	1.8984	590–770	500
S550Q	1.8904	640–820	550
S550QL	1.8926	640–820	550
S550QL1	1.8986	640–820	550
S620Q	1.8914	700–890	620
S620QL	1.8927	700–890	620
S620QL1	1.8987	700–890	620
S690Q	1.8931	770–940	690
S690QL	1.8928	770–940	690
S690QL1	1.8988	770–940	690
S890Q	1.8940	940–1100	890
S890QL	1.8983	940–1100	890
S890QL1	1.8925	940–1100	890
S960Q	1.8941	980–1150	960
S960QL	1.8933	980–1150	960

Notes:
- Q indicates quenched steel.
- L indicates steel may be specified with a minimum impact energy at temperatures not below −40°C.
- L1 indicates steel may be specified with a minimum impact energy at temperatures not below −60°C.
- Data is for thicknesses between 3 mm and 50 mm.

Standard sections

Standard steel sections include flat plate, I-sections, channels, angles, universal columns and beams, as well as cellular and castellated beams. Long-span cellular and plated beams spanning 8–24 m as floor beams and up to 60 m as roof rafters are routine customised production.

Hollow sections

Circular, oval, semi-elliptical, square and rectangular hollow sections are usually made from flat sections which are progressively bent until almost round. They are then passed through a high-frequency induction coil to raise the edges to fusion temperature, when they are forced together to complete the tube. Excess metal is removed from the surface. The whole tube may then be reheated to normalising temperature (850–950°C), and hot rolled into circular, oval, rectangular or square sections. For smaller sizes, the tube is heated to 950–1050°C, and stretch reduced to appropriate dimensions. The standard steel grades to BS EN 10210–1: 2006 are S275J2H and S355J2H (Table 5.6). Cold-formed hollow sections differ in material characteristics from the hot-finished sections and conform to BS EN 10219: 2006. The lowest grade S235, with a minimum yield strength of 235 MPa, is imported, but the standard non-alloy grades are S275 and S355. Grades S420 and S460 are designated as alloy special steels (Table 5.7).

Fig. 5.9 Structural steelwork – Wembley Stadium, London. *Architects*: Foster + Partners. *Photograph*: Arthur Lyons

Table 5.6 Steel designations for hot finished structural hollow sections to BS EN 10210: 2006 (hot finished structural sections of non-alloy and fine grain structural steels)

Designation		Properties	
BS EN 10027–1: 2005 and pr EN 10027–2 : 2013		BS EN 10210–1: 2006 limits	
Grade	Number	Tensile strength (MPa)	Minimum yield strength (MPa)
S235JRH	1.0039	360–510	235
S275J0H	1.0149	410–560	275
S275J2H	1.0138	410–560	275
S355J0H	1.0547	470–630	355
S355J2H	1.0576	470–630	355
S355K2H	1.0512	470–630	355
S275NH	1.0493	370–510	275
S275NLH	1.0497	370–510	275
S355NH	1.0539	470–630	355
S355NLH	1.0549	470–630	355
S420NH	1.8750	520–680	420
S420NLH	1.8751	520–680	420
S460NH	1.8953	540–720	460
S460NLH	1.8956	540–720	460

Notes:
– H refers to hollow sections.
– Sub-grades JR, J0 and J2 indicate impact resistance at room temperature, 0°C and -20°C, respectively.
– K2 refers to higher impact energy than J2.
– Sub-grade N (normalised or normalised rolled) relates to the physical state of the steel and L (low temperature impact) to high-impact resistance.
– The standard UK production grades are the S275J2H and S355J2H designations.
– Data is for thicknesses between 3 mm and 16/40 mm.

For larger hollow sections, the rotary forge process is used to produce seamless tubes. A hot tapered ingot is pierced by a hydraulic ram, and the central void is opened up by the action of rollers and a rotating mandrel. The steel subsequently passes through a series of eccentric rollers which elongate the tube, reducing the section to the required dimensions.

Bending of structural sections

Castellated beams, rolled, hollow and other sections may be bent into curved forms by specialist metal-bending companies. The minimum radius achievable depends on the metallurgical properties, thickness and the cross-section. Generally, smaller sections may be curved to smaller radii than the larger sections, although for a given cross-section size the heavier-gauge sections may be bent to smaller radii than the thinner-gauge sections. Normally, universal sections may be bent to tighter radii than hollow sections of the same dimensions. Elegant structures, such as Merchants Bridge, Manchester (Fig. 5.10), can be produced with curved standard sections and also curved tapered beams. The cold-bending process work hardens the steel, but without significant loss of performance within the elastic range appropriate to structural steelwork. Tolerances on units may be as low

Table 5.7 Steel designations for cold-formed structural hollow sections to BS EN 10219–1: 2006 (cold-formed welded structural hollow sections of non-alloy and fine grain steels)

Designation		Properties	
BS EN 10027–1: 2005 and pr EN 10027–2: 2013		BS EN 10219–1: 2006 limits	
Grade	Number	Tensile strength (MPa)	Minimum yield strength (MPa)
S235JRH	1.0039	360–510	235
S275J0H	1.0149	410–560	275
S275J2H	1.0138	410–560	275
S355J0H	1.0547	470–630	355
S355J2H	1.0576	470–630	355
S355K2H	1.0512	470–630	355
S275NH	1.0493	370–510	275
S275NLH	1.0497	370–510	275
S355NH	1.0539	470–630	355
S355NLH	1.0549	470–630	355
S460NH	1.8953	540–720	460
S460NLH	1.8956	540–720	460
S275MH	1.8843	360–510	275
S275MLH	1.8844	360–510	275
S355MH	1.8845	450–610	355
S355MLH	1.8846	450–610	355
S420MH	1.8847	500–660	420
S420MLH	1.8848	500–660	420
S460MH	1.8849	530–720	460
S460MLH	1.8850	530–720	460

Notes:
- H refers to hollow sections.
- Sub-grades JR, J0 and J2 indicate impact resistance at room temperature, 0°C and −20°C, respectively.
- K2 refers to higher impact energy than J2.
- Sub-grades M (thermomechanical rolled) and N (normalised or normalised rolled) relate to the physical state of the steel and L (low temperature impact) to high-impact resistance.
- The standard UK production grades are the S275J2H and S355J2H designations.
- Data is for thicknesses between 3 mm and 16/40 mm.

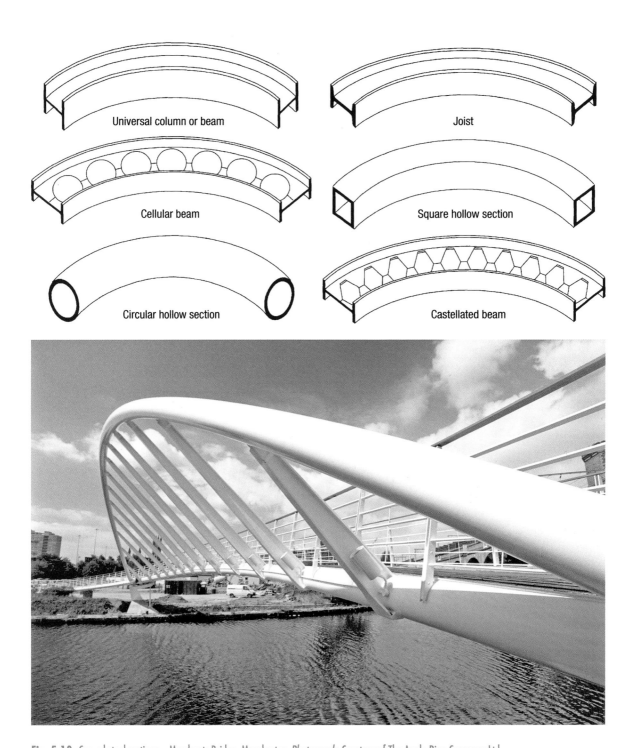

Universal column or beam

Joist

Cellular beam

Square hollow section

Circular hollow section

Castellated beam

Fig. 5.10 Curved steel sections — Merchants Bridge, Manchester. *Photograph:* Courtesy of The Angle Ring Company Ltd

as ± 2 mm, with multiple bends, reverse curvatures and bends into three dimensions all possible. Increasingly, cold bending is replacing induction or hot bending which requires subsequent heat treatment to regain the initial steel properties.

Cast sections

Components such as the nodes for rectangular and circular hollow-section constructions and large pin-joint units are manufactured directly as individual castings. They may then be welded to the standard milled steel sections to give continuity of structure. Structural steels appropriate for casting are listed in the standard BS EN 10340: 2007. Alloy steels for casting may be subjected to quenching and tempering, normalising or annealing according to the specific component requirements.

Steel frame construction

The use of light steel frame construction is an alternative to timber-frame technology. Most light framing is manufactured from cold-formed galvanised steel channel sections, factory assembled, ready for subsequent bolting together on site. Panel units are typically 600 mm wide, storey-height modules, lined with a vapour check layer and plasterboard. This modern method of construction (MMC) considerably reduces the time on site compared to building traditional load-bearing masonry.

Bi-steel

Bi-steel panels consist of two steel plates held apart by an array of welded steel bar connectors (Fig. 5.11). The panels are usually assembled into larger modules for delivery to site, where they are erected and the void space filled with concrete. The combination of permanent steel formwork and concrete fill acts as reinforced concrete, with the steel providing resistance to in-plane and bending forces, and the concrete offering resistance to compression and shear. Units are manufactured up to 2 m wide and 18 m long in S275 or S355 steel to thicknesses of between 200 and 700 mm and may be flat or curved. Adjacent panels may be bolted or fixed with proprietary connectors, giving fast erection times on site.

FIRE PROTECTION OF STRUCTURAL STEEL

The two approaches to the design of fire precautions in buildings rely on periods of fire resistance as described within the Building Regulations Approved Document B and in BS 9999: 2008, or according to a risk profile based on occupancy, fire growth rate, ventilation conditions and building geometry also described in BS 9999: 2008. This standard includes consideration of more complex spaces such as auditoria and atria as well as less sophisticated building configurations.

The fire protection of structural steel may be approached either by the traditional method involving the application of insulation materials with standard

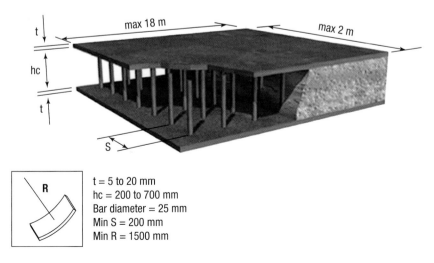

t = 5 to 20 mm
hc = 200 to 700 mm
Bar diameter = 25 mm
Min S = 200 mm
Min R = 1500 mm

Fig. 5.11 Bi-steel unit. *Photograph:* Courtesy of Tata Steel

fire resistance periods (Fig. 5.12) or by a structural fire engineering method, which predicts the potential rate of rise of temperature of exposed steel members in each situation, based on the calculated fire load and particular exposure of the steel.

Applied protection to structural steel

Intumescent coatings

Thin film intumescent coatings, which do not seriously affect the aesthetic of exposed structural steelwork, offer up to 120 minutes' fire protection. Typical expansion rates are 50:1, such that a 1 mm thick coating will expand to about 50 mm when affected by fire. A full colour range for application by spray, brush or roller may be used on steel and also for remedial work on old cast iron or wrought iron structures. The three standards of finish usually specified are basic, normal decorative finish and high decorative finish. The normal finish has a slight orange peel effect, whereas the high decorative finish is smooth and even. Thin film intumescent coatings account for 75% of the market share in fire protection for steel-framed buildings. The standard pr BS EN 16623: 2013 relates to reactive coatings for fire protection.

Sprayed coatings

Sprayed coatings are either low density based on mineral fibre cement or higher density based on vermiculite cement. The materials may be applied directly to steel to give up to 240 minutes' fire protection. The process is particularly appropriate for structural steel in ceiling voids, where the over-spray onto other materials is less critical. The finish, which may be adjusted to the required thickness, is heavily textured, and the products are relatively cheap.

Boarded systems

Boxed boards around steel sections offer between 30 and 240 minutes' fire protection according to their thicknesses. Products are generally based on vermiculite or mineral fibres within cement, calcium silicate or gypsum binders. Boarded systems are screwed directly onto the structural steel, to light-gauge steel fixings or to a box configuration. Lightweight boards are generally not suitable for decorative finishes. Heavyweight boards are either pre-coated or decorated in situ.

Preformed casings

Preformed sheet-steel casings, which encase lightweight vermiculite plaster, give a high-quality appearance and

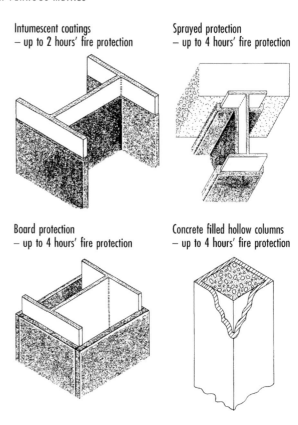

Intumescent coatings
– up to 2 hours' fire protection

Sprayed protection
– up to 4 hours' fire protection

Board protection
– up to 4 hours' fire protection

Concrete filled hollow columns
– up to 4 hours' fire protection

Fig. 5.12 Structural steelwork, typical fire-protection systems

up to 240 minutes' fire resistance. The calculated fire resistance is based solely on the thickness of insulation and does not take into account any additional protection afforded by the sheet steel.

Masonry and concrete

Structural steel may be fully encased with masonry or suitably reinforced lightweight concrete in which non-spalling aggregates should be used. Hollow steel columns may be filled with plain, fibre-reinforced or bar-reinforced concrete to give up to 120 minutes' fire resistance. For plain or fibre-reinforced concrete a minimum section of 140 × 140 mm or 100 × 200mm is required and 200 × 200 mm or 150 × 250 mm for bar-reinforced concrete filling.

Water-filled systems

Interconnecting hollow steel sections can be given fire protection by filling with water as part of a gravity feed or pumped system. Water loss is automatically replaced from a tank, where corrosion inhibitor and anti-freeze agents are added to the system as appropriate. However, this system is rarely used in modern construction.

Fire engineering

The heating rate of a structural steel section within a fire depends upon the severity of the fire and the degree of exposure of the steel. Where a steel section has a low 'Section Factor' (i.e. surface area per unit length A/volume per unit length V (A/V); [formerly known as the surface/cross-sectional area (Hp/A) ratio] (Fig. 5.13)), its temperature will rise at a slower rate than a section with a high A/V or Hp/A ratio. Performance-based fire engineering solutions calculate the severity of a potential fire based on the enclosure fire loads, ventilation rates and thermal characteristics, and then predict temperature rises within the structural steel based on exposure. The stability of the structural member may therefore be predicted, taking into consideration its steel grade, loading and any structural restraint. From these calculations it may be determined whether additional fire protection is required and at what level to give the required fire resistance period. Section factors range from 25 m^{-1} for very large sections to 300 m^{-1} for small, slender sections.

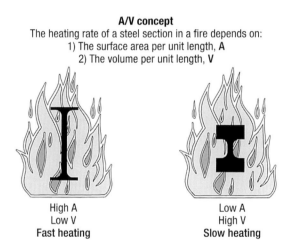

A/V concept
The heating rate of a steel section in a fire depends on:
1) The surface area per unit length, **A**
2) The volume per unit length, **V**

High A	Low A
Low V	High V
Fast heating	**Slow heating**

Fig. 5.13 A/V ratios and rates of heating in fire

Depending on the particular circumstances, a fully loaded unprotected column with a section factor (A/V) of less than 50 m^{-1} may offer 30 minutes' structural fire resistance; similarly, lighter columns with lightweight concrete blocks in the web can achieve 30 minutes' fire resistance. Shelf angle floors of suitable section, and in which a high proportion of the steel is encased by the concrete floor construction, can achieve 60 minutes' fire resistance (Fig. 5.14).

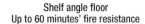
Shelf angle floor
Up to 60 minutes' fire resistance

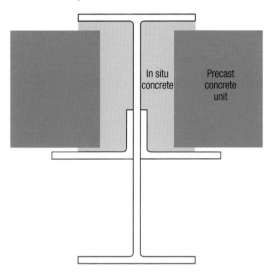
In situ concrete

Precast concrete unit

'Slimflor' beam
Up to 60 minutes' fire resistance

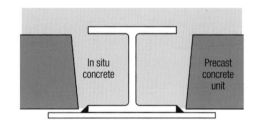
In situ concrete

Precast concrete unit

Column with blocked in web
Up to 30 minutes' fire resistance

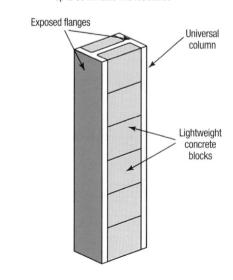

Exposed flanges

Universal column

Lightweight concrete blocks

Fig. 5.14 Fire resistance of structural-steel systems

STEEL PRODUCTS

Profiled steel sheeting

The majority of profiled sheet steel is produced by shaping the pre-coated strip through a set of rollers, which gradually produce the desired section without damage to the applied coating. The continuous profiled sheet is then cut and packaged to customer requirements. The standard sections have a regular trapezoidal profile, with the depth of the section dependent on the loading and required span (Fig. 5.15). In cases where there is the risk of buckling, stiffeners are incorporated into the profile. Curved profiled sheets for eaves and soffits are manufactured by brake pressing from the same coated strip. Trapezoidal profiles may be crimped in this process, although sinusoidal sheets and shallow trapezoidal sections may be curved without this effect. The rigidity of curved sections reduces their flexibility and thus the tolerances of these components.

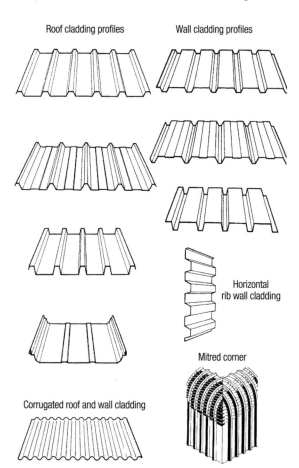

Fig. 5.15 Typical profiles for sheet-steel roofing and cladding

Proprietary spring-clip fixings may be used when concealed fixings are required for certain profiled sheet sections. The standard BS EN 14782: 2006 requires a minimum thickness of 0.4 mm (excluding any coating) for self-supporting profiled sheet steel in construction.

Steel cables

Steel cables are manufactured by drawing annealed thin steel rods through a series of lubricated and tapered tungsten carbide dies, producing up to a 10-fold elongation. The drawing process increases the strength and reduces the ductility of the steel; thus, the higher carbon steels, required for the production of high tensile wires, need special heat treatment before they are sufficiently ductile for the sequence of drawing processes.

In order to manufacture steel cables for suspended structures or pre-stressed concrete, a set of individual wires are twisted into a strand, then a series of strands are woven around a central core of steel or fibre strand to produce a rope. A series of ropes are then woven to produce cable to the required specifications. Typical configurations use 7 or 19 strands or ropes to form heavy-duty cables. Stainless steel (grade 1.4401) may be used for exposed structures; however, for bridge suspension cables the higher grades of steel are necessary.

Perforated steel sheets, steel meshes and nets

Perforated steel sheets are manufactured in mild steel, galvanised steel and stainless steel for use in architectural features, sunshades, balustrades, as well as wall and ceiling panels. Perforated sheets, also available in aluminium, copper, brass and bronze, may have round, square or slotted holes within a wide range of sizes and spacings to produce the desired aesthetic effect. Metal sheets are either punched or plasma profile cut.

Stainless steel meshes are available as flexible or rigid self-supporting weaves, each in a wide variety of patterns offering choice for use as external façades and sun-screening, as well as internal space dividers, balustrades, wall coverings and suspended ceilings. Patterns range from traditional weave and expanded metal to chain-mail, with a wide variation in texture and transparency. Some patterns are available in mild steel and also in non-ferrous metals. Stainless steel netting may be used creatively to form open tent and canopy structures.

Ferrous alloys

WEATHERING STEEL

Weathering steels are structural steels which have been alloyed with small proportions of copper, usually between 0.25 and 0.55%, together with silicon, manganese, chromium and phosphorus. Additional minor constituents are combinations of aluminium, niobium, vanadium and titanium (BS 7668: 2004 and pr EN 10025–5: 2011). The alloying has the effect of making the naturally formed brown rust coating adhere tenaciously to the surface, thus preventing further loss by spalling. All weathering steel must be carefully detailed to ensure that the rainwater run-off does not impinge upon other materials, particularly concrete or glass where it will cause severe staining during the first few years of exposure to the elements. *Cor-Ten* is the commercial name for weathering steels. Table 5.4 (page 198) gives the steel specification and steel number to the European Standards. Weathering steels are used for structural and cladding applications as well as for sculptural works of art (Fig. 5.16).

STAINLESS STEELS

Stainless steels are a range of alloys containing at least 10.5% chromium. The corrosion resistance of the material is due to the natural passive film of chromium oxide which immediately forms over the material in the presence of oxygen; thus, if the surface is subsequently

Fig. 5.16 Corten weathering steel – Copenhagen. *Photograph*: Arthur Lyons

Table 5.8 Stainless steel compositions and grades to pr EN 10088–1: 2011 for different environmental conditions

Designation		Suitable environments	
Type	Name (indicating composition of alloying components)	Number	
Austenitic	X5CrNi18–10	1.4301	Rural and clean urban
	X5CrNiMo17–12–2	1.4401	Urban, industrial and marine
	X2CrNiMo18–14–3	1.4435	Marine
Ferritic	X6Cr17	1.4016	Interior
Duplex	X2CrNiMoN22-5-3	1.4462	Severe industrial and marine

Notes:
– Cr, Ni, Mo and N refer to chromium, nickel, molybdenum and nitrogen, respectively.
– X2, X5 and X6 refer to the carbon contents of 0.02, 0.05 and 0.06%, respectively.

scratched or damaged the protective film naturally re-forms. The corrosion resistance is increased by the inclusion of nickel and molybdenum as additional alloying components. The standard (austenitic) grades used within construction are 18% chromium, 10% nickel (1.4301), and 17% chromium, 12% nickel, 2.5% molybdenum (1.4401). The 18/10 alloy is suitable for use in rural and lightly polluted urban sites, while the 17/12/2.5 higher specification alloy is more appropriate for use within normal urban, marine and industrial environments. The marine environment of the façade to the Copenhagen Opera House required the use of Grade 1.4435 stainless steel. The elegant cantilevered steel roof (Fig. 5.17) is constructed as a closed steel box, typical of modern steel bridge design. For certain aggressive environments, the high-alloy (duplex) stain-less steel (number 1.4462) should be used. Ferritic stainless steel (1.4016) containing only chromium, with a reduced corrosion resistance, is appropriate for internal building use where corrosion is a less critical factor. Standard grades to pr EN 10088–1: 2011 for stainless steels are given in Table 5.8. Extended listings of grades for construction are detailed in BS EN 10088–4: 2009 for flat products and in BS EN 10088–5: 2009 for sections, bars and rods.

Stainless steel is manufactured by a three-stage process. Scrap is melted in an electric arc furnace, then refined in an argon-oxygen decarburiser and alloyed to the required composition in a ladle furnace by the addition of the minor constituents. Most molten metal is continuously cast into billets or slabs for subsequent forming. Stainless steel is hot rolled into plate, bar and

Fig. 5.17 Stainless steel façade – Copenhagen Opera House. *Architects*: Henning Larsens Architects. *Photograph*: Arthur Lyons

Fig. 5.18 Stainless steel finishes – Courtesy of Rimex Metals Group. *Photograph*: Arthur Lyons

sheet, while thin sections may be cold rolled. Heavy universal sections are made up from plates. Stainless steel may be cast or welded and is readily formed into small components such as fixings and architectural ironmongery. Polished, brushed, matt, patterned and profiled finishes are available (Fig. 5.18); additionally, the natural oxide film may be permanently coloured by chemical and cathodic treatment to bronze, blue, gold, red, purple or green according to its final thickness. Approximately 50% of stainless steel derives from recycled scrap steel.

Stainless steel is available in square, rectangular, oval and circular hollow sections as well as the standard sections for structural work. Its durability is illustrated by the Lloyd's Building (Fig. 5.19), which maintains its high-quality finish within the urban context of the City of London. Stainless steel is widely used for roofing, cladding, interior and exterior trim owing to its combined strength, low maintenance and visual impact. Stainless steels for self-supporting profile sheet and tile roofing systems are listed in BS EN 508–3: 2008. A minimum thickness of 0.4 mm is specified in BS EN 14782: 2006 and long strip roofing to 15 m is possible due to the fatigue resistance and moderate coefficient of expansion. The standard also lists optional organic finishes including polyester, silicone-modified polyester, polyurethane, polyvinylidene fluoride and PVC plastisol. The corrosion resistance of stainless steel also makes it eminently suitable for masonry fixings, such as corbels, anchor bolts, cavity wall ties and for concrete reinforcement. Austenitic stainless steels are used for the manufacture of pipework, catering and drainage products where durability and corrosion resistance are critical. Exposed exterior stainless steel should be washed regularly to retain its surface characteristics, particularly where a brushed rather than polished finish has been used. Pitting corrosion causing surface pin-point attack, crevice corrosion under tight-fitting washers and stress corrosion cracking, where the material is under high tensile load, may occur where inappropriate grades are used in aggressive and marine environments.

High-performance (superduplex) stainless steel (1.4507) wire ropes, which have 50% more tensile strength than standard (austenitic) stainless steel (1.4401), are appropriate for architectural support and restraint systems. These alloys containing around 1.6% copper have a lower elastic stretch than standard stainless steels and a higher fatigue resistance, which makes them appropriate for architectural tensile elements including those within marine and swimming pool environments.

Fig. 5.19 Stainless steel construction – Lloyd's Building, London. *Architects*: Richard Rogers. *Photograph*: Arthur Lyons

In addition to organic coatings, stainless steel may be coated with terne or electroplated with $\geqslant 10$ g/m² tin to BS EN 508–3: 2008 (see tin and terne-coated steel, page 212).

HEAT-TREATED STEELS

The size effect, which causes a reduced yield strength in large sections due to their slower cooling rates than the equivalent thin sections, may be ameliorated by the addition of small quantities of alloying elements such as chromium, manganese, molybdenum and nickel.

Coated steels

To inhibit corrosion, steel may be coated with metallic or organic finishes. Metallic finishes are typically zinc (Z), zinc-iron alloy (ZF), zinc-aluminium alloy (ZA), zinc-magnesium alloy (ZM), aluminium-zinc alloy (AZ), aluminium-silicon-alloy (AS) and aluminium (A), all of which may be applied by hot-dipping of steel into the molten metal (Table 5.9). These metallic coatings, except for pure aluminium, are covered by the standard pr EN 10346: 2013. Aluminium-coated steel is described in BS EN 505: 2013. The standard BS EN ISO 14713–1: 2009 indicates an anticipated annual corrosion loss of between 0.7 and 2 μm of zinc per year from galvanised steel in urban and coastal UK locations. Organic coatings (BS EN 10169: 2010) may be divided into liquid paints, powder coatings and films. Certain products are suitable only for interior applications.

Table 5.9 Typical examples for zinc- and aluminium-based coatings on steel

Coating	Designation	Coating mass (g/m^2) (total mass)	Nominal thickness (μm) (on each face)
Zinc	Z275	275	20
Zinc-iron	ZF120	120	8
Zinc-aluminium	ZA255	255	20
Zinc-magnesium	ZM250	250	19
Aluminium	A195	195	32
Aluminium-zinc	AZ150	150	20
Aluminium-silicon	AS100	100	17

Notes:
- The standard pr EN 10346: 2013 gives a range of coating thicknesses for coatings other than aluminium.
- The standard BS EN 505: 2013 refers to aluminium coated steel.

METAL-COATED STEEL

Zinc-coated steel

The zinc coating of steel has for many years been a standard method for its protection against corrosion. The zinc coating may be applied by hot-dipping or spraying with the molten metal, sheradising in heated zinc powder (BS EN 13811: 2003) or electrolytically (ZE) to BS EN 10152: 2009. In hot-dip galvanising the steel is cleaned by pickling in acid followed by immersion in molten zinc at 450°C. The zinc coating (Z) protects the steel by acting as a physical barrier between the steel and its environment, and also by sacrificially protecting the steel where it is exposed by cutting or surface damage. If annealed, the pure zinc coating becomes alloyed from the substrate steel producing an alloy of 8–12% iron (ZF) which gives a better surface for painting or welding.

The durability of the coated steel is dependent on the thickness of the coating and the environment. Coating designations to pr EN 10346: 2013 (Table 5.9) indicate the total mass and nominal thickness of applied metallic coatings.

Coastal situations and industrial environments with high concentrations of salt and sulphur dioxide, respectively, may cause rapid deterioration. The alkalis in wet cement, mortar and plaster etch zinc coatings, but once dry, corrosion is slow; however, calcium chloride used as an accelerator in plaster is aggressive and should only be used sparingly. Fixings for zinc-coated sheet should be carefully chosen to avoid the formation of bimetallic couples, which can cause accelerated corrosion. In particular, no copper or brass should make contact with either zinc or iron-zinc alloy-coated steel. Other metals, such as lead, aluminium and stainless steel, have less serious effects in clean atmospheres, but generally all fixings should be sealed and insulated by rubber-faced washers. Where zinc-coated steel is to be fixed to unseasoned timber or timber impregnated with copper-based preservatives, the wood should be coated with bitumen paint. Where damaged in cutting, fixing or welding, the zinc coating should be repaired with the application of zinc-rich paint.

Zinc-coated steel may be painted for decoration or improved corrosion resistance. However, the *normal spangle* zinc finish will show through paint and the *minimised spangle* or iron-zinc alloy uniform matt grey finish is more appropriate for subsequent painting.

Aluminium-zinc alloy-coated steel

The two distinct zinc/aluminium alloys used for coating steel contain 5% (ZA) and 55% (AZ) aluminium. The ZA coating to BS ISO 14788: 2011 is usually a substrate for an organic finish. For profile sheet metal roofing to BS EN 508–1: 2008, the minimum coating mass totals are 255 and 185 g/m^2, respectively, compared to 350 g/m^2 for pure zinc. Steel coated with the AZ alloy of aluminium (55%), zinc (43.4%) and silicon (1.6%) is more durable than that coated with an equivalent thickness of pure zinc, and may be used without further protection in non-aggressive environments. It is also used as the substrate for certain organic coatings. The finish is a metallic lustre due to crystal formation.

Aluminium-coated steel

Hot-dip aluminium on steel for profiled roofing sheet or tile to BS EN 508–1: 2008 and BS EN 505: 2013 may be applied to three grades with total mass levels of 195, 230 or 305 g/m². However, the 195 g/m² product is only suitable for subsequent organic coating. Usually, an alloy containing 8–11% silicon is used, which produces a durable alloy finish of aluminium, iron and silicon.

Tin and terne-coated steel

Tin and terne, an alloy of lead (80–90%) and tin (10–20%), are used as finishes to steel and stainless steel for cladding and roofing units. Terne may be applied to sheet stainless steel by immersion in the molten alloy. Terne-coated stainless steel does not suffer from bimetallic corrosion and may normally be used in contact with lead, copper, aluminium or zinc. Thermal movement is similar to stainless steel, allowing for units of up to 9 m in length to be used for roofing and cladding. Terne weathers to a patina similar to that produced by pure lead. Tin-coated stainless steel manufactured by electrodeposition is also used for roofing where an appearance similar to lead is required, but without the associated risk of lead theft. Specifications for the various terne-coated steels are listed in pr BS ISO 4999: 2013. Coating thicknesses range from 17 to 114 μm depending upon the designated application. A minimum coating mass of 20 g/m² for terne and 10 g/m² for tin is recommended. The standard grades of the substrate stainless steel are 1.4301 and 1.4401 (Table 5.8). For environmental reasons tin coating is becoming more prevalent than terne coating.

ORGANIC-COATED STEEL

Since the 1960s, a range of heat-bonded organic coatings for steel have been developed, including PVC plastisol (Colorcoat), polyvinylidene fluoride (PVDF), polyesters and PVC film (Stelvetite). Within this product range the PVC plastisol currently has the largest market share for cladding and roofing within the UK. The standard BS EN 10169: 2010 lists the full range of liquid paints, coating powders and films as well as the designation system for organic coated steels to assist appropriate specification within the environmental context.

PVC plastisol coating

PVC plastisol is applied to aluminium/zinc-coated steel to a thickness of 0.2 mm. It has a tough, lightly textured finish and is available in a wide range of colours, referenced to the British Standard or RAL colour systems. The reverse side is usually coated with a grey corrosion-resistant primer and polyester finish, although PVC plastisol may be specified for unusually aggressive internal environments. Careful site storage and handling is required to prevent physical damage to the surface. For non-marine environments the most durable colours will give a period to first maintenance of greater than 40 years in Northern Europe. In coastal locations the periods to first repainting are reduced.

Polyvinylidene fluoride coating

Polyvinylidene fluoride (PVDF), an inert fluorocarbon, when applied as a 0.027 mm coating to zinc-coated steel, has good colour stability at temperatures up to 120°C, making it suitable for worldwide use and for buildings which are likely to be extended at a later date. The finish is smooth and self-cleaning, although considerable care is required on site to prevent handling damage. A period to first maintenance of 15 years is typical within the UK for non-coastal locations. The wide colour range includes metallic silver. Polyvinylidene fluoride finished galvanised or stainless steel is also available in a range of colour finishes which include copper, copper patina and stainless steel. The material may be used for cladding and roofing; the solar-reflective paint finish reduces excessive solar gain.

Polyester and polyester/acrylic bead coatings

Polyester and silicone polyester-coated galvanised steels are economic products, but offer only medium-term life in non-aggressive environments. Externally the period to first maintenance will be typically 10 years in unpolluted inland locations, but they are suitable for internal use. However, polyester coatings on aluminium/zinc-coated steel have a greater resistance to marine environments. Silicone polyester should not be used in marine or hot, humid environments. Polyester and silicone polyester coatings are smooth and typically 0.025 mm in thickness. Polyester with acrylic bead finishes is durable with a 0.050 mm application typically lasting up to 40 years. Finishes are metallic and non-metallic, matt or gloss in a wide range of colours. They are suitable for both marine and inland environments and offer good scratch resistance.

Polyurethane coating

Polyurethane is applied as a 0.05 mm finish to zinc/aluminium alloy-coated steel as a primer and top coat. It is available in a range of solid and metallic colours with a 30-year period to first maintenance for roofs and cladding. Matt colours have a 25-year period to first maintenance in non-marine environments. The product is approved for use in conjunction with rainwater harvesting systems, as well as active solar air-heating systems and for the direct attachment of frameless photovoltaic panels.

Enamel coating

Organic enamel-coated steels offering good light reflectance are suitable for internal use as wall and roof linings. Coatings, usually 0.022 mm thick, are typically applied to hot-dip zinc/aluminium alloy-coated steel and are easily cleaned. The standard colour is brilliant white, but a range of light colours is also available.

PVC film coating

PVC film (0.02 mm) in a range of colours, decorative patterns and textured finishes is calendered to zinc-coated steel strip. The product is suitable only for internal applications.

Paints

The wide range of paints used to protect steel from corrosion fall into several categories listed in BS EN ISO 12944–5: 2007. Paints including solvent-borne chlorinated rubber, acrylic polymers and vinyl chloride copolymers dry by solvent evaporation which may be reversed by solvent action. The majority of other products dry irreversibly by a loss of solvent or coalescence from the water emulsion. Alkyds, urethane alkyds and epoxy esters dry by a combination of a loss of solvent and aerial oxidation. Water-borne acrylics, vinyls and urethanes coalesce as the water evaporates. Two-component paints such as epoxys and polyurethanes set through chemical processes activated upon mixing.

Paint is normally applied in three coats. The primer adheres to the cleaned surface and gives corrosion protection. The second coat builds up thickness and the finishing coat provides protection from the environment and the required aesthetic finish. The standard BS EN ISO 12944: 2007 gives guidelines on the suitability of a range of paint systems in relation to environmental conditions and durability.

Steel tiles and slates

Lightweight steel tile and slate units, manufactured from galvanised or aluminium zinc alloy hot-dipped steel, coated with acrylic or polyester resin and a granular finish, give the appearance of traditional slate or pantile roofs. The products have the advantage, particularly for refurbishment work, of lightness in comparison to the traditional materials. Units may typically be used for roof pitches between 12° and 90°. A span of 1200 mm allows for wider spacing of roof trussed rafters. Units in a range of traditional material colours are available with appropriate edge and ventilation accessories. The standard BS EN 14782: 2006 requires a minimum thickness of 0.4 mm (excluding any coating) for self-supporting sheet steel in construction, but a typical 0.7 mm polyester-coated galvanised sheet steel system will span 1 m.

Aluminium

Aluminium has only been available as a construction material for about 100 years. Possibly the most well-known early use of the metal was for the cast statue of Eros, which has stood in Piccadilly Circus, London since 1893. Because of its durability, it is used widely in construction, particularly for secondary components, as illustrated in the permanent shading devices on the Faculty of Divinity building of the University of Cambridge (Fig. 5.20). Aerofoil-shaped shading systems manufactured from extruded aluminium, either fixed or linked to an active building management system, are standard features in many new buildings.

MANUFACTURE

Aluminium, the most common metallic element in the earth's crust, is extracted from the ore bauxite, an impure form of aluminium oxide or alumina. Large reserves of bauxite are available in Australia, China and Brazil. India, Indonesia, Jamaica and Russia

Fig. 5.20 Aluminium shading devices – Faculty of Divinity, University of Cambridge. *Architects*: Edward Cullinan Architects. *Photograph*: Arthur Lyons

also produce significant quantities of bauxite. The ore is dissolved in caustic soda, filtered, reprecipitated to remove impurities and dried. The pure alumina is then dissolved in fused cryolite (sodium aluminium fluoride) within a carbon-lined electrolytic cell. Electrolysis of the aluminium oxide produces oxygen and the pure aluminium, which is tapped off periodically and cast. The process is highly energy intensive, and worldwide production of 1 tonne of aluminium requires 15000 kWh of electrical energy, although new systems have reduced this to less than 13000 kWh per tonne. Globally, 50% of this energy is from renewable hydroelectric power. However, recycling requires only 5% of the energy input compared to primary production, and recycling from building and construction in Europe is at least 92%. Cast ingots or slabs are hot rolled at 500°C into 5 mm coiled sheet, which may subsequently be cold rolled into thinner sheet or foil.

Due to the ductility of aluminium, the metal may be extruded into complex shapes or drawn into wire. Forming and machining processes are generally easier than with steel. Aluminium components may also be formed by casting.

PROPERTIES

Aluminium is one of the lightest metals with a density of 2700 kg/m^3 compared to steel 7900 kg/m^3. Standard-grade aluminium (99% pure) has a tensile strength between 70 and 140 MPa, depending on temper; however, certain structural aluminium alloys (e.g. alloy 5083) achieve 345 MPa comparable to the 410–560 MPa for S275 steel. This compares favourably on a strength-to-weight basis, but the modulus of elasticity for aluminium is only one-third that of steel, so deflections will be greater unless deeper sections are used. For an aluminium section to have the same stiffness as an equivalent steel member, the aluminium section must be enlarged to approximately half the weight of the steel section.

Durability

The durability of aluminium as a construction material is due to the protection afforded by natural oxide film, which is always present on the surface of the metal. The aluminium oxide film, which is immediately produced when the surface of the metal is cut or scratched, is naturally only 0.01 μm thick, but may be thickened by the process of anodisation.

Fire

The strength of aluminium is halved from its ambient value at a temperature of 200°C, and for many of the alloys it is minimal by 300°C.

Contact with other building materials

While dry cement-based materials do not attack aluminium, the alkalinity of wet cement, concrete and mortar causes rapid corrosion. Thus, where these materials make contact during the construction process, the metal should be protected by a coating of bitumen paint. Furthermore, anodised and particularly coloured sections, such as glazing units, can be permanently damaged by droplets of wet cement products, and should be protected on site by a removable lacquer or plastic film. Under dry conditions aluminium is unaffected by contact with timber; however, certain timber preservatives, particularly those containing copper compounds, may cause corrosion under conditions of high humidity. Where this risk is present the metal should be protected with a coating of bitumen.

Although aluminium is highly resistant to corrosion in isolation, it can be seriously affected by corrosion when in contact with other metals. The most serious effects occur with copper and copper-based alloys, and rainwater must not flow from a copper roof or copper pipes into contact with aluminium. Except in marine and industrial environments it is safe to use stainless steel fixings or lead with aluminium, although zinc and zinc-coated steel fixings are more durable. Unprotected mild steel should not be in electrical contact with aluminium.

ALUMINIUM ALLOYS

Aluminium alloys fall into two major categories, either cast or wrought. In addition, the wrought alloys may be subjected to heat treatment. The majority of aluminium used in the construction industry is wrought, the content and degree of alloying components being directly related to the physical properties required, with the pure metal being the most malleable. BS EN 573–1: 2004 designates aluminium alloys into categories according to their major alloying components (Table 5.10).

Table 5.10 Broad classification of aluminium alloys to BS EN 573–1: 2004

Alloy series	Major alloying components
1000	Greater than 99% aluminium
2000	Copper alloys
3000	Manganese alloys
4000	Silicon alloys
5000	Magnesium alloys
6000	Magnesium and silicon alloys
7000	Zinc alloys
8000	Other elements

Notes:
- In many cases minor alloying components are also present.
- The standard BS EN 573–3: 2013 details the full range of alloy compositions.

The following example illustrates the coding system for aluminium and its alloys:

Structural aluminium is alloy EN AW-6082.

Where EN refers to the European Norm, A is for aluminium and W is for wrought products. 6082 is within the 6000 series of magnesium and silicon alloys, and the final three digits refer to the exact chemical composition as listed in BS EN 573–3: 2013.

For flashings where on-site work is necessary, 99.8% pure aluminium (alloy EN AW-1080A) or 99.5% (EN AW-1050A) offer the greatest malleability, although the standard commercial grade 99% pure aluminium (alloy EN AW-1200) is suitable for insulating foils and for continuously supported sheet roofing. The minimum nominal thickness for fully supported aluminium sheet roofing and cladding is 0.6 mm to BS EN 14783: 2013. A minimum roof fall of 1° and bay lengths of up to 10 m are possible for standing seam and batten roll fully supported roof systems using 0.7 mm aluminium.

Profiled aluminium for roofing and cladding, requiring additional strength and durability, is alloyed with 1.25% manganese (alloy EN AW-3103). It is produced from the sheet by roll forming, and may be manufactured into curved sections to increase design flexibility. Preformed rigid flashings to match the profile sheet are manufactured from the same alloy and finish. The alloy with 2% magnesium (alloy EN AW-5251) is more resistant to marine environments. The standard BS EN 508–2: 2008 details the full range of alloys for self-supporting roofing systems and illustrates examples of profiles applicable to aluminium sheets and tiles.

Aluminium rainscreen cladding panels up to 2.8 × 1.5 m in size may also be shaped using the superplastic forming (SPF) process, which relies on the high extensibility of the alloy EN AW-5083SPF. Sheet alloy, typically 2 mm in thickness, is heated to 380–500°C, and forced by air pressure into the three-dimensional form of the mould. Horizontal or vertical ribs are frequently manufactured to give enhanced rigidity, but cladding panels may be formed to individual designs including curvature in two directions. Coloured finishes are usually polyester powder or polyvinylidene fluoride coatings.

Extruded sections for curtain walling, doors and windows require the additional strength imparted by alloying the aluminium with magnesium and silicon (alloy EN AW-6063). Thermal insulation within such extruded sections is achieved by a hidden thermal break or by an internal plastic or timber-insulating cladding.

Structural aluminium for load-bearing sections and space frames typically contains magnesium, silicon and manganese (alloy EN AW-6082). Tempering increases the tensile strength to the range 270–310 MPa, which is more comparable to the standard grade of structural steel S275 (minimum tensile strength 410 MPa).

FINISHES FOR ALUMINIUM

Anodising

The process of anodising thickens the natural aluminium oxide film to typically 5–25 μm. The component is immersed in sulphuric acid and electrolytically made anodic which converts the surface metal into a porous aluminium oxide film, which is then sealed by boiling in water. The anodising process increases durability and may be used for trapping dyes within the surface to produce a wide range of coloured products (Fig. 5.21). Some dyes fade with exposure to sunlight, the most durable colours being gold, blue, red and black. Exact colour matching for replacement or extensions to existing buildings may be difficult, and manufacturers will normally produce components within an agreed band of colour variation. If inorganic salts of tin are incorporated into the surface during the anodising process then colour-fast bronzes are produced. Depending upon the period of exposure to the electrolytic anodisation process, a range of colours from gold and pale bronze to black may be produced. Different aluminium alloys respond differently to the anodising treatment. Pure aluminium produces a silver mirror finish, whereas the aluminium-silicon alloys (e.g. alloy EN AW-6063) produce a grey finish. The standard BS EN ISO 7599: 2010 specifies the range of clear, coloured and textured finished produced by the anodic oxidation of aluminium and its alloys. Light-reflecting properties and colour fastness should be specified by the customer.

Surface textures

A range of surface textures is achieved by mechanical and chemical processing. Finishes include bright polished, matt, etched and pattern-rolled according to the pre-treatments applied, usually before anodising, and also the particular alloy used. The aluminium discs on the façade of the Selfridges building in Birmingham give an innovative decorative finish to the steel-frame store, forming an elegant contrast to the blue-painted rendered surface (Fig. 5.22).

Coatings

Metallic
Zinc-coated aluminium is a foldable roofing and cladding material combining the durability of aluminium with the appearance of pre-weathered zinc. The standard sheet thickness is 1.0 mm. Matching rainwater goods including half-round gutters and downpipes are available. An equivalent pre-patinated or pre-oxidised titanium-coated aluminium with the appearance of bright steel is under development.

Organic
Polyester coatings, predominantly white, but with a wide range of colour options, are used for double-glazing systems, cladding panels and rainwater goods.

Fig. 5.21 Anodized aluminium tiles – Ravensbourne College , London. *Architects*: Foreign Office Architects. *Photographs*: Arthur Lyons

Fig. 5.22 Aluminium discs – Selfridges Store Birmingham. *Architects*: Future Systems. *Photographs*: Arthur Lyons

Electrostatically applied polyester powder is heat-cured to a smooth self-cleaning finish. PVC and PVDF simulated wood-grain and other pattern film finishes may also be applied to aluminium extrusions and curtain-wall systems. The range of factory-applied organic coatings, including polyester, acrylic, polyvinylidene fluoride (PVDF), alkyd and polyurethane, suitable for self-supporting roofing systems including tiles, is described in BS EN 508–2: 2008.

Paint

Where aluminium is painted for decorative purposes it is important that the appropriate primer is used. The aluminium should be abraded or etched to give a good key to the paint system, although cast aluminium normally has a sufficiently rough surface. Oxide primers are appropriate but red lead should be avoided.

Maintenance of finished aluminium

For long-term durability, all external aluminium finishes should be washed regularly, at intervals not normally exceeding three months, with a mild detergent solution. Damaged paint coatings may be touched up on site, but remedial work does not have the durability of the factory-applied finishes.

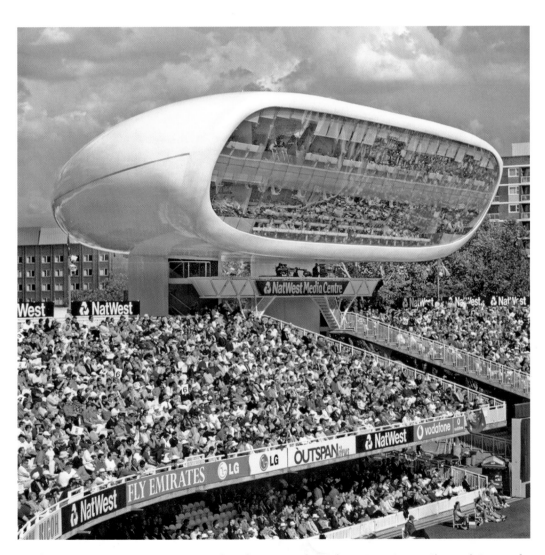

Fig. 5.23 Aluminium semi-monocoque construction – Lord's Media Centre, London. *Architects:* Future Systems. *Photograph:* Courtesy of Richard Davies

ALUMINIUM IN BUILDING

Typical applications for aluminium and its alloys in building include roofing and cladding, curtain-wall and structural glazing systems, flashings, rainwater goods, vapour barriers, and internally ceilings, panelling, luminaires, ducting, architectural hardware and walkways. Minimum sheet thicknesses for self-supporting sheet aluminium roofing and other applications to BS EN 14782: 2006 are 0.6 and 0.4 mm, respectively. For fully supported aluminium roofing and cladding to BS EN 14783: 2013, the nominal minimum sheet thickness is 0.6 mm.

Monocoque construction

The Lord's Cricket Ground Media Centre (Fig. 5.23) was the world's first semi-monocoque building in aluminium. The media centre is a streamlined pod raised 14 m off the ground on two concrete support towers, giving journalists and commentators an uninterrupted view over the cricket ground. The structure consists of a curved 6 and 12 mm aluminium-plate skin welded to a series of ribs. Thus acting together, the skin and the ribs provide both the shape and the structural stability, a system typically used in the boat-building and aircraft industries. The building was made in 26 sections and transported to the site for assembly.

Thermal breaks in aluminium

In order to overcome thermal bridging effects, where aluminium extrusions are used for double-glazing systems, thermal breaks are inserted between the aluminium in contact with the interior and exterior spaces. These may be manufactured from pre-formed polyamide strips, or alternatively the appropriate extrusions are filled with uncured polymer, then the bridging aluminium is milled out once the plastic has set.

Jointing methods

Aluminium components may be joined mechanically with aluminium bolts or rivets; non-magnetic stainless steel bolts are also appropriate. If aluminium is to be electric arc welded, the use of an inert gas shield, usually argon, is necessary to prevent oxidation of the metal surface. A filler rod, compatible with the alloy to be welded, supplies the additional material to make up the joint. Strong adhesive bonding of aluminium components is possible, provided that the surfaces are suitably prepared.

Copper

Copper was probably one of the first metals used by man, and evidence of early workings suggests that the metal was smelted as early as 8700 BC. Later it was discovered that the addition of tin to copper improved the strength of the material and by 2300 BC the Bronze Age had arrived. The Romans made extensive use of copper and bronze for weapons, utensils and ornaments. Brass from the alloying of copper and zinc emerged from Egypt during the first century BC. By the mid-eighteenth century South Wales was producing 90% of the world's output of copper, with the ore from Cornwall, but now the main sources are from Chile, the United States, Peru, China, Australia and Indonesia. The traditional visual effect of copper is illustrated in a modern context by the millennium project, Swan Bells in Perth, Australia (Fig. 5.24). The copper was initially clear-coated to prevent gradual oxidation and patinisation within the marine environment of the harbour.

MANUFACTURE

The principal copper ores are the sulphides (e.g. chalcocite), and sulphides in association with iron (e.g. chalcopyrite). Ores typically contain no more than 1% copper and therefore require concentration by flotation techniques before the copper is extracted through a series of furnace processes. The ores are roasted, then smelted to reduce the sulphur content and produce *matte*, which contains the copper and a controlled proportion of iron sulphide. The molten matte is refined in a converter by a stream of oxygen. This initially oxidises the iron which concentrates into the slag and is discarded; sulphur is then burnt off to sulphur dioxide, leaving 99% pure metal, which on casting evolves the remaining dissolved gases and solidifies to *blister* copper. The blister copper is further refined in a furnace to remove remaining sulphur with air and then oxygen with methane or propane. Finally, electrolytic purification produces 99.9% pure metal. Approximately 41% of copper and the majority of brass and bronze used within the UK are recycled from scrap. Recycling requires approximately 25% of the energy used in the primary production of copper depending upon the level of impurities present.

Fig. 5.24 Copper cladding – Swan Bells Tower, Perth Australia. *Architects*: Hames-Sharley Architects. *Photographs*: Arthur Lyons

GRADES AND FORMS OF COPPER

Grades of copper and copper alloys are designated by both a symbol and a number system to the current standards BS EN 1172: 2011 and BS EN 1412: 1996.

Thus, for example, phosphorus deoxidised non-arsenical copper, typically used for roofing and cladding, is defined as:

Symbol system:
Cu-DHP
(Deoxidised high-residual phosphorus)

Number system:
CW024A-R240
(C refers to copper, W refers to wrought products, 024A identifies the unique composition (Table 5.11). The subsequent letters and numbers define a specific requirement such as tensile strength or hardness, in this case a minimum tensile strength of 240 MPa, which is half-hard temper.)

Only four of the numerous grades of copper are commonly used within the construction industry.

Electrolytic tough pitch high-conductivity copper (Cu-ETP or CW004A)

Electrolytic tough pitch high-conductivity copper is used mainly for electrical purposes; however, the sheet material is also used for fully supported traditional and long strip copper roofing. It contains approximately 0.04% dissolved oxygen which is evolved as steam if the copper is heated to 400°C in a reducing flame, thus rendering the metal unsuitable for welding or brazing.

Fire refined tough pitch copper (Cu-FRHC or CW005A)

Fire refined tough pitch copper has a similar specification to Cu-ETP, but with marginally more impurities.

Tough pitch non-arsenical copper (Cu-FRTP or CW006A)

Tough pitch non-arsenical copper is used for general building applications. It is suitable for sheet roofing.

Phosphorus deoxidised non-arsenical copper (Cu-DHP or CW024A)

Phosphorus deoxidised non-arsenical copper is the standard grade for most building applications including roofing, but not for electrical installations. The addition of 0.04% phosphorus to refined tough pitch copper isolates the oxygen rendering the metal suitable for welding and brazing. It is therefore used for plumbing applications where soldering is inappropriate.

The detailed chemical compositions of these and all other grades of copper are listed in PD CEN/TS 13388: 2013.

Copper forms and sizes

Copper is available as wire, rod, tube, foil, sheet and plate. Standard sheet thicknesses to BS EN 1172: 2011 are 0.4, 0.5, 0.6, 0.7, 0.8 and 1.0 mm with typical roofing grades of 0.6 and 0.7 mm. (The minimum nominal thickness for self-supporting sheet copper to BS EN 14782: 2006 and for the fully supported material to BS EN 14783: 2013 is 0.5 mm.) Curtain walling is usually in the range 1.5–3.0 mm. The metal is supplied dead soft (fully annealed), one-eighth or one-quarter hard, half-hard or full-hard. It rapidly work hardens on bending, but this may be recovered by annealing at red heat. Copper may be worked at any temperature, since, unlike zinc, it is not brittle when cold. The standard grade of copper used for roofs, pipes and domestic water-storage cylinders is phosphorus deoxidised non-arsenical copper CW024A, although the other tough pitch grades CW004A, CW005A and CW006A may also be used for roofs. Copper for pipework is supplied in annealed coils for mini/microbore systems, in 6 m

Table 5.11 Broad classification of copper alloys

Number series	Letters	Materials
000–099	A or B	Copper
100–199	C or D	Copper alloys (less than 5% alloying elements)
200–299	E or F	Copper alloys (more than 5% alloying elements)
300–349	G	Copper – aluminium alloys
350–399	H	Copper – nickel alloys
400–449	J	Copper – nickel – zinc alloys
450–499	K	Copper – tin alloys
500–599	L or M	Copper – zinc alloys
600–699	N or P	Copper – zinc – lead alloys
700–799	R or S	Copper – zinc alloys, complex

Note:
The three-digit number designates each material and the letter indicates copper or the alloy group.

lengths half-hard and hard for general plumbing work. The hard temper pipes cannot be bent. Plastic-coated tubes, colour coded to identify the service (e.g. yellow-gas), are available. New forms of copper façade materials include textured surfaces, profiled, pressed or perforated sheets and woven mesh.

WEATHERING

Patina

The green patina of basic copper sulphate or chloride on exposed copper gradually develops according to the environmental conditions. On roofs within a marine or industrial environment the green patina develops within eight years; under heavy pollution it may eventually turn dark brown or black. Within a town environment, the patina on roofs will typically develop over a period of 12 years. However, vertical copper cladding will normally remain a deep brown, due to the fast rainwater run-off, except in marine environments when the green colour will develop. On-site treatment to accelerate the patinisation process is unreliable, but pre-patinised copper sheet is available if the effect is required immediately.

Pre-oxidised copper is either mid- or dark brown. Pre-patinised copper is available in a range of colours from green to blue, and either uniform or flecked showing the brown metallic under-layer (Fig. 5.25). Pre-patinised copper sheet should not be welded, brazed or soldered, as heat treatment causes discoloration of the patina. The factory-generated bright-green patina will weather according to the local environmental conditions, often turning quickly to a blue-green. However, if the original metallic appearance is to be retained externally, then a copper/aluminium/zinc alloy which resists weathering and retains its bright gold colour may be used (Fig. 5.25). Friar Gate Square, Derby (Fig. 5.26) illustrates the impact of bright copper cladding, while Maggie's Highlands Cancer Caring Centre Raigmore Hospital, Inverness (Fig. 5.27) shows the coordination between pre-patinised copper and natural timber.

Corrosion

Generally, copper itself is resistant to corrosion; however, rainwater run-off may cause staining on adjacent materials and severe corrosion to other metals. Zinc, steel, galvanised steel and non-anodised aluminium should not be used under copper, although

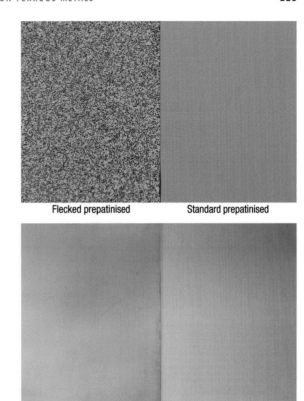

Flecked prepatinised   Standard prepatinised

Natural copper   Copper/aluminium/zinc alloy

Fig. 5.25 Prepatinised, natural and alloyed copper colours. Courtesy of Aurubis. *Photograph*: Arthur Lyons

in this respect lead, stainless steel and brass are un-affected. Copper may cause corrosion to steel or anodised aluminium in direct contact, if moisture is present. Specifically, copper should not be installed below exposed bitumen, bitumen paint, or cedarwood shingles where leaching action-producing acid solutions may cause localised attack on the metal. In addition, some corrosion may arise from the acid produced by algae on tiled roofs. The accidental splashing of lime or cement mortar onto copper causes a blue-green discoloration; however, this can readily be removed with a soft brass brush. Some corrosion of copper pipework may be caused by soft water, particularly if high levels of dissolved carbon dioxide are present; hard waters generally produce a protective film of calcium compounds, which inhibits corrosion. Pitting corrosion has been reported in rare cases associated with either hard, deep-well waters or hot, soft waters with a significant manganese content. In addition, excessive acidic flux residues not removed by flushing the system may cause corrosion. Within heating systems in which oxygen in the primary

Fig. 5.26 Copper cladding – Friar Gate Square, Derby. *Architects*: Panter Hudspith. *Photographs*: Courtesy of TECU®Consulting, KME UK

circulating water is constantly being replenished through malfunction or poor design, bimetallic corrosion will occur between steel radiators and copper pipework. This will result in the buildup of iron oxide residues at the bottom of the radiators. The use of appropriate inhibitors will reduce this effect.

Protective coatings for copper

A range of clear coatings is available for external and internal application to copper where the original colour and surface finish is to be retained. Air-drying acrylic thermosetting resins and soluble fluoropolymers are suitable for exterior use. The acrylic resins, which incorporate benzotriazole to prevent corrosion if the coating is damaged, have a life expectancy of up to eight years. However, in the long term external copper will progress to a patina and this should be anticipated in

the design. For internal use, polyurethane, vinyls, epoxy and alkyd resins are appropriate. Silicones are necessary for high temperature applications.

COPPER IN BUILDING

Traditional and long-strip systems

Copper roofing systems may be categorised as traditional or long strip. The latter has the advantage that bays between 8.1 and 14.6 m, depending on the pitch, may be constructed without the necessary cross-welts on sloping roofs or drips on flat roofs appropriate to the traditional system. This has significant cost benefits in terms of installation costs. The long-strip copper roof system (Fig. 5.28) with bays up to approximately 600 mm wide may be laid on roofs with pitches between 3° and 90°, and uses one-quarter or half-hard temper

Fig. 5.27 Patina copper- Maggie's Highlands Cancer Caring Centre Raigmore Hospital, Inverness. *Architects:* Page / Park. *Photographs:* Courtesy of TECU®Consulting, KME UK

0.6 or 0.7 mm copper strip. The system requires specified areas of the roof to be fixed with conventional welted joint clips, and the remainder with expansion clips, which allow for the longitudinal expansion of the bays, but ensure a secure fixing to the substructure. Lateral thermal movement is accommodated by a space at the base of the standing seams. Long-strip copper is laid on a breather membrane which allows free movement between the metal and the structure, while isolating the copper from any ferrous fixings in the structure and providing some sound reduction from the effects of wind and rain. All fixings should be made from the same copper as the roof. Nails with minimum 6 mm diameter heads should be copper or brass. Prefabricated roofing trays up to 8 m long may be supplied direct to site to reduce on-site workmanship.

Within the traditional copper roofs (Fig 5.29), standing seams or batten roll jointing systems are used depending on the pitch and appearance required. For pitches of 5° or less, batten rolls (square or tapered) are appropriate, as standing seams are vulnerable to accidental flattening and subsequent failure by capillary action. Cross-welts may be continuous across roofs where batten rolls are used, but should be staggered where standing seams are used. Bays should not exceed approximately 1700 mm in length. Either soft or one-quarter hard temper copper is normally used. The substructure, breather membrane and fixed clips are as used in long-strip roofing. These differences in articulation within the traditional roofing systems and particularly by contrast to the smooth line of the long-strip system offer alternative visual effects to the designer of copper roofs. Copper rainwater systems are available with a range of standard components. Copper shingles offer an additional aesthetic to traditional copper roofing and cladding systems.

Position of fixed clips in relation to roof pitch

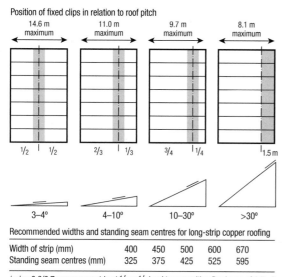

Fig. 5.28 Long-strip copper roofing

Self-supporting systems

Typical sections for profiled sheet copper and copper tiles are illustrated in the standard BS EN 506: 2008. Phosphorus deoxidised copper (Cu-DHP) is specified for this purpose. The standard BS EN 14782: 2006 defines a minimum thickness of 0.5 mm for self-supporting copper sheet for roofing and cladding. Self-supporting façade panels, up to 300 mm × 3 m, may be arranged vertically, horizontally or diagonally to give a striated effect. Larger rectangular cassettes up to 900 mm × 3 m and preformed on all four edges produce an alternative aesthetic.

Bonded copper system

One proprietary system offers visual effects similar to that of traditional copper, as well as aluminium and stainless steel, by bonding metal foil to reinforced bitumen roofing membrane (see Chapter 6 on bitumen and flat roofing materials, page 255).

COPPER ALLOYS

Copper may be alloyed with zinc, tin, aluminium, nickel or silicon to produce a range of brasses and bronzes. The full range of copper alloys is listed in PD CEN/TS 13388: 2013.

Brass

Brass is an alloy of copper and zinc, most commonly with zinc content between 10% and 45%. It is used for small components such as architectural ironmongery, door and window furniture, handrails and balustrades. It may be lacquered to prevent deterioration of the polished finish, although externally and in humid environments the lacquer fails, requiring the brass to be cleaned with metal polish to remove the tarnish. Brass plumbing fittings manufactured from 60/40 brass may corrode in soft, high-chloride content waters by dezincification. The process produces insoluble zinc corrosion products and ultimately porous metal fittings which may cause failure of the system. In situations where this problem is likely, dezincification-resistant (DZR) fittings made from alloy CW602N containing 2% lead should be used. Such components are marked with the 'ℝ' dezincification-resistant symbol.

Bronze

Bronze is an alloy of copper and tin, used for high-quality door furniture and as a woven fabric cladding material, as at the Theatre Royal rehearsal centre, Plymouth. Bronzes are usually harder and more durable than the equivalent brasses and exhibit a greater resistance to corrosion. Phosphor bronze contains up to 0.5% phosphorus in an 8% tin bronze. Because of its load-bearing properties and durability, it is frequently used as corbel plates, as well as fixings for stone and precast concrete cladding panels. Aluminium bronze (copper and aluminium), silicon bronze (copper and silicon) and gunmetals (copper-tin-zinc alloys) are also used for masonry fixings and cast components by virtue of their strength and durability. Nickel bronze alloys (copper, nickel and zinc) may be manufactured to highly polished *silver* finishes, particularly appropriate for interior fittings.

ANTI-MICROBIAL SURFACES

It is now recognised that copper and copper alloys are naturally anti-microbial materials and therefore have a vital role to play in the reduction of hospital-acquired infections such as MRSA and *E.coli*. When cleaned regularly, antimicrobial copper kills 99.9% of germs within two hours of exposure. It is known that a large proportion of infections are spread by touch, so it is appropriate that touch surfaces, such as door handles, push plates, taps and light switches within hospitals and other healthcare facilities should be manufactured from copper, brass or bronze rather than stainless steel or other inactive materials. A range of 300 antimicrobial alloys give a choice of colours and textures, but surfaces must not be lacquered or varnished, as this impairs the antimicrobial efficacy.

Batten roll

Standing seam

Copper nails are
ring shanked,
25 mm × 2.8 mm

Underlay

Supporting
substance

Recommended centres for batten roll
copper roofing
Width of sheet (mm)   600
Batten roll centres (mm)   500
(using 0.6 mm copper strip to a maximum
1.8 m length in soft or 1/4 hard temper)

Recommended centres for standing seam
copper roofing
Width of sheet (mm)   600
Batten roll centres (mm)   500
(using 0.6 mm copper strip to a maximum
1.8 m length in soft or 1/4 hard temper)

Copper shingles

97.5 mm

160 mm

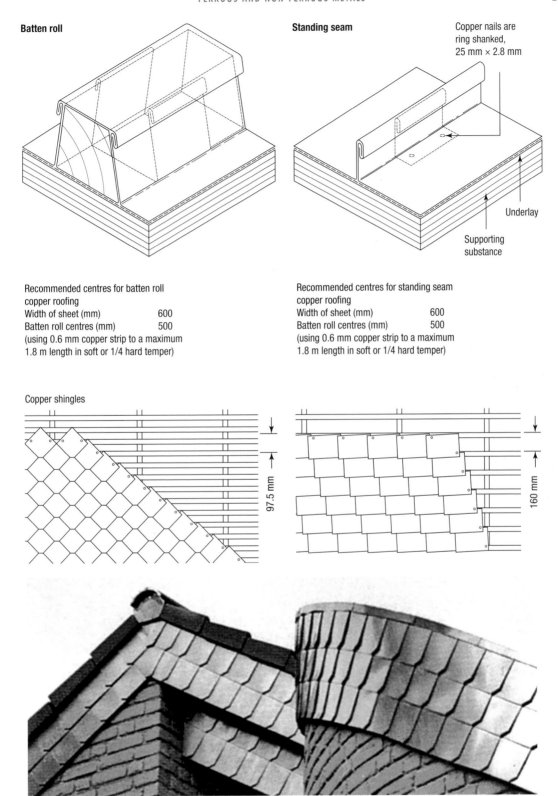

Fig. 5.29 Traditional copper roofing and copper shingles. *Photograph*: Courtesy of Aurubis

Lead

The Egyptians used lead in the glazing of their pottery and for making solder by 5000 BC. It was mined in Spain by the Phoenicians around 2000 BC.

MANUFACTURE

Lead occurs naturally as the sulphide ore, galena. The manufacturing process involves the concentration of the ore by grinding and flotation. The sulphide is converted to the oxide by roasting, then reduced to the metal in a blast furnace charged with limestone and coke. Further refining removes impurities which would otherwise reduce the softness of the metal. The building industry takes approximately 6% of the lead currently used within the UK. Overall some 90% of all lead used is recycled. UK consumption is approximately 120,000 tonnes per year.

Lead sheet

The majority of lead for roofing, cladding, flashings and gutter linings is produced as rolled sheet to the required thickness. A 2-tonne billet is repeatedly rolled to produce an 18 m sheet of the required thickness. Continuous machine-cast lead, which accounts for approximately 10% of the UK market, is manufactured by immersing a rotating water-cooled metal drum in a bath of molten lead at constant temperature. The lead solidifies on the surface of the drum and is peeled off as it emerges from the melt. The thickness may be adjusted by altering the rotation speed of the drum, the temperature and the depth of immersion. The sheet produced is without the anisotropic directional grain structure associated with the standard rolling process, but has variable thickness. Sand-cast lead sheet (Fig. 5.30) is still manufactured by the traditional method, which involves pouring molten lead onto a prepared bed of damp sand. The sheet thickness is controlled by the temperature of the molten lead and by drawing a piece of timber across the molten metal surface to remove the excess material. Sand-cast lead is normally only used for conservation work on key historic buildings, when much of the old lead may be recycled in the process. A typical cast lead sheet size is 6 × 1.2 m.

Fig. 5.30 Traditional lead casting – Courtesy of Norman and Underwood. *Photograph:* Arthur Lyons

Table 5.12 Lead sheet thicknesses and colour codes to BS EN 12588: 2006 with typical applications

	green	yellow	blue	red	black	white	orange
European designation							
Colour code	green	yellow	blue	red	black	white	orange
Nominal thickness (mm)	1.25	1.50	1.75	2.00	2.50	3.00	3.50
UK designation							
Lead codes	3		4	5	6	7	8
Nominal thickness (mm)	1.32	1.59	1.80	2.24	2.65	3.15	3.55
Nominal weight (kg/m^2)	15.0	18.0	20.4	25.4	30.1	35.7	40.3
Typical application:							
Flat roofing				✓	✓	✓	✓
Pitched roofing				✓	✓	✓	✓
Vertical cladding			✓	✓			
Soakers	✓		✓				
Hip and ridge flashings			✓	✓			
Parapets, box/tapered valley gutters					✓	✓	✓
Pitched valley gutters			✓	✓			
Weatherings to parapets			✓	✓			
Apron and cover flashings			✓	✓			
Chimney flashings			✓	✓			

The standard BS EN 12588: 2006 which relates only to rolled lead, defines lead sheet by thickness rather than by the code system still sometimes used within the UK (Table 5.12). The minimum nominal thickness for fully supported lead sheet roofing and cladding is 1.25 mm to BS EN 14783: 2013. The handling of lead requires appropriate on-site care, particularly due to the weight of the material and the toxic fumes produced during lead burning. The use of lead in the EU is registered under the REACH (Regulation, Evaluation, Authorisation and restriction of Chemicals) regulations.

WEATHERING

Corrosion

Freshly cut lead has a bright finish, but it rapidly tarnishes in the air with the formation of a blue-grey film of lead carbonate and lead sulphate. In damp conditions a white deposit of lead carbonate is produced, and in cladding this can be both aesthetically unacceptable and cause some staining of the adjacent materials. The effect can be prevented by the application of patination oil or acrylic copolymer emulsion after the lead has been fixed. Lead is generally resistant to corrosion due to the protection afforded by the insoluble film; however, it is corroded by organic acids. Acidic rainwater run-off from mosses and lichens may cause corrosion, and contact with damp timbers, particularly oak, teak and western red cedar, should be avoided by the use of building paper or bitumen paint. Trapped condensation under sheet lead may cause significant corrosion, so consideration must be given to the provision of adequate ventilation underneath the decking which supports the lead. Dew points must be checked to ensure that condensation will not occur and be trapped under the lead sheet in either new work or renovation. Generally, lead is stable in most soils; however, it is attacked by the acids within peat and ash residues. Electrolytic corrosion rarely occurs when lead is in contact with other metals, although within marine environments aluminium should not be used in association with lead. Corrosion does occur between wet Portland cement or lime products and lead during the curing process; thus, in circumstances where the drying out will be slow, the lead should be isolated from the concrete with a coat of bitumen paint.

Fatigue and creep

In order to prevent fatigue failure due to thermal cycling or creep, that is, the extension of the metal under its own weight over extended periods of time, it is necessary to ensure that sheet sizes, thicknesses and fixings are in accordance with the advice given by the Lead Sheet Association in their technical manuals. The metal must be relatively free to move with temperature changes, so that alternating stresses are not focused in small areas leading to eventual fatigue fracture. A geotextile separating underlay may be used. The addition of 0.06% copper to 99.9% pure lead refines the crystal structure, giving increased fatigue resistance without significant loss of malleability. The composition of lead sheet is strictly controlled by BS EN 12588: 2006.

LEAD IN BUILDING

Roofing

Lead roofing requires a smooth, continuous substrate. Generally, the bay sizes depend on the roof geometry

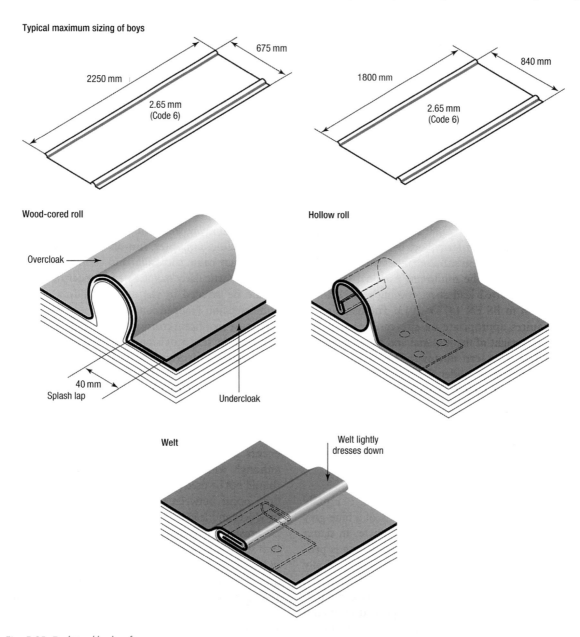

Fig. 5.31 Traditional lead roofing

Fig. 5.32 Traditional lead roof. *Contractor:* Norfolk Sheet Lead. *Photograph:* Courtesy of Lead Contractors Association

and the thickness of lead to be used (Fig. 5.31). For flat roofs (from 1 in 80 [approximately 1°] to 10°), joints are generally wood-cored rolls down the fall and drips across. For pitched roofs (10° to 80°), joints in the direction of the fall may be wood-cored or hollow rolls, with laps across the fall, unless, for aesthetic reasons, the bays are to be divided by drips. Roofs with hollow rolls are less vulnerable to lead theft than wood-cored rolls, as the sheets are more difficult to remove. Hollow rolls of 3.0 mm thickness are not normally damaged by maintenance work on flat roofs.

For steep pitches, welts are used and over 80°, standing seams are appropriate (Fig. 5.32). Fixings are copper or stainless steel nails and clips within the rolls, welts or standing seams. Lead as a highly malleable material may be formed or *bossed* into shape with the specialist tools including the bossing stick and bossing mallet. Welding or lead burning involves the joining of lead to lead using additional material to make the joint thicker by one-third than the adjacent material.

The David Mellor Cutlery Factory, Hathersage, Derbyshire (Fig. 9.15, Chapter 9, page 337), illustrates a traditionally detailed lead roof. The wood-cored roll-jointed lead is supported on a stepped deck manufactured from prefabricated stressed-skin insulated plywood boxes, tapered to fit the radial design. These units are supported on a series of lightweight steel trusses, tied at the perimeter by a steel tension ring and at the centre lantern by a ring-truss. Around the perimeter, the lead is burnt to ensure a vertical seal. A steeper roof without the need for drips is illustrated in Figure 5.32.

Lead sheet cladding

For cladding, the thickness of lead to be used dictates the maximum spacing between vertical joints and distance between laps. Vertical joints may be wood-cored rolls or welts and occasionally standing seams or hollow rolls, where the risk of physical damage from ladders is negligible. The lead is hung by nailing at the head, with allowance for up to 6 mm thermal movement to occur within the lap joints.

An alternative form of lead cladding is the use of preformed lead-faced cladding panels, which are then fixed to the building façade (Fig. 5.33). Typically, 25 mm exterior grade plywood covered with 1.80 or 2.24 mm (Code 4 or Code 5) lead is used. The panels are set against a lead-faced timber structural support leaving 25 mm joints for thermal movement. Standard details are illustrated in the relevant Code of Practice BS 6915: 2001.

Lead tiles

An imaginative use of lead as a roofing material is illustrated by the Haberdashers' Hall in London (Fig. 5.34). The two-storey building, constructed around a courtyard, features a roof clad with diamond-shaped lead tiles. The individual units are formed from 1×1.5 m marine plywood diamonds each dressed with lead sheet, and incorporating a flashing on two edges to seal under the panels above.

Flashings

Lead, because of its malleability and durability, is an ideal material from which to form gutters and gutter linings, ridge and hip rolls, and the full range of

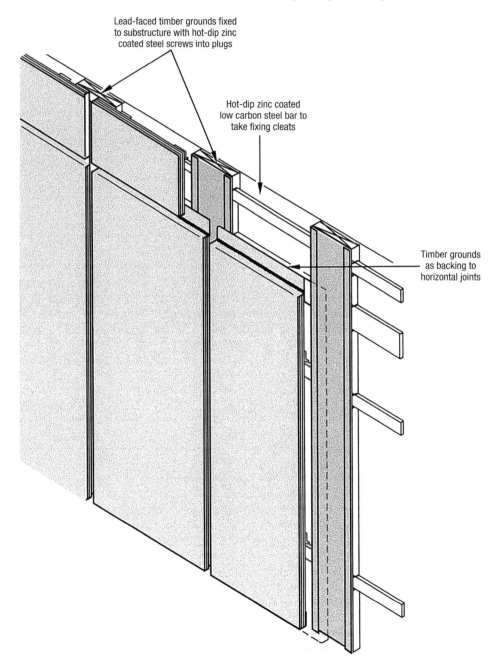

Lead-faced timber grounds fixed to substructure with hot-dip zinc coated steel screws into plugs

Hot-dip zinc coated low carbon steel bar to take fixing cleats

Timber grounds as backing to horizontal joints

Fig. 5.33 Lead-faced timber cladding panels. Permission to reproduce extracts from BS 6915: 2001 is granted by the British Standards Institute

Fig. 5.34 Lead tile roofing – Haberdashers' Hall, London. *Architects*: Hopkins Architects. *Photograph*: Courtesy of Lead Contractors Association

standard and specialist flashings, including ornamental work to enhance design features. For most flashing applications, lead sheet of 1.32, 1.80 or 2.24 mm (Codes 3–5) are used, fixed with copper or stainless steel and occasionally lead itself.

Acrylic-coated lead sheet

Acrylic-coated lead sheet is produced to order and in the standard colours of white, slate grey, terracotta and dark brown for use in colour-coordinated flashings. The colour-coated 1.80 mm (Code 4) rolled lead is produced in widths of 250, 300, 450 and 600 mm. The material is moulded by bossing as for standard lead sheet, and where welded, the exposed grey metal may be touched up if necessary.

Zinc

Zinc was known to the Romans as the alloy, brass, but it was not produced industrially until the mid-eighteenth century, and was not in common use on buildings until the nineteenth century. The cut surface tarnishes quickly to a light grey due to the formation of a patina of basic zinc carbonate. The metal is hard at ambient temperatures and brittle when cold. It should therefore not be worked at metal temperatures below 10°C without prior warming, and heavy impacts should not be used within the forming processes of bending and folding.

MANUFACTURE

Zinc occurs naturally as the sulphide ore, zinc blende. The main deposits are in China, Peru, Australia, North America and India. The ore is first concentrated and then roasted to produce zinc oxide. The oxide is dissolved in sulphuric acid and the solution electrolysed between lead alloy anodes and aluminium cathodes. The zinc is stripped off the cathodes and cast into ingots for subsequent forming. Zinc is classified according to its purity, as specified in the standard BS EN 1179: 2003 (Table 5.13). Approximately 3610 kW hours of energy is consumed in the primary production of one tonne of sheet zinc, although a large proportion is recycled.

Zinc sheet

Zinc sheet is manufactured by continuous casting and rolling in a range of thicknesses (Table 5.14) to a

Table 5.13 Colour codes for grades of zinc

Grade classification	Z1	Z2	Z3	Z4	Z5
Zinc content (%)	99.995	99.99	99.95	99.5	98.5
Colour code	White	Yellow	Green	Blue	Black

Table 5.14 Zinc/copper/titanium sheet thicknesses and weights

Nominal thickness (mm)	0.6	0.65	0.7	0.8	1.0	1.20	1.50
Nominal weight (kg/m²)	4.3	4.7	5.0	5.8	7.2	8.6	10.8

maximum coil width of 1000 mm. The two standard products are the pure metal (99.995% zinc) and its alloy with small additions of titanium and copper (e.g. 0.06% and 0.08% minima, respectively). The rolling process modifies the grain structure, particularly in the pure metal; however, this does not affect the working of the sheets. The alloy has improved performance with respect to strength and creep resistance but also a reduced coefficient of thermal expansion which enables the construction of roof bays up to 10 m, or in certain cases up to 16 m in length, depending on design considerations including bay width. The zinc-copper-titanium alloy (BS EN 988: 1997) may be folded or curved to produce interlocking cladding panels for vertical, horizontal or diagonal installation. Both the pure metal and the titanium alloy may be worked by hand at room temperature and do not work-harden. The minimum nominal thickness for self-supporting zinc alloy to BS EN 14782: 2006 and for fully supporting zinc roofing and cladding to BS EN 14783: 2013 is 0.6 mm. The zinc-copper-titanium alloy is used to add a unity to the suite of elegant buildings forming the University of Cambridge, Centre for Mathematical Sciences, by Edward Cullinan Architects (Fig. 5.35).

WEATHERING

Patina

Bright zinc tarnishes in the air with the production of a thin oxide film, which is rapidly converted into basic zinc carbonate by the action of water and carbon dioxide. The patina then prevents further degradation of the surface. Ordinary zinc has a lighter blue-grey patina than the alloyed sheet, so the two materials should not be mixed within the same construction. The factory-produced patina is grey or black, but mineral pigments rolled into the surface and coated with a protective film produce subtle red, blue, green or brown pre-weathered finishes. The lifetime of zinc depends directly on the thickness. A 0.8 mm roof should last for 40 years in urban conditions, whereas the same sheet as cladding, washed clean by rain, could last for 60

years. The titanium alloy with considerably improved durability has a predicted life of up to 100 years in a rural environment depending upon the pitch of the application.

Lacquered zinc sheet

A factory-applied 25 μm heat-treated polyester lacquer finish to zinc gives a range of colour options through white, brown, gold, terracotta, red, green, grey and blue. Alternative organic coatings include acrylic, silicone-polyester, polyvinylidene fluoride and PVC plastisol to BS EN 506: 2008.

Corrosion

Zinc should not be used in contact with copper or where rainwater draining from copper or copper alloys would discharge onto zinc. It may, however, be used in association with aluminium. In contact with steel or stainless steel, the zinc must be the major component to prevent significant corrosion effects. Unprotected cut edges of galvanised steel located above zinc can cause unsightly rust stains and should be avoided. If the underside of zinc sheet remains damp due to condensation for extended periods of time then pitting corrosion will occur, causing eventual failure. It is therefore necessary to ensure that the substructure is designed appropriately with vapour barrier, insulation and ventilation to prevent interstitial condensation. Sulphur dioxide within polluted atmosphere prevents the formation of the protective carbonate film and causes corrosion.

Zinc is not affected by Portland cement mortars or concrete, although it should be coated with acrylic resin paint where it will be in contact with soluble salts from masonry or cement additives. Zinc may be laid directly onto seasoned softwoods, unless impregnated with copper-salt preservatives, which have a slight corrosion-promoting effect. However, zinc should not be used on acidic timbers such as oak, chestnut

Fig. 5.35 Zinc alloy roofing – Centre for Mathematical Sciences, University of Cambridge. *Architects*: Edward Cullinan Architects. *Photograph*: Arthur Lyons

and western red cedar. Furthermore, zinc should not be used in association with western red cedar shingles, which generate an acidic discharge. The acidic products from the effect of ultraviolet radiation on bitumen can cause corrosion in zinc. If the bitumen is not protected from direct sunlight by reflective chippings, then any zinc must be separated from the bitumen with an impermeable material.

ZINC IN BUILDING

Fixings

Fixings for zinc should be of galvanised or stainless steel. Clips are made of zinc, cut along the rolled direction of the sheet and folded across the grain. Watertight joints may be made by soldering using tin/lead solder in conjunction with zinc chloride flux.

Roofing and cladding

Both the roll cap and standing seam systems (Fig. 5.36) are appropriate for fully supported zinc and zinc alloy roofing (BS EN 501: 1994). Welted joints are standard practice across the bays at pitches steeper than 15°; below 15° drips are necessary. A minimum fall of 3° is recommended, although 1.5° is possible for shorter bays; however, a pitch in excess of 7° will ensure self-cleaning, preventing the accumulation of dirt, which reduces service life. Where the bay length is greater than 3 m, a section 1.3 m in length is fixed rigidly, while the remaining area is secured to the substructure with sliding clips which accommodate the thermal movement. Timber roof boarding, oriented strand board or plywood form the ideal substructures for zinc, but chipboard is inappropriate except cement-bonded particleboard for cladding. Where concrete is used it must be sealed against trapped moisture. For cladding,

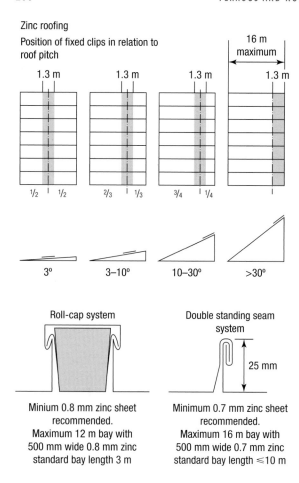

Zinc roofing

Position of fixed clips in relation to roof pitch

Roll-cap system

Minium 0.8 mm zinc sheet recommended.
Maximum 12 m bay with 500 mm wide 0.8 mm zinc standard bay length 3 m

Double standing seam system

Minimum 0.7 mm zinc sheet recommended.
Maximum 16 m bay with 500 mm wide 0.7 mm zinc standard bay length ≤10 m

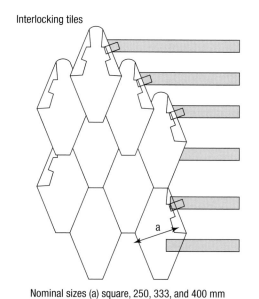

Interlocking tiles

Nominal sizes (a) square, 250, 333, and 400 mm diamond 200, 250, and 285 mm

Fig. 5.36 Zinc roofing and interlocking tiles

the vertical joints may be welted, standing seam or roll cap with the horizontal joints welted. The maximum bay length for cladding is 6 m, although 3 m is more practical on site. Titanium zinc rainwater systems are available with an appropriate range of standard components.

Zinc may also be used as the surface material for interlocking façade panel systems, with recessed horizontal or vertical joints, fixed to a sub-frame of timber, aluminium or stainless steel. A range of colours is available with lacquered finish.

Titanium zinc interlocking square and diamond tiles are appropriate for vertical hanging and roof pitches down to 25°. They are fixed to timber battens with soldered and sliding clips. A range of sizes is available to give a choice of scale. Tiles are either pre-weathered or bright and manufactured in 0.7 or 0.8 mm alloy. To muffle the sound of rain, a full-surface substructure is advisable.

Recycling of zinc

Worldwide, approximately 40% of newly formed zinc is from recycled sources, and within Europe approximately 90% of sheet material and rainwater goods zinc is recycled, although sheet material accounts for only a small percentage of zinc consumption, which is predominantly used for galvanizing.

Titanium

Titanium ore is abundant in the earth's crust, with reserves well exceeding currently anticipated demands. The main producing countries are China, Japan, Russia, Kazakhstan and Ukraine, although the ores, rutile (titanium oxide) and ilmenite (iron-titanium oxide), are also found in Europe, Australia, America and South America. Originally isolated in 1887, it was developed for use in the aerospace industry in the 1950s but more recently has become a significant building cladding material. The Glasgow Science Centre (Fig. 5.37) illustrates titanium's eye-catching appearance as a modern construction material.

MANUFACTURE

The ore is treated with chlorine and coke at 900°C to produce titanium tetrachloride, which is then purified to remove other unwanted elements. Treatment with

Fig. 5.37 Titanium cladding – The Glasgow Science Centre. *Architects*: BDP – Building Design Partnership. *Photograph*: Courtesy of Don Clements

metallic magnesium or sodium reduces the titanium tetrachloride to a sponge of titanium metal, which is then melted under vacuum to produce solid ingots. Ingots are then forged into slabs and rolled out into sheet. Where required an embossed finish may be applied during the final rolling process. Other sections and forms may be produced by hot rolling or cold forming as for steel. Titanium has a high embodied energy; however, this is to some extent balanced against its life cycle costing and ultimate full recycling.

PROPERTIES AND USES

Titanium is an appropriate material for construction due to its corrosion resistance. It is resistant to acids and alkalis as well as industrial and marine environments. Titanium has a density of 4510 kg/m³, intermediate between aluminium (2700 kg/m³) and steel (7900 kg/m³), giving it the advantage of a good strength-to-weight ratio. It is less ductile than steel, so hot forming is required for severe bending. The metal has a modulus of elasticity half that of steel. Titanium has a low coefficient of expansion (8.9×10^{-6}), half that of stainless steel and copper and one-third of that for aluminium. This reduces the risk of thermal stress, and enables titanium sheet roofing to be laid in longer lengths than other metals. The use of relatively thin roofing and cladding panels (0.3–0.4 mm) minimises both the dead load and the supporting structural system. Titanium with its very high melting point of 1660°C can withstand fire tests at 1100°C. Further applications include fascias, panelling, protective cladding for piers and columns, and three-dimensional artwork.

Durability

The corrosion resistance of titanium arises from its self-healing and tenacious protective oxide film. However, rainwater run-off from zinc, lead or copper roofs should be avoided. The Guggenheim Museum in Bilbao, clad in 32,000 m² of commercially pure 0.3–0.4 mm titanium sheet panels, shows some staining due to lack of protection during the construction process and also rainwater run-off. Although initially expensive, on a life cycle basis, due to its low maintenance costs, titanium may prove to be a highly competitive cladding and roofing material. Already, one manufacturer is offering a 100-year guarantee against corrosion failure in roofing applications. Titanium can cause the corrosion of contact aluminium, steel or zinc, but austenitic stainless steel (grade 1.4401) is not affected.

Finishes

The normal oxide film may be thickened by heat treatment or anodising, giving permanent colours ranging from blue and mauve to cream and straw. Control is necessary to ensure the absence of colour variations within a project. Surface finishes range from reflective bright to soft matte and embossed, as used on the Glasgow Science Centre buildings (Fig. 5.37). In this case, the rolling grain direction was maintained over the building façades to ensure no visible variation of the embossed stipple effect.

Welding titanium

Titanium may be arc welded, but this requires the exclusion of air usually by the use of argon gas shielding. Other welding technologies such as plasma arc and laser or electron beam are used for more specialist applications.

TITANIUM ALLOYS

Titanium is available as a wide range of alloys classified according to increased corrosion resistance, higher strength or higher temperature resistance. However, their current use is mainly confined to aerospace, industrial and medical applications. The standard architectural cladding material is 99% pure titanium (grade 1 or grade 2).

Process of metallic corrosion

Corrosion is an electro-chemical process, which can only occur in the presence of an electrolyte; that is, moisture containing some dissolved salts. The process may be understood by considering the action of a simple Daniell cell as shown in Figure 5.38.

When the cell operates, two key processes occur. At the anode, the zinc gradually dissolves, generating zinc ions in solution and electrons which flow along the wire and light up the lamp as they move through its filament. At the copper cathode, the electrons are received on the

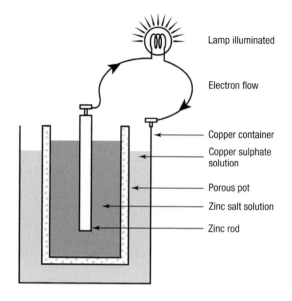

Fig. 5.38 Daniell cell

surface of the metal, and combine with copper ions in solution to plate out new shiny metal on the inside of the copper container.

Anode

$$Zn \longrightarrow Zn^{2+} + 2e^-$$
zinc   zinc ions   electrons

Cathode

$$Cu^{2+} + 2e^- \longrightarrow Cu$$
copper ions   electrons   copper

An equivalent process takes place in the dry Leclanché cell – the standard torch battery (Fig. 5.39). However, in this case the central carbon rod replaces the copper and the liquid is replaced by an aqueous paste. The anode process is the same as in the Daniell cell with the gradual dissolution of the zinc container.

Anode

$$Zn \longrightarrow Zn^{2+} + 2e^-$$
zinc   zinc ions   electrons

At the cathode the carbon rod is surrounded by manganese dioxide, which oxidises the hydrogen gas that would otherwise have been produced there by the reaction between water and the electrons.

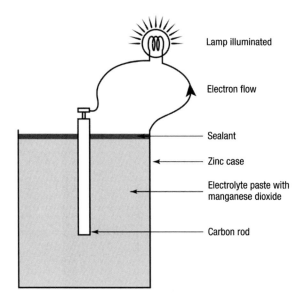

Fig. 5.39 Dry Leclanché cell

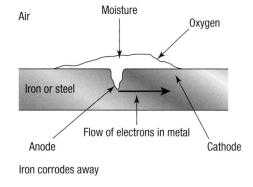

Fig. 5.40 Corrosion of iron

Cathode

$$H_2O \quad + \quad O \quad + \quad 2e^- \quad \longrightarrow \quad 2OH^-$$
water   oxygen   electrons   hydroxyl
        (from manganese   ions
             dioxide)

This sequence is similar to that seen in the corrosion of iron (Fig. 5.40). In this case the presence of both an electrolyte and oxygen is necessary for corrosion to occur.

Anode

$$Fe \quad \longrightarrow \quad Fe^{2+} \quad + \quad 2e^-$$
iron   iron ions   electrons

Cathode

$$2H_2O \quad + \quad O_2 \quad + \quad 4e^- \quad \longrightarrow \quad 4OH^-$$
water   oxygen   electrons   hydroxyl
        (from the air)   ions

Rust formation

$$Fe^{2+} \quad + \quad 2OH^- \quad \rightarrow \quad Fe(OH)_2 \quad \rightarrow \quad Fe_2O_3 \cdot H_2O$$
iron   hydroxyl   iron   rust
ions   ions   hydroxide

Overall summary

$$4Fe \quad + \quad 3O_2 \quad + \quad 2H_2O \quad \longrightarrow \quad 2Fe_2O_3 \cdot H_2O$$
iron   oxygen   water   rust

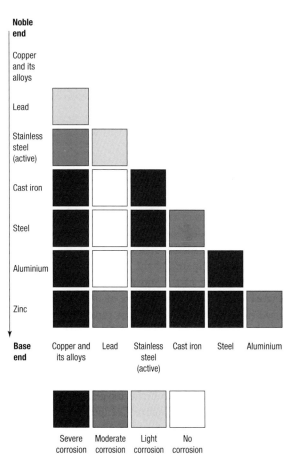

Indication of typical corrosion rates between pairs of metals in contact in the presence of moisture. The metal lower down the series copper (noble end) to zinc (base end) corrodes. The rate of corrosion will be generally increased within more aggressive environments.

Fig. 5.41 Bimetallic corrosion between pairs of metals in building applications

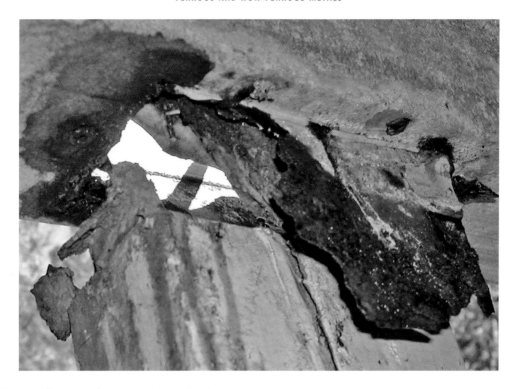

Fig. 5.42 Bimetallic corrosion between aluminium and steel

FACTORS AFFECTING THE RATE OF CORROSION

The key factors which accelerate the rate of corrosion are the presence of two dissimilar metals in mutual contact and the degree of pollution within any moisture surrounding the metals. If the more base metal is small in quantity compared to the more noble metal, then rapid corrosion of the more base metal will occur. Figure 5.41 shows which pairs of metals commonly used in construction should not generally be allowed into contact. Within a single metal, minor surface variations, such as crystal grain boundaries, the effects of cold working or welding, the presence of impurities or alloying components within the metal, variable cleanliness or access to aerial oxygen, may all cause accelerated corrosion. Figure 5.42 illustrates the effect of corrosion between an extruded aluminium gutter and a steel rainwater pipe. The aluminium corroded producing a white deposit near the point of contact between the two metals; this was followed by rapid corrosion of the steel.

References

FURTHER READING

Allbury, K., Abbe, O. and Anderson, J. 2013: *Environmental impact of metals.* Bracknell: IHS BRE.

Association for Specialist Fire Protection. 2007: *Fire protection for structural steel in buildings*, 4th edn. Aldershot: Association for Specialist Fire Protection.

British Gypsum. 2011: *The fire book. Passive fire protection solutions.* East Leake: British Gypsum.

English Heritage. 2012: *Metals. Practical building conservation.* Farnham: Ashgate Publishing Ltd.

Foley, G. 2009: *Houses of steel.* Victoria: Images Publishing.

Fröhlich, B. and Schulenburg, S. (eds) 2003: *Metal architecture: Design and construction.* Basel: Birkhäuser.

Hayward, A., Weare, F. and Oakhill, A.C. 2011: *Steel designers' manual*, 3rd edn. Chichester: Wiley-Blackwell.

Lead Sheet Association. 2007: *Rolled lead sheet manual. The complete manual.* London: Lead Sheet Association.

LeCuyer, A. 2003: *Steel and beyond; New strategies for metals in architecture.* Basel: Birkhäuser.

Lennon, T. 2011: *Structural fire engineering.* Tonbridge: ICE.

Leyens, C. and Peters, M. 2003: *Titanium and titanium alloys, fundamentals and applications.* Germany: Wiley.

Links International. 2011: *Architecture and construction in metal.* Barcelona: Links.

Mazzolani, F.M. 2003: *Aluminium structural design.* Berlin: Springer-Verlag.

Müller, U. 2011: *Introduction to structural aluminium design.* Caithness: Whittles Publishing.

National Building Specification. 2007: *Galvanising opinion. An examination of bi-metallic corrosion.* NBS Shortcut 25. Newcastle-upon-Tyne: NBS.

Reichel, A., Ackermann, P., Hentschel, A. and Hochberg, A. 2007: *Detail practice. Building with steel.* Basel: Birkhäuser.

Rheinzink. 2002: *Rheinzink: Applications in architecture.* Datteln: Rheinzink.

Schulitz, H., Sobek, W. and Habermann, K. 2000: *Steel construction manual.* Basel: Birkhäuser.

Scott, D.A. and Eggert, G. 2009: *Iron and steel. Corrosion, colourants and conservation.* Los Angeles: Archetype Publications.

Steel Construction Institute. 2012: *Fire resistant design of steel framed buildings in accordance with Eurocodes and UK National Annexes.* Publication No. 375. Ascot: The Steel Construction Institute.

Steel Construction Institute. 2012: *Steel designers manual,* 7th edn. Chichester: Wiley-Blackwell.

Trebilcock, P. and Lawson, M. 2003: *Architectural design in steel.* London: Taylor and Francis.

Vargel, C. 2004: *Corrosion of aluminium.* Oxford: Elsevier.

Zahner, L.W. 2005: *Architectural metal surfaces.* New Jersey: John Wiley and Sons.

STANDARDS

BS 4
Structural steel sections:
Part 1: 2005   Specification for hot-rolled sections.

BS ISO 209: 2007
Aluminium and aluminium alloys. Chemical composition.

BS ISO 404: 2013
Steel and steel products. General technical delivery requirements.

BS 405: 1987
Specification for uncoated expanded metal carbon steel sheets for general purposes.

BS 416
Discharge and ventilating pipes and fittings sand-cast or spun in cast-iron:
Part 1: 1990   Specification for spigot and socket systems.

BS 417
Galvanised mild steel cisterns and covers, tanks and cylinders:
Part 2: 1987   Metric units.

BS 437: 2008
Specification for cast iron drain pipes, fittings and their joints.

BS 460: 2002
Cast-iron rainwater goods. Specification.

BS 493: 1995
Airbricks and gratings for wall ventilation.

pr BS ISO 630–5: 2013
Structural steels. Technical delivery conditions for structural steels with improved atmospheric corrosion resistance.

BS 1161: 1977
Specification for aluminium alloy sections for structural purposes.

BS 1189: 1986
Specification for baths made from porcelain enamelled cast iron.

BS 1202
Nails:
Part 1: 2002   Steel nails.
Part 2: 1974   Copper nails.
Part 3: 1974   Aluminium nails.

BS 1245: 2012
Pedestrian doorsets and door frames made from steel sheet. Specification.

BS 1390: 1990
Baths made from vitreous enamelled sheet steel.

BS 1449
Steel plate, sheet and strip:
Part 1.1: 1991 Specification for carbon and carbon/manganese plate, sheet and strip.

BS 1566
Copper indirect cylinders for domestic purposes:
Part 1: 2002   Open ventilated copper cylinders.
Part 2: 1984   Specification for single feed indirect cylinders.

BS 3083: 1988
Specification for hot-dip zinc coated and hot-dip aluminium/zinc coated corrugated steel sheets for general purposes.
BS 3198: 1981
Specification for copper hot water storage combination units for domestic purposes.
BS ISO 3575: 2011
Continuous hot-dip zinc-coated carbon steel sheet of commercial and drawing qualities.
BS 3987: 1991
Anodic oxide coatings on wrought aluminium for external architectural applications.
BS 4449: 2005
Steel for the reinforcement of concrete. Weldable reinforcing steel. Specification.
BS 4482: 2005
Steel wire for the reinforcement of concrete products. Specification.
BS 4483: 2005
Steel fabric for the reinforcement of concrete. Specification.
BS 4513: 1969
Specification for lead bricks for radiation shielding.
BS 4842: 1984
Specification for liquid organic coatings for application to aluminium alloy extrusions, sheet and preformed sections for external architectural purposes.
BS 4868: 1972
Profiled aluminium sheet for building.
BS 4873: 2009
Aluminium alloy windows and door sets. Specification.
BS 4921: 1988
Specification for sheradized coatings on iron or steel.
pr BS ISO 4995: 2014
Hot-rolled steel sheet of structural quality.
pr BS ISO 4996: 2014
Hot-rolled steel sheet of high yield stress structural quality.
pr BS ISO 4997: 2014
Cold reduced carbon steel sheet of structural quality.
pr BS ISO 4998: 2013
Continuous hot-dip zinc-coated carbon sheet steel of structural quality.
pr BS ISO 4999: 2013
Continuous hot-dip terne (lead alloy) coated

cold-reduced carbon steel sheet of commercial, drawing and structural qualities.
BS 5427–1: 1996
Code of practice for the use of profiled sheet for roof and wall cladding on buildings. Design.
BS 5896: 2012
High tensile steel wire and strand for the prestressing of concrete. Specification.
pr BS ISO 5951: 2012
Hot-rolled steel sheet of higher yield strength with improved formability.
BS ISO 5952: 2011
Continuously hot-rolled steel sheet of structural quality with improved atmospheric corrosion resistance.
BS ISO 5954: 2007
Cold reduced carbon steel sheet according to hardness requirements.
BS 5977
Lintels:
Part 1: 1981   Method for assessment of load.
BS 6496: 1984
Specification for powder organic coatings for application and stoving to aluminium alloy extrusions, sheet and preformed sections for external architectural purposes.
BS 6510: 2010
Specification for steel windows, sills, window boards and doors.
BS 6582: 2000
Specification for continuously hot-dip lead alloy (terne) coated cold reduced carbon steel flat rolled products.
BS 6722: 1986
Recommendations for dimensions of metallic materials.
BS 6744: 2001
Stainless steel bars for the reinforcement of and use in concrete.
BS 6915: 2001
Design and construction of fully supported lead sheet roof and wall coverings.
pr BS ISO 6929: 2012
Steel products. Vocabulary.
BS 7364: 1990
Galvanised steel studs and channels for stud and sheet partitions and linings using screw fixed gypsum wallboards.
BS 7543: 2003
Guide to durability of buildings and building elements, products and components.

BS ISO 7583: 2013
  Anodizing of aluminium and its alloys. Terms and definitions.

BS 7668: 2004
  Weldable structural steels. Hot finished structural hollow sections in weather resistant steels. Specification.

BS 8202
  Coatings for fire protection of building elements:
  Part 1: 1995   Code of practice for the selection and installation of sprayed mineral coatings.
  Part 2: 1992   Code of practice for the use of intumescent coating systems to metallic substrates for providing fire resistance.

BS 8537: 2010
  Copper and copper alloys. Plumbing fittings.

BS 9999: 2008
  Code of practice for fire safety in the design, management and use of buildings.

BS ISO 13270: 2013
  Steel fibres for concrete. Definitions and specifications.

BS ISO 13887: 2011
  Cold reduced steel sheet of higher yield strength with improved formability.

BS ISO 14788: 2011
  Continuous hot-dip zinc 5% aluminium alloy coated steel sheet.

BS ISO 15510: 2014
  Stainless steels. Chemical composition.

BS ISO 16020: 2005
  Steel for the reinforcement and prestressing of concrete. Vocabulary.

BS ISO 16143
  Stainless steels for general purposes:
  Part 1: 2014   Corrosion-resistant flat products.
  Part 2: 2014   Corrosion-resistant semi-finished products, bars, rods and secctions.
  Part 3: 2014   Wire.

BS ISO 16160: 2012
  Hot-rolled steel sheet products. Dimensional and shape tolerances.

BS ISO 16162: 2012
  Cold-rolled steel sheet products. Dimensional and shape tolerances.

BS ISO 16163: 2012
  Continuously hot-dipped coated steel sheet products. Dimensional and shape tolerances.

BS ISO 17615: 2007
  Aluminium and aluminium alloys. Alloyed ingots for remelting. Specifications.

pr BS EN 124 Parts 1–6: 2012
  Gully tops and manhole tops for vehicular and pedestrian areas.

BS EN ISO 225: 2010
  Fasteners. Bolts, screws, studs and nuts. Symbols and description of dimensions.

BS EN 485
  Aluminium and aluminium alloys. Sheet, strip and plate:
  Part 1: 2008   Technical conditions for inspection and delivery.
  Part 2: 2013   Mechanical properties.
  Part 3: 2003   Tolerances on shape and dimensions for hot-rolled products.
  Part 4: 1994   Tolerances on shape and dimensions for cold-rolled products.

BS EN 486: 2009
  Aluminium and aluminium alloys. Extrusion ingots. Specifications.

BS EN 487: 2009
  Aluminium and aluminium alloys. Rolling ingots. Specifications.

BS EN 501: 1994
  Roofing products from metal sheet. Specification for fully supported roofing products of zinc sheet.

BS EN 502: 2013
  Roofing products from metal sheet. Specification for fully supported products of stainless steel sheet.

BS EN 504: 2000
  Roofing products from metal sheet. Specification for fully supported roofing products of copper sheet.

BS EN 505: 2013
  Roofing products from metal sheet. Specification for fully supported products of steel sheet.

BS EN 506: 2008
  Roofing products from metal sheet. Specification for self-supporting products of copper or zinc sheet.

BS EN 507: 2000
  Roofing products from metal sheet. Specification for fully supported products of aluminium sheet.

BS EN 508
Roofing products from metal sheet. Specification for self-supported products of steel, aluminium or stainless steel sheet:
Part 1: 2008   Steel.
Part 2: 2008   Aluminium.
Part 3: 2008   Stainless Steel.

BS EN 515: 1993
Aluminium and aluminium alloys. Wrought products. Temper designations.

BS EN 545: 2010
Ductile iron pipes, fittings, accessories and their joints for water pipelines.

BS EN 573
Aluminium and aluminium alloys. Chemical composition and form of wrought products:
Part 1: 2004   Numerical designation system.
Part 2: 1995   Chemical symbol based designation system.
Part 3: 2013   Chemical composition and form of products.
Part 5: 2007   Codification of standardized wrought products.

BS EN 586
Aluminium and aluminium alloys. Forgings:
Part 1: 1998   Technical conditions.
Part 2: 1994   Mechanical properties.
Part 3: 2001   Tolerances on dimensions and form.

BS EN 598: 2007
Ductile iron pipes, fittings, accessories and their joints for sewerage applications. Requirements and test methods.

BS EN 754
Aluminium and aluminium alloys:
Parts 1–8: 2008   Cold drawn rod/bar and tube. (Part 2: 2013).

BS EN 755
Aluminium and aluminium alloys:
Parts 1–9: 2008   Extruded drawn rod/bar, tube and profiles. (Part 2: 2013).

BS EN 845
Specification for ancillary components for masonry:
Part 1: 2003   Ties, tension straps, hangers and brackets.
Part 2: 2003   Lintels.
Part 3: 2003   Bed joint reinforcement of steel meshwork.

BS EN 877: 1999
Cast iron pipes and fittings, their joints and accessories.

BS EN 969: 2009
Specification for ductile iron pipes, fittings, accessories and their joints for gas applications. Requirements and test methods.

BS EN 988: 1997
Zinc and zinc alloys. Specification for rolled flat products for building.

BS EN 1057: 2006
Copper and copper alloys. Seamless round copper tubes for water and gas in sanitary and heating applications.

BS EN 1090
Execution of steel structures and aluminium structures:
Part 1: 2009   Requirements for conformity assessment of structural components.
Part 2: 2008   Technical requirements for the execution of steel structures.
Part 3: 2008   Technical requirements for aluminium structures.
pr Part 4: 2014   Technical requirements for thin-guage, cold-formed steel elements and structures for roof, ceiling, floor and wall applications.

BS EN 1172: 2011
Copper and copper alloys. Sheet and strip for building purposes.

BS EN 1173: 2008
Copper and copper alloys. Material condition designation.

BS EN 1179: 2003
Zinc and zinc alloys. Primary zinc.

BS EN 1254 Parts 1–6 & 8
Copper and copper alloys. Plumbing fittings.

BS EN 1412: 1996
Copper and copper alloys. European numbering system.

BS EN 1559–1: 2011
Founding. Technical conditions of delivery. General.

BS EN 1560: 2011
Founding. Designation system for cast iron. Material symbols and material numbers.

BS EN 1561: 2011
Founding. Grey cast irons.

BS EN 1562: 2012
Founding. Malleable cast irons.

BS EN 1563: 2011
Founding. Spheroidal graphite cast irons.
BS EN 1564: 2011
Founding. Ausferritic spheroidal graphite ductile cast irons.
BS EN 1774: 1997
Zinc and zinc alloys. Zinc for foundry purposes. Ingot and liquid.
BS EN 1976: 2012
Copper and copper alloys. Cast unwrought copper products.
BS EN 1977: 2013
Copper and copper alloys. Copper drawing stock.
BS EN 1982: 2008
Copper and copper alloys. Ingots and castings.
BS EN 1993
Eurocode 3: Design of steel structures:
Part 1.1: 2005	General rules and rules for buildings.
Part 1.2: 2005	Structural fire design.
Part 1.3: 2006	Cold formed members and sheeting.
Part 1.4: 2006	Stainless steels.
Part 1.5: 2006	Plated structural elements.
Part 1.6: 2007	Strength and stability of shell structures.
Part 1.7: 2007	Plated steel structures subject to out of plane loading.
Part 1.8: 2005	Design of joints.
Part 1.9: 2005	Fatigue.
Part 1.10: 2005	Material toughness.
Part 1.11: 2006	Design of structures with tension components.
Part 1.12: 2007	Additional rules for steel grades S 700.

BS EN 1994
Eurocode 4: Design of composite steel and concrete structures:
Part 1.1: 2004 General rules and rules for buildings.
Part 1.2: 2005 Structural fire design.
BS EN 1999
Eurocode 9. Design of aluminium structures:
Part 1: 2007 General structural rules.
Part 2: 2007 Structural fire design.
Part 3: 2007 Structures susceptible to fatigue.
Part 4: 2007 Cold formed structural sheeting.
Part 5: 2007 Shell structures.
pr EN ISO 6509: 2012
Corrosion of metals and alloys. Determination of dezincification resistance of copper alloys with zinc.

BS EN ISO 7441: 1995
Corrosion of metals and alloys. Determination of bimetallic corrosion in outdoor exposure corrosion tests.
BS EN ISO 7599: 2010
Anodizing of aluminium and its alloys. General specifications for anodic oxidation coatings for aluminium.
BS EN ISO 8044: 2000
Corrosion of metals and alloys. Basic terms and definitions.
BS EN ISO 8504
Preparation of steel substrates before application of paints and related products. Surface preparation methods:
Part 1: 2001 General principles.
Part 2: 2001 Abrasive blast cleaning.
Part 3: 2001 Hand and power tool cleaning.
BS EN ISO 9444
Continuously cold-rolled stainless steel. Tolerances on dimensions:
Part 2: 2010 Wide strip and sheet/plate.
BS EN ISO 9445
Continuously cold-rolled stainless steel. Tolerances on dimensions:
Part 1: 2010 Narrow strip and cut lengths.
Part 2: 2010 Wide strip and plate/sheet.
BS EN 10020: 2000
Definition and classification of grades of steel.
BS EN 10021: 2006
General technical delivery conditions for steel.
BS EN 10025
Hot rolled products of structural steels:
pr Part 1: 2011	General technical delivery conditions.
pr Part 2: 2011	Non-alloy structural steels.
pr Part 3: 2011	Normalized/normalized rolled weldable fine grain structural steels.
pr Part 4: 2011	Thermomechanical rolled weldable fine grain structural steels.
pr Part 5: 2011	Structural steels with improved atmospheric corrosion resistance.
pr Part 6: 2011	High yield strength structural steels.

BS EN 10027
Designation systems for steels:
Part 1: 2005 Steel names.
pr Part 2: 2013 Numerical system.

BS EN 10029: 2010
Hot-rolled steel plates 3 mm thick. Tolerances on dimensions and shape.

BS EN 10034: 1993
Structural steel I and H sections. Tolerances on shape and dimensions.

BS EN 10051: 2010
Continuously hot-rolled strip and plate/sheet cut from wide strip of non-alloy and alloy steels. Tolerances on dimensions and shape.

BS EN 10052: 1994
Vocabulary of heat treatment terms for ferrous products.

BS EN 10056
Specification for structural steel equal and unequal leg angles:

Part 1: 1999	Dimensions.
Part 2: 1993	Tolerances, shape and dimensions.

BS EN 10079: 2007
Definition of steel products.

BS EN 10080: 2005
Steel for the reinforcement of concrete. Weldable reinforcing steel.

BS EN 10083
Steels for quenching and tempering:

Part 1: 2006	General technical delivery conditions.
Part 2: 2006	Technical delivery conditions for non alloy steels.
Part 3: 2006	Technical delivery conditions for alloy steels.

BS EN 10088
Stainless steels:

pr Part 1: 2011	List of stainless steels.
Part 2: 2005	Technical delivery conditions for sheet, plate and strip for general purposes.
pr Part 3: 2011	Technical delivery conditions for semi-finished products, bars, rods and sections for general purposes.
Part 4: 2009	Technical delivery conditions for sheet/plate and strip of corrosion resisting steels for construction purposes.
Part 5: 2009	Technical delivery conditions for bars, rods, wire, sections and bright products of corrosion resisting steels for construction purposes.

BS EN 10095: 1999
Heat resisting steels and nickel alloys.

BS EN 10130: 2006
Cold rolled low carbon steel flat products for cold forming. Technical delivery conditions.

BS EN 10131: 2006
Cold rolled uncoated and zinc or zinc-nickel electrolytically coated low carbon and high yield strength steel flat products for cold forming. Tolerances on dimensions and shape.

BS EN 10140: 2006
Cold rolled narrow steel strip. Tolerances on dimensions and shape.

BS EN 10143: 2006
Continuously hot-dip coated steel sheet and strip. Tolerances on dimensions and shape.

BS EN 10149
Hot-rolled products made of high yield strength steels for cold forming:

Part 1: 2013	General technical delivery conditions.
Part 2: 2013	Delivery conditions for thermomechanically rolled steels.
Part 3: 2013	Delivery conditions for normalized or normalized rolled steels.

BS EN 10152: 2009
Electrolytically zinc coated cold rolled steel flat products. Technical delivery conditions.

BS EN 10169: 2010
Continuously organic coated (coil coated) steel flat products. Technical delivery conditions.

BS EN 10210
Hot finished structural hollow sections of non-alloy and fine grain structural steels:

Part 1: 2006	Technical conditions.
Part 2: 2006	Tolerances, dimensions and sectional properties.

BS EN 10219
Cold-formed welded structural hollow sections of non-alloy and fine grain steels:

Part 1: 2006	Technical delivery requirements.
Part 2: 2006	Tolerances, dimensions and sectional properties.

BS EN 10220: 2002
Seamless and welded steel tubes. Dimensions and masses per unit length.

BS EN 10223 Parts 1–7: 2012
Steel wire and wire products for fencing and netting.

Part 8: 2013   Welded mesh gabion products.

BS EN 10242: 1995

Threaded pipe fittings in malleable cast iron.

BS EN 10244

Steel wire and wire products. Non-ferrous metallic coatings on steel wire:

Part 1: 2009   General principles.

Part 2: 2009   Zinc or zinc alloy coatings.

BS EN 10250

Open steel die forgings for general engineering purposes:

Part 1: 1999   General requirements.

Part 2: 2000   Non-alloy quality and special steels.

Part 3: 2000   Alloy special steels.

Part 4: 2000   Stainless steels.

BS EN 10264

Steel wire and wire products. Steel wire for ropes:

Part 1: 2012   General requirements.

Part 4: 2012   Stainless steel wire.

BS EN 10279: 2000

Hot rolled steel channels. Tolerances on shape, dimension and mass.

BS EN 10283: 2010

Corrosion resistant steel castings.

BS EN 10340: 2007

Steel castings for structural uses.

BS EN 10343: 2009

Steels for quenching and tempering for construction purposes. Technical delivery conditions.

pr EN 10346: 2013

Continuously hot-dip coated steel flat products. Technical delivery conditions.

BS EN 10349: 2009

Steel castings. Austenitic manganese steel castings.

PD ISO/TR 10809

Cast irons:

Part 1: 2009   Materials and properties for design.

Part 2: 2011   Welding.

BS EN 12020: 2008

Aluminium and aluminium alloys. Extruded precision profiles in alloys.

BS EN 12060: 1998

Zinc and zinc alloys. Method of sampling. Specification.

BS EN 12165: 2011

Copper and copper alloys. Wrought and unwrought forging stock.

BS EN 12258–1: 2012

Aluminium and aluminium alloys. Terms and definitions. General terms.

pr EN 12420: 2012

Copper and copper alloys. Forgings.

BS EN 12449: 2012

Copper and copper alloys. Seamless, round tubes for general purposes.

BS EN 12513: 2011

Founding. Abrasion resistant cast irons.

BS EN 12588: 2006

Lead and lead alloys. Rolled lead sheet for building purposes.

BS EN 12844: 1999

Zinc and zinc alloys. Castings. Specifications.

BS EN ISO 12944

Paints and varnishes. Corrosion protection of steel structures by protective paint systems:

Part 1: 1998   General information.

Part 2: 1998   Classification of environments.

Part 3: 1998   Design considerations.

Part 4: 1998   Types of surface and surface preparation.

Part 5: 2007   Protective paint systems.

BS EN 13283: 2002

Zinc and zinc alloys. Secondary zinc.

PD CEN/TS 13388: 2013

Copper and copper alloys. Compendium of compositions and products.

BS EN 13438: 2013

Paints and varnishes. Powder organic coatings for galvanized or sheradized steel products for construction purposes.

BS EN 13501

Fire classification of construction products and building elements:

Part 1: 2007   Classification using test data from reaction to fire tests.

Part 2: 2007   Classification using data from fire resistance tests.

Part 5: 2005   Classification using data from external fire exposure to roof tests.

BS EN 13811: 2003

Sheradizing. Zinc diffusion coatings on ferrous products. Specification.

BS EN 13830: 2003

Curtain walling. Product standard.

BS EN 13835: 2012

Founding. Austenitic cast irons.

BS EN 13858: 2006
Corrosion protection of metals. Non-electrolytically applied zinc flake coatings on iron or steel components.

BS EN 14351: 2006
Windows and doors. Product standard, performance characteristics.

BS EN 14431: 2004
Vitreous and porcelain enamels. Coatings applied to steel panels intended for architecture.

BS EN ISO 14713
Zinc coatings. Guidelines for the protection against corrosion of iron and steel in structures:
Part 1: 2009   General principles of design and corrosion resistance.
Part 2: 2009   Hot-dip galvanizing.
Part 3: 2009   Sheradizing.

BS EN 14782: 2006
Self-supporting metal sheet for roofing, external cladding and internal lining. Product specification and requirements.

BS EN 14783: 2013
Fully supported metal sheet and strip for roofing, external cladding and internal lining. Product specification and requirements.

BS EN 15088: 2005
Aluminium and aluminium alloys. Structural products for construction works.

BS EN 15530: 2008
Aluminium and aluminium alloys. Environmental aspects of aluminium products. General guidelines for inclusion in their standards.

BS EN ISO 15630
Steel for the reinforcement and prestressing of concrete. Test methods:
Part 1: 2010   Reinforcing bars, wire rod and wire.
Part 2: 2010   Welded fabric.
Part 3: 2010   Prestressing steel.

BS EN 16079: 2011
Founding. Compacted (vermicular) graphite cast irons.

pr EN 16623: 2013
Paints and varnishes. Reactive paints for fire protection of metallic substrates. Definitions, requirements, characteristics and marking.

BS EN 18286: 2010
Hot-rolled stainless steel plates. Tolerances on dimensions and shape.

BS EN ISO 28722: 2011
Vitreous and porcelain enamels. Characteristics of enamel coatings applied to steel panels intended for architecture.

CP 143
Code of practice for sheet roof and wall coverings:
Part 5: 1964   Zinc.
Part 10: 1973   Galvanized corrugated steel. Metric units.
Part 12: 1970   Copper. Metric units.
Part 15: 1973   Aluminium. Metric units.

BUILDING RESEARCH ESTABLISHMENT PUBLICATIONS

BRE Digests

BRE Digest 444: 2000
Corrosion of steel in concrete (Parts 1, 2 and 3).

BRE Digest 455: 2001
Corrosion of steel in concrete – service life design and prediction.

BRE Digest 461: 2001
Corrosion of metal components in walls.

BRE Digest 462: 2001
Steel structures supporting composite floor slabs – design for fire.

BRE Digest 487 Part 2: 2004
Structural fire engineering design: materials behaviour – steel.

BRE Information papers

BRE IP 11/00
Ties for masonry walls: A decade of development.

BRE IP 10/02
Metal cladding – assessing thermal performance of built up systems with use of Z spacers.

BRE IP 6/12
Passive and reactive fire protection to structural steel.

ALUMINIUM FEDERATION PUBLICATIONS

Fact Sheet 5
Aluminium recycling.

Fact Sheet 8
Aluminium in building and construction.

BRITISH CONSTRUCTIONAL STEELWORK ASSOCIATION PUBLICATION

Tata Steel: 2012
  The whole story from cradle to grave.

COPPER DEVELOPMENT ASSOCIATION

Publication 154: 2006
  The guide to copper in architecture.
Publication 205: 2010
  Antimicrobial copper. Introducing a new category of touch surface material for architects and designers.

ADVISORY ORGANISATIONS

Aluminium Federation Ltd, National Metalforming Centre, 47 Birmingham Road, West Bromwich, West Midlands B70 6PY (0121 601 6363).

Aluminium Rolled Products Manufacturers Association, National Metalforming Centre, 47 Birmingham Road, West Bromwich, West Midlands B70 6PY (0121 601 6363).

British Constructional Steelwork Association Ltd, 4 Whitehall Court, Westminster, London SW1A 2ES (0207 839 8566).

British Non-Ferrous Metals Federation, 10 Greenfield Crescent, Birmingham B15 3AU (0121 456 3322).

British Stainless Steel Association, Broomgrove, 59 Clarkehouse Road, Sheffield, South Yorkshire S10 2LE (0114 267 1260).

Cast Iron Drainage Development Authority, 72 Francis Road, Edgbaston, Birmingham, West Midlands B16 8BR (0121 693 9909).

Confederation of British Metal Forming, National Metalforming Centre, 47 Birmingham Road, West Bromwich, West Midlands B70 6PY (0121 601 6350).

Copper Development Association, 5 Grovelands Business Centre, Boundary Way, Hemel Hempstead, Hertfordshire HP2 7TE (01442 275705).

Corus Research, Development and Technology, Teeside Technology Centre, Eston Road, Middlesbrough TS6 6UB (01642 467144).

Council for Aluminium in Building, Bank House, Bond's Mill, Stonehouse, Gloucestershire GL10 3RF (01453 828851).

Galvanizers Association, Wren's Court, 56 Victoria Road, Sutton Coldfield, West Midlands B72 1SY (0121 355 8838).

Lead Sheet Association, Unit 10, Archers Park, Branbridges Road, East Peckham, Tonbridge, Kent TN12 5HP (01622 872432).

Metal Cladding and Roofing Manufacturers Association, 106 Ruskin Avenue, Rogerstone, Newport, South Wales NP10 0BD (01633 895633).

Steel Construction Institute, Silwood Park, Ascot, Berkshire SL5 7QN (01344 636525).

Zinc Information Centre, 6 Wrens Court, 56 Victoria Road, Sutton Coldfield, West Midlands B72 1SY (0121 362 1201).

6

BITUMEN AND FLAT ROOFING MATERIALS

CONTENTS

Introduction

Flat roofing materials, which form an impermeable water barrier, include reinforced bitumen membrane systems, mastic asphalt, single-ply plastic membranes and liquid coatings. All of these materials require continuous support on an appropriate roof decking system. Green roofs are considered as an extension of the standard roofing systems. Metal roofing systems are described in Chapter 5.

FIRE EXPOSURE OF ROOFS

All materials used as finishes for roofs, both pitched and flat, are classified with respect to external fire exposure. The classification system (BS 476–3: 2004) indicates whether the roof is flat or pitched followed by a two-letter coding on fire performance.

Roof system: EXT. F. (flat) or EXT. S. (sloping)

Fire penetration (first letter): A, B, C or D

    A = no penetration in 1 hour,
    B = no penetration in 30 minutes,
    C = penetration within 30 minutes,
    D = penetrated by preliminary flame test

Spread of flame (second letter): A, B, C or D

    A = no spread of flame,
    B = spread of flame less than 533 mm,

C = spread of flame more than 533 mm,

D = specimens that continue to burn after the test flame was removed or which had a spread of flame more than 381 mm in the preliminary test

A suffix 'X' is added where the material develops a hole or suffers mechanical failure.

Thus, a flat roof material classified as EXT. F. AA suffers no fire penetration or spread of flame during the standard one-hour fire test.

The European fire test for roofs is detailed in BS EN 13501: 2005 + A1: 2009, with test methodology in DD CEN/TS 1187: 2012 and extended application in PD CEN/TS 16459: 2013. Classification is according to four tests.

t1   Burning brands
t2   Burning brands and wind
t3   Burning brands, wind and radiant heat
t4   Burning brands, wind and radiant heat in two stages

Results lead to a B to F classification according to the response to each of these standard test procedures. For example, Class B_{ROOF}(t4) corresponds to the external fire performance for no penetration of the roof system within one hour in test t4 and with limited burning on the surface during the initial part of the two-stage test. This corresponds approximately to the BS 476–3: 2004 rating of EXT. F. AA.

However, the standard BS EN 13956: 2012 states that currently flexible sheets for waterproofing alone are categorised as Class F_{ROOF} to BS EN 13501–5: 2005 (external fire exposure to roofs tests), as their external fire performance is dominated by the build-up and not the individual sheet material.

Warm, cold and inverted roofs

WARM ROOFS

In warm roof construction the thermal insulation is laid between the roof deck and the weatherproof covering (Fig. 6.1). This ensures that the roof deck and its supporting structure are insulated from extremes of temperature, thus limiting excessive thermal movement which may cause damage. As the insulating material is directly under the waterproof layer, it must be sufficiently strong to support any foot traffic associated with maintenance of the roof. The waterproof and insulation layers will require mechanical fixing or ballasting to prevent detachment in strong winds. Surface condensation on the underside of a roof deck within warm roof construction would normally indicate insufficient thermal insulation. Warm roof construction is the preferred method for lightweight roofs.

COLD ROOFS

In cold roof construction, the weatherproof layer is applied directly onto the roof decking, usually particleboard or plywood, and this is directly supported by the roof structure, frequently timber joists (Fig. 6.1). Thermal insulation is laid over the gypsum plasterboard ceiling, leaving cold void spaces between the structural timbers or steel. In this form of roof construction, there is a significant risk of condensation forming on the underside of the decking, and this may cause deterioration of the structure. Precautions must be taken to ensure adequate ventilation of the cold voids; in addition, the underside of the deck must not cool below the dew point when the external temperature is –5°C. Any vapour check under the insulation layer is vulnerable to leakage around electrical service cables. In remedial work on cold roofs, if adequate ventilation cannot be achieved, then conversion to a warm roof or inverted roof system may be advantageous. Cold roof construction is not recommended for new building work in the Code of Practice (BS 8217: 2005).

INVERTED ROOFS

In inverted roof construction, both the structural deck and the weatherproof membrane are protected by externally applied insulation (Fig. 6.1). This ensures that the complete roof system is insulated from extremes of hot and cold, as well as from damage by solar radiation and maintenance traffic. The insulation layer is usually ballasted with gravel, paving slabs or roof garden finishes. Disadvantages of inverted roof construction are the greater dead-weight, and the difficulty in locating leaks under the insulation layer. Inverted roof construction is the preferred method for concrete and other heavyweight roof systems, but additional insulation is required to allow for the cooling effect of water draining beneath the insulation layer.

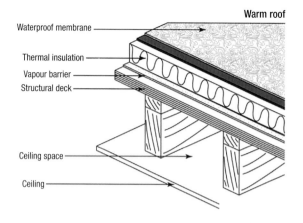

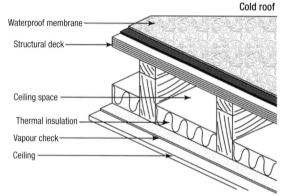

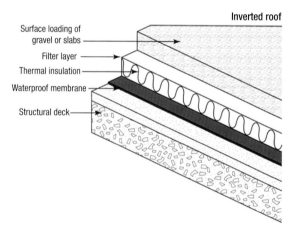

Fig. 6.1 Warm roof, cold roof and inverted roof

Reinforced bitumen membranes

Reinforced bitumen membrane (RBM) roofs (formerly built-up roofing) consist of two or more layers of bitumen sheets bonded together with self-adhesive or hot bitumen. The individual bitumen sheets are manufactured from a base carrier of either glass or fibre matting which is impregnated and coated with bitumen. Some products also have an applied adhesive backing and/or a surface protective finish.

Bitumen is the residual material produced after the removal by distillation of all volatile products from crude oil. The properties of bitumen are modified by controlled oxidation, which produces a more rubbery material suitable for roofing work. In the manufacture of bitumen sheets, the continuous base layer of organic fibres, glassfibre or bitumen-saturated polyester is passed through molten oxidised bitumen containing inert filler; the material is then rolled to the required thickness. The bitumen sheet is coated with sand to prevent adhesion within the roll or with mineral chippings to produce the required finish. A range of thicknesses is available, and for ease of recognition of the base fibre layer, the rolls are colour-coded along one edge.

TYPES OF REINFORCED BITUMEN MEMBRANES

Reinforced bitumen membranes for roofing systems are classified by BS EN 13707: 2013 according to a range of physical properties but without specific reference to the materials of manufacture. Guidance given in BS 8747: 2007 relates the key physical properties of tensile strength (S class) and puncture resistance (P class) to the types of product. Puncture resistance is derived from a combination of resistance to impact (D subclass) and resistance to static loading (L subclass). Both the S and P classes have five criteria levels of increasing standard from 1 to 5. The classification does not apply to venting and partial bonding base layers or underlay for mastic asphalt. The polyester-based products have greater strength and durability at a higher initial cost than the glassfibre-based products. Fine finishes are appropriate for the underlayers or where surface protection is applied, and granular finishes may be used for the exposed layer. Table 6.1 shows the typical relationship between generic products and the derived membrane S and P classes.

The requirement for a particular membrane performance (e.g. S2P3) is determined from a risk

Table 6.1 Reinforced bitumen membranes for roofing to BS 8747: 2007

Description	Type (previous coding)	Colour code	Tensile class S	Resistance to impact — subclass D	Static loading — subclass L	Puncture class P (derived from D and L)
Glass fibre – fine granule surface	3B	Red	S1	D1	L1	D1L1 = P1
Polyester – fine granule underlayer	5U	Blue	S2	D2	L2	D2L2 = P2
Polyester – fine granule surface	5B / 180	Blue	S3	D2	L3	D2L3 = P3
Polyester – fine granule surface	5B / 250	Blue	S4	D2	L4	D2L4 = P4
Polyester – fine granule surface	5B / 350	Blue	S5	D3	L4	D3L4 = P5

Notes:
- These are typical combinations, but specifiers must check the manufacturers' specifications as different combinations may apply.
- There are five levels of tensile strength from S1 (lowest) to S5 (highest).
- There are three levels of resistance to impact from D1 (lowest) to D3 (highest).
- There are four levels of resistance to static loading from L1 (lowest) to L4 (highest).
- Numbers 180, 250 and 350 refer to the mass g/m^2 of the base layer (not including any granular finish).

analysis, including the geometry of the roof (slope and membrane substrate), access (maintenance, pedestrian and vehicle) and degree of protection (self-protection, ballast, pavers and gardens). BS 8747: 2007 includes a matrix as guidance for specification of the appropriate membrane classification required.

Examples:

- Flat concrete roof, 0–5° pitch, gravel ballast, light maintenance access only – membrane classification required S2P3.

- Plywood substrate, > 5° pitch, self-protected membrane, light maintenance access only – membrane classification required S3P2.

The classification applies to both single-layer and multi-layer reinforced bitumen membrane systems. For multi-layer systems, one layer (usually the under-layer) should have a minimum S2P2 classification.

The Annex to BS 8747: 2007 describes Type 3G glassfibre-reinforced bitumen perforated venting layer used as a first layer of built-up RBM roofing systems where partial bonding or venting is required.

Underlay materials for discontinuous roofing such as slates and tiles are described in BS EN 13859–1: 2014. They are characterised by three classes of resistance to water penetration, namely class W1 (highest resistance) to class W3 (least resistance). The Annex to BS 8747: 2007 describes Type 1F fibre-based bitumen underlay reinforced with hessian and possibly incorporating an aluminium foil heat-reflecting layer on the underside. Type 5U polyester-reinforced bitumen roofing underlay has greater mechanical strength with a higher resistance to tearing than the Type 1F underlay. Generally, where the risk of condensation is high, sarking membranes, which are waterproof but vapour permeable, should be used. Alternative breather materials for tile underlay include a range of polyolefin laminates and modified PVC products. The colour codes for the bitumen-based membranes are listed in Table 6.2.

Table 6.2 Reinforced bitumen membranes for roofing to Annex of BS 8747: 2007

Class	Base material	Type	Use	Colour code
Class 1	Organic fibres and jute hessian base	Type 1F	Pitched roof underslating sheet	White
Class 3	Glass fibre base	Type 3G	RBM roof venting base layer	None
Class 5	Polyester base	Type 5U	Pitched roof underslating sheet	Blue

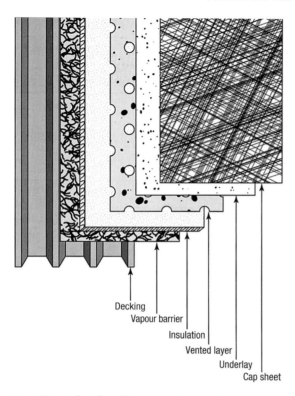

Decking
Vapour barrier
Insulation
Vented layer
Underlay
Cap sheet

Fig. 6.2 Typical reinforced bitumen membrane roofing system

A typical reinforced bitumen membrane system is illustrated in Figure 6.2. All oxidised bitumen sheets, when exposed to ultraviolet light and ozone, gradually age harden and become less resistant to fatigue failure.

ROOFING SYSTEMS

Reinforced bitumen roofing systems of substrate, insulation, membrane and finishes are described in the Code of Practice BS 8217: 2005 and in BS EN 13707: 2013. Roofing sheets may be applied to roofs constructed from precast or in situ reinforced concrete, plywood (exterior grade), timber (19 mm tongued and grooved, with preservative treatment and conditioning), oriented strand board, wood wool slabs or profile metal decking (galvanised steel or aluminium). Certain proprietary composite decking systems, such as units comprising plywood, rigid urethane foam and aluminium foil, are also appropriate, except in areas of high humidity. In warm roof applications a vapour check is applied over the decking. Appropriate vapour check materials include lapped and bonded bitumen membranes, lapped polythene sheets or 12 mm mastic asphalt on glassfibre tissue depending on the structural material.

Insulation materials include cork, rigid mineral (MW) or glass wool, perlite (EPB), cellular foamed glass (CG), rigid polyurethane foam (RUP), extruded (XPS) or expanded moulded (EPS) polystyrene, phenolic foam (PF), polyisocyanurate (PIR), bitumen-impregnated fibreboard and various proprietary composite systems. Most manufacturers are now able to supply CFC-free insulation products. The heat-sensitive expanded plastics are frequently supplied pre-bonded to a cork, perlite or fibreboard layer to receive the *pour and roll* hot bitumen or *torch-on* sheet systems. Where suitable falls are not incorporated into the roof structure, the insulation can be supplied ready cut-to-falls. A minimum in service fall of 1 in 80 is required to prevent *ponding*; this requires a design fall of 1 in 40 to allow for settlement.

On sloping roofs the first layer of roofing sheet is applied down the slope, but on flat roofs (less than 10°) the direction of the first layer need not relate to the falls. The first layer is either partly or fully bonded depending upon the substrate. Perforated sheet is frequently laid loose as the first layer, and becomes spot bonded as the hot bitumen for the second layer is applied. On timber the first layer is nailed. Partial bonding permits some thermal movement between the sheet system and the decking, and also allows for the escape of any water vapour trapped in the decking material. The use of proprietary breather vents on large roofs allows the escape of this entrapped air from the roof structure by migration under the partially bonded layer. Side laps of 50 mm and end laps of typically 100 mm should all be staggered between layers. On sloping roofs the first layer should be nailed at the top of each sheet at 50 mm centres and higher melting point 115/15 bitumen should be used for bonding the subsequent layers to prevent slippage. Protection from the effects of ultraviolet light is afforded either by the factory-applied mineral surface finish to the *cap* sheet, by the application of reflective paint, or on flat roofs, typically by a 12 mm layer of reflective white spar stone chippings.

POLYMER-MODIFIED BITUMEN MEMBRANES

High-performance bitumen membranes based on glassfibre or polyester bases for toughness, and polymer-modified bitumen coatings for increased flexibility, strength and fatigue resistance, offer considerably enhanced durability over the standard oxidised bitumen sheets. The two types are based on styrene-butadiene-styrene- (SBS) and atactic polypropylene- (APP) modified bitumen. These systems are covered by BS 8747: 2007.

Elastomeric SBS high-performance membranes

Styrene-butadiene-styrene polymer-modified bitumen membranes have greater elasticity than standard oxidised bitumen sheets. They are laid either by the traditional pour and roll technique which is used for standard bitumen sheets or by torching-on (Fig. 6.3). In the pour and roll process, bonding bitumen is heated to between 200 and 250°C and poured in front of the sheet as it is unrolled, giving continuous adhesion between the layers. In the torching-on process as the sheet is unrolled, the backing is heated to the molten state with propane burners or hot air. Alternatively, cold adhesive or mechanical fastening may be used. Additional fire protection is afforded by the incorporation of expandable graphite crystals into the membrane. The crystals absorb heat by an endothermic intumescent reaction, creating an insulating layer and reducing the risk of dripping from the bitumen coating. Ecological membranes incorporate titanium dioxide catalyst systems which under UV radiation convert oxides of nitrogen and sulphur into environmentally neutral substances.

Plastomeric APP high-performance membranes

Atactic polypropylene polymer-modified bitumen contains typically 25% atactic polymer in bitumen with some inert filler. The product is more durable than oxidised bitumen and has enhanced high-temperature resistance and low-temperature flexibility. Sheets are manufactured with a polyester and/or glassfibre core. Some additionally have a glassfibre-reinforced weathering surface. The APP polymer-modified bitumen sheets are bonded with cold adhesive or by torching the heat-sensitive backing, as the temperature of hot-poured bitumen is too low to form a satisfactory bond.

METAL-FACED ROOFING SHEETS

Copper, stainless steel and aluminium-faced high-performance SBS membranes give visual quality and enhanced durability compared with standard mineral surfaced reinforced bitumen membrane roofing systems. The small squared pattern of indentations (Fig. 6.4) allows for thermal movement between the bonded sheet and metal finish. The copper finish weathers similarly to traditional copper roofing

Hot bitumen bonding roofing sheet

Torching-on high-performance roofing sheet

Fig. 6.3 Laying reinforced bitumen membrane roofing with poured hot bitumen and by torching-on

systems, producing a green patina. Aluminium is available in the natural finish and stove-enamel white, grey, green or red. The aluminium and copper foil thicknesses are typically 0.08 mm and stainless steel 0.05 mm.

INVERTED ROOFS

In the inverted roof the reinforced bitumen waterproofing membrane is laid directly onto the roof deck. Non-absorbent insulation, such as extruded polystyrene, is laid onto the membrane, which is then covered with a filter sheet to prevent the ingress of excessive organic material. River-washed ballast or

Fig. 6.4 Copper-faced bitumen membrane roofing

pavings on supports protect the system from mechanical and wind damage. Inverted roofs have the advantage that the waterproof membrane is protected from thermal stress by the insulating layer. This in turn is protected from damage by the paving or ballast finish. High-performance reinforced bitumen membrane systems are suitable for inverted roofs.

Mastic asphalt

TYPES OF MASTIC ASPHALT

Mastic asphalt is a blended bitumen-based product. It is manufactured either from the bitumen produced by the distillation of crude oil, or from lake asphalt, a naturally occurring blend of asphalt containing 36% by weight of finely divided clay, mainly imported from Trinidad. The bitumen is blended with limestone powder and fine limestone aggregate to produce the standard roofing types listed in BS 6925: 1988 (Table 6.3). Most frequently specified are the BS 988T25 and the polymer-modified grades.

The effect of the finely divided clay particles within lake asphalt type BS 988T confers better laying characteristics and enhanced thermal properties; these are advantageous when the material is to be exposed to wide temperature changes, particularly in warm roof construction systems.

Mastic asphalt is usually delivered as blocks for melting on site prior to laying, although hot molten asphalt is occasionally supplied for larger contracts. Laid mastic asphalt is brittle when cold but softens in hot, sunny weather. The hardness is increased by the re-melting process, and also by the addition of further limestone aggregate. Polymer-modified mastic asphalts, usually containing styrene-butadiene-styrene block copolymers, are more durable and have enhanced flexibility and extensibility at low temperatures, allowing for greater building movements and better resistance to thermal shock. Where mastic asphalt roofs are subjected to foot or vehicle access, paving-grade mastic asphalt should be applied as a wearing layer over

Table 6.3 Mastic asphalt grades to BS 6925: 1988

Type	Composition
BS 988B	100% bitumen
BS 988T25	75% bitumen, 25% lake asphalt
BS 988T50	50% bitumen, 50% lake asphalt
Specified by manufacturers	Polymer-modified grades

the standard roofing grade material. Two key grades S and H are available; the softer grade is suitable for footways and rooftop car parks, and the harder grade for heavily stressed areas. For standard flooring, type F1076 mastic asphalt is required, for coloured flooring type F1451, and for tanking and damp-proof courses type T1097 is necessary. The flooring types of mastic asphalt are available in four grades (hard, light, medium and heavy duty) according to the required wearing properties.

ROOFING SYSTEMS

Mastic asphalt may be laid over a wide variety of flat or pitched roof decking systems to the Code of Practice BS 8218: 1998. As mastic asphalt is a brittle material it requires continuous firm support. Appropriate decks are concrete (in situ or precast), plywood (exterior grade), timber boarding (19 mm minimum thickness), particleboard, wood wool slabs (50 mm minimum thickness) and profiled metal decking. A typical concrete warm roof system is illustrated in Figure 6.5.

In dense concrete construction, a sand and cement screed is laid to falls over the in situ slab. The falls should be designed such that even with inevitable variations on site, they are never less than 1 in 80, as this is essential to ensure the immediate removal of the surface water and to prevent ponding. Plywood, profiled metal and other decking systems would be similarly laid to falls. A layer of bitumen-bonded Type 3B glassfibre roofing sheet is applied over the structure to act as a vapour check.

Insulation

Thermal insulation, to provide the necessary roof U-value, is bonded with hot bitumen. A wide variety of insulation boards or blocks, including compressed cork, high-density mineral wool, fibreboard, perlite, cellular foamed glass, high-density extruded polystyrene and polyisocyanurate, are suitable, although where insufficient rigidity is afforded by the insulation material, or if it would be affected by heat during application of the hot asphalt, it must be overlaid with firm heat-resistant boards to prevent damage during

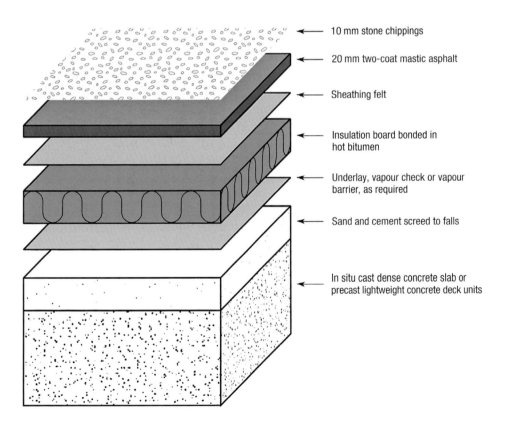

10 mm stone chippings

20 mm two-coat mastic asphalt

Sheathing felt

Insulation board bonded in hot bitumen

Underlay, vapour check or vapour barrier, as required

Sand and cement screed to falls

In situ cast dense concrete slab or precast lightweight concrete deck units

Fig. 6.5 Typical mastic asphalt roofing system

maintenance or construction. Where falls are not provided by the structure, the insulation may be set appropriately, provided that the thinnest section gives the required thermal properties. A separating layer of loose-laid geotextile material or sheathing membrane is then applied to allow differential thermal movement between the decking system and the mastic asphalt waterproof finish.

Mastic asphalt application

Mastic asphalt is laid to 20 mm in two layers on roofs up to 10° and in three layers to 20 mm on slopes greater than 10°. Horizontal surfaces, such as roof gardens, require three layers to 30 mm. Upstands of 150 mm are required to masonry, rooflights, pipes, etc., where they penetrate the roof membrane. Where adhesion on vertical surfaces is insufficient, expanded metal lathing should be used to support the mastic asphalt. The top of the upstand should be tucked into a chase or protected with an apron flashing. A layer of sharp sand is rubbed into the top of the final layer while it is hot to break up the skin of bitumen-rich material which forms at the worked surface.

Surface protection

Mastic asphalt gradually hardens over a period of a few years and should be protected from softening under bright sunlight by the application of surface protection. Reflective paint coatings (for example, titanium oxide in polyurethane resin or aluminium-pigmented bitumen) are effective until they become dirty, but for vertical surfaces they are the only appropriate measure. Reflective coatings are available in a range of colours giving differing levels of solar reflectivity. For horizontal surfaces and pitches up to 10°, a layer of 10–14 mm white stone chippings will give better protection not only from sunlight but also from ultraviolet light, which gradually degrades bitumen products. In addition, a layer of stone will act to reduce the risk of thermal shock during very cold periods. Where traffic is anticipated, the mastic asphalt should be protected with glassfibre-reinforced concrete (GRC) tiles or concrete pavings.

INVERTED ROOFS

Mastic asphalt forms a suitable waterproofing system for protected membrane inverted roofs. The application of the insulating layer over the mastic asphalt has the advantage that it protects the waterproof layer from thermal shock, impact damage and degradation by ultraviolet light. The insulation, usually extruded polystyrene boards, is held down by either gravel or precast concrete paving slabs.

Single-ply roofing systems

Single-ply roofing systems consist of a continuous membrane usually between 1 and 3 mm thick, covering any form of flat or pitched roof (Fig. 6.6). As waterproofing is reliant on the single membrane, a high quality of workmanship is required, which is normally provided by the specialist installer. In refurbishment work where the substrate may be rough, a polyester fleece may be used to prevent mechanical damage to the membrane from below. Life expectancies are typically quoted as 25 years. The wide range of membrane materials used may generally be categorised into thermoplastic and elastomeric products. In many cases the single-layer membrane is itself a laminate, incorporating either glassfibre or polyester to improve strength and fatigue resistance or dimensional stability, respectively. Both thermoplastic and elastomeric products are resistant to ageing under the severe conditions of exposure on roofs. Fixings offered by the proprietary systems include fully bonded, partially bonded, mechanically fixed and loose laid with either ballast or concrete slabs. Joints are lapped and either heat or solvent welded, usually with tetrahydrofuran (THF). A final seal of the plastic in solvent may be applied to the joint edge after the lap joint has been checked for leaks. Most manufacturers provide a range of purpose-made accessories such as preformed corners, rainwater outlet sleeves and fixings for lightning protection. The standard BS EN 13956: 2012 Annex E lists an extensive range of flexible sheet materials (plastics, rubbers and thermoplastic rubbers) used for roofing systems across Europe, but the market in the UK for non-bitumen-based products is dominated by plasticised PVC and EPDM.

THERMOPLASTIC SYSTEMS

Thermoplastic systems, made from non-cross-linked plastics, can be joined by solvent or heat welding. They generally exhibit good weathering properties and chemical resistance. The dominant thermoplastic systems are based on plasticised polyvinyl chloride (PVC) which is normally available in a range of colours. Certain PVC products contain up to 35% by weight of

Mechanically fastened

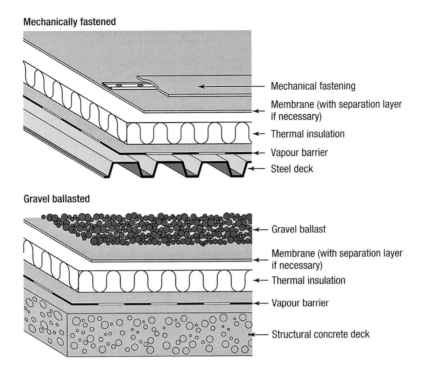

← Mechanical fastening

← Membrane (with separation layer if necessary)

← Thermal insulation

← Vapour barrier

← Steel deck

Gravel ballasted

← Gravel ballast

← Membrane (with separation layer if necessary)

← Thermal insulation

← Vapour barrier

← Structural concrete deck

Fig. 6.6 Typical single-ply roofing system, mechanically fastened and gravel ballasted

plasticisers, which can migrate to adjacent materials, leaving the membrane less flexible and causing incompatibility with extruded polystyrene insulation or bitumen products. Anticipated service lives range between 25 and 35 years. The Imperial War Museum hanger at Duxford by Foster and Partners (Fig. 6.7) is finished with a PVC roofing membrane.

Products manufactured from vinyl ethylene terpolymer (VET, a blend of 35% ethyl vinyl acetate and 65% PVC with only 4% plasticiser added as lubricant) are more compatible with bitumen- and polystyrene-based insulation products. Other products include chlorinated polyethylene (CPE) which has enhanced chemical resistant properties and chlorosulphonated

Fig. 6.7 PVC roofing membrane – Imperial War Museum, Duxford. *Architects*: Foster + Partners. *Photograph*: Arthur Lyons

polyethylene (CSM) which is highly weather resistant. Thermoplastic polyolefin (TPO) membranes, based on an alloy of ethylene propylene rubber (EPR) and polypropylene, are halogen-free and combine the temperature, chemical resistance and weldability of PVC with the flexibility of the elastomeric single-ply systems. The thermoplastic polyolefin elastomer (TPE) single-ply roofing membranes are similar to the well-established TPOs but have enhanced properties of elasticity and flexibility resulting from the particular catalytic process which generates very pure highly elastic polyolefins. They are highly resistant to UV and ozone, giving an anticipated service life in excess of 20 years.

ELASTOMERIC SYSTEMS

Elastomeric systems are dominated by ethylene propylene diene monomer (EPDM) which is a cross-linked or cured polymer. It is characterised by high elongation and good weathering resistance to ultra-violet light and ozone. The standard material is black or grey in colour but white is also available. Most products are seamed with adhesives or applied tapes as EPDM cannot be softened by solvents or heat; however, EPDM laminated with thermoplastic faces can be heat welded on site. Products may be mechanically fixed, ballasted or adhered to the substrate. Systems usually incorporate polyester or glassfibre reinforcement for increased dimensional stability. A service life in excess of 20 years may be expected. Polyisobutylene (PIB) is a relatively soft material which is easily joined using solvent and tape systems. Both PIB and EPDM are blended with carbon black to enhance stability.

Liquid coatings

A range of bitumen- and polymer-based materials is used in the production of liquid-roof waterproofing membranes. While some products are installed in new work, many products are used for remedial action on failed existing flat roofs as an economic alternative to re-roofing (Fig. 6.8). They may be appropriate when the exact location of water ingress cannot be located or when re-roofing is not practicable due to the disruption it would cause. It is essential that the nature of the existing roof is correctly determined so that an appropriate material may be applied; also failures in the substrate must be identified and rectified. The surface of the existing material must be free of loose material and dust to ensure good adhesion with the liquid coating which may be applied by brush, roller or air-less spray. While achieving a uniform thickness is difficult, the systems have the advantage of being seamless. Solar-reflective white and a range of colours are available. Without further protection, roofs should only be subjected to light maintenance pedestrian traffic. Installation should normally be carried out by specialist roofing contractors. Systems include glassfibre-reinforced polyester and reinforced elastomeric polyurethane.

BITUMEN-BASED SYSTEMS

Most bitumen systems require a primer to seal the existing roof membrane and provide a base for the liquid coating. Two or three coats of bitumen solution or emulsion will normally be required for the waterproofing layer, and a solar-reflective finish should be applied after the membrane has fully dried. A layer

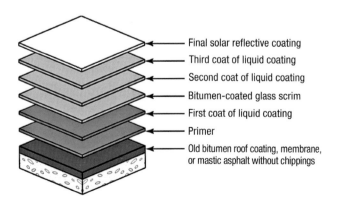

Final solar reflective coating
Third coat of liquid coating
Second coat of liquid coating
Bitumen-coated glass scrim
First coat of liquid coating
Primer
Old bitumen roof coating, membrane, or mastic asphalt without chippings

Fig. 6.8 Typical refurbishment of a membrane or mastic asphalt roof with a liquid coating system

of glassfibre reinforcement is usually incorporated during application of the waterproofing membrane to give dimensional stability. Two-component systems mixed during the spraying process cure more rapidly, allowing a seamless 4 mm elastomeric coat to be built up in one layer on either flat or pitched roofs. Where the material is ultraviolet light resistant, a solar-reflective layer may not be necessary.

Hot applied polymer-modified bitumen systems are laid at around 200°C on a conditioned substrate in two 3 mm layers with intermediate polyester reinforcement, producing a monolithic waterproof layer. A protective finish, usually 50 mm of loose aggregate, is finally applied.

POLYMER-BASED SYSTEMS

The range of polymers used for liquid roof finishes is extensive, including acrylic resins, polyurethanes, polyesters, silicones, rubber copolymers and modified bitumens. Some manufacturers offer a wide range of colours incorporating the necessary solar-reflecting properties. Glassfibre or polyester mat is used as reinforcement within the membrane layer, which is applied in a minimum of two coats. Additives to improve fire resistance, such as antimony trioxide and bromine compounds, may be incorporated into the formulations, and good fire ratings in respect of flame penetration and surface spread may be achieved with some products. Solvent-based products have the advantage of rapid drying times, while solvent-free products have 'greener' credentials and some are odour-free, reducing the disruption caused during the refurbishment of occupied buildings. A typical polyurethane system requires a primer appropriate to the particular substrate (concrete, plywood, steel, mastic asphalt or roofing membrane) followed by one-component polyurethane in which a glassfibre mat is embedded, producing the elastomeric membrane. A one-component UV-resistant polyurethane finish is applied, giving an anticipated service life of 25 years. Typical GRP systems incorporate a heavy-duty glass-fibre mat where foot traffic is present and double-layer glass fibre reinforcement for green roofs and water containment. A standard system with a 450 g/m² glassfibre mat offers a service life of 25 years.

Green roofs

Green roofs are flat or low-pitched roofs landscaped over the waterproofing layer. The landscaping may include some hard surfaces and have access for leisure and recreational functions as well as the necessary routine maintenance. Green roofs offer not only increased life expectancy for the waterproofing layer by protecting it from physical damage, ultraviolet light and temperature extremes, but also increased usable space. Environmental advantages include reduced and delayed rainwater run-off as a Sustainable Drainage

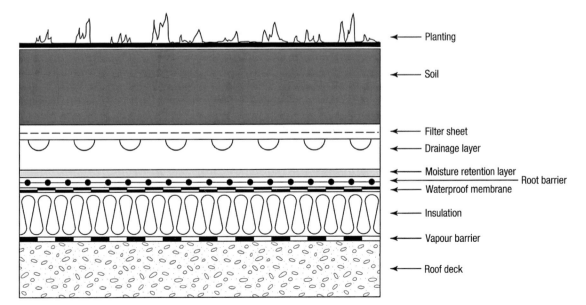

Fig. 6.9 Typical green roof system

Fig. 6.10 Extensive green roof – Westonbirt Arboretum, Gloucestershire. *Photograph:* Arthur Lyons

System (SuDS), as well as considerable environmental noise, thermal control, air quality and wildlife habitat benefits. Green roofs (Fig. 6.9) may be waterproofed using modified-bitumen high-performance membrane systems, single-ply membrane systems or mastic asphalt. Under planting, T-grade mastic asphalt should be laid to 30 mm in three layers rather than the usual two layers to 20 mm thickness. Green roofs are divided between the *intensive* and the *extensive* systems.

EXTENSIVE GREEN ROOFS

Extensive green roofs are designed to be lighter in weight, relatively cheap, not open to recreational use and requiring the minimum of maintenance. Their prime purpose is either ecological or for the environmental masking of buildings. Planting should be of drought-tolerant, wind- and frost-resistant species, such as sedums, herbs and grasses. Instant cover can be created by the installation of pre-cultivated vegetation blankets where the immediate visual effect is required. Alternatively, a mixture of seeds, plant cuttings, mulch and fertiliser is sprayed onto the growing medium, and this will mature into the finished green roof over a period of between one and two years. The complete system with planting, soil, filter sheet, drainage, moisture-retention layer and root barrier will add between 60 and 200 kg/m² loading to the roof structure, which must be capable of this additional imposed load. Typical soil depth is up to 100 mm. Limited maintenance is required to remove unwanted weeds, fill bare patches and apply organic fertilisers in the spring, and to remove dead plants and weeds in the autumn. An alternative approach to the fully planted extensive green roof is the biodiverse green roof which has some initial planting and incorporates natural features such as logs and boulders but which is then left to nature to develop with the local flora and fauna. Figure 6.10 illustrates a typical low-maintenance green roof at the Westonbirt Arboretum, Gloucestershire; a similar effect is achieved at the former Earth Centre, Doncaster, which is planted with sedum (Fig. 17.1, page 460).

Biodiverse brown roofs

Biodiverse brown roofs are another form of extensive green roof which make use of recycled materials (e.g. crushed brick and concrete, etc.) in the growing medium. The medium may be seeded initially to create growth, but subsequently the surface is essentially unmanaged to permit the natural development of flora and fauna.

INTENSIVE GREEN ROOFS

Intensive green roofs are generally designed to accept recreational activity and to include the wider range of vegetation from grass and herbaceous plants to shrubs. Depths of soil are typically between 200 and 300 mm, which, together with the necessary minimum 50 mm of water reservoir and drainage systems, generate an additional imposed load of typically 400 kg/m² on the existing or proposed structural system. Intensive green roofs may incorporate both soft and hard landscaping, and slopes of up to 20° are practicable. To conform to Health and Safety requirements edge protection (e.g. handrails) or a fall-arrest system (e.g. harness attachment points) must be incorporated into the design.

A typical intensive green roof system requires the following construction:

- soil, compost and planting – hard and soft land-scaped areas;
- filter fleece to prevent soil from blocking the drainage system;
- moisture-retention material and drainage system;
- protection mat (to prevent damage to root barrier and waterproofing layers);
- polyethylene foil (isolating layer);
- root barrier;
- waterproof layer;
- insulation;
- vapour barrier.

Where trees are required, soil depths may need to be increased to 750–1000 mm, with the associated increase in overall weight. Care must be taken to ensure that the roof membrane is not damaged by gardening implements. Intensive green roofs will require careful choice of planting to cope with the exposed conditions and regular long-term maintenance including weeding, trimming and pest control.

GREEN WALLS

Green walls (living walls) as illustrated by the market façade in Avignon, France (Fig. 6.11) are constructed with the appropriate growing medium contained in matting or blocks fixed directly to the building façade. Provision for watering and maintenance is necessary and the whole system requires refurbishment at regular intervals.

Green walls derived from climbing plants supported on stainless steel rods, ropes and mesh are more easily irrigated and maintained.

Fig. 6.11 Green wall – Les Halles, Avignon. *Photograph*: Arthur Lyons

References

FURTHER READING

Blanc, P. 2012: *The vertical garden. From nature to the city.* London: W.W. Norton.

CIBSE. 2007: *Green roofs.* Knowledge series KS11. London: Chartered Institution of Building Services Engineers.

CIRIA. 2007: *The SUDS manual.* Publication C697. London: CIRIA.

CIRIA. 2007: *Building greener. Guidance on the use of green roofs, green walls and complementary features on buildings.* Publication C644. London: CIRIA.

Dakin, K., Benjamin, L.L. and Pantiel, M. 2013: *The professional design guide to green roofs.* London: Timber Press.

Grant, G. 2006: *Green roofs and facades.* EP74. Watford: BRE.

Guertin, M. 2003: *Roofing with asphalt shingles.* USA: Taunton Press.

IKO. 2012: *Flat roofing design guide.* Wigan: IKO.

Mastic Asphalt Council. 2006: *Roofing.* Technical Guide. Hastings: Mastic Asphalt Council.

Mastic Asphalt Council. 2013: *Green roofs.* Technical Guide. Hastings: Mastic Asphalt Council.

McDonough, W. 2005: *Green roofs, ecological design and construction.* USA: Schiffer Publishing.

National Federation of Roofing Contractors. 2005: *Benefits of reinforced bitumen membrane (RBM) roofing in new build and refurbishment.* Information Sheet 29. London: National Federation of Roofing Contractors.

National Federation of Roofing Contractors. 2012: *Pitched roof underlay.* London: National Federation of Roofing Contractors.

NBS. 2009: *Sustainable drainage systems.* Newcastle-upon-Tyne: NBS.

NBS. 2010: *Green roofs. An introduction.* Newcastle-upon-Tyne: NBS.

NHBC. 2013: *NHBC Standards 2013. Part 7 – Roofs.* Milton Keynes: NHBC.

RIBA. 2008: *Roofing – technical review.* London: RIBA Publishing.

Sedlbauer, K. 2010: *Flat roof construction. Design applications.* Basel: Birkhäuser.

Single Ply Roofing Association. 2012: *Design guide for single ply roofing.* London: Single Ply Roofing Association.

STANDARDS

BS 476
   Fire tests on building materials and structures:
      Part 3: 2004   Classification and method of test for external fire exposure to roofs.

BS 743: 1970
   Materials for damp-proof courses.

BS 1446: 1973
   Mastic asphalt (natural rock asphalt fine aggregate) for roads and footways.

BS 4841
   Rigid polyurethane (PUR) and polyisocyanurate (PIR) foam for building end-use applications:
      Part 3: 2006   Specification for laminated board (roofboards) with auto-adhesively or separately bonded reinforcing facings for use as roofboard thermal insulation under built-up bituminous roofing membranes.
      Part 4: 2006   Specification for laminated board (roofboards) with auto-adhesively or separately bonded reinforcing facings for use as roofboard thermal insulation under built-up non-bituminous single-ply roofing membranes.

BS 5250: 2011
   Code of practice for control of condensation in buildings.

BS 5534: 2003
   Code of practice for slating and tiling (including shingles).

BS 6229: 2003
   Code of practice for flat roofs with continuously supported coverings.

BS 6398: 1983
   Specification for bitumen damp-proof courses for masonry.

BS 6925: 1988
   Mastic asphalt for building and engineering (limestone aggregate).

BS 8204
   Screeds, bases and in situ floorings:
      Part 5: 2004   Code of practice for mastic asphalt underlays and wearing surfaces.

BS 8217: 2005
    Reinforced bitumen membranes for roofing. Code
    of practice.
BS 8218: 1998
    Mastic asphalt roofing. Code of practice.
BS 8747: 2007
    Reinforced bitumen membranes (RBMs) for
    roofing. Guide to selection and specification.
BS ISO TR 11925–1: 1999
    Reaction to fire tests. Ignitability of building
    products subjected to direct impingement of
    flame. Guidance on ignitability.
BS 594987: 2010
    Asphalt for roads and other paved areas.
BS EN 495–5: 2013
    Flexible sheets for waterproofing. Plastic and
    rubber sheets for roof waterproofing.
BS EN 544: 2011
    Bitumen shingles with mineral and/or synthetic
    reinforcements.
BS EN 1107
    Flexible sheets for waterproofing:
        Part 1: 2000   Bitumen sheets for roof
                        waterproofing.
        Part 2: 2001   Plastic and rubber sheets for
                        roof waterproofing.
BS EN 1108: 2000
    Flexible sheets for waterproofing. Bitumen sheets
    for roof waterproofing. Determination of form
    stability.
BS EN 1109: 2013
    Flexible sheets for waterproofing. Bitumen sheets
    for roof waterproofing. Determination of
    flexibility at low temperatures.
BS EN 1110: 2010
    Flexible sheets for waterproofing. Bitumen sheets
    for roof waterproofing. Determination of flow
    resistance.
DD CEN/TS 1187: 2012
    Test methods for external fire exposure to roofs.
BS EN 1297: 2004
    Flexible sheets for roofing. Bitumen, plastic and
    rubber sheets for roof waterproofing. Method of
    artificial ageing.
BS EN 1848
    Flexible sheets for waterproofing. Determination
    of length, width and straightness:
        Part 1: 2000   Bitumen sheets for roof
                        waterproofing.
        Part 2: 2001   Plastic and rubber sheets for
                        roof waterproofing.

BS EN 1849
    Flexible sheets for waterproofing. Determination
    of thickness and mass:
        Part 1: 2000   Bitumen sheets for roof
                        waterproofing.
        Part 2: 2009   Plastic and rubber sheets for
                        roof waterproofing.
BS EN 1850
    Flexible sheets for waterproofing. Determination
    of visible defects:
        Part 1: 2000   Bitumen sheets for roof
                        waterproofing.
        Part 2: 2001   Plastic and rubber sheets for
                        roof waterproofing.
BS EN 1931: 2000
    Flexible sheets for waterproofing. Bitumen, plastic
    and rubber sheets. Determination of vapour
    transmission.
BS EN ISO 11925–2: 2010
    Reaction to fire tests. Ignitability of products
    subjected to direct impingement of flame. Single-
    flame source test.
BS EN 12039: 2000
    Flexible sheets for waterproofing. Bitumen sheets
    for roof waterproofing. Adhesion of granules.
BS EN 12310
    Flexible sheets for waterproofing. Determination
    of resistance to tearing:
        Part 1: 2000   Bitumen sheets for roof
                        waterproofing.
        Part 2: 2000   Plastic and rubber sheets for
                        roof waterproofing.
BS EN 12311
    Flexible sheets for waterproofing. Determination
    of tensile properties:
        Part 1: 2000   Bitumen sheets for roof
                        waterproofing.
        Part 2: 2013   Plastic and rubber sheets for
                        roof waterproofing.
BS EN 12316
    Flexible sheets for waterproofing. Determination
    of peel resistance of joints:
        Part 1: 2000   Bitumen sheets for roof
                        waterproofing.
        Part 2: 2013   Plastic and rubber sheets for
                        roof waterproofing.
BS EN 12317
    Flexible sheets for waterproofing. Determination
    of shear resistance of joints:
        Part 1: 2000   Bitumen sheets for roof
                        waterproofing.

Part 2: 2010   Plastic and rubber sheets for roof waterproofing.

BS EN 12591: 2009
Bitumen and bituminous binders. Specifications for paving grade bitumens.

BS EN 12597: 2014
Bitumen and bituminous binders. Terminology.

BS EN 12691: 2006
Flexible sheets for waterproofing. Bitumen, plastic and rubber sheets for roof waterproofing. Determination of resistance to impact.

BS EN 12697: 2012
Bituminous mixtures. Test methods for hot mix asphalt.

BS EN 12730: 2001
Flexible sheets for waterproofing. Bitumen, plastic and rubber sheets for roof waterproofing. Determination of resistance to static loading.

BS EN 13055
Lightweight aggregates:
     Part 2: 2004   Lightweight aggregates for bituminous mixtures.

BS EN 13108
Bituminous mixtures. Material specification:
     pr Part 1: 2013   Asphalt concrete.
     pr Part 2: 2013   Asphalt concrete for very thin layers.
     pr Part 3: 2013   Soft asphalt.
     pr Part 4: 2013   Hot rolled asphalt.
     pr Part 5: 2013   Stone mastic asphalt.
     pr Part 6: 2013   Mastic asphalt.
     pr Part 7: 2013   Porous asphalt.
     pr Part 8: 2013   Reclaimed asphalt.

BS EN 13111: 2010
Flexible sheets for waterproofing. Underlays for discontinuous roofing and walls.

BS EN 13238: 2010
Reaction to fire tests for building products. Conditioning procedures and general rules for selection of substrates.

BS EN 13416: 2001
Flexible sheets for waterproofing. Bitumen, plastic and rubber sheets for roof waterproofing. Rules for sampling.

BS EN 13501
Fire classification of construction products and building elements:
     Part 1: 2007   Classification using data from reaction to fire tests.
     Part 2: 2007   Classification using data from fire resistance tests.

     Part 3: 2005   Classification using data from fire resistance tests on products and elements used in building service installations.
     Part 4: 2007   Classification using data from fire resistance tests on components of smoke control systems.
     Part 5: 2005   Classification using data from external fire exposure to roofs tests.

BS EN 13583: 2012
Flexible sheets for waterproofing. Bitumen, plastic and rubber sheets for roof waterproofing. Hail resistance.

BS EN 13707: 2013
Flexible sheets for waterproofing. Reinforced bitumen sheets for roof waterproofing. Definitions and characteristics.

BS EN 13859
Flexible sheets for waterproofing. Definitions and characteristics of underlays:
     Part 1: 2014   Underlays for discontinuous roofing.
     Part 2: 2014   Underlays for walls.

BS EN 13924
Bitumen and bituminous binders. Specification framework for special paving grade bitumen:
     Part 1: 2014   Hard paving grade bitumens.
     Part 2: 2014   Multigrade paving grade bitumens.

BS EN 13948: 2007
Flexible sheets for waterproofing. Bitumen, plastic and rubber sheets for roof waterproofing. Determination of resistance to root penetration.

BS EN 13956: 2012
Flexible sheets for waterproofing. Bitumen, plastic and rubber sheets for roof waterproofing. Definitions and characteristics.

BS EN 13969: 2004
Flexible sheets for waterproofing. Bitumen damp-proof sheets.

BS EN 13970: 2004
Flexible sheets for waterproofing. Bitumen vapour control layers.

BS EN 13984: 2013
Flexible sheets for waterproofing. Plastic and rubber vapour control layers.

BS EN 14023: 2010
Bitumen and bituminous binders. Specification framework for polymer modified bitumens.

BS EN 14695: 2010
Flexible sheets for waterproofing. Reinforced bitumen sheets for waterproofing of concrete bridge decks and other trafficked areas of concrete.

BS EN 15812: 2011
Polymer modified bituminous thick coatings for waterproofing. Determination of crack bridging ability.

BS EN 15813: 2011
Polymer modified bituminous thick coatings for waterproofing. Determination of flexibility at low temperatures.

BS EN 15814: 2011
Polymer modified bituminous thick coatings for waterproofing. Definitions and requirements.

BS EN 15815: 2011
Polymer modified bituminous thick coatings for waterproofing. Resistance to compression.

BS EN 15816: 2011
Polymer modified bituminous thick coatings for waterproofing. Resistance to rain.

BS EN 15817: 2011
Polymer modified bituminous thick coatings for waterproofing. Water resistance.

BS EN 15818: 2011
Polymer modified bituminous thick coatings for waterproofing. Determination of dimensional stability at high temperature.

BS EN 15819: 2011
Polymer modified bituminous thick coatings for waterproofing. Reduction of the thickness of the layer when fully dried.

BS EN 15820: 2011
Polymer modified bituminous thick coatings for waterproofing. Determination of watertightness.

BS EN 15976: 2011
Flexible sheets for waterproofing. Determination of emissivity.

BS EN 16002: 2010
Flexible sheets for waterproofing. Determination of the resistance to wind load of mechanically fastened flexible sheets.

PD CEN/TS 16459: 2013
External fire exposure of roofs and roof coverings. Extended application of test results from CEN/TS 1187.

PD 476–3: 2012
Classification rules for the end-use application of test results arising from BS 476–3.

PD 6484: 1979
Commentary on corrosion at bimetallic contacts and its alleviation.

BUILDING RESEARCH ESTABLISHMENT PUBLICATIONS

BRE Digests

BRE Digest 486: 2004
Reducing the effects of climate change by roof design.

BRE Digest 493: 2005
Safety considerations in designing roofs.

BRE Digest 499: 2006
Designing roofs for climate change. Modifications to good practice guidance.

BRE Digest 528: 2013
External fire performance of roofs. A guide to test methods and classification.

BRE Good building guides

BRE GBG 36: 2000
Building a new felted flat roof.

BRE GBG 43: 2000
Insulated profiled metal roofs.

BRE GBG 51: 2002
Ventilated and unventilated cold pitched roofs.

BRE Information papers

BRE IP 7/04
Designing roofs with safety in mind.

BRE IP 13/10
Cool roofs and their application in the UK.

BRE Report

BR 504: 2009
Roofs and roofing. Performance, diagnosis, maintenance, repair and the avoidance of defects.

ADVISORY ORGANISATIONS

Flat Roofing Alliance, Roofing House, 31 Worship Street, London EC2A 2DY (0207 638 7663).

Institute of Asphalt Technology, PO Box 17399, Edinburgh EH12 1FR (01506 238 397).

Liquid Roofing and Waterproofing Association, Roofing House, 31 Worship Street, London EC2A 2DY (0207 448 3859).

Mastic Asphalt Council, PO Box 77, Hastings, East Sussex TN35 4WL (01424 814400).

Metal Cladding and Roofing Manufacturers Association, 106 Ruskin Avenue, Rogerstone, Newport, South Wales NP10 0BD (01633 895633).

National Federation of Roofing Contractors Ltd, Roofing House, 31 Worship Street, London EC2A 2DY (020 7638 7663).

Single Ply Roofing Association, Roofing House, 31 Worship Street, London EC2A 2DY (0115 914 4445).

GLASS

CONTENTS

Introduction

Glass, as exemplified by The Shard in London (Fig. 7.1), is now the fashionable architectural cladding material of choice for the twenty-first century. The Shard, designed by Renzo Piano and completed in 2012, is 309 m high and its totally glazed façade comprises 56,000 m² of glass. It is currently the tallest building in the European Union.

The term *glass* refers to materials, usually blends of metallic oxides, predominantly silica, which do not

crystallise when cooled from the liquid to the solid state. It is the non-crystalline or amorphous structure of glass (Fig. 7.2) that gives rise to its transparency.

Glass made from sand, lime and soda ash has been known in Egypt for 5000 years, although it probably originated in Assyria and Phoenicia. The earliest man-made glass was used to glaze stone beads, later to make glass beads (*circa* 2500 BC), but it was not until about 1500 BC that it was used to make hollow vessels.

Fig. 7.1 'The Shard', London. *Architect:* Renzo Piano. *Photograph:* Nick Hilder

For many centuries glass was worked by drawing the molten material from a furnace. The glass was then rolled out or pressed into appropriate moulds and finally fashioned by cutting and grinding. Around 300 BC the technique of glass blowing evolved in Assyria, and the Romans developed this further by blowing glass into moulds. Medieval glass produced in the Rhineland contained potash from the burning of wood rather than soda ash. Together with an increase in lime

content this gave rise to a less durable product which has caused the subsequent deterioration of some church glass dating from that period.

The various colours within glass derived from the addition of metallic compounds to the melt. Blue was obtained by the addition of cobalt, while copper produced blue or red and iron or chromium produced green. In the fifteenth century white opaque glass was produced by the addition of tin or arsenic, and by the seventeenth century ruby-red glass was made by the addition of gold chloride. Clear glass could only be obtained by using antimony or manganese as a decolouriser to remove the green coloration caused by iron impurities within the sand.

The Faculty of Law building at the University of Cambridge, (Fig. 7.3) illustrates the movement in the late twentieth century towards fully glazed façades, resulting in the construction industry becoming a major consumer of new glass, and a proactive force in the development of new products. The wide range of glass materials used within the building industry is classified in the standard BS 952–1: 1995.

Within the UK, thousands of tonnes of glass are recycled each year, but this is mainly domestic waste, which cannot be used for the production of window glass as this requires pure materials. Even architectural waste glass is of variable composition with contamination from wire, sealants and special glasses, making it unusable as cullet in the manufacturing process without careful sorting. However, clean cullet from glass-processing companies is recycled, and new float glass may incorporate up to 30% of recycled material, significantly reducing overall energy and CO_2 emissions. Re-melting clean cullet uses 25% less energy than making the equivalent quantity of new glass from raw materials. In addition, there is a saving on the extraction of raw materials. Excess recycled glass, not required for remaking bottles, has been used for making decorative paving surfaces and incorporated into the manufacture of bricks.

Recent experimentation with pulverised domestic glass waste has shown that when expanded by CO_2 from added calcium or magnesium carbonate or carbon black, the resulting material may be used as a lightweight aggregate in concrete. Expansion of 15–20 times gives a granular product with a density range 200–900 kg/m^3.

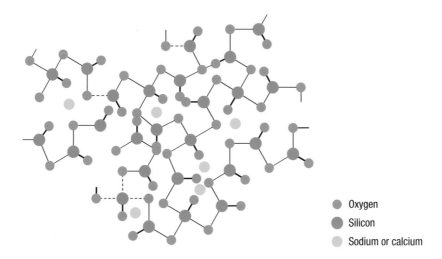

Fig. 7.2 Structure of glass (after Button D. and Pye B. (ed.) 1993. *Glass in building*. Butterworth Architecture)

Fig. 7.3 Glazed façade – Faculty of Law, University of Cambridge. *Architects*: Foster + Partners. *Photograph*: Arthur Lyons

Manufacture

COMPOSITION

Soda lime silicate glass

Modern glass is manufactured from sand (silica), soda ash (sodium carbonate) and limestone (calcium carbonate), with small additions of salt-cake (calcium sulphate) and dolomite (magnesian limestone). This gives a final composition of typically 70–74% silica, 12–16% sodium oxide, 5–12% calcium oxide, 2–5% magnesium oxide with small quantities of aluminium, iron and potassium oxides. The addition of 25% broken glass or *cullet* to the furnace mix accelerates the melting process and recycles the production waste. Most raw materials are available within the UK, although some dolomite is imported. The production process is relatively energy intensive at 15,000 kWh/m^3 (cf. concrete: 625 kWh/m^3), but the environmental pay back arises from its appropriate use in energy-conscious design. Soda lime silicate glass may be chemically strengthened by an ion exchange process which replaces small surface ions by larger ones, thus putting the surfaces and edges into compression.

Alkaline earth silicate, borosilicate and ceramic glass

Other products used within construction include the alkaline earth silicate and borosilicate glasses; these have significantly different chemical compositions giving rise to their particular physical properties. The composition of alkaline earth silicate glass is typically 55–70% silica, 5–14% potassium oxide, 3–12% calcium oxide, 0–15% aluminium oxide with quantities of zirconium, strontium and barium oxides. Borosilicate glass is typically 70–87% silica, 0–8% sodium oxide, 0–8% potassium oxide, 7–15% boron oxide, 0–8% aluminium oxide with small quantities of other oxides. A particular characteristic of borosilicate glass is that it has a coefficient of expansion one-third that of standard soda lime silicate glass, making it significantly more resistant to thermal shock in case of fire.

Ceramic glass is characterised by a near-zero coefficient of expansion, making it highly resistant to thermal shock. The composition of ceramic glass is complex, comprising typically 50–80% silica, 15–27% alumina, with small quantities of sodium oxide, potassium oxide, barium oxide, calcium oxide, magnesium oxide, titanium oxide, zirconium oxide, zinc oxide, lithium oxide and other minor constituents. The glass is initially produced by a standard float or rolling technique but subsequent heat treatment converts part of the normal glassy phase into a fine-grained crystalline form, giving rise to the particular physical properties. Ceramic glass has a high softening point compared to other glass products used in construction.

FORMING PROCESSES

Early methods

Early crown glass was formed by spinning a 4 kg cylindrical gob of molten glass on the end of a blowpipe. The solid glass was blown, flattened out and then transferred to a solid iron rod or *punty*. After reheating, it was spun until it opened out into a 1.5 m diameter disc. The process involved considerable wastage including the bullion in the centre, which nowadays is the prized piece. An alternative process involved the blowing of a glass cylinder which was then split open and flattened out in a kiln. This process was used for the manufacture of the glass for the Crystal Palace in 1851.

Subsequently, in a major development, a circular metal bait was lowered into a pot of molten glass and withdrawn slowly, dragging up a cylindrical ribbon of glass 13 m high, the diameter of the cylinder being maintained with compressed air. The completed cylinder was then detached, opened up and flattened out to produce flat window glass.

It was only by the early twentieth century with the development of the Fourcault process in Belgium and the Colburn process in America that it became possible to produce flat glass directly. A straight bait was drawn vertically out of the molten glass to produce a ribbon of glass, which was then drawn directly up a tower, or in the Colburn process turned horizontally, through a series of rollers; finally, appropriate lengths were cut off. However, such drawn sheet glass suffered from manufacturing distortions. This problem was overcome by the production of plate glass, which involved horizontal casting and rolling, followed by grinding to remove distortions and polishing to give a clear, transparent but expensive product. The process was ultimately fully automated into a production line in which the glass was simultaneously ground down on both faces. The plate glass manufacturing process is now virtually obsolete having been replaced by the *float process*, which was invented in 1952 by Pilkington and developed into commercial production by 1959. Drawn glass is only manufactured for conservation work, where gaseous and solid inclusions in the glass are required to emulate the historic material.

Float glass

A furnace produces a continuous supply of molten glass at approximately 1100°C, flowing across the surface of a large shallow bath of molten tin contained within an atmosphere of hydrogen and nitrogen, which prevents oxidation of the surface of the molten metal (Fig. 7.4). The glass ribbon moves across the molten metal, initially at a sufficiently high temperature for the irregularities on both surfaces to become evened out, leaving a flat and parallel ribbon of glass. The temperature of the glass is gradually reduced as it moves forward until, at the end of the molten tin, it is sufficiently solid at 600°C not to be distorted when supported on rollers. Thickness is controlled by the speed at which the glass is drawn from the bath. Any residual stresses are removed as the glass passes through the 200 m annealing lehr or furnace, leaving a fire-polished material. The glass is washed and substandard material discarded for recycling. The computer-controlled cutting, first across the ribbon and then the removal of the edges, is followed by stacking, warehousing and dispatch. A typical float glass plant will manufacture 5000 tonnes of glass per week, operating continuously for several years.

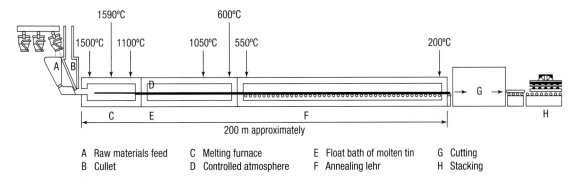

A   Raw materials feed   C   Melting furnace   E   Float bath of molten tin   G   Cutting
B   Cullet   D   Controlled atmosphere   F   Annealing lehr   H   Stacking

Fig. 7.4 Float glass process

Float glass for the construction industry is made within the thickness range 2–25 mm, although 0.5 mm is available for the electronics industry. Many surface-modified glasses are produced by incorporating metal ions into the glass within the float process, or under vacuum by magnetically enhanced cathodic sputtering; alternatively, by spraying the surface with metal oxides or silicon. Body-tinted glass, which is of uniform colour, is manufactured by blending additional metal oxides into the standard melt. The use of all-electric melting processes offers higher quality control and less environmental pollution than previously produced from earlier oil- or gas-fired furnaces.

Non-sheet products

GLASS FIBRES

Continuous filament

Continuous glass fibres are manufactured by constantly feeding molten glass from a furnace into a fore-hearth fitted with 1600 accurately manufactured holes through which the glass is drawn at several thousand metres per second. The fibres (as small as 9 microns in diameter) pass over a size applicator and are gathered together as a bundle prior to being wound up on a collet. The material may then be used as rovings, chopped strand or woven strand mats for the production of glassfibre-reinforced materials such as glassfibre-reinforced polyester (GRP), glassfibre-reinforced cement (GRC), or glassfibre-reinforced gypsum (GRG) (see Chapter 11).

Glass wool

Glass wool is made by the Crown process, which is described in Chapter 13.

CAST GLASS

Glass may be cast and pressed into shape for glass blocks and extruded sections.

Profiled sections

Profile trough sections in clear or coloured 6 or 7 mm cast glass are manufactured in sizes ranging from 232 to 498 mm wide, 41 and 60 mm deep and up to 7 m long, with or without stainless steel longitudinal wires (Fig. 7.5) (BS EN 572–7: 2012). The system may be used horizontally or vertically, as single or double glazing, and as a roofing system spanning up to 3 m. A large radius curve is possible as well as the normal straight butt jointed system and the joints are sealed with one-part translucent silicone. The standard double-glazed system has a U-value of 2.8 W/m^2 K, but this may be enhanced to 1.8 W/m^2 K by the use of low-emissivity coated glass. A variety of surfaces and colours including amber and blue are available for solar control or aesthetic reasons. The double-glazed system produces a sound reduction within the 100–3200 Hz range of typically 40 dB.

The incorporation of 16 mm-thick translucent aerogel insulated polycarbonate panels into profiled trough sections significantly increases the thermal insulating properties of the units. Airborne sound insulation is enhanced, particularly at frequencies below 500 Hz. Light transmission through the translucent aerogel panels is approximately 50%, but UV transmission is zero. (Aerogels are described in Chapter 13, Insulation materials, page 400.)

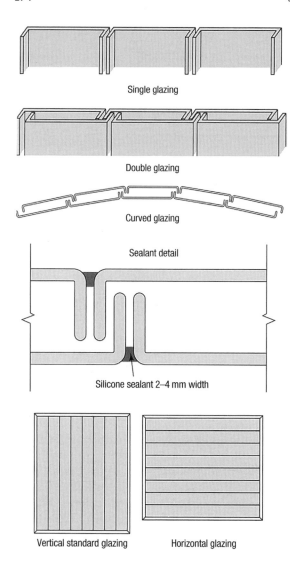

Single glazing

Double glazing

Curved glazing

Sealant detail

Silicone sealant 2–4 mm width

Vertical standard glazing   Horizontal glazing

Fig. 7.5 Profiled glass sections

Glass blocks

Glass blocks for non-load-bearing walls and partitions are manufactured by casting two half-blocks at 1050°C, joining them together at 800°C, followed by annealing at 560°C. The standard blocks (Fig. 7.6) are 115, 190, 240 and 300 mm square with thicknesses of 80, 100 or 150 mm, although rectangular and circular blocks are also available. A standard 190 × 190 × 80 mm block has a U-value of 2.9 W/m² K, but this is reduced to 1.6 W/m² K for the 150 mm-thick blocks. Walls may be curved, as illustrated in Figure 7.6. The variety of patterns, offering differing degrees of privacy, include clear, frosted, Flemish, reeded and crystal designs with

colours ranging from blue, green and grey to pink and bronze. Blocks with solar-reflective glass or incorporating white glass fibres offer additional solar control; colour may be added to either the edge coating or the glass itself. Special blocks are also available to form corners and ends; also for ventilation.

For exterior and fire-retarding applications natural or coloured mortar (2 parts Portland cement, 1 part lime, 8 parts sand) is used for the jointing. Walls may be straight or curved; in the latter case, the minimum radius varies according to the block size and manufacturer's specification. Vandal- and bullet-proof blocks are available for situations requiring higher security. For interior use, blocks may be laid with sealant rather than mortar. Glass blocks are readily available in preformed fire-rated panels for speedier installation.

Glass blocks, jointed with mortar, give a fire resistance of 60 minutes with respect to integrity (G60) and either 15 (F15), 30 (F30) or 60 (F60) minutes with respect to thermal isolation according to pr EN 1364–1: 2011. Fire stop blocks (TF30) are manufactured from 26 mm rather than 8 mm glass to offer the increased fire protection. Fire stop construction requires the incorporation of stainless steel bars within the mortar joints. Sound reduction over the 100–3150 Hz range is typically 40–42 dB for standard blocks, but up to 45–49 dB for the TF60 blocks. Visible light transmission ranges downwards from 80–60% depending upon the pattern and block size, but this is reduced significantly for coloured glass.

Glass pavers

Glass pavers are cast as either single-layer lenses or shells, and as hollow blocks where insulation is required. The standard sizes are 120, 150 and 190 mm square, with a depth of 55, 60 or 80 mm for single-layer shells and 80, 100 and 160 mm for double-layer insulating blocks. Single-layer lenses range in size from 100–300 mm with thicknesses typically between 18 and 38 mm. Surfaces may be clear or sand-blasted for non-slip action. On-site installation requires reinforced concrete construction allowing a minimum 30 mm joint between adjacent blocks and appropriate expansion joints around panels. Precast panels offer a high standard of quality control and rapid installation. Fire-rated glass block floors are usually G60 (60 minutes' integrity) and F15, F30 or F60 for 15, 30 or 60 minutes of thermal isolation, preventing significant heat transfer through the floor.

Fig. 7.6 Glass Blocks. *Photograph:* Courtesy of Glass Block Technology (www.glassblocks.co.uk)

CELLULAR OR FOAMED GLASS

The manufacture of cellular or foamed glass as an insulation material is described in Chapter 13.

Sheet products

STANDARD FLOAT GLASS SIZES

The standard thicknesses for float glass are 3, 4, 5, 6, 8, 10, 12, 15, 19 and 25 mm to typical maximum sheet sizes of 3.2 × 9 m, but sizes of up to 3.3 × 18 m have been manufactured for triple 10 mm-layer units weighing 4.5 tonnes. Units of this size are subject to the limitations of transport and on-site vacuum lifting equipment. (The U-value for standard 6 mm float glass is 5.7 W/m^2 K.)

TRADITIONAL BLOWN AND DRAWN GLASSES

Traditional blown and drawn glasses are available commercially both clear and in a wide range of colours.

Drawn glass up to 1600 × 1200 mm is optically clear but varies in thickness from 3–5 mm. It is suitable for conservation work where old glass requires replacement. Blown glass contains variable quantities of air bubbles and also has significant variations in thickness, giving it an antique appearance. Where laminated glass is required, due to variations in thickness, these traditional glasses may only be bonded to float glass with resin.

CURVED SHEET GLASS

Curved glass is defined by BS ISO 11485–1: 2011 and BS 952–2: 1980 refers to standard glass curvatures. Curved glass can be manufactured by heating annealed glass to approximately 600°C, when it softens and sag bends to the shape of the supporting mould. Sheets up to 3 × 4 m or 2 × 5 m can be bent with curvature in either one or two directions. Thicknesses between 4 and 19 mm are standard and tight curvatures can be produced for architectural feature glass. Patterned, textured, tinted, clear white and pyrolytic-coated solar-control glasses can all be curved by this technique.

Bent glass can subsequently be sand-blasted, toughened or laminated, incorporating coloured interlayers if required.

Glass laminated with PVB can also be cold bent on site and mechanically anchored to its shape-fixing substructure, leaving it under strain within the framing. Alternatively, the use of shear resistant interlayers can hold laminated glass in the required curvature after modest heating and forming in an autoclave.

SELF-CLEANING GLASS

Self-cleaning glass has an invisible hard coating 15 μm thick, which incorporates two special features. The surface incorporating titanium dioxide is photo-catalytic, absorbing ultraviolet light, which, with oxygen from the air, breaks down or loosens any organic dirt on the surface. In addition, the surface is hydrophilic, causing rainwater to spread evenly over the surface, rather than running down in droplets, thus uniformly washing the surface and preventing any unsightly streaks or spots from appearing when the surface dries. Self-cleaning glass has a slightly greater mirror effect than ordinary float glass, with a faint blue tint. It is available as normal annealed glass as well as in toughened or laminated form. The surface coating, which reduces the transmittance of the glass by about 5%, is tough, but as with any glass can be damaged by scratching. Blue and neutral solar-control versions, suitable for conservatory roofs and façades, reduce the summer solar heat gain and winter heat loss when combined with low-emissivity glass. Self-cleaning glass may be used at pitches down to 10° and combined in an Insulating Glazing Unit (IGU) with other specialist glasses.

CLEAR WHITE GLASS

Standard float glass is slightly green due to the effect of iron oxide impurities within the key raw material sand. However, clear white glass may be produced, at greater cost, by using purified ingredients. The light transmittance of clear white glass is approximately 2% greater than that of standard glass for 4 and 6 mm glazing. Unlike standard glass, which appears green at exposed polished edges, clear white glass is virtually colourless. Clear white glass is normally required for the external leaf in double-glazing units to maximise passive solar gain for an A or A+ rating in the Window Energy Rating (WER) classification. It is also beneficial when using thick float glass.

PATTERNED GLASS

A wide range of 3, 4, 6, 8 and 10 mm patterned glass is commercially available, offering a range of obscuration factors depending on the depth and design of the pattern, as illustrated in Figure 7.7. The classification of privacy levels differs between manufacturers, with 1 (lowest) to 5 (highest) for Pilkington and 1 (highest) to 10 (lowest) for Saint Gobain. The degree of privacy afforded by the various glasses is dependent not only on the pattern but also on the relative lighting levels on either side and the proximity of any object to the glass. The maximum stock sheet sizes are 2140 × 1320 mm and 2160 × 1650 mm.

Patterned glass is manufactured to BS EN 572–5: 2012, from a ribbon of molten cast glass, which is passed through a pair of rollers, one of which is embossed. Certain strong patterns require client choice of orientation, whereas the more flowing designs may need appropriate matching. Patterned glasses may be toughened, laminated or incorporated into double-glazing units for thermal, acoustic or safety considerations; a limited range is available in bronze tinted glass. Wired patterned glass (BS EN 572–6: 2012) with a thickness range of 6 to 9 mm is available with a 12.5 or 25 mm square mesh.

SCREEN-PRINTED GLASS

White or coloured ceramic frit is screen printed onto clear or tinted float glass, which is then toughened and heat soaked, causing the ceramic enamel to fuse permanently into the glass surface. Standard patterns (Fig. 7.8) or individual designs may be created, giving the required level of solar transmission and privacy. Screen-printed glass, which is colourfast and abrasion-resistant, is usually installed with the printed side as the inner face of conventional glazing. The lace pattern on the John Lewis, Leicester store is printed on two layers of glass to give views out when viewed frontally from within the retail outlet, but when viewed obliquely from street level the patterns are displaced, increasing privacy. The pattern (Fig. 7.9) draws its inspiration from Leicester's history of textiles and weaving, the John Lewis tradition of quality fabrics and the translucency of Indian saris worn by the local Asian population.

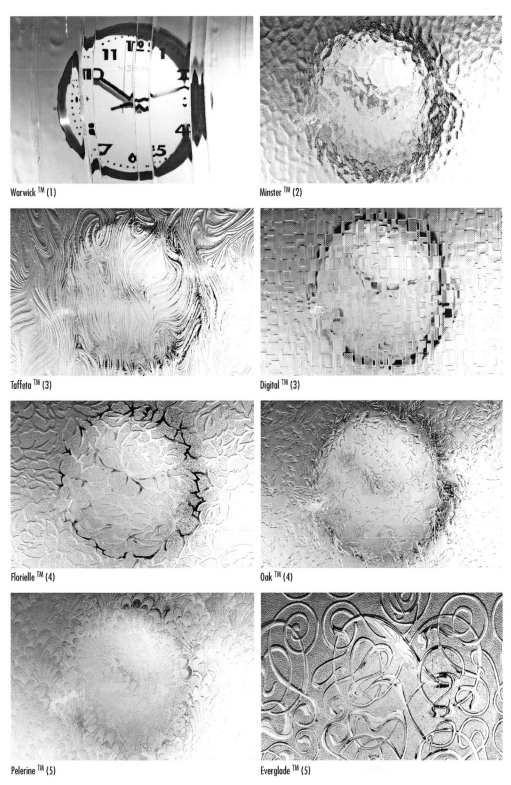

Fig. 7.7 Selection of patterned (texture) glasses with appropriate obscuration factors. Copyright © images reproduced by permission of Pilkington plc.

20% dots, 2 mm dia. 2 mm apart   40% dots, 3.6 mm dia. 1.5 mm apart

60% mesh, 3.6 mm dia. 1.5 mm apart   80% mesh, 2 mm dia. 2 mm apart

10 mm   3 mm   10 mm   3 mm

6 mm Single clear float

	20%	40%	60%	80%
Percentage coverage	20%	40%	60%	80%
Percentage light transmission	75	62	49	37
Shading co-efficient	0.86	0.76	0.66	0.56

Fig. 7.8 Screen printed glass

DECORATIVE ETCHED AND SAND-BLASTED GLASS

Plain acid etched glass, with a range of thicknesses, is available to a stock maximum sheet size of 3210 × 2400 mm as well as with a small range of patterns on 4 and 6 mm sheet (Fig. 7.10). These glasses have a low obscuration factor and should not be used in areas of high humidity, as condensation or water causes temporary loss of the pattern. Etched glasses need to be handled carefully on site, as oil, grease and finger marks are difficult to remove completely. Etched glasses may be toughened or laminated; when laminated, the etched side should be outermost to retain the pattern effect, and when incorporated into double glazing, the etched glass forms the inner leaf with the etched face towards the air gap. As with embossed glass, pattern matching and orientation is important. Similar visual effects may be achieved by sand-blasting techniques, although the surface finish is less smooth. Patterns may be clear on a frosted background, or the reverse, depending on the aesthetic effect and level of privacy required. In addition, glass with both a textured and etched finish is available.

DECORATIVE COLOURED GLASS

Traditional coloured glass windows constructed with lead cames, soldered at the intersections and wired to saddle bars, are manufactured from uniform pot, surface flashed or painted glasses. For new work, additional support is afforded by the use of lead cames with a steel core, and non-corroding saddle bars of bronze or stainless steel should be used. The basilica at Brioude, France (Fig. 7.11) is enhanced by a complete set of modernistic coloured glass windows, installed in 2008, ranging in colours from soft blues to brilliant red and orange.

A three-dimensional effect is achieved by fixing with ultraviolet-sensitive adhesive, coloured bevelled glass to clear or coloured sheet glass, the thin edges (1.5–2.5 mm) being covered with adhesive lead strip. Such effects may also be simulated by the use of coloured polyester or vinyl film and lead strip acrylic bonded to a single sheet of clear glass. The base glass may be toughened or laminated as appropriate, and the decorative coloured glass laminate may be incorporated into standard double-glazing units. The effect of coloured glass may also be achieved by using a coloured polyvinyl butyral interlayer within laminated glass. Various colour effects and tones may be achieved by combining up to four different-coloured interlayers and, if required, a variety of patterns including geometric designs, photographs, text or logos may be incorporated. These laminated coloured and patterned glasses have the same impact resistance and acoustic insulation as standard clear laminated glass of the same dimensions.

Painted glass

Painted glass is available in a range of colours including black and white. Painted glass is only suitable for internal applications, where the lacquered surface is to the rear to be protected from damage.

Fig. 7.9 Screen printed glass façade — John Lewis Department Store, Leicester. *Architects*: Foreign Office Architects. *Photographs*: Arthur Lyons

Smooth   Textured

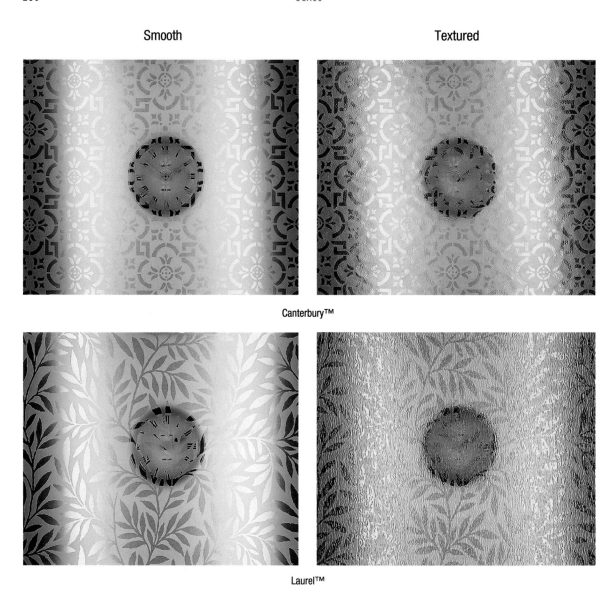

Canterbury™

Laurel™

Fig. 7.10 Etched glasses – *Oriel*™ collection. Copyright © images reproduced by permission of Pilkington plc.

GEORGIAN WIRED GLASS

Georgian wired glass (BS EN 572–3: 2012) is produced by rolling a sandwich of 13 mm electrically welded steel wire mesh between two ribbons of molten glass. This produces the standard cast 7 mm sheet, suitable when obscuration is required. For visual clarity the cast product is subsequently ground with sand and water, and then polished with jeweller's rouge to 6 mm sheet (Fig. 7.12). Both the cast and polished grades have a light transmission of 80%. Wired glass is no

stronger than the equivalent thickness of annealed glass; however, when cracked, the pieces remain held together.

On exposure to fire, the wire mesh dissipates some heat, but ultimately Georgian glass will crack, particularly if sprayed with water when hot. However, the wire mesh holds the glass in position, thus retaining its integrity and preventing the passage of smoke and flame. Accidental damage may cause breakage of the glass, but again it is retained in position by the mesh, at least until the wires are affected by corrosion.

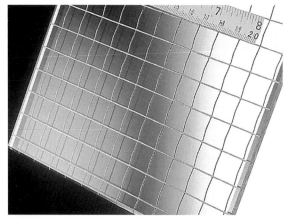

Polished

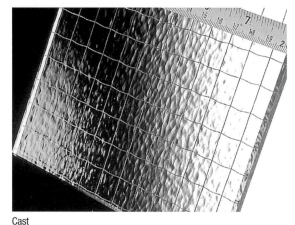

Cast

Fig. 7.12 Georgian wired glass

Fig. 7.11 Contemporary coloured glass – Basilique Saint-Julien, Ville de Brioude, France. *Artist*: Kim En Joong. *Photograph*: Arthur Lyons

Georgian wired glass is available in sheet sizes up to 1829 × 3353 mm (cast) and 1980 × 3300 mm (polished). It is easily cut and may be laminated to other glasses but cannot be toughened. Standard Georgian glass is not considered to be a *safety glass* to BS 6206: 1981, which defines three classes with decreasing impact resistance down from Class A to Class C. However, certain laminates or products with increased wire thickness do achieve the impact resistance standards for safety glass to BS 6206: 1981 and should be marked accordingly. They may therefore be used in locations requiring safety glass according to Part N of the Building Regulations and BS 6262–4: 2005.

TOUGHENED GLASS

Toughened glass (Fig. 7.13) is up to four or five times stronger than standard annealed glass of the same thickness. It is produced by subjecting pre-heated annealed glass at 650°C to rapid surface cooling by the application of jets of air. This causes the outer faces to be set in compression with balancing tension forces within the centre of the glass. As cracking within glass commences with tensile failure at the surface, much greater force may be withstood before this critical point is reached.

Toughened glass cannot be cut or worked, and therefore all necessary cutting, drilling of holes and grinding or polishing of edges must be completed in advance of the toughening process. In the *roller hearth* horizontal toughening process, some bow, roller wave distortions and end edge sag may develop but these will be within narrow tolerances; however, they may be observed in the more reflective glasses when viewed from the outside of a building. In the vertical toughening process the sheet is held by tongs, which leave slight distortions where they have gripped the glass.

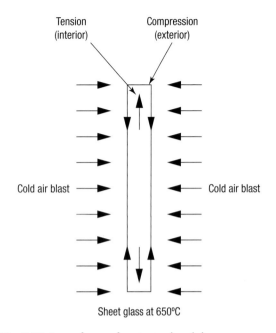

Tension   Compression
(interior)   (exterior)

Cold air blast   Cold air blast

Sheet glass at 650°C

Fig. 7.13 Process for manufacturing toughened glass

Toughened glass will withstand considerable extremes of temperature and sudden shock temperatures. If broken, it shatters into small granules which are not likely to cause the serious injuries associated with the accidental breakage of annealed glass. The form of breakage for toughened glass is defined in BS EN 12600: 2002. To be classified as a *safety glass*, toughened glass must be tested and marked according to required standard BS 6206: 1981. When toughened glass is specified for roof glazing, balustrades and spandrel panels, it is subjected to a heat soaking at 290°C, a process which is destructive to any substandard units. This removes the low risk of spontaneous breakage of toughened glass on site, caused by the presence of nickel sulphide inclusions within the material. All standard float, coated, rough-cast and some patterned glasses may be toughened. Toughened glass is defined by the standard pr EN 12150–1: 2012.

HEAT-STRENGTHENED GLASS

Heat-strengthened glass is manufactured by a process similar to toughening, but with a slower rate of cooling which produces only half the strength of toughened glass. On severe impact, heat-strengthened glass breaks into large pieces like annealed glass, and therefore alone is not a safety glass. It does not require heat soaking to prevent the spontaneous breakage that

occurs occasionally with toughened glass. Heat-strengthened glass is frequently used in laminated glass where the residual strength after fracture gives some integrity to the laminate. Typical applications include locations where resistance to wind pressure is necessary, such as the upper storeys and corners of high buildings, and also in spandrels where there is an anticipated higher risk of thermal cracking. Modern applications of laminated heat-strengthened glass include its use in roofing panels where the residual strength of the laminate prevents the glazing from falling out of the frame if broken, reducing the hazard to users below.

LAMINATED GLASS

Laminated glass (Fig. 7.14) is produced by bonding two or more layers of glass together with a plastic inter-layer of polyvinyl butyral (PVB) sheet or a polymethyl methacrylate low-viscosity resin. The low-viscosity resin is more versatile, as it allows for the manufacture of curved laminates or the incorporation of patterned glasses. The lamination process greatly increases the impact resistance over annealed glass of the same thickness. Furthermore, on impact, while the glass laminations crack, they do so without splintering or disintegration, being held together by the interlayer. The form of breakage for laminated glass is defined in BS EN 12600: 2002. Laminated glass may be defined as a *safety glass*, provided that it achieves the appropriate class standard to BS 6206: 1981.

The all-glass façade of the Prada Store in Tokyo by Herzog & de Meuron (Fig. 7.15) is manufactured from rhomboidal panes of PVB-bonded laminated glass; some units are flat, but others are curved with either the convex or concave surface to the outside, giving rise to varied colours by reflection. The perimeters of the double-glazed units are necessarily flat to ensure correct positioning and sealing.

The impact resistance of laminated glass may be increased by the use of thicker interlayers, increasing the numbers of laminates, or by the inclusion of polycarbonate sheet. Typically, glazing resistant to manual attack (BS EN 356: 2000) has two or three glass laminates, while bullet-resistant glass (BS EN 1063: 2000) has four or more glass laminates depending on the anticipated calibre and muzzle velocity. To prevent spalling, the rear face of bullet-resistant glass may be sealed with a scratch-resistant polyester film, and for fire protection Georgian wired glass may be incorporated. Laminated glasses made from annealed glass may be cut and worked after manufacture.

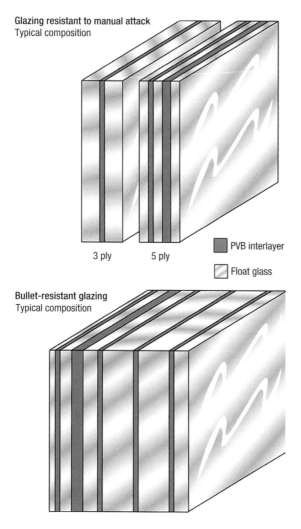

Fig. 7.14 Laminated glass – Glazing resistant to manual attack and bullet-resistant glazing

Fig. 7.15 Curved rhomboidal glass panes – Prada Store, Tokyo. *Architects*: Herzog and de Meuron. *Photograph*: Courtesy of Sam Glynn

Specialist properties for X-ray or ultraviolet light control may be incorporated into laminated glasses by appropriate modifications to the standard product. The latter reduces transmissions in the 280–380 nm wavelength ranges, which cause fading to paintings, fabrics and displayed goods.

The incorporation of specialist film interlayers within laminated glass offers further diversity. Interlayers may have variable optical and thermal transmission properties, by incorporating photochromic, thermochromic, thermotropic or electrochromic materials. Alternatively, the interlayer may diffract the incident light through specified angles, as within prisms and holograms. Thus, within a deep room, natural light can be refracted up to a white ceiling for dispersion further back within the space.

PLASTIC FILM LAMINATES

A range of transparent and translucent plastic films is readily applied internally or externally to modify the properties of glass. These include patterned films to create privacy, manifestation films to prevent people from accidentally walking into clear glass screens or doors, and reflective films to reduce solar gain and glare. Safety films applied to overhead glazing reduce the risk of injury from falling glass when nickel sulphide inclusions cause the spontaneous failure of toughened glass. Similarly, security films ensure that glass damaged by accidental impact or vandalism remains in place.

FIRE-RESISTANT GLASS

The ability of a particular glass to conform to the criteria of integrity and insulation within a fire is a measure of its fire resistance (Table 7.1). However, to achieve a specified performance in fire it is necessary to ensure that the appropriate framing, fixings and glass have all been used, as fire resistance is ultimately dependent on the whole glazing system and not the glass alone.

The European specification (BS EN 13501–2: 2007) for the fire resistance of a material or an assembly is classified by its performance against the criteria:

integrity (E), insulation (I), radiation (W) and also, not normally relevant to glass, load-bearing capacity (R). The standard time periods are 15, 20, 30, 45, 60, 90, 120, 180 and 240 minutes.

Typical classifications:

E 30	integrity only 30 minutes
EW 30	integrity and radiation protection for 30 minutes
EI 30	integrity and insulation for 30 minutes
E 60 EI 30	integrity 60 minutes and insulation 30

Integrity

Integrity-only non-insulating glass products will prevent the passage of flame, hot gases and smoke,

Table 7.1 Typical fire-resistance properties of glass

Fire resistance of non-insulating glass – integrity only

Type	Thickness (mm)	E integrity (min)
Georgian wired glass – cast	7	30–120
Georgian wired glass – polished	6	30–120
Toughened standard glass	6–10	30–60
Toughened borosilicate glass	6–12	30–120
Glass blocks	80	30–60

Fire resistance of enhanced integrity and radiation glass

Type	Thickness (mm)	EW integrity and radiation (min)
Laminated intumescent layer	7–18	30
	10–27	60
	37	90
Laminated gel interlayer	13	30
	14	60

Fire resistance of insulating glass – integrity and insulation

Type	Thickness (mm)	EI integrity and insulation (min)
Laminated intumescent glass	23	30
	32	60
Laminated gel interlayer glass	16	30
	25	60
	38	90
	58	120

Notes:
- Fire resistance data are significantly dependent upon glass thickness, glazing size, aspect ratio and the metal, timber, plastic or butt-jointed glazing system.
- Product specifications vary between manufacturers.

but will allow heat transmission by radiation and conduction; thus, ultimately further fire spread may occur through the ignition of secondary fires. Intense radiation through glass areas may render adjacent escape routes impassable. Commercial fire-resistant glass can give 60 minutes of integrity (E 60) depending upon thickness.

Georgian wired glass offers up to a 120 minutes' fire resistance rating with respect to integrity, depending on the panel size and fixings. If the glass cracks in a fire its integrity is retained, as the wire mesh prevents loss of the fractured pieces. Georgian wired glass is cheaper than insulated fire-resistant glasses and may be cut to size on site.

Toughened calcium-silica-based glasses can achieve 90 minutes' fire resistance with respect to integrity. The glazing remains intact and transparent, but will break up into harmless granules upon strong impact if necessary for escape. Toughened glass cannot be cut or worked after manufacture.

Boro-silicate glass, with a low coefficient of expansion, is more resistant to thermal shock than standard annealed glass and does not crack on exposure to fire. It may be thermally strengthened to increase its impact resistance. Certain ultra-heat-resistant ceramic glasses have a zero coefficient of thermal expansion. As a result, they can resist temperatures of up to 1000°C and the thermal shock of a cold water spray when heated by fire.

Integrity and radiation

Enhanced integrity glass (EW 30 to EW 60) gives full integrity but also reduced radiant heat transfer for 30 to 120 minutes. It is suitable for use in internal partitions and smoke-proof doors, subject to appropriate regulations. Partially insulating glass typically consists of two sheets of toughened float glass with one intumescent or gel layer.

Integrity and insulation

Insulating glasses are manufactured from float glass laminated with either intumescent or gel materials. Intumescent laminated glass has clear interlayers which, on exposure to fire, expand to a white opaque material, inhibiting the passage of conductive and radiant heat (Fig. 7.16). The glass layers adjacent to the fire crack but retain integrity owing to their adhesion with the interlayers. The fire resistance, ranging between 30 and

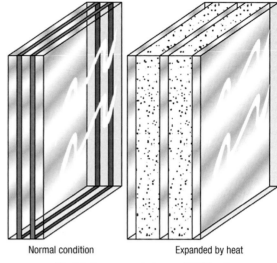

Normal condition   Expanded by heat

The interlayers expand at a round 120°C and transform into a rigid and opaque shield

Fig. 7.16 Fire-resistant glass with intumescent material laminates

180 minutes for insulation and integrity, depends on the number of laminations, usually between 3 and 5. To avoid the green tint associated with thick laminated glass, a reduced-iron-content glass may be used to maintain optimum light transmission. For exterior use, the external grade has an additional glass laminate with a protective ultraviolet filter interlayer. Laminates may be manufactured with tinted glass or combined with other patterned or solar-control glasses. Insulating glass is supplied cut to size and should not be worked on site. Double-glazed units with two leaves of intumescent laminated glass give insulation and integrity ratings of 120 minutes; alternatively, units may be formed with one intumescent laminate in conjunction with specialist solar-control glasses. The fire-resistant laminated glasses conform to the requirements of BS 6206: 1981 in respect of Class A impact resistance.

Gel-insulated glasses are manufactured from laminated toughened glass with the gel layer sandwiched between two or more glass layers. In the event of fire, the gel interlayer, which is composed of a polymer-containing aqueous inorganic salt solution, absorbs heat by the evaporation of water and produces an insulating crust. The process is repeated layer by layer. Depending on the thickness of the gel layer, fire resistance times between 30 and 120 minutes are achieved.

ENERGY CONSERVATION

The Building Regulations (Approved Document Part L1A: Conservation of fuel and power in new dwellings [2013 edition]) requires that each new dwelling, for its particular plot, has a Dwelling Fabric Energy Efficiency (DFEE) which does not exceed its mandatory Target Fabric Energy Efficiency (TFEE), measured in kWh/(m^2.year). In addition, the predicted carbon dioxide emissions, the Dwelling Emission Rate (DER), expressed as the mass of CO_2 in kilograms emitted per square metre of floor area per year, must not be greater than the Target Emission Rate (TER) determined for the equivalent notional building of the same size and shape. The TER and TFEE are calculated using the Standard Assessment Procedure (SAP 2012).

The following elemental data sets the notional dwelling specification, although developers are permitted to deviate from these specifications provided that the same level of carbon dioxide emissions and fabric energy efficiency performance is achieved or exceeded. However, no element must be worse than the limiting fabric parameters as listed in Chapter 2, page 44.

Concurrent notional specification for new dwellings:

Wall	0.18 W/m^2 K
Party wall	0.00 W/m^2 K
Floor	0.13 W/m^2 K
Roof	0.13 W/m^2 K
Window/roof window/door	1.40 W/m^2 K (whole window value)
Opaque doors	1.0 W/m^2 K
Semi-glazed doors	1.2 W/m^2 K
Airtightness	5.0 m^3/hr/m^2 at 50Pa
Ventilation type	Natural (with extract fans)
Air conditioning	None

Note: The full specification is in the Approved Document Part L1A (2013 edition).

The Building Regulations also require provision to prevent excessive summer solar gains causing high internal temperatures. This may be achieved by the appropriate use of window size and orientation, solar control with shading or specialist glazing systems, ventilation and high thermal capacity. The Building Regulations set limits on services performance and envelope airtightness (5 m^3/hr/m^2 at 50 Pa), and, in addition, heat losses from internal hot water pipes should be limited. Each new building owner should be supplied with an energy performance certificate and instructions on how to operate the building with no more fuel and power than is reasonable in the circumstances.

The Building Regulations (Approved Document Part L2A: Conservation of fuel and power in new buildings other than dwellings [2013 edition]) requires the predicted carbon dioxide emissions – Building Emission Rate (BER) to be no greater than the Target Emission Rate (TER) calculated by the Simplified Building Energy Model (SBEM) or other approved software tools.

The following are the notional specifications for new buildings other than dwellings to Building Regulations Part L2A (2013 edition), although these are flexible, provided that the limiting values (Chapter 2, page 45) are not exceeded.

Concurrent notional specification for new buildings other than dwellings:

Wall	0.26 W/m^2 K
Floor	0.22 W/m^2 K
Roof	0.18 W/m^2 K
Window (not top lit)	1.60 W/m^2 K
Window (top lit)	1.80 W/m^2 K
Airtightness	5.0 m^3/hr/m^2 at 50Pa (area $\leqslant$ 250 m^2) 3.0 m^3/hr/m^2 at 50Pa (area $\geqslant$ 250 m^2)
Heat recovery efficiency	70%

Note: The full specification is in the Approved Document Part L2A (2013 edition).

Building Regulations Approved Documents Part L1B and L2B refer to work on existing dwellings and other buildings, respectively. Guidance is quoted on the reasonable provision and standards for new extensions and replacement of existing thermal elements.

Standards for new elements in existing dwellings and other buildings to Building Regulations Part L1B and Part L2B (England):

Wall	0.28 W/m^2 K
Floor	0.22 W/m^2 K

Pitched roof – insulation at ceiling level	0.16 W/m² K
Pitched roof – insulation at rafter level	0.18 W/m² K
Flat roof – integral insulation	0.18 W/m² K
Window/roof window/ roof light for buildings of domestic character	WER Band C or better or U-value 1.6 W/m² K
All other windows/roof windows/roof lights for non-domestic buildings	U-value 1.8 W/m² K for the whole unit
Doors in domestic buildings	DSER Band E or better or U-value 1.8 W/m² K
Pedestrian doors in non-domestic buildings	U-value 1.8 W/m² K

Notes: WER refers to Window Energy Rating (Fig. 7.18). DSER refers to Door Set Energy Rating (Fig. 7.18).

Full specifications are in the Building Regulations 2010 edition Approved Documents Part L1B and L2B as amended in 2013.

Double and triple glazing

Whenever the internal surface of exterior glazing is at a lower temperature than the mean room surface temperature and the internal air temperature, heat is lost by a combination of radiation exchange at the glass surfaces, air conduction and air convection currents inside and out, as well as by conduction through the glass itself. This heat loss can be reduced considerably by the use of multiple glazing with air, partial vacuum or inert gas fill (Fig. 7.17).

Gas-filled glazing units
Double glazing reduces the direct conduction of heat by the imposition of an insulating layer of air between the two panes of poorly insulating glass. The optimum air gap is between 16 and 20 mm, as above this value convection currents between the glass panes reduce the insulating effect of the air. Argon is now normally used as the filling agent, as it has a lower thermal conductivity than air, thus reducing heat transfer by conduction. The standard is 90% argon, because of the difficulties in ensuring that all air has been flushed out of the void space. The use of krypton or even xenon within a 16 mm double-glazing gap in conjunction with low-emissivity glass can further enhance U-values. These rarer inert gases are now commercially available, although they are very expensive compared to argon (thermal conductivities λ air = 0.025, λ argon = 0.017, λ krypton = 0.009, λ xenon = 0.005 W/m K; BS EN ISO 10456: 2007). Similar, further reductions in conducted heat may be achieved by the incorporation of an additional air or argon space within triple glazing. Thin low-emissivity films suspended within the cavity can further reduce the U-values of double-glazing units to as low as 0.6 W/m² K. Typical centre pane U-values are shown in Table 7.2.

Vacuum glazing units
Vacuum glazing units incorporating a gap of only 0.2 mm have the same level of thermal efficiency as standard gas-filled low-emissivity systems. The sheets of clear float glass and low-emissivity glass are kept apart by 0.5 mm diameter microspacers at 20 mm centres, which are only visible upon close inspection. The vacuum is created through a hole in the inner pane, which is permanently sealed with a 15 mm cover. The 6.5 mm-thick units may be manufactured up to 2.4 × 1.35 m, and may be fitted as replacements for existing single glazing. The product is especially relevant to heritage building refurbishment where normal double-glazing systems would be inappropriate. The centre pane U-value for this vacuum system is 1.1 W/m² K.

Window energy ratings
The British Fenestration Rating Council (BFRC) system of window energy rating (WER) bands for complete units, based on the range A+ (best) to G (poorest), gives guidance to specifiers on energy efficiency. The ratings take into account a combination of the three key factors which affect performance; namely thermal efficiency for heat retention (U-value), solar heat gain (g-value between zero and 1) and heat loss by air leakage (L-factor). The solar heat gain and U-values relate to the whole unit, not just to the glazed areas. The leakage rate is taken for average conditions. It is determined as m³/hr/m² at a pressure of 50 Pa and then scaled down by a factor of 20 to a realistic level. It is then converted into an equivalent heat loss (W/m² K) to be added to the heat gains and losses. The window energy rating (WER) bands are colour coded from purple (A+) through light green (C) to red (F) and brown (G) and they measure the overall energy performance of the windows as the total annual energy

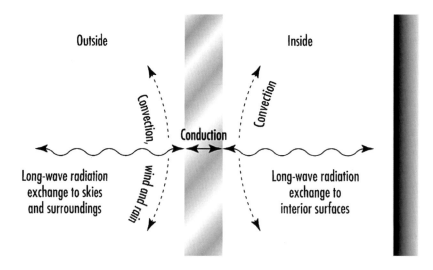

Mechanism for heat loss, through single glass

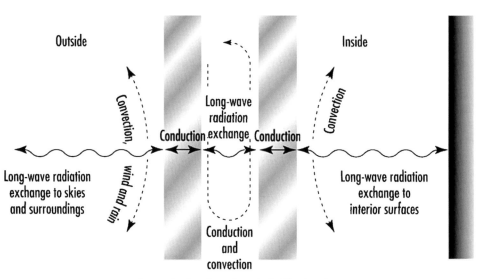

Mechanism for heat loss, through double glazed units

Fig. 7.17 Mechanism of heat loss through single and double glazing (after Button D. and Pye B. (ed.) 1993. *Glass in building*. Butterworth Architecture)

flow (kWh/(m².year)). Windows in the A band will give a small positive energy contribution to the building, while C band windows contribute an energy loss of the order of 15 kW hour/m²/year. The scheme, supported by the Glass and Glazing Federation, is based on a standard window of 1.40 × 1.23 m with a central mullion, one fixed light and one opening light to ensure equity of comparison between different products.

The target for domestic windows in new-build housing corresponds to the B rating. However, demand for the A and A+ rated products is increasing as energy costs increase. A major advancement in double-glazing units has been the development of warm edge spacer bar technology and the increased thermal efficiency of frames. Materials used for spacers include aluminium, thermoplastics and GRP. Preformed rigid GRP units, foil sealed to prevent leakage, may be gas filled for increased thermal resistance. In some cases steel reinforcement previously used in PVC-U frames is omitted or reduced to improve thermal performance.

Table 7.2 U-values for single and multiple glazing systems – typical centre pane values

Glass system	U-value (W/m² K)
Single clear glass	5.8
Double clear glass	2.8
Double clear glass with argon fill	2.6
Double clear glass with hard low-emissivity coating	1.6
Double clear glass with hard low-emissivity coating and argon fill	1.5
Double clear glass with soft low-emissivity coating	1.4
Double clear glass with soft low-emissivity coating and argon fill	1.1
Double glazed with one soft low-emissivity coating, and hard coating on surface #4 with argon fill	0.9
Triple clear glass with two low-emissivity coatings and two argon fills	0.8
Triple glazing with two low-emissivity coatings and one hard coating on inner surface with argon fills	0.5
Vacuum 0.5 mm spacing and two 3 mm clear glass	1.4

Notes:
- The data relate to 4 mm standard glass and 16 mm spacing.
- Argon is 90% filling.
- Overall window ratings will differ considerably according to the perimeter materials and construction system.
- Most double- and triple-glazing units incorporate solar control glasses as standard.
- Surface #4 is the inside of the inner leaf of double glazing.

Windows for domestic housing are dominated by PVC-U and timber systems, but aluminium and steel frames also have a significant market share. In addition, composite windows, typically with aluminium for the exterior and timber on the interior, combine the advantages of both materials including aesthetic and maintenance properties; they also have the potential for the installation of a wide range of sealed double- or triple-glazing units for thermal and acoustic efficiency. By 2011 over 25,000 tonnes of PVC-U windows and profiles were being recycled each year in the UK.

An appropriately insulated and sealed PVC-U framed system with double glazing, using 4 mm clear white glass for the outer pane, 4 mm vacuum sputtered soft-coated low-emissivity glass as the inner pane, PVC-U thermal inserts, 20 mm spacing with 90% argon fill, will achieve the A rating. The typical combination of factors to give the A rating would be an overall window U-value of 1.4 W/m² K with a solar factor (g)

of 0.45 and effective air leakage (L) of 0.0 W/m² K. An A+ rating (U-value 0.8 W/m² K) may be achieved with inert gas-filled triple glazing set in multi-chamber reinforced PVC-U frames. Normally, to achieve an overall window U-value of 0.8 W/m² K would require a centre pane U-value of 0.5–0.6 W/m² K, depending upon the performance of the frame and spacer bar.

The highest specification triple-glazed units typically offer the following U-values: timber/aluminium frame 0.63 W/m² K, timber/fibre glass frame 0.65 W/m² K, timber frame 0.69 W/m² K and PVC-U/aluminium frame 0.69 W/m² K depending on the size and exact construction details of the unit.

Relationship between BFRC window ratings and annual heat loss per year:

WER rating	Energy loss per year (kWh/(m².year))	
Band A (and A+)	greater than zero	(positive energy gain)
Band B	−10 to less than 0	(small energy loss)
Band C	−20 to less than -10	
Band D	−30 to less than -20	
Band E	−50 to less than -30	
Band F	−70 to less than -50	
Band G	more than −70	(high energy loss)

Figure 7.18 illustrates the British Fenestration Rating Council energy rating labels for windows and doors.

Door set energy ratings
The British Fenestration Rating Council (BFRC) rates the energy efficiency of doors under the Door Set Energy Ratings scheme (DSER), on an equivalent basis to that used for windows. Ratings range from A to G (Fig. 7.18), but the F and G ratings will only be appropriate for listed buildings and architecturally sensitive applications. The scheme covers all pedestrian doors manufactured in PVC-U, aluminium, timber or composite materials. Glazed patio and sliding folding doors use the window formula and rating scale. Newly installed glazed doors in existing buildings will normally require a minimum E rating to comply with the 2014 regulation. Higher standards are required for new-build and many products achieve a C rating or better.

The Door Set Energy Rating (DSER) is derived from a numerical formula incorporating the U-value of the door materials and an air leakage value. Solar gain is not included for glazed pedestrian doors.

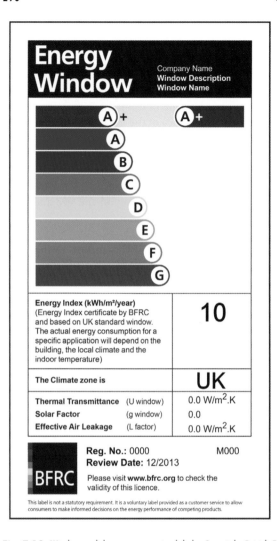

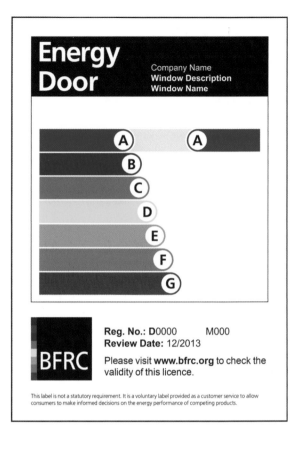

Fig. 7.18 Window and door energy rating labels. *Copyright:* British Fenestration Rating Council

BFRC door set energy rating:

DSER rating

Band A	DSER ⩾ -70
Band B	−70 < DSER ⩽ −85
Band C	−85 < DSER ⩽ −100
Band D	−100 < DSER ⩽ −115
Band E	−115 < DSER ⩽ −130
Band F	−130 < DSER ⩽ −145
Band G	< DSER ⩽ −145

For example, a Band E rating would be equivalent to a door U-value of 1.8 W/m² K.

Low-emissivity glass

Low-emissivity glasses are manufactured from float glass by the application of a transparent low-emissivity coating on one surface. The coating may be applied either on-line, within the annealing lehr at 650°C, as a pyrolytic hard-coat, or off-line after glass manufacture by magnetic sputtering under vacuum which produces a softer coat. On-line manufactured low-emissivity glasses and some off-line products may be toughened after coating; also off-line low-emissivity coatings may be applied to previously toughened glass. The on-line surface coating is more durable and is not normally damaged by careful handling.

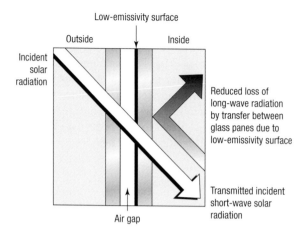

Fig. 7.19 Mechanism of heat loss control with low-emissivity glass

Low-emissivity glass functions by reflecting back into the building the longer wavelength heat energy associated with the building's occupants, heating systems and internal wall surfaces, while allowing in the transmission of the shorter wavelength solar energy (Fig. 7.19). The incoming solar energy is absorbed by the internal walls and re-radiated as longer wavelength energy, which is then trapped by the low-emissivity coating on the glass.

Low-emissivity coatings can reduce by three-quarters the radiant component of the thermal transfer between the adjacent surfaces within double glazing. The reduction in emissivity gives a decrease in U-value from 2.8 W/m² K for standard double glazing to a centre pane value of 1.0 W/m² K with low-emissivity glass in an argon-filled 4/16/4 mm double-glazing unit and 0.4 W/m² K in triple glazing. Normally the low-emissivity surface is on the cavity face of the inner leaf within sealed double-glazed units, but further gain is achieved by a low-emissivity surface on the inside of the inner leaf (surface #4). The outer leaf in the double-glazing system may be clear or any other specialist glass for security or solar control. Pyrolytic low-emissivity coatings are suitable for incorporation into secondary glazing for existing windows. The emissivities of low-E coatings range for hard coats from 0.15–0.20 and for soft coats from 0.04–0.10.

To achieve the requirements of current building legislation, low-emissivity glass double glazing has become the standard for all new building works, as in the Swiss Re ('Gherkin') building in central London (Fig. 7.20).

Fig. 7.20 Glazing and detail – Swiss Re Building 'The Gherkin', London. *Architects*: Foster + Partners. *Photographs*: Arthur Lyons

Double-glazing units

Hermetically sealed Insulated Glazing Units (IGU) are frequently manufactured with aluminium or thin stainless steel spacers which incorporate moisture-adsorbing molecular sieve or silica gel and are sealed with polyisobutylene, polyurethane, polysulphide or epoxysulphide. The primary seal is backed up by a secondary seal, usually a two-part silicone to prevent leakage, and is covered with a protective cap (Fig. 7.21). Warm edge spacers which reduce cold bridging and

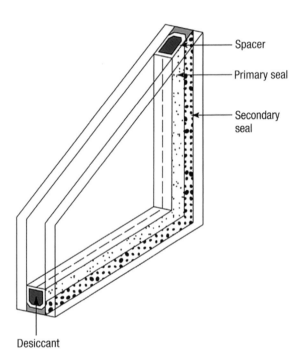

Fig. 7.21 Typical double glazing unit

the risk of condensation are manufactured in thermoplastic or GRP with some products coated with a thin foil of aluminium or stainless steel to give additional protection against gas loss. Timber frames offer good insulation. PVC frames use multi-chamber systems with minimal steel reinforcement to reduce thermal conduction. Aluminium and steel frames require the inclusion of thermal breaks to reduce the risk of surface condensation and significant heat loss

SOLAR-CONTROL GLASSES

Solar-control glasses offer a modified passage of light and heat energy compared to clear glass of the same thickness (Fig. 7.22). A descriptive code indicates the relative quantities of light and heat transmitted for a single sheet of a particular glass (e.g. 49/58 for 6 mm bronze glass) and this can be related to the equivalent data for clear float glass (87/83 for 6 mm clear glass). However, solar-control glasses are always used within an insulating glass unit (IGU), so the overall transmittance of light and heat is important in specification (Table 7.3). The method of solar control is a combination of heat absorption and heat reflection by the outer pane of the IGU. The absorption of some of the incident solar radiation by the outer pane causes the glass to warm. The glass then dissipates this absorbed heat mainly to the outside due to external air movement. The outer pane of the IGU also reflects some solar energy dependent upon any coatings applied on-line during the float process or subsequently as sputtered surface treatment. Solar-control glasses are typically grey, bronze, green, blue or silver, and may be toughened or laminated for use in hazardous areas. In addition, double-glazed units may be augmented by

Table 7.3 Characteristics of typical solar control glasses

Colour	Manufacturer's code	Light transmittance	Light reflectance	Solar radiant heat transmission	Shading coefficient	U-value
Clear	69/37	0.69	0.16	0.37	0.43	1.0
Silver	49/31	0.49	0.39	0.31	0.36	1.0
Blue	49/28	0.49	0.19	0.28	0.32	1.1
Blue/green	51/37	0.51	0.21	0.37	0.43	1.6
Green	43/27	0.43	0.17	0.27	0.31	1.6
Grey	29/31	0.29	0.10	0.31	0.36	1.6
Bronze	34/35	0.34	0.13	0.35	0.40	1.6

Note:
6 mm solar control glass, 16 mm space with 90% argon fill and 6 mm clear white glass inner pane.

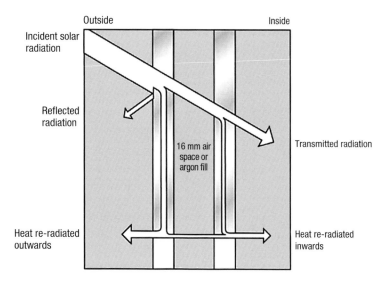

Fig. 7.22 Action of solar control glass within double glazing units

adjustable blinds or louvres. One system incorporates magnetically adjustable Venetian blinds into the 18 or 20 mm sealed double-glazing units. Solar-control glasses may be combined in double- or triple-glazing units with other specialist functions such as security and fire protection. Standard toughened glass IGU maximum sizes are typically 4800 × 2400 mm. Solar-control reflective film is available in a range of grades, thickness and colours for retrofitting onto existing glazing.

ACOUSTIC CONTROL

The level of sound reduction by glazing is influenced by the mass of the glass and the extent of air leakage around the opening lights. Sound insulation for single glazing follows the mass law – doubling the glass thickness reduces sound transmission by approximately 4 dB. Toughened, patterned and wired glasses of the same thickness respond as for plain glass, but laminated glasses based on a 1 mm layer of polymethyl methacrylate (PMMA) or a thick polyvinyl butyral (PVB) interlayer (one or two 0.4 mm layers) have enhanced sound insulation properties. The plastic interlayers, because they are soft material, change the frequency response of the composite sheet in comparison with the same weight of ordinary glass and also absorb some of the sound energy. Double glazing for sound insulation should be constructed with the component glasses differing in thickness by at least 30% to reduce sympathetic resonances; typically 6 mm and 10 mm would be effective. Polyvinyl butyral-laminated glass is

available in thicknesses of 6.8, 8.8, 9.1, 10.8, 12.8 and 13.1 mm and may be combined with other specialist function glasses in insulated glazing units.

Typical sound reduction:

8.8 mm laminated glass	R_w 37 dB
8.8 mm laminated glass/	R_w 42 dB
16 mm 90% argon/	
8 mm clear glass	
13.1 mm laminate/	R_w 50 dB
12 mm krypton/	
6 mm clear glass/	
12 mm krypton/	
9.1mm laminate	

For further enhanced sound insulation an air gap of at least 100 mm is required, with the economical optimum being 200 mm. The reveals should be lined with sound-absorbing material such as fibreboard, to reduce reverberation within the air space. All air gaps must be fully sealed with opening lights closed by multipoint locking systems and compressible seals. An air gap corresponding to only 1% of the window area can reduce the efficiency of sound insulation by 10 dB.

COLOURED ENAMELLED GLASS

Opaque or translucent enamelled glass spandrel panels for curtain walling may be manufactured to match or contrast with the range of vision area solar-control glasses. Manufactured from toughened heat-soaked glass for impact and thermal shock resistance, they may be single or double glazed, with integral glassfibre or polyurethane foam insulation and an internal finish. Panels are colourfast and scratch-resistant, and may be manufactured from plain, screen-printed or decorative glass. A standard colour range including black and white is available.

SPECIALIST GLASSES

One-way observation glass

Where unobserved surveillance is required, one-way observation mirror glass can be installed. In order to maintain privacy, the observer must be at an illumination level no greater than one-eighth that of the observed area and wear dark clothing. From the observed area the one-way observation glass has the appearance of a normal mirror. One-way observation glass is available annealed, toughened or laminated.

Mirror glass

Standard mirror glass is manufactured by the chemical deposition onto float glass of a thin film of silver, from aqueous silver and copper salt solutions. The film is then protected with two coats of paint or a plastic layer. A recent development is the production of mirror glass by chemical vapour deposition within the float glass process. Mirror glass is produced by the on-line application of a three-layer coating of silicon-silica-silicon, which acts by optical interference to give the mirror effect. Mirror glass manufactured by this process is less prone to deterioration and may be more easily toughened, laminated or bent than traditional mirror glass.

Anti-reflection glass

Treatment of standard float glass can reduce surface reflection from 8% to 2%, thus increasing the transmittance. The coating is applied equally to both faces. Although used mostly for the protection of displayed artworks, this material may be used for interior display windows and dividing screens, as well as to reduce multiple reflections from the surfaces of double-glazed units, where both sheets of glass must be anti-reflective.

Anti-condensation glass

Anti-condensation glass is an on-line low-emissivity product, available in 4 mm sheets which have a specialist external coating to keep the outer surface warmer, thus delaying the onset of external condensation.

Alarm glass

Glass containing either a ceramic loop or a series of straight wires may be incorporated into an intruder alarm system, which is activated when the glass is broken. Usually, the alarm glass conductive circuit is incorporated into laminated glass on the inside of the outer ply, or fixed to the inner face of the toughened outer pane within a double-glazing system.

Heated glass

Electrically heated laminated glass incorporates a conducting pyrolytic low-emissivity metal oxide coating which may be switched on for space heating or when there is the risk of condensation. Typical applications are in areas of high humidity, such as swimming pools, kitchens and glass roofs, particularly when there are significant differences between the internal and external temperatures. Power consumption ranges from 100 W/m^2 for homes to 1000 W/m^2 for commercial applications, depending on the internal environment and external ambient conditions. The double-glazed units, incorporating low-emissivity glass to minimise heat loss, act as a radiant heating system which can raise room temperatures up to appropriate comfort levels.

Dichroic glass

Dichroic glass has a series of metal oxide coatings which create optical interference effects. These cause the incident light to be split into the spectral colours which, depending on the angle of incidence of the light, are either reflected or transmitted. This effect may be used to create interesting colour patterns, which vary with both the movement of the sun and the observer. One material produces a range of iridescent colours from blue through purple to violet.

Cracked ice glass

Cracked ice glass is a feature material manufactured from a laminate of toughened glass between layers of annealed glass. After curing, the laminate is machined, causing the toughened glass to crack, giving the desired aesthetic effect. It is available in a range of colours and mirror finish.

Antibacterial glass

Antibacterial glass kills the majority of bacteria which fall onto its surface. The action is due to the presence of silver ions in the surface layer of glass. The silver ions interact with the bacteria and inhibit their division metabolism. Antibacterial glass may be used in locations where hygiene requirements are paramount as well as in damp environments prone to bacterial and fungal development. The material is also manufactured as mirror glass.

Electromagnetic radiation-shielding glass

Electromagnetic radiation-shielding glass may be used to protect building zones containing magnetically stored data from accidental or deliberate corruption by external electric fields. For maximum security, the conducting laminates within the composite glass should be in full peripheral electrical contact with the metal window frames and the surrounding wall surface screening.

X-ray protection glass

X-ray protection glass contains 70% lead oxide, which produces significant shielding against ionising radiation. The glass is amber in colour due to the high lead content. A 6 mm sheet of this lead glass has the equivalent shielding effect of a 2 mm lead sheet against χ and γ radiation.

Sound-generating glass

Terfenol-D is a magnetostrictive material, which when stimulated by a magnetic field expands and contracts rapidly, producing a large physical force. If a device containing Terfenol-D is attached to the smooth surface of glass and an audio input is fed into the system, then the whole sheet of glass will vibrate, acting as a loudspeaker. Thus, shop windows can be turned into loudspeakers, producing across their surface a uniform sound which can be automatically controlled to just greater than the monitored street noise level, thus avoiding sound pollution. Two devices, appropriately positioned, will generate stereo sound.

Magnetostrictive devices will operate similarly on any flat, rigid surfaces such as tabletops, work surfaces and rigid partitions. Terfenol-D is named after the metallic elements iron, terbium and dysprosium from which it is manufactured.

Light-emitting diode (LED) illuminated glass

Light-emitting diodes laminated between two sheets of glass and a PVB interlayer may be illuminated by the electrical conductivity of the surface oxide film on special glass, thus eliminating any visible electrical connections. White and colour points of light activated by complex switching sequences are all attainable. Similar products based on 5–8 mm acrylic sheets are also produced.

Manifestation of glass

Where there is a risk that glazing may be unseen, and thus cause a hazard to the users of a building, particularly large areas at entrances and in circulation spaces, the presence of the glass should be made clear

Fig. 7.23 Manifestation of glass. *Photograph:* Arthur Lyons

with solid or broken lines, decorative features or company logos at heights between 850 and 1000 mm as well as between 1400 and 1600 mm above floor level. In such circumstances the risk of impact injury must be reduced by ensuring that the glass is either robust, protected, in small panes, or breaks safely. Critical locations with a risk of human impact (BS 6262–4: 2005) are clear glazed panels from floor level to 800 mm; also floor level to 1500 mm for glazed doors and glazed side panels within 300 mm of doors. Figure 7.23 illustrates a typical example of glass manifestation.

VARIABLE TRANSMISSION (SMART) GLASSES

Variable transmission or *smart* glasses change their optical and thermal characteristics under the influence of light (photochromic), heat (thermochromic) or electric potential (electrochromic). These glasses offer the potential of highly responsive dynamic climate control to building façades. These smart materials, including thermotropic products, are also available as plastic laminates for incorporation into laminated glass systems.

Photochromic glass

Photochromic glasses incorporate silver halide crystals, which are sensitive to ultraviolet or shortwave visible light. The depth of colour is related directly to the intensity of the incident radiation and is fully reversible. For use in buildings, these materials have the disadvantage that they respond automatically to changes in solar radiation, rather than to the internal environment within the building.

Thermochromic glass

Thermochromic glasses change in transmittance in response to changes in temperature. Like photochromic glass, these materials have the disadvantage of responding to local conditions rather than to the requirements of the building's internal environment.

Electrochromic glass

Electrochromic glasses change their transmittance in response to electrical switching and are therefore the basis of *smart* windows. One system uses multilayer thin-films which become coloured in response to an applied low voltage, and are then cleared by reversal of the electric potential. The depth of coloration is dependent upon the magnitude of the applied d.c. voltage. Both manual and automatic control systems are available. Optically stable materials, which exhibit electrochromism, are the oxides of tungsten, nickel and vanadium. Electrochromic thin film systems may be laminated to any flat sheet glass. An alternative system uses a ceramic coated glass in which level of tint depends on the magnitude of the applied electric current.

Electro-optic laminates

Electrically operated vision-control glass (*Priva-lite*) consists of laminated systems of glass and polyvinyl butyral layers containing a polymer-dispersed liquid crystal layer, which can be electrically switched from transparent to white/translucent for privacy. The system uses only a low current and when switched off the glass is translucent.

Intelligent glass

Conventional glass coatings reduce both light and heat transmission. However, a coating based on tungsten-modified vanadium dioxide allows visible light through at all times, but reflects infrared radiation at temperatures above 29°C. Thus, at this temperature, further heat penetration through the glass is blocked. Therefore the intelligent glass, which has a slight yellow/green colour, admits useful solar gain in cooler conditions but cuts out excessive infra-red solar gain under hot conditions.

An alternative system uses micro-lamella layers of transparent steel in the glass. The steel layers are positioned so that they reflect sun infiltration from high angles during the hot summer months, but allow sunlight from low angles to enter relatively unimpeded, maintaining internal light levels during the winter months.

Intelligent glass façades

An intelligent glass building façade changes its physical properties in response to sensors detecting the external light and weather conditions, thus reducing the energy consumption necessary to maintain the appropriate internal environment. Therefore intelligent façades have ecological significance in reducing global greenhouse emissions and also in reducing operational building costs to clients and users.

Truly intelligent façades capitalise on the incident solar energy striking the façade of a building, adapt the skin functionality to the appropriate thermal control and solar protection, and, in addition, may generate electricity through photovoltaic cell systems. Solar control may be provided by switched electrochromic glass, or by using laminated prismatic or holographic films which deflect the solar radiation according to its angle of incidence. In addition, intelligent façades respond to air flows or ground heat sources to ensure appropriate and responsive ventilation. This function is usually achieved by the use of a double-skin façade, which acts as a ventilation cavity. During the heating season the double skin can prewarm the incoming fresh air, and when cooling is required it can remove, by convection, built-up excess heat from the double-glazing unit. Furthermore, excess heat energy may be stored for redistribution when required.

Smart shading systems respond to reduce excess incident solar radiation. Electrically controlled louvres or blinds located between two glass panes open and close according to either a solar detector or a range of weather-sensing devices. Alternatively, a computer-controlled external shading device, like a traditional Arabic *mashrabiya* lattice, will open and close according to sun exposure and changing angles of incidence throughout the year.

Fig. 7.24 Glass façade – 55 Baker Street, London. *Architects*: Make Architects. *Photograph*: Zander Olssen

Glass supporting systems

Modern glazing façades are sophisticated, as illustrated in Figure 7.24. The fixing of glazing and particularly solar-control glasses should be sufficiently flexible to allow for tolerances and thermal movements. A minimum edge clearance of at least 3 mm is required for single-glazing and 5 mm for double-glazing units. Edge cover should be sufficient to cope with the design wind loading, with a minimum normally equal to the glass or unit thickness to ensure a neat sight line. Glass thickness should be checked for suitability against predicted wind speeds, modified appropriately by consideration for the effects of local topography, building height and size of the glazing component.

The Pilkington *Planar System* offers the designer a flush and uninterrupted façade of glass. The system, which may be used for single, double or triple glazing, vertically or sloping, is designed such that each glazing unit is separately supported so that there is no restriction on the height of the building (Fig.7.25). Countersunk stainless steel bolts are embedded in the

inner leaf of the glass laminate, using the strength of the stiff polymer interlayer to spread the load. This leaves the outer face of the laminated glass free from countersunk bolt heads. The structure is supported by steel framing, glass mullions, tensile wires, or a combination of these schemes. Toughened glass is heat soaked to prevent spontaneous shattering. Thermal and wind movement can be absorbed by spring plates or small rotations in the link rod between the glazing and primary structure. Glass-to-glass butt joints are sealed with silicone.

The *Planar Triple Glazing System*, with one solar control, two low-emissivity glazings and two 16 mm air spaces, can achieve a U-value of 0.8 W/m² K. Insulating glass units are manufactured from three panes of toughened and heat-soaked glass, with thicknesses of between 4 and 19 mm according to the unit size.

The new entrance to Blackfriars Station, London (Fig. 7.26) illustrates the finely engineered Pilkington *Planar*™ system which creates a delicate façade with few visual obstructions.

Single glazed – steelwork fixing   Single glazed – glass mullion

Triple glazed – steelwork fixing   Triple glazed – glass mullion

Fig. 7.25 *Planar*™ glazing system. *Diagram:* Copyright © images reproduced by permission of Pilkington plc.

STRUCTURAL GLASS

Glass columns are frequently used as fins to restrain excessive deflection caused by wind and other lateral loads to glass façades. The fixings between the façade glazing and fin units are usually stainless steel clamps bolted through preformed holes in the toughened or laminated glass, although silicone adhesives may also be used. BS EN 13022: 2014 indicates that where exposed to UV light the outer seal of any structural glazing unit should be UV resistant. Furthermore, BS EN 15434: 2006 indicates that where exposed to UV light, silicone adhesive is the only structural

adhesive appropriate for fixing glazing units. Typically, storey-height fins are 200–300 mm wide in 12–15 mm toughened glass, fixed into aluminium or stainless steel shoes to the floor and/or glazing head. A soft interlayer between the metal fixing and the glass is incorporated to prevent stress concentrations on the glass surface and to allow for differential thermal movement between the glass and metal.

Glass is strong in compression and therefore an appropriate material for load-bearing columns and walls, provided that the design ensures sufficient strength, stiffness and stability. Generally, consideration

of buckling is the critical factor, although safety factors must be considered in relation to robustness and protection against accidental damage. Column sections need not be rectangular, as, for example, a cruciform section manufactured from laminated toughened glass gives both an efficient and elegant solution.

Glass beams are usually manufactured by laminating toughened glass. Typically, a 4 m × 600 mm-deep beam manufactured from three 15 mm toughened glass laminates could carry a load of over 5 tonnes, thus supporting at 2 m centres a 4 m-span glass roof. Glass beams may be jointed to glass columns by mortice and tenon jointing fixed with adhesive.

Single-storey all-glass structures, such as the small pavilions at the Gateshead Millennium Bridge (Fig. 7.27), entrance foyers and cantilevered canopies have been constructed using a combination of laminated and toughened glass walls, columns and beams. Usually metal fixings have been used, but where the purity of an all-glass system is required, high-modulus structural adhesives are used for invisible fixing.

Where clear double-glazed structural units are required, standard aluminium spacers may be replaced by glass spacers sealed with clear silicone, although the edges may require etching to conceal the necessary desiccant. It should be remembered that all-glass constructions require careful design consideration in relation to excessive solar gain and other environmental factors.

Fig. 7.26 Glazing system – Blackfriars Station, London. *Architects*: Pascall + Watson. *Photographs*: Arthur Lyons

Fig. 7.27 Structural glazing detail – Gateshead Millennium Bridge Pavilions. *Architects*: Wilkinson Eyre Architects. *Photograph*: Courtesy of Wilkinson Eyre Architects

Where glass is used as a load-bearing element it must conform to the structural and fire performance requirements of the relevant parts of BS EN 1365. The requirement is usually for 60 minutes of integrity and insulation, but this may be reduced to EI 30 for residential buildings no higher than 5 m.

Structural double-glazing units filled with aerogel may be considered as alternatives to façades and roofing. A 70 mm-wide cavity filled with aerogel gives a U-value of 0.25 W/m^2 K, and a 30 mm-filled cavity produces a U-value of 0.54 W/m^2 K. Solar and light transmittance through 70 mm of nanogel is 19% and 49% for a 30 mm-filled cavity. Structural requirements would include a toughened laminated inner pane.

Glass staircases, balustrades and floors

Other structural applications for glass include stairs, walkways, floors, balustrades and canopies. For stairs, laminated glass, either annealed or toughened, with a polymer interlayer may be used. The overall thickness depends upon the span and fixings, but a typical system would be a laminate of 25 and 10 mm, with the thicker layer uppermost. Glass is slippery when wet however, even if treated by sand-blasting, chemically bonded polymer resin or grooved, and should not normally be considered for external locations. All-glass staircases (Fig. 7.28) may be manufactured with glass risers and treads joined with structural silicone. The complete stair unit may be free standing, supported by vertical structural glass to which it is fixed only with metal brackets and structural silicone, or alternatively

supported on one side by masonry or concealed steel work. The steps usually incorporate sand-blasted dots or texturing to prevent slipping. Anti-slip ratings to DIN 51130: 2004 range from R9 (most slippery) to R13 (least slippery). Anti-slip glass for glass floors and stair treads is usually rated R10 or R11.

Glass flooring, usually in panels of up to a maximum of 1 m^2, is used for effect in domestic and commercial locations. Coloured or clear panels may be illuminated from below to create interesting features. Floor thicknesses depend upon the anticipated loading and fire regulations, but are normally within the range from 29 mm (19 + 10 mm) for domestic environments to 50 mm (25 + 25 mm) for commercial situations. Surfaces will normally be etched or sand-blasted for safety. The Chicago Sear's Tower has 40 mm-thick glass floor balconies projecting 1.2 m out from the 103rd floor, to give a view down onto the city.

Free-standing glass balustrades are usually manufactured from heat-soaked toughened glass ranging in thickness from 12–25 mm depending upon their height and anticipated loading. Features may include curvature and fired-on screen print decoration.

Glass canopies for commercial buildings may be manufactured from laminates of toughened glass,

Fig. 7.28 Glass staircase. *Architects and Photograph*: McInnes Gardner Architects

Fig. 7.29 Glazed roof – British Museum, London. General interior view of the Great Court and Reading Room.
Architects: Foster + Partners. *Photograph*: Courtesy of Nigel Young, Foster and Partners

which gives double security against the risk of danger-ous failure. Recent innovations also include the struc-tural use of glass rods in web compression members, and glass tubes in structural compression elements.

Structural glass must conform to the relevant parts of BS EN 1365 in relation to load-bearing properties and fire performance.

Glazing checklist

Perhaps more than any other building component, glazing is expected to perform many functions. It is therefore necessary to ensure that all factors are taken into consideration in the specification of glasses. It is evident that many of the environmental control factors are closely interrelated and the specifier must check the consequences of a design decision against all the parameters. Many of these factors are illustrated by

the feature roof to the Great Court of the British Museum in London (Fig. 7.29).

Key functions:

- view in and out by day and night;
- visual appearance by day and by night – colour and reflectivity;
- energy-conscious balance between daylight and artificial lighting;
- sky and reflected glare;
- overheating and solar control;
- shading;
- passive solar gain and energy efficiency;
- thermal comfort, U-values and condensation;
- ventilation;
- acoustic control;
- security – impact damage, vandalism and fire spread.

References

FURTHER READING

Bahamón, A. 2006: *Glass houses.* New York: Collins Design.

Behling, S. and Behling, S. (eds) 2000: *Glass, structure and technology in architecture.* München: Prestel.

Bell, M. and Kim, J. (eds) 2009: *Engineering transparency. The technical, visual and spatial effects of glass.* New York: Princeton Architectural Press.

Boubekri, M. 2008: *Daylighting, architecture and health. Design strategies.* Oxford: Architectural Press.

Bricknell, D. 2010: *Float. Pilkingtons' glass revolution.* Lancaster: Crucible Books.

Centre for Accessible Environments. 2006: *Glass in buildings. Specifiers' handbook for inclusive design.* London: CAE.

CIBSE. 2006: *Designing for improved solar shading control.* TM37. London: Chartered Institution of Building Services Engineers.

Compagno, A. 2002: *Intelligent glass façades. Material,* 5th edn. Boston: Birkhäuser.

Crosbie, M.J. 2005: *Curtain walls: Recent developments by Cesar Pelli.* Basel: Birkhäuser.

DECC. 2009: *Standard Assessment Procedure.* London: Department of Energy & Climate Change.

English Heritage. 2012: *Glass and glazing. Practical building conservation.* Farnham: Ashgate Publishing Ltd.

Evergreen. 2008: *Architecture materials. Glass.* Köln: Taschen GmbH.

Gingko Press. 2012: *Materials in architecture. Glass, stone, concrete, steel.* Hamburg: Gingko Press.

Glass and Glazing Federation. 2006: *Marking of installed safety glass.* London: Glass and Glazing Federation.

Glass and Glazing Federation. 2008: *Project green good glazing guide. Environmentally green working practices in the glass and glazing industry.* London: Glass and Glazing Federation.

Glass and Glazing Federation. 2011: *Window installation safety. Code of practice.* London: Glass and Glazing Federation.

Glass and Glazing Federation. 2011: *Guide to best practice in the specification and use of fire-resistant glazed systems,* 3rd edn. London: Glass and Glazing Federation.

Glass and Glazing Federation. 2012: *Non-vertical overhead glazing. Guide to the selection of glass from the point of view of safety.* London: Glass and Glazing Federation.

Hyatt, P. and Hyatt, J. 2004: *Designing with glass: Great glass buildings.* Australia: Images Publishing Group.

Juracek, J.A. 2006: *Architectural surfaces: Details for architects, designers and artists.* London: Thames and Hudson.

Kaltenbach, F. 2004: *Translucent materials: Glass, plastics, metals.* Basel: Birkhäuser.

Kristal, M. 2011. *Immaterial world.* New York: Monacelli Press.

Links International. 2012: *Architecture and construction in glass.* S.A.: Leading International Key Services.

McLeod, V. 2011: *Detail in contemporary architecture.* London: Laurence King.

Murray, S. 2009: *Contemporary curtain wall architecture.* Princeton, NJ: Princeton Architectural Press.

Murray, S. 2012: *Translucent building skins. Material innovations.* Abingdon: Routledge.

Pilkington. 2012: *Global glass handbook. Architectural products.* Ormskirk: Pilkington Group Ltd.

Richards, B. and Gilbert, D. 2006: *New glass architecture.* London: Laurence King.

Ryan, P., Otlet, M. and Ogden, R.G. 1998: *Steel supported glazing systems.* SCI publication 193. Ascot: Steel Construction Institute.

Saint Gobain. 2000: *Glass guide.* Goole: Saint Gobain.

Schittich, C., Staib, G., Balkow, D., Schuler, M. and Sobek, W. 2007: *Glass construction manual,* 2nd edn. Basel: Birkhäuser.

Slessor, C. 2011: *Glass houses. Inspirational homes and features in glass.* London: Ryland Peters & Small.

Steel Window Association. 2010: *Specifiers guide to steel windows,* 10th edn. London: Steel Window Association.

Uffelen, C. Van. 2009: *Clear glass. Creating new perspectives.* Hamburg: Braun.

Watts, A. 2004: *Modern construction. Facades.* Austria: Springer-Verlag.

Watts, A. 2006: *Facades. Technical review.* London: RIBA Publishing.

Weller, B. 2009: *Detail practice. Glass in building.* Basel: Birkhäuser.

Wigginton, M. 2002: *Glass in architecture,* 2nd edn. London: Phaidon.

Wurm, J. 2007: *Glass structures. Design and construction of self-supporting skins.* Basel: Birkhäuser.

STANDARDS

BS 476
Fire tests on building materials and structures.

BS 644: 2012
Timber windows and doorsets. Fully finished factory-assembled windows and doorsets of various types.

BS 952
Glass for glazing:
Part 1: 1995   Classification.
Part 2: 1980   Terminology for work on glass.

BS 3447: 1962
Glossary of terms used in the glass industry.

BS 4255
Rubber used in preformed gaskets for weather exclusion from buildings:
Part 1: 1986   Specification for non-cellular gaskets.

BS 4873: 2009
Aluminium alloy window and door sets. Specification.

BS 4904: 1978
Specification for external cladding for building purposes.

BS 5051
Bullet-resistant glazing:
Part 1: 1988   Specification for glazing for interior use.

BS 5252: 1976
Framework for colour co-ordination for building purposes.

BS 5357: 2007
Code of practice for installation of security glazing.

BS 5516
Patent glazing and sloping glazing for buildings:
Part 1: 2004   Code of practice for sloping and vertical patent glazing.
Part 2: 2004   Code of practice for sloping glazing.

BS 5821
Methods for rating the sound insulation in buildings and of building elements:
Part 3: 1984   Method for rating the airborne sound insulation of façade elements and façades.

BS 6100
Building and civil engineering terms:
Part 1: 2004   Vocabulary. General terms.

BS 6180: 2011
Barriers in and about buildings. Code of practice.

BS 6206: 1981
Specification for impact performance requirements for flat safety glass and safety plastics for use in buildings.

BS 6262
Glazing for buildings:
Part 1: 2005   General methodology for the selection of glazing.
Part 2: 2005   Code of practice for energy, light and sound.
Part 3: 2005   Code of practice for fire, security and wind loading.
Part 4: 2005   Code of practice for safety related to human impact.
Part 6: 2005   Code of practice for special applications.
Part 7: 2005   Code of practice for the provision of information.

BS 6375
Performance of windows and doors:
Part 1: 2009   Classification for weather-tightness and guidance on selection and specification.
Part 2: 2009   Classification for operation and strength characteristics and guidance on selection and specification.
Part 3: 2009   Classification for additional performance characteristics and guidance on selection and specification.

BS 6510: 2010
Steel-framed windows and glazed doors. Specification.

BS 7412: 2007
Specification for windows and door sets made from unplasticised polyvinyl chloride (PVC-U) extruded hollow profiles.

BS 8000
Workmanship on building sites:
Part 7: 1990   Code of practice for glazing.

BS 8206–2: 2008
Lighting for buildings. Code of practice for daylighting.

BS 8213
Windows, doors and rooflights:
Part 1: 2004   Design for safety in use and during cleaning of windows.

Part 4: 2007   Code of practice for the survey and installation of windows and external door sets.

BS 8214: 2008
Code of practice for fire door assemblies.

pr BS ISO 9050: 2014
Glass in building. Determination of light transmittance, solar direct transmittance, total solar energy transmittance, ultraviolet transmittance and related glazing factors.

BS ISO 11485
Glass in building. Curved glass:
Part 1: 2011   Terminology and definitions.
Part 2: 2011   Quality requirements.
pr Part 3: 2012 Requirements for thermally tempered and laminated curved glass.

pr BS ISO 18178: 2014
Glass in building. Laminated solar PV glass.

BS ISO 18292: 2011
Energy performance of fenestration systems for residential buildings. Calculation procedure.

pr BS ISO 29584: 2014
Glass in building. Pendulum impact testing and classification of safety glass.

BS EN 356: 2000
Glass in building. Security glazing. Classification of resistance to manual attack.

BS EN 357: 2004
Glass in building. Fire resistant glazed elements with transparent or translucent products.

BS EN 410: 2011
Glass in building. Determination of luminous and solar characteristics of glazing.

BS EN 572
Glass in building. Basic soda lime silicate glass products:
Part 1: 2012   Definitions and general properties.
Part 2: 2012   Float glass.
Part 3: 2012   Polished wired glass.
Part 4: 2012   Drawn sheet glass.
Part 5: 2012   Patterned glass.
Part 6: 2012   Wired patterned glass.
Part 7: 2012   Wired or unwired channel shaped glass.
Part 8: 2012   Supplied and final cut sizes.
Part 9: 2004   Evaluation of conformity/ product standard.

BS EN 673: 2011
Glass in building. Determination of thermal transmittance (U-value). Calculation method.

BS EN 674: 2011
Glass in building. Determination of thermal transmittance (U-value). Guarded hot plate method.

BS EN 675: 2011
Glass in building. Determination of thermal transmittance (U-value). Heat flow meter method.

BS EN 1026: 2000
Windows and doors. Air permeability. Test method.

BS EN 1036
Glass in building. Mirrors from silver-coated float glass for internal use:
Part 1: 2007   Definitions, requirements and test methods.
Part 2: 2008   Evaluation of conformity. Product standard.

BS EN 1051
Glass in building. Glass blocks and glass pavers:
Part 1: 2003   Definitions and description.
Part 2: 2007   Evaluation of conformity. Product standard.

BS EN 1063: 2000
Glass in building. Security glazing. Classification of resistance against bullet attack.

BS EN 1096
Glass in building. Coated glass:
Part 1: 2012   Definitions and classification.
Part 2: 2012   Class A, B and S coatings.
Part 3: 2012   Class C and D coatings.
pr Part 4: 2012 Evaluation of conformity. Product standard.
pr Part 5: 2011 Test method and classification for the self-cleaning performances of coated glass surfaces.

BS EN 1279
Glass in building. Insulating glass units:
Part 1: 2004   Generalities, dimensional tolerances.
Part 2: 2002   Requirements for moisture penetration.
Part 3: 2002   Requirements for gas leakage rate.
Part 4: 2002   Methods of test for the physical attributes of edge seals.
Part 5: 2005   Evaluation of conformity.
Part 6: 2002   Factory production control and periodic tests.

BS EN 1364
Fire resistance tests for non-loadbearing elements:
pr Part 1: 2011 Walls.

Part 2: 1999   Ceilings.
Part 3: 2014   Curtain walling. Full configuration.
Part 4: 2014   Curtain walling. Part configuration.

BS EN 1365

Fire resistance tests for loadbearing elements:

Part 1: 2012   Walls.
pr Part 2: 2012 Floors and roofs.
Part 3: 2000   Beams.
Part 4: 1999   Columns.
Part 5: 2004   Balconies and walkways.
Part 6: 2004   Stairs.

BS EN 1522: 1999

Windows, doors, shutters and blinds. Bullet resistance. Requirements and classification.

BS EN 1634

Fire resistance and smoke control tests for door, shutter and openable window assemblies:

Part 1: 2008   Fire resistance tests for doors, shutters and openable windows.
Part 2: 2008   Fire resistance characterisation test for elements of building hardware.
Part 3: 2004   Smoke control test for door and shutter assemblies.

BS EN 1748

Glass in building. Special basic products:

Part 1.1: 2004   Borosilicate glasses. Definition.
Part 1.2: 2004   Borosilicate glasses. Conformity.
Part 2.1: 2004   Glass ceramics. Definitions.
Part 2.2: 2004   Glass ceramics. Conformity.

BS EN 1863

Glass in building. Heat strengthened soda lime silicate glass:

Part 1: 2011   Definition and description.
Part 2: 2004   Evaluation of conformity.

BS EN ISO 10077

Thermal performance of windows, doors and shutters:

Part 1: 2006   Calculation of thermal transmittance. General.
Part 2: 2012   Calculation of thermal transmittance. Numerical method for frames.

BS EN ISO 10456: 2007

Building materials and products. Hygrothermal properties. Tabulated design values and procedures for determining the declared and design thermal values.

BS EN 12150

Glass in building. Thermally toughened soda lime silicate safety glass:

pr Part 1: 2012   Definition and description.
Part 2: 2004   Evaluation of conformity.

BS EN 12337

Glass in building. Chemically strengthened soda lime silicate glass:

Part 1: 2000   Definition and description.
Part 2: 2004   Evaluation of conformity/ product standard.

BS EN 12464–1: 2011

Light and lighting. Lighting of work places. Indoor work places.

BS EN 12519: 2004

Windows and pedestrian doors. Terminology.

BS EN ISO 12543

Glass in building. Laminated glass and laminated safety glass:

Part 1: 2011   Definitions and descriptions.
Part 2: 2011   Laminated safety glass.
Part 3: 2011   Laminated glass.
Part 4: 2011   Test methods for durability.
Part 5: 2011   Dimensions and edge finishing.
Part 6: 2011   Appearance.

BS EN ISO 12567

Thermal performance of windows and doors. Determination of thermal transmittance:

Part 1: 2010   Complete windows and doors.
Part 2: 2005   Roof windows and other projecting windows.

BS EN 12578: 2011

Glass in building. Glazing and airborne sound insulation. Product description and determination of properties.

BS EN 12600: 2002

Glass in building. Impact test method and classification for glass.

BS EN 12608: 2003

Unplasticised polyvinyl chloride (PVC-U) profiles for the fabrication of windows and doors.

BS EN 12758: 2011

Glass in building. Glazing and airborne sound insulation.

BS EN 12898: 2001

Glass in building. Determination of the emissivity.

BS EN 13022

Glass in building. Structural sealant glazing:

Part 1: 2014   Glass products for structural sealant glazing systems for supported and unsupported monolithic and multiple glazing.

Part 2: 2014　Assembly rules.

BS EN 13024

Glass in building. Thermally toughened borosilicate safety glass:

Part 1: 2011　Definition and description.
Part 2: 2004　Evaluation of conformity.

BS EN 13119: 2007

Curtain walling. Terminology.

BS EN 13363

Solar protection devices combined with glazing:

Part 1: 2003　Calculation of solar and light transmission. Simplified method.
Part 2: 2005　Calculation of total solar energy transmittance and light transmittance. Detailed calculation method.

BS EN 13501

Fire classification of construction products and building elements:

Part 1: 2007　Classification using data from reaction to fire tests.
Part 2: 2007　Classification using data from fire resistance tests.

BS EN 13541: 2012

Glass in building. Security glazing. Testing and classification.

pr EN 13830: 2013

Curtain walling. Product standard.

BS EN 13947: 2006

Thermal performance of curtain walling. Calculation of thermal transmittance.

BS EN 14178

Glass in building. Basic alkaline earth silicate glass products:

Part 1: 2004　Float glass.
Part 2: 2004　Evaluation of conformity.

BS EN 14179

Glass in building. Heat soaked thermally toughened soda lime silicate safety glass:

Part 1: 2005　Definition and description.
Part 2: 2005　Evaluation of conformity. Product standard.

BS EN 14321

Glass in building. Thermally toughened alkaline earth silicate safety glass:

Part 1: 2005　Definition and description.
Part 2: 2005　Evaluation of conformity. Product standard.

BS EN 14351

Windows and doors. Product standard, performance characteristics:

Part 1: 2006　Windows and external pedestrian doorsets.
Part 2: 2014　Internal pedestrian doorsets.

BS EN ISO 14438: 2002

Glass in building. Determination of energy balance value. Calculation method.

BS EN 14449: 2005

Glass in building. Laminated glass and laminated safety glass. Evaluation of conformity.

BS EN 14600: 2005

Doorsets and openable windows with fire resisting and/or smoke control characteristics. Requirements and classification.

BS EN 14963: 2006

Roof coverings. Continuous rooflights of plastics. Classification.

BS EN 15193: 2007

Energy performance of buildings. Energy requirements for lighting.

BS EN 15254

Extended application of results from fire-resistance tests. Non-loadbearing walls:

Part 4: 2008　Glazed constructions.
Part 6: 2014　Curtain walling.

BS EN 15269–1: 2010

Extended application of test results from fire-resistance and/or smoke control for door, shutter and openable window assemblies. General requirements.

BS EN 15434: 2006

Glass in building. Product standard for structural and/or UV resistant sealant.

BS EN 15682

Glass in building. Heat soaked thermally toughened alkaline earth silicate safety glass:

Part 1: 2013　Definition and description.
Part 2: 2013　Evaluation of conformity. Product standard.

BS EN 15683

Glass in building. Thermally toughened soda lime silicate channel shaped safety glass:

Part 1: 2013　Definition and description.
Part 2: 2013　Evaluation of conformity. Product standard.

BS EN 15998: 2010

Glass in building. Safety in case of fire, fire resistance. Glass testing methodology for the purpose of classification.

pr EN 16477
    Glass in building. Painted glass for internal use:
        Part 1: 2012   Requirements.
        Part 2: 2012   Evaluation of conformity.

pr EN 16759: 2014
    Structural sealant glazing systems (SSGS).

PAS 24: 2012
    Enhanced security performance requirements for doorsets and windows in the UK.

PD 6512
    Use of elements of structural fire protection:
        Part 3: 1987   Guide to the fire performance of glass.

BUILDING RESEARCH ESTABLISHMENT PUBLICATIONS

BRE Digests

BRE Digest 453: 2000
    Insulating glazing units.

BRE Digest 457: 2001
    The Carbon Performance Rating for offices.

BRE Digest 497: 2005
    Factory glazed windows (Parts 1 and 2).

BRE Digest 498: 2006
    Selecting lighting controls.

BRE Good building guide

BRE GBG 61: 2004
    Lighting (Parts 1, 2 and 3).

BRE Information papers

BRE IP 2/02
    Control of solar shading.

BRE IP 3/02
    Whole life performance of domestic automatic window controls.

BRE IP 11/02
    Retrofitting solar shading.

BRE IP 17/03
    Impact of horizontal shading devices on peak solar gains through windows.

BRE IP 1/05
    Impact standards for glass.

BRE IP 2/06
    Rain noise from glazed and lightweight roofing.

BRE IP 10/07
    Safer floors.

BRE Reports

BR 443: 2006
    Conventions for U-value calculations.

FB 55: 2013
    Security glazing. Is it all it's cracked up to be?

FB 66: 2014
    Environmental impact of windows.

ADVISORY ORGANISATIONS

British Glass Manufacturers Confederation, 9 Churchill Way, Chapeltown, Sheffield, South Yorkshire S35 2PY (0114 2901850).

Glass & Glazing Federation, 54 Ayres Street, London SE1 1EU (0207 939 9101).

Plastics Window Federation, Federation House, 85–87 Wellington Street, Luton, Bedfordshire LU1 5AF (01582 456147).

Steel Window Association, 42 Heath Street, Tamworth, Staffordshire B79 7JH (0844 249 1355).

Wood Window Alliance, The Building Centre, 26 Store Street, London WC1E 7BT (0844 209 2610).

8

CERAMIC MATERIALS

CONTENTS

Introduction

Ceramic materials, manufactured from fired clay, have been used in construction since at least 4000 BC in Egypt, and represent the earliest manufactured building materials. While the strict definition of ceramics includes glass, stone and cement, this chapter deals only with the traditional ceramics based on clays. The variety of traditional ceramic products used within the building industry arises from the wide range of natural and blended clays used for their production. Manufacturers' energy consumption and landfill waste is reduced where possible by the recycling of substandard products. The roof of the spectacular Sydney Opera House (Fig. 8.1) is surfaced with white ceramic tiles which reflect the changing light associated with the time of day.

The Equality Act 2010 requires accessible environments for disabled persons in places of work or public spaces. In respect of ceramic materials this requires a good visual contrast for the visually impaired between various elements of a building by the use of Light Reflectance Values (LRV), measured on a scale of 0 to 100. A minimum of 30 LRV points between adjacent surfaces is advocated. This is particularly vital between floors and walls, and where provided, the skirting should visually contrast with the floor. Tactile floor tiles, usually with a corduroy profile, are used as warning features in advance of stairs, ramps and platforms.

CLAY TYPES

Clays are produced by the weathering of igneous rocks, typically granite, which is composed mainly of feldspar, an alumino-silicate mineral. Clays produced within the vicinity of the parent rock are known as *primary clays*. They tend to be purer materials, less plastic and more vulnerable to distortion and cracking on firing. Kaolin ($Al_2O_3 \cdot 2SiO_2 \cdot 2H_2O$), which is the purest clay, comes directly from the decomposition of the feldspar

Fig. 8.1 Ceramic tile roof and detail – Sydney Opera House. *Architects*: Jørn Utzon and Ove Arup. *Photographs*: Arthur Lyons

in granite. Secondary clays, which have been transported by water, have a higher degree of plasticity, and fire to a buff or brown colour depending on the nature and content of the incorporated oxides. Generally, secondary clays, laid down by the process of sedimentation, have a narrower size distribution and their particulate structure is more ordered.

The most common clay minerals used in the manufacture of building materials are kaolin, illite (a micaceous clay) and montmorillonite, a more plastic clay of variable composition. Clay crystals are generally hexagonal in form, and in pure kaolin the crystals are built up of alternating layers of alumina and silica (Fig. 8.2). However, in illite and montmorillonite clays,

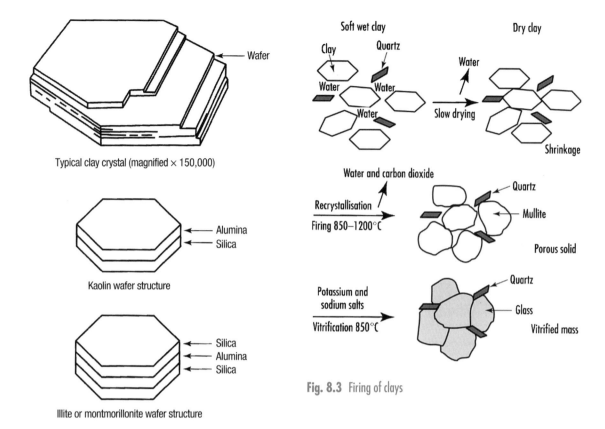

Typical clay crystal (magnified × 150,000)

Kaolin wafer structure

Illite or montmorillonite wafer structure

Fig. 8.2 Structure of clays

Fig. 8.3 Firing of clays

the variable composition produced by sedimentation yields more complex crystal structures.

Ball clays are secondary clays containing some organic matter which is burnt off during the firing process; they tend to have a fine grain size which makes them plastic. When fired alone they have a high shrinkage and produce a light-grey or buff ceramic, but they are usually blended into other clays such as kaolin to make a workable clay. Terracotta clays contain significant proportions of iron oxide which gives rise to the characteristic red colour on firing. While the major clay materials used in the manufacture of ceramics are kaolin, illite, feldspar and ball clay, chalk, quartz and other minor constituents are frequently incorporated to produce the required ceramic properties on firing.

Water in clay

Moist clay contains both chemically and physically bonded water. It is the latter which permeates between the clay particles, allowing them to slide over each other during the wet forming processes. As the formed clay slowly dries out before firing, a small proportion of the residual physically bonded water holds the clay in shape. On firing, the last of the physically bonded water is removed as the temperature exceeds 100°C.

MANUFACTURING PROCESSES

Clay products are formed by either wet or dry processes. In the former case, the artefacts after extrusion must be dried slowly prior to firing, allowing for shrinkage without cracking. Where a high level of dimensional accuracy is required, as in most wall and floor tiles, a dry process is used in which powdered clay is compressed into the required form. The standard BS EN 14411: 2012 defines all ceramic tiles according to whether they are produced by extrusion (method A) or dry pressing (method B), in addition to their water absorption properties, irrespective of their end use.

During firing of the clay, as the temperature is gradually increased, the majority of the chemically bonded water is removed by 500°C. At 800°C, carbonaceous matter has been burnt off as carbon dioxide, and the sintering process commences, at first producing a highly porous material. As the temperature is further raised towards 1200°C, the alumina and silica components recrystallise to form mullite. With an additional increase in firing temperature, a more glassy ceramic is produced due to further recrystallisation, and if the firing temperature reaches 1300°C, any remaining free silica is recrystallised. In the presence of potassium or sodium salts vitrification occurs giving an impervious product (Fig. 8.3).

Ceramic products

FIRECLAY

A range of clays, predominantly blends of alumina and silica, high in silica (40–80%) and low in iron oxide (2–3%), produce fireclay refractory products, which will withstand high temperatures without deformation. Dense products have high flame resistance, while the insulating lower-density products are suitable for flue linings. White glazed fireclay is typically used for urinals, floor channels, and industrial and laboratory sinks.

BRICK CLAYS

Bricks and roofing tiles

Bricks can be manufactured from a wide range of clays, the principal ones being Keuper marl, Etruria marl, Oxford clay, London clay, Coal measure shale, and Weald and Gault clays, with some production from alluvial and fireclay deposits. The composition of the clay varies widely depending on the type, but typically contains 40–65% silica, 10–25% alumina and 3–9% iron oxide. The loss on firing may reach 17% in the case of clay containing high levels of organic matter. The production of bricks is described in detail in Chapter 1.

Glazed bricks are manufactured in a wide range of high-gloss, uniform or mottled colours. Colour-fast glazed bricks offer a low-maintenance, frost- and vandal-resistant material suitable for light-reflecting walls. Standard and purpose-made specials may be manufactured to order. Normal bricklaying techniques are appropriate but to reduce the visual effect of the mortar joints they may be decreased from the standard 10 to 6 mm. For conservation work, in order to match

new to existing, it may be necessary to fire the glazed bricks a second time at a reduced temperature to simulate the existing material colour.

Roofing tiles are made from similar clays to bricks, such as Etruria marl, but for both hand- and machine-made tiles, the raw materials have to be screened to a finer grade than for brick manufacture. Traditional red, brown, buff, brindled or 'antique' ceramic roofing tiles are unglazed with a plain or sanded finish. Most plain tiles have a minimum pitch of 30°. While most interlocking clay tiles may be used to a minimum pitch of 22.5°, a limited selection with full interlocking may be used down to only 15°. Where bright colours are required, high- and low-gloss pantiles are available in a range of strong colours, or to individual specification. For plain tiles, a range of standard fittings are produced for hips, valleys, eaves, ridges, verges, and internal and external angles, as shown in Figure 8.4. Some flat, single-lap interlocking tiles are compatible with photovoltaic tiles. Roofing tiles are usually shrink-wrapped for protection and ease of handling on site. Clay roofing tiles and fittings are specified in BS EN 1304: 2013.

Marl floor tiles

Certain floor tiles are also manufactured from Etruria marl. Firing to 1130°C produces sufficient vitrification to give a highly durable, chemical- and frost-resistant product, with water absorption limited to less than 3%. This corresponds to the lowest category (Group I) of water absorption to BS EN 14411: 2012. Medium water absorption (Group II) is subdivided (II_a is–6% and II_b is 6–10%), and Group III refers to water absorption greater than 10% (Table 8.1). Where high slip resistance is required, a studded profile or carborundum (silicon carbide) grit may be incorporated into the surface (Fig. 8.5). Figure 8.6 illustrates the appropriate use of ceramic floor tiles.

TERRACOTTA

In order to produce intricately detailed terracotta building components, the clay has to be more finely divided than is necessary for bricks and roof tiles. The presence of iron oxide within the clay causes the buff, brown or red coloration of the fired product. During the latter part of the nineteenth century many civic buildings were constructed with highly decorative terracotta blocks. The material was used because it was cheaper than stone, durable and could be readily

Standard titles

Feature titles

Plain tile fittings

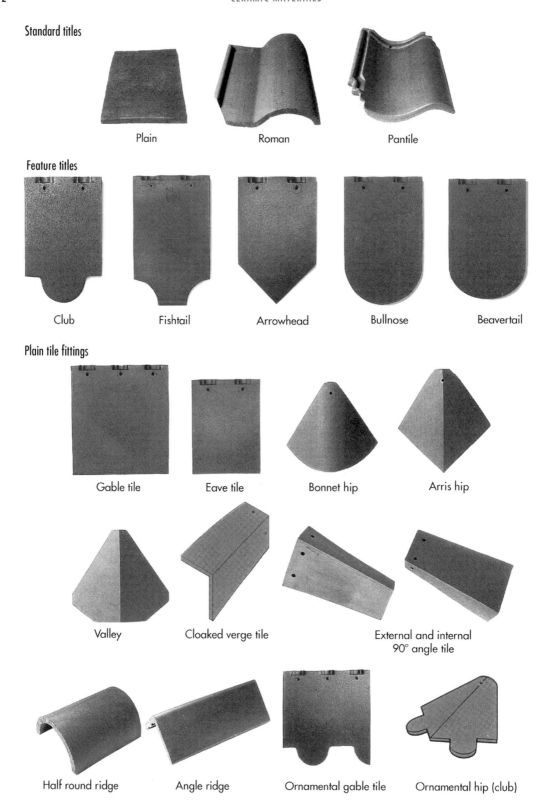

Fig. 8.4 Roof tiles – feature tiles and plain tile fittings

Table 8.1 Classification of ceramic tiles by manufacturing process and water absorption to BS EN 14411: 2012

Manufacturing process	Water absorption (%)			
	Group I ≤ 3%	Group IIa 3–6%	Group IIb 6–10%	Group III > 10%
Extruded tiles, Method A	Group AI$_a$ ≤ 0.5% Group AI$_b$ 0.5–3%	Group AII$_a$	Group AII$_b$	Group AIII
Dry-pressed tiles, Method B	Group BI$_a$ ≤ 0.5% Group BI$_b$ 0.5–3%	Group BII$_a$	Group BII$_b$	Group BIII

Notes:
- Groups AII$_a$ and AII$_b$ are subdivided into two parts due to specification variation of products manufactured around Europe.
- Group III covers glazed ceramic tiles only.

moulded. The blocks, which were usually partly hollowed out to facilitate drying and firing, were filled with concrete during construction.

Modern terracotta blocks may still be supplied for new work or refurbishment as plain ashlar, profiled or with sculptural embellishments. Terracotta may be used as the outer skin of cavity wall construction or as 25–40 mm-thick cladding hung with stainless steel mechanical fixings. The production of terracotta blocks requires the manufacture of an oversize model (to allow for shrinkage), from which plaster moulds are made. Prepared clay is then pushed into the plaster mould, dried under controlled conditions and finally fired. Traditional colours together with greens and

blues and various textures are produced. For refurbishment work existing terracotta may, subject to natural variations, usually be colour matched. In addition to cladding units, terracotta clay is also used in the manufacture of terracotta floor tiles and an extensive range of decorative ridge tiles and finials (Fig. 8.7). Standard terracotta building blocks made by extrusion are described in Chapter 2.

Terracotta rainscreen cladding

Rainscreen cladding is the external weathering element to multi-layer rainscreen wall systems. The rainscreen façade is drained and back-ventilated to protect the

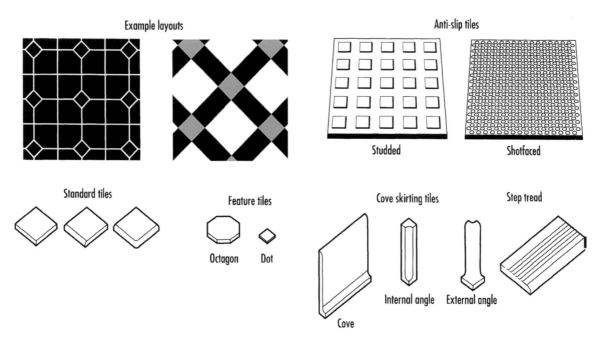

Fig. 8.5 Floor tiles – textured, smooth and specials

Fig. 8.6 Floor tiles. *Photograph:* Courtesy of Johnson-Tiles

structural wall from the adverse effects of the sun, wind and rainwater. The design of the joints and the cavity between the façade and the structure result in an equalisation of air pressure between the cavity and the exterior, thus inhibiting the drive of airborne moisture across the cavity. In some systems a vertical EPDM gasket seal is located in the vertical joints. A breather membrane is usually fixed to the structure before the rainscreen system is applied.

Rainscreen systems are appropriate for masonry, concrete, timber frame and concrete frame construction. A grid of vertical or horizontal aluminium extrusions is fixed to the façade, creating a minimum air gap of 25 mm. The rainscreen units are then clipped to the support system. The rainscreen cladding units may be manufactured in terracotta or from a wide range of other materials including stone laminate, stainless steel, copper, aluminium or zinc. Rainscreen units are shaped to shed water out of the open-drained joints, and individual units may be removed for maintenance or repair. A range of colours and dimensions is available in terracotta units to create the required aesthetic effect. Terracotta rainscreen cladding (Fig. 8.8) is fire resistant and durable, requiring virtually no maintenance except for occasional cleaning.

FAIENCE

Faience is glazed terracotta, used either as structural units or in the form of decorative slabs applied as cladding. It was popular in the nineteenth century and was frequently used in conjunction with polychrome brickwork on the façades of buildings such as public houses. Either terracotta may be glazed after an initial firing to the *biscuit* condition, or the *slip* glaze may be applied prior to a single firing. The latter has the advantage that it reduces the risk of the glaze crazing, although it also restricts the colour range. Faience, with an orange-peel texture, is available with either a matt or gloss finish and in plain or mottled colours. It is a highly durable material unaffected by weathering, frost or ultraviolet light, but strong impacts can chip the surface, causing unsightly damage.

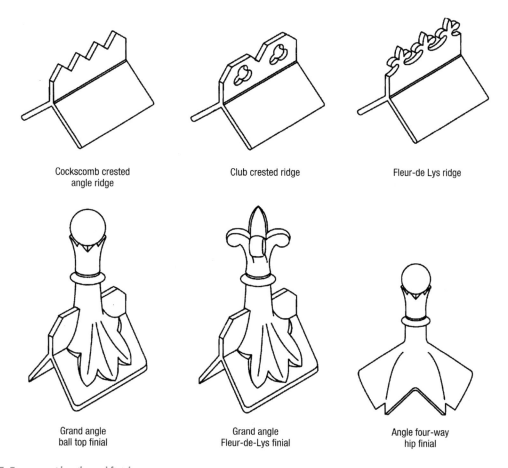

Cockscomb crested angle ridge

Club crested ridge

Fleur-de Lys ridge

Grand angle ball top finial

Grand angle Fleur-de-Lys finial

Angle four-way hip finial

Fig. 8.7 Terracotta ridge tiles and finials

STONEWARE

Stoneware is manufactured from secondary plastic clays, typically fireclays blended with an added flux such as feldspar. On firing to between 1200°C and 1300°C, the material vitrifies, producing an impermeable ceramic product with high chemical resistance. The majority of unglazed vitrified clay pipes are stoneware. For most purposes push-fit polypropylene couplings are used, which allow flexibility to accommodate ground movement; however, if required, traditional jointed socket/spigot drainage goods are also available in stoneware.

Large stoneware ceramic panels up to 1.2 m square and 8 mm in thickness are manufactured as cladding units for façades. The units, which are colour-fast, frost- and fire-resistant, may be uniform in colour or flecked and glazed or unglazed. Fixing systems are exposed or hidden; the open joint system offers rear ventilation, allowing any moisture diffusing

from the supporting wall to be dissipated by natural air movement.

Stoneware floor tiles

Stoneware is also used in the manufacture of some floor tiles. The high firing temperature gives a product of low porosity, typically around 3%. In one manufacturing process a granular glaze is applied to the tiles within the kiln to produce an impervious vitreous finish. The standard BS EN 14411: 2012 classifies six levels of abrasion resistance for glazed floor tiles. Class 0 are not recommended for use on floors. Class 5 tiles have the maximum resistance to heavy pedestrian traffic over sustained periods. Slip resistance is coded according to the *pendulum* test (BS 7976: 2002) which simulates walking over a floor. The test is available for dry surfaces (e.g. 4S Slider 96) or wet surfaces (e.g. TRRL Slider 55). For wet surfaces high slip potential is graded 24 and below; low slip potential is graded 36 and above.

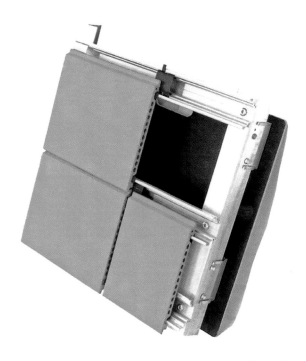

Fig. 8.8 Terracotta rainscreen cladding. *Image*: Courtesy of EH Smith

EARTHENWARE

Earthenware is produced from a mixture of kaolin, ball clay and flint, with, in some cases, feldspar as a flux. The material when fired at 1100°C is porous and requires a glaze to prevent water absorption. In the manufacture of traditional glazed drainage goods, the salt glaze is produced by adding damp common salt to the kiln during the firing process. The salt decomposes to form sodium oxide, which then reacts with silica and alumina on the surface of the clay component to produce the salt glaze which is impermeable to moisture.

Earthenware wall tiles

Wall tiles (Fig. 8.9) are generally manufactured from earthenware clay to which talc (magnesium silicate) or limestone (calcium carbonate) is added to ensure a white burning clay. To prepare the clay for manufacturing wall tiles by the dry process, the components, typically a blend of china clay (kaolin), ball clay and silica sand together with some ground recycled tiles, are

Fig. 8.9 Wall tiles. *Photograph*: Courtesy of Johnson-Tiles

mixed with water to form a slip. This is sieved, concentrated to a higher density slip and then dried to a powder by passage down a heated tower at 500°C. The clay dust, which emerges with a moisture content of approximately 8%, is then pressed into tiles. A glaze is required to both decorate and produce an impermeable product and this may be applied before a single firing process or after the tiles have been fired at 1150°C to the biscuit stage in a tunnel kiln. Either the unfired or biscuit tiles are coated with a slip glaze followed by firing under radiant heat for approximately 16 hours. Damaged tiles are rejected for recycling; the quality-checked tiles are packaged for dispatch. Standard sizes are 108 × 108 mm, 150 × 150 mm, 150 × 200 mm, 200 × 200 mm and 200 × 250 mm, although many larger format sizes are available.

VITREOUS CHINA

Vitreous china, used for the manufacture of sanitary ware, has a glass-like body which limits water absorption through any cracks or damage in the glaze to a mean maximum of 0.5%. It is typically manufactured from a blend of china clay (20–30%), ball clay (20–30%), feldspar or nepheline syenite (10–20%), quartz or flint (30–40%) and talc (0–3%). For large units such as WCs and wash-basins, a controlled drying-out period is required before firing to prevent cracking. Glaze containing metallic oxides for coloration is applied before firing to all visually exposed areas of the components.

Vitreous china floor and wall tiles

Vitreous china is also used in the manufacture of some floor and wall tiles due to its impermeable nature. Porcelain which has a similar composition to vitreous china has a water absorption of less than 0.5% (Table 8.1). Unglazed tiles may be smooth or polished, alternatively textured, ribbed or bush hammered to give additional non-slip properties. Thickness is usually in the range 8–13 mm. Sizes range from 200 × 200 mm and 300 × 300 mm square format to rectangular and extra-large format up to 3000 × 1000 mm for certain product ranges. Matching skirtings (typically 75–100 mm) are usually available. For lining swimming pools, additional protection against water penetration is given by the application of a glaze.

Anti-pollutant and anti-bacterial tiles are based on incorporated titanium dioxide which catalyses the conversion of nitrogen oxides to nitrates and reduces bacterial concentrations. Most commercial floor tiles are rated B to the BRE Green Guide, but slim porcelain products (4 mm thick) may achieve an A rating.

REPRODUCTION DECORATIVE TILES

Reproduction moulded ceramic wall tiles, encaustic tiles with strong colours burnt into the surface, and geometrical floor tiles can be manufactured to match existing units with respect to form, colour and texture for restoration work. Some manufacturers retain both the necessary practical skills and appropriate detailed drawings to ensure high-quality conservation products, which may be used to replace lost or seriously damaged units. There is also an increasing demand for reproduction decorative tiles in new-build work.

MOSAICS

Mosaics in glazed or unglazed porcelain are hard-wearing, frost-proof and resistant to chemicals. Unglazed mosaics may be used for exterior use and other wet areas such as swimming pools, where good slip resistance is important. Mosaics are usually supplied attached to paper sheets for ease of application. Mosaics are also available in glass, stone and stainless steel. Figure 8.10 illustrates a formal mosaic wall, while Figure 8.11 shows the broken tile mosaic finish used by Calatrava on the Tenerife Concert Hall, following the technique developed by Gaudi.

Fig. 8.10 Mosaic. *Photograph:* Courtesy of Architectural Ceramics

Fig. 8.11 Ceramic mosaic finish and detail – Concert Hall, Tenerife. *Architect*: Santiago Calatrava. *Photographs*: Arthur Lyons

CERAMIC GRANITE

Ceramic granite is a blend of ceramic and reconstituted stone, manufactured from a mixture of feldspar, quartz and clay. The components are crushed, graded, mixed and compressed under very high pressure, followed by firing at 1260°C. The material is produced in 20 and 30 mm slabs, which can be cut and polished to produce a hard, shiny finish with the appearance of natural marble or granite, suitable for worktops. Colours range from ochre, off-white and grey to green and blue, depending on the initial starting materials.

References

FURTHER READING

CIRIA. 2010: *Safer surfaces to walk on. Reducing the risk of slipping.* C652. London: Construction Industry Research and Information Association.

Creative Publishing International. 2003: *The complete guide to ceramic and stone tile.* USA: Creative Publishing International.

Daab. 2007: *Ceramic design.* Cologne: Daab.

DfES Publications. 2007: *Floor finishes in schools.* Nottingham: Department for Education and Skills.

Durbin, L. 2004: *Architectural tiles: Conservation and restoration.* Oxford: Elsevier.

Dutfield, A., Mundy, J. and Anderson, J. 2011: *Environmental impact of materials. Floor finishes.* Bracknell: IHS BRE Publications.

Lemmen, H.V. 2001: *Architectural ceramics.* Princes Risborough: Shire Publications.

NHBC. 2013: *NHBC Standards Part 8. Services and internal finishing.* Milton Keynes: NHBC.

Ripley, J. 2005: *Ceramic and stone tiling.* Marlborough: Crowood Press.

Taylor, K. 2008: *Roof tiling and slating.* Marlborough: Crowood Press.

Wilhide, E. 2003: *Materials; A directory for home design.* London: Quadrille Publishing.

STANDARDS

BS 65: 1991
Specification for vitrified clay pipes, fittings, and ducts; also flexible mechanical joints for use solely with surface water pipes and fittings.

BS 493: 1995
Airbricks and gratings for wall ventilation.

BS 1125: 1987
Specification for WC flushing cisterns.

BS 1188: 1974
Ceramic washbasins and pedestals.

BS 1196: 1989
Clayware field drain pipes and junctions.

BS 1206: 1974
Fireclay sinks: Dimensions and workmanship.

BS 3402: 1969
Quality of vitreous china sanitary appliances.

BS 5385
Wall and floor tiling:

Part 1: 2009	Code of practice for the design and installation of internal ceramic and natural stone wall tiling and mosaics in normal conditions.
Part 2: 2006	Design and installation of external ceramic and mosaic wall tiling in normal conditions. Code of practice.
Part 3: 2007	Design and installation of internal and external ceramic floor tiles and mosaics in normal conditions. Code of practice.
Part 4: 2009	Design and installation of ceramic and mosaic tiling in special conditions. Code of practice.
Part 5: 2009	Design and installation of terrazzo, natural stone and agglomerated stone tile and slab flooring. Code of practice.

BS 5506
Specification for wash basins:

Part 3: 1977	Wash basins (one or three tap holes), materials, quality, design and construction.

pr BS 5534: 2013
Code of practice for slating and tiling.

BS 7976 Parts 1–3: 2002
Pendulum testers.

BS 8000
Workmanship on building sites:

Part 6: 2013	Code of practice for slating and tiling of roofs and walls.
Part 11: 2011	Internal and external wall and floor tiling. Ceramic and

agglomerated stone tiles, natural stone and terrazzo tiles and slabs and mosaics. Code of practice.

BS ISO 13007

Ceramic tiles. Grouts and adhesives:

pr Part 1: 2013	Terms, definitions and specifications for adhesives.
Part 2: 2013	Test methods for adhesives.
Part 3: 2010	Terms, definitions and specifications for grouts.
Part 4: 2013	Test methods for grouts.

BS EN 295

Vitrified clay pipes for drains and sewers:

Part 1: 2013	Requirements for pipes, fittings and joints.
Part 2: 2013	Evaluation of conformity and sampling.
Part 3: 2012	Test methods.
Part 4: 2013	Requirements for adaptors, connectors and flexible couplings.
Part 5: 2013	Requirements for perforated pipes and fittings.
Part 6: 2013	Requirements components of manholes and inspection chambers.
Part 7: 2013	Requirements for pipes and joints for pipe jacking.

BS EN 538: 1994

Clay roofing tiles for discontinuous laying. Flexural strength test.

BS EN 539

Clay roofing tiles for discontinuous laying. Determination of physical characteristics:

Part 1: 2005	Impermeability test.
Part 2: 2013	Test for frost resistance.

BS EN 997: 2012

WC pans and WC suites with integral trap.

BS EN 1304: 2013

Clay roofing tiles and fittings. Product definitions and specifications.

BS EN 1308: 2007

Adhesives for tiles. Determination of slip.

BS EN 1324: 2007

Adhesives for tiles. Determination of shear adhesion strength of dispersion adhesives.

BS EN 1346: 2007

Adhesives for tiles. Determination of open time.

BS EN 1347: 2007

Adhesives for tiles. Determination of wetting capability.

BS EN 1348: 2007

Adhesives for tiles. Determination of tensile adhesion strength for cementitious adhesives.

BS EN 1457

Chimneys. Clay/ceramic flue liners:

Part 1: 2012	Flue liners operating under dry conditions. Requirements and test methods.
Part 2: 2012	Flue liners operating under wet conditions. Requirements and test methods.

BS EN 1806: 2006

Chimneys. Clay/ceramic flue blocks for single wall chimneys. Requirements and test methods.

BS EN ISO 10545

Ceramic tiles:

pr Part 1: 2013	Sampling and basis for acceptance.
Part 2: 1997	Determination of dimensions and surface quality.
Part 3: 1997	Determination of water absorption, apparent porosity, relative and bulk density.
pr Part 4: 2013	Determination of modulus of rupture and breaking strength.
Part 5: 1998	Determination of impact resistance.
Part 6: 2012	Determination of resistance to deep abrasion for unglazed tiles.
Part 7: 1999	Determination of resistance to surface abrasion for glazed tiles.
pr Part 8: 2013	Determination of linear thermal expansion.
Part 9: 2013	Determination of resistance to thermal shock.
Part 10: 1997	Determination of moisture expansion.
Part 11: 1996	Determination of crazing resistance.
Part 12: 1997	Determination of frost resistance.
Part 13: 1997	Determination of chemical resistance.
Part 14: 1997	Determination of resistance to stains.
Part 15: 1997	Determination of lead and cadmium given off by glazed tiles.

Part 16: 2012   Determination of small colour differences.

BS EN 12002: 2008
Adhesives for tiles. Determination of transverse deformation for cementitious adhesives and grouts.

BS EN 12003: 2008
Adhesives for tiles. Determination of shear adhesive strength of reaction resin adhesives.

BS EN 12004: 2007
Adhesives for tiles. Requirements, evaluation of conformity, classification and designation.

BS EN 12808
Grouts for tiles. Requirements, evaluation of conformity, classification and designation:

Part 1: 2008   Determination of chemical resistance of reaction resin mortars.

Part 2: 2008   Determination of resistance to abrasion.

Part 3: 2008   Determination of flexural and compressive strength.

Part 4: 2009   Determination of shrinkage.

Part 5: 2008   Determination of water absorption.

BS EN 13502: 2002
Chimneys. Requirements and test methods for clay/ceramic flue terminals.

BS EN 13888: 2009
Grout for tiles. Requirements, evaluation of conformity.

BS EN 14411: 2012
Ceramic tiles. Definitions, classification, characteristics and marking.

BS EN 14437: 2004
Determination of the uplift resistance of installed clay or concrete tiles for roofing.

DD CEN/TS 15209: 2008
Tactile paving surface indicators from concrete, clay and stone.

BUILDING RESEARCH ESTABLISHMENT PUBLICATIONS

BRE Digests

BRE Digest 467: 2002
Slate and tile roofs: avoiding damage from aircraft wake vortices.

BRE Digest 486: 2004
Reducing the effect of climate change by roof design.

BRE Digest 499: 2006
Designing roofs for climate change. Modifications to good practice guidance.

BRE Good building guide

BRE GBG 64: 2005
Tiling and slating pitched roofs (Parts 1, 2 and 3).

ADVISORY ORGANISATIONS

British Ceramic Confederation, Federation House, Station Road, Stoke-on-Trent ST4 2SA (01782 744631).

CERAM Research, Queens Road, Penkhull, Stoke-on-Trent, Staffordshire ST4 7LQ (01782 764428).

Clay Roof Tile Council, Federation House, Station Road, Stoke-on-Trent, Staffordshire ST4 2SA (01782 744631).

Tile Association, Forum Court, 83 Copers Cope Road, Beckenham, Kent BR3 1NR (020 8663 0946).

STONE AND CAST STONE

CONTENTS

Introduction

The term *stone* refers to natural rocks after their removal from the earth's crust. The significance of stone as a building material is illustrated by widespread prehistoric evidence and its sophisticated use in the early civilisations of the world, including the Egyptians, the Incas of Peru and the Mayans of Central America.

Geologically, all rocks may be classified into one of three groups: igneous, metamorphic or sedimentary, according to the natural processes by which they were produced within or on the earth's surface.

For construction work, stone specification to BS EN 12440: 2008 requires the scientific petrological family as defined within the standard BS EN 12670: 2002, the typical colour range and the place of origin. The latter should be as detailed as possible, including the location of the quarry, its nearest town, the region and the country. The new Derby visual arts and media centre, the Derby Quad (Fig. 9.1), designed by Fielden Clegg Bradley Studios, illustrates the use of banded sandstone in contemporary architecture.

Types of stone

IGNEOUS ROCKS

Igneous rocks are the oldest, having been formed by the solidification of the molten core of the earth or *magma*. They form about 95% of the earth's crust, which is up to 16 km thick. Depending upon whether solidification

Fig. 9.1 Stancliffe Stanton Moor sandstone – Derby Quad. *Architects*: Fielden Clegg Bradley Studios. *Photograph*: Courtesy of Stancliffe Stone

occurred slowly within the earth's crust or rapidly at the surface, the igneous rocks are defined as plutonic or volcanic, respectively. In the plutonic rocks, slow cooling from the molten state allowed large crystals to grow, which are characteristic of the granites. Volcanic rocks such as pumice and basalt are fine-grained and individual crystals cannot be distinguished by eye; thus, the stones are visually less interesting. Dolerites, formed by an intermediate rate of cooling, exhibit a medium-grained structure.

Apart from crystal size, igneous rocks also vary in composition according to the nature of the original magma, which is essentially a mixture of silicates. High-silica-content magma produces acid rocks (e.g. granite) while low-silica content forms basic rocks (e.g. basalt and dolerite). Granites are mainly composed of feldspar (white, grey or pink), which determines the overall colour of the stone but they are modified by the presence of quartz (colourless to grey or purple), mica (silver to brown) or horneblende (dark coloured). The basic rocks such as dolerite and basalt in addition to feldspar contain augite (dark green to black) and sometimes olivine (green). Although basalt and dolerite have not been used widely as building stones they are frequently used as aggregates, and cast basalt is now being used as a reconstituted stone.

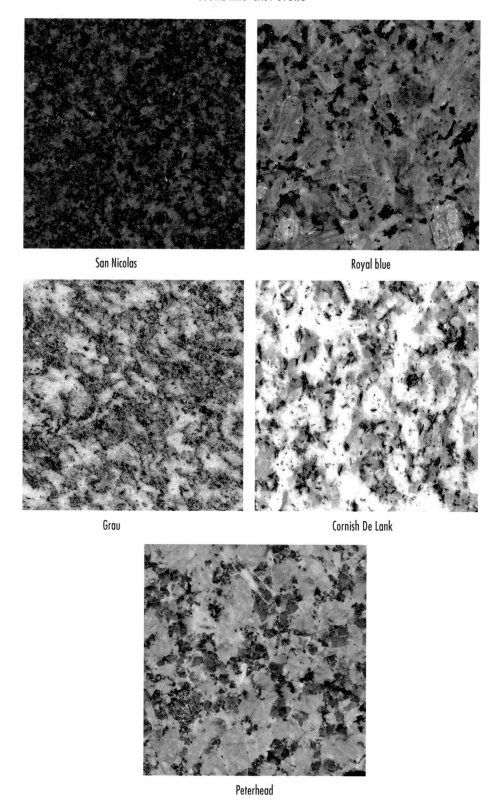

Fig. 9.2 Selection of granites

Granites

Most granites are hard and dense, and thus form highly durable building materials, virtually impermeable to water, resistant to impact damage and stable within industrial environments. The appearance of granite is significantly affected by the surface finish which may be sawn, rough punched, picked, fine-tooled, honed or polished. It is, however, the highly polished form of granite which is most effective at displaying the intensity of the colours and reflectivity of the crystals. In addition, granites may be flamed to a spalled surface, produced by the differential expansion of the various crystalline constituents. Many buildings have combined the polished and flamed material to create interesting contrasts in depth of colour and texture. Grey and pink granites are quarried in Scotland, the North of England, Devon and Cornwall, but a wide variety of colours including black, blue, green, red, yellow and brown are imported from other countries (Fig. 9.2; Table 9.1). Because of the high cost of quarrying and finishing granite, it is frequently used as a cladding material (40 mm externally or 20 mm internally), or alternatively cast directly onto concrete cladding units. Granite is available for flooring and for hard landscaping including pavings, setts and kerbs. Polished granite is also used as a kitchen countertop material due to its strength, durability and high-quality finish.

Cast basalt

Basalt is a fine-grained stone which is nearly as hard as granite. It can be melted at 2400°C and cast into tile units which are deep steel grey in colour. A slightly patterned surface can be created by swirling the molten basalt within the mould. Annealing in a furnace produces a hard, virtually maintenance-free, shiny textured surface flecked with shades of green, red and bronze. Larger cast units for worktops, in either a honed or polished finish, may be cut to size.

SEDIMENTARY ROCKS

Sedimentary rocks are produced by the weathering and erosion of older rocks. In the earliest geological time these would have been the original igneous rocks, but subsequently other sedimentary and metamorphic rocks too will have been reworked. Weathering action by water, ice and wind breaks the rocks down into small fragments which are then carried by rivers and sorted into size and nature by further water action. Most

Table 9.1 UK and imported granites

Colour	Name	Country of origin
UK		
Light grey	Merrivale, Devon	England
Silver grey	De Lank & Hantergantick, Cornwall	England
Light and dark pink to brownish red	Shap	England
Pink	Peterhead	Scotland*
Pale to deep red	Ross of Mull	Scotland
Grey	Aberdeen	Scotland*
Black	Hillend	Scotland*
Black	Beltmoss	Scotland*
Imported		
Red with black	Balmoral Red	Finland
Red	Bon Accord Red	Sweden
Black	Bon Accord Black	Sweden
Red	Virgo Granite	Sweden
Dark red with blue to purple quartz	Rose Swede	Sweden
Grey	Grey Royal	Norway
Grey	Sardinian Grey	Sardinia
Yellow	Nero Tijuca	Brazil
Beige/brown	Juparana	Brazil
Blue	Blue Pearl	Norway
Green/black	Emerald Pearl	Norway
Pink to red	Torcicoda	Brazil
Beige/brown	Giallio Veneziano	Brazil

Note:
* Available only in limited quantities.

deposits are laid down in the oceans as sedimentary beds of mud or sand, which build up in layers, become compressed and eventually are cemented together by minerals such as calcium carbonate (calcite), quartz (silica), iron oxide or dolomite (magnesium and calcium carbonate) remaining in the groundwater. The natural bedding planes associated with the formation of the deposits may be thick or thin but are potentially weak; this is used to advantage in the quarrying process. In masonry, to obtain maximum strength and durability, stones should be laid to their natural bed except for cornices, cills and string courses which should be edge-bedded. Face-bedded stones will tend to delaminate (Fig. 9.3). When quarried, stones contain *quarry sap* and may be worked and carved more easily than after exposure to the atmosphere.

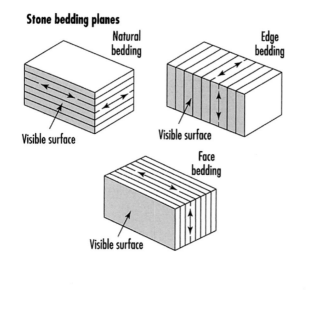

Stone bedding planes

Natural bedding

Edge bedding

Face bedding

Visible surface

Appropriate application of bedding planes

Natural or edge on thick copings

Edge

Natural

Lead flashing

Edge on cornice

Natural on cornerstone

Natural

Lintel

Voussoirs

Fig. 9.3 Natural stone bedding planes

Sandstones

Deposits of sand cemented together by calcium carbonate, silica, iron oxide and dolomite produce calcareous, siliceous, ferruginous and dolomitic sandstones respectively. Depending on the nature of the original sand deposit, the sandstones may be fine or coarse in texture. Sandstones range in colour from white, buff and grey through to brown and shades of red depending on the natural cement; they are generally frost-resistant. Some common UK sandstones are listed in Table 9.2, and examples are illustrated in

Figure 9.4. The Chester Song School (Fig. 9.5) was built in Locharbriggs red sandstone to empathise with the old red sandstone of the cathedral. Some sandstones imported from India have fern fossils embedded within the strata (Fig. 9.6). Typical finishes are sawn, split faced and clean rubbed, although a range of tooled finishes, including broached and droved, may also be selected (Fig. 9.7). For cladding, sandstone is normally 75–100 mm thick and fixed with non-ferrous cramps and corbels. Sandstones are quarried in Scotland, the North of England, Yorkshire and Derbyshire; they

Table 9.2 Typical UK sandstones and their characteristics

Name	Colour	Source	Characteristics
Doddington	Purple/pink	Northumberland	Fine to medium grained
Darley Dale – Stancliffe	Buff	Derbyshire	Fine grained
Birchover gritstone	Pink to buff	Derbyshire	Medium to coarse grained
York Stone	Buff, fawn, grey, light brown	Yorkshire	Fine grained
Mansfield Stone	Buff to white	Nottinghamshire	Fine grained
Hollington	Pale pink, dull red, pink with darker stripe	Staffordshire	Fine to medium grained
St Bees	Dark red	Cumbria	Fine grained
Blue Pennant	Dark grey/blue	Mid-Glamorgan	Fine grained

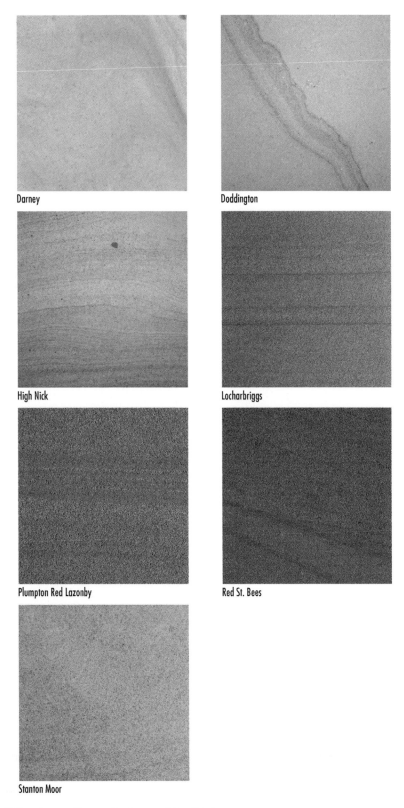

Fig. 9.4 Examples of UK sandstones. *Photographs*: Courtesy of Stancliffe Stone

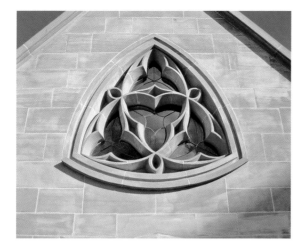

Fig. 9.5 Locharbriggs red sandstone – Chester Song School. *Architects*: Arrol and Snell. *Photograph*: Courtesy of Stancliffe Stone

Fig. 9.6 Indian fossil sandstone

Broached with draughted margin

Stugged or punched face

Droved

Herringbone

Bats

Fig. 9.7 Typical tooled stone finishes

include the old and new red sandstones, York Stone and Millstone Grit. Sandstone is also imported from Spain and Italy from where *Pietra Serena* is sourced.

Calcareous sandstone
Calcareous sandstones are not durable in acid environments, which may cause the slow dissolution of the natural calcium carbonate cement of the stone. Pure calcite is white, so these sandstones are generally white in colour.

Siliceous sandstone
Siliceous sandstones are predominantly grains of silica (sand) cemented with further natural silica, and are therefore durable even in acid environments. Siliceous sandstones are generally grey in colour.

Ferruginous sandstone
Ferruginous sandstones are bound with oxides of iron which may be brown, ochre or red. They are generally durable.

Dolomitic sandstone
Dolomitic sandstones are bound with a mixture of magnesium and calcium carbonates, and therefore do not weather well in urban environments. They are generally off-white and buff in colour.

Limestones

Limestones consist mainly of calcium carbonate, either crystallised from solution as calcite or formed from accumulations of fossilised shells deposited by various sea organisms (Fig. 9.8). They are generally classified according to their mode of formation. Many colours are available, ranging from off-white, buff, cream, to grey and blue. Limestones are found in England in a belt from Dorset, the Cotswolds, Oxfordshire and Lincolnshire to Yorkshire. Limestone is also imported from Ireland, France and Portugal to widen the palette of colours. Some common UK limestones are listed in Table 9.3, and examples of limestones and

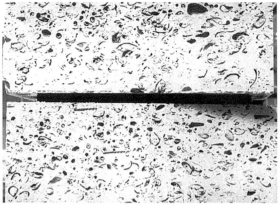
Roach limestone

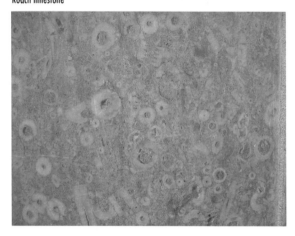
Fig. 9.8 Portland fossil limestones

Table 9.3 Typical UK limestones and their characteristics

Name	Colour	Source	Characteristics
Ancaster	Cream to buff	Lincolnshire	Oölitic limestone – variable shell content; freestone available
Bath Stone	Pale brown to	Avon	Oölitic limestone
– Westwood Ground	light cream		Coarse-grained – buff coloured
– Monks Park			Fine-grained – buff coloured
Clipsham	Buff to cream	Rutland	Medium-grained oölitic limestone with shells; some blue stone; best quality stone is durable
Doulting	Pale brown	Somerset	Coarse textured; fossils uncommon
Hopton Wood	Cream or grey	Derbyshire	Carboniferous limestone containing many attractive fossils; may be polished
Ketton	Pale cream to buff and pink	Lincolnshire	Medium-grained oölitic limestone; even-textured; durable stone
Portland Stone	White	Dorset	Exposed faces weather white, protected faces turn black.
– Roach			Coarse open-textured shelly stone; weathers very well
– Whitbed			Fine grained – some shells; fragments; durable stone
– Basebed			Fine grained with few shells; suitable for carving
Purbeck	Blue/grey to buff	Dorset	Some shells; durable stone

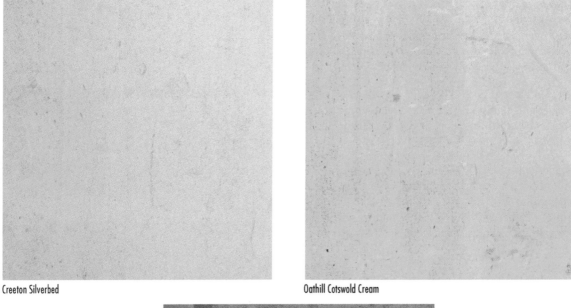

Creeton Silverbed

Oathill Cotswold Cream

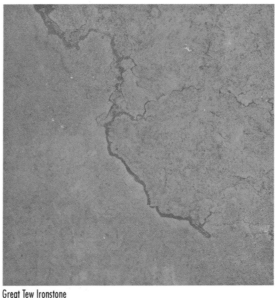

Great Tew Ironstone

Fig. 9.9 Examples of UK limestones. *Photographs:* Courtesy of Stancliffe Stone

ironstone are illustrated in Figure 9.9. The standard finishes are fine rubbed, fine dragged and split faced, although tooled finishes are also appropriate. Externally, limestones must not be mixed with or located above sandstones, as this may cause rapid deterioration of the sandstone.

Oölitic limestone

Oölitic limestones are formed by crystallisation of calcium carbonate in concentric layers around small fragments of shell or sand, producing spheroidal grains or oöliths. The oöliths become cemented together by the further deposition of calcite to produce the rock. Typically, the oöliths are up to 1 mm in diameter, giving a granular texture to the stone, which may also incorporate other fossils. Oölitic limestones are very workable, and include Bath stone and Portland stone. Clipsham and Ketton Stone have been widely used at Oxford and Cambridge, respectively, including the Queen's Building of Emmanuel College, Cambridge (Fig. 9.10), which is built of load-bearing Ketton limestone, with appropriately massive columns and flat voussoir arches to the colonnade and window openings. Lime mortar is used to ensure an even spread of the load between stones. In the case of Foundress Court, Pembroke College, Cambridge (Fig. 9.11), the Bath stone (Monks Park) is built up three storeys from ground level as a well-detailed cladding, with restraint back to the load-bearing blockwork inner skin. The flexibility of lime mortar is used to reduce the number of visible movement joints.

Organic limestone

Organic limestones are produced in bedded layers from the broken shells and skeletal remains of a wide variety of sea animals and corals. Clay is frequently incorporated into organic limestones and this adversely affects the polish which may otherwise be achieved on the cut stone.

Crystallised limestone

When water containing calcium bicarbonate evaporates, it leaves a deposit of calcium carbonate. In the case of hot springs the material produced is travertine, and in caves, stalactites and stalagmites or *onyx-marble* result.

Fig. 9.10 Ketton Limestone – The Queen's Building, Emmanuel College, Cambridge. *Architects*: Hopkins Architects. *Photograph*: Arthur Lyons

Fig. 9.11 Bath Stone – Foundress Court, Pembroke College, Cambridge. *Architects*: Eric Parry Architects. *Photograph*: Arthur Lyons

Dolomitic limestone
Dolomitic limestones have had the original calcium carbonate content partially replaced by magnesium carbonate. In general, this produces a more durable limestone, although it is not resistant to heavily polluted atmospheres.

METAMORPHIC ROCKS

Metamorphic rocks are formed by the recrystallisation of older rocks, when subjected to intense heat or pressure or both, within the earth's crust. Clay is metamorphosed to slate, limestone to marble and sandstone to quartzite.

Slate

Slate is derived from fine-grained sand-free clay sediments. The characteristic cleavage planes of slate were produced when the clay was metamorphosed, and frequently they do not relate to the original bedding planes. Slate may be split into thin sections (typically 4–10 mm for roofing slates) giving a natural riven finish, or it may be sawn, sanded, fine rubbed, honed, polished, flame textured or bush hammered. A range of distinctive colours is available: blue/grey, silver grey and green from the Lake District, blue, green, grey and plum red from North Wales, and grey from Cornwall. Slate is also imported from Ireland (grey/green), Canada (blue/grey), France (blue/grey), India (blue/grey), China (blue/green/grey), Brazil (grey/green/plum) and blue/black from Spain, which is the world's largest producer of the material. Slate is strong, acid- and frost-resistant, lasting up to 400 years as a roofing material. The minimum recommended pitch for slate roofing is 20° under sheltered or moderate exposure and 22.5° under severe exposure, and these situations require the use of the longest slates (460, 560 or 610 mm). Where thick slates (up to 20 mm in thickness) are used for a roof pitch of less than 25°, it should be noted that the slates lie at a significantly lower pitch than the rafters. Fixing nails should be of copper or aluminium. Slate is also used for flooring, cladding, copings, cills and stair treads. When used as a cladding material it should be fixed with non-ferrous fixings or cast directly onto concrete cladding units.

Roofing and external cladding slates satisfy the requirements for the Class A1 characteristic reaction to fire performance, without the need for testing.

Recycled roofing slates, particularly Welsh slate, are generally available in a range of sizes and are appropriate for both conservation work and new build where

Welsh slate – uniform size

Swithland slate – graded size

Fig. 9.12 Slate roofs

an immediate weathered appearance is required. Welsh slates have a good reputation for durability, making the recycled product a viable option. Certain regional slates, such as Swithland in Leicestershire, are only available as recycled products. This particular type of slate has a single top nail fixing and, unlike most roofing slates which are of a uniform size, is graded from large slates at the eaves to smaller units at the ridge (Fig. 9.12). The standard pr EN 5534: 2013 gives the Code of Practice for slating and tiling.

Reconstituted slate

Reconstituted slate for roofing is manufactured from slate granules and inert filler, mixed with a thermo-setting resin and cast into moulds to give a natural riven slate finish. Certain products incorporate glass-fibre reinforcement, and offer a wider range of colours than are available in natural slate. Some interlocking slates may be used down to a pitch of 17.5°, while double-lap simulated natural slates may be used down to a pitch of 20° depending upon the degree of exposure. Reconstituted slate is also manufactured in glassfibre-reinforced cement (GRC) as described in Chapter 11.

Marble

Marble is metamorphosed limestone in which the calcium carbonate has been recrystallised into a mosaic of approximately equal-sized calcite crystals. The process, if complete, will remove all traces of fossils, the size of the crystals being largely dependent on the duration of the process. Some limestones which can be polished are sold as marble, but true marble will not contain any fossilised remains. Calcite itself is white, so a pure marble is white and translucent. The colours and veining characteristics of many marbles are associated with impurities within the original limestone; they range from red, pink, violet, brown, green, beige, cream and white to grey and black. Marble is attacked by acids; therefore, honed, rather than highly polished surfaces, are recommended for external applications. Marbles are generally hard and dense, although fissures and veins sometimes require filling with epoxy resins. Most marbles used within Britain are imported from Europe, as indicated in Table 9.4; a selection is illustrated in Figure 9.13.

For external cladding above first-floor level 40 mm-thick slabs are used, although 20 mm may be appropriate for internal linings and external cladding up to first-floor level. Fixing cramps and hooks should be

Table 9.4 A selection of imported marbles

Colour	Name	Country of origin
White	White Carrara/Sicilian	Italy
White	White Pentelicon	Greece
Cream	Perlato	Sicily
Cream	Travertine	Italy
Beige	Botticino	Sicily
Pink	Rosa Aurora	Portugal
Red	Red Bilbao	Spain
Brown	Napoleon Brown	France
Green	Verde Alpi	Italy
Black	Belgian Black	Belgium
Black with white veins	Nero Marquina	Algeria

in stainless steel, phosphor bronze or copper. Floor slabs, to a minimum thickness of 30 mm, should be laid on a minimum 25 mm bed. Marble wall and bathroom floor tiles are usually between 7 and 10 mm in thickness.

Reconstituted marble

Reconstituted marble is manufactured from marble chippings and resin into tiles and slabs for use as floor and wall finishes. The material has the typical colours of marble but without the veining associated with the natural material.

Quartzite

Quartzite is metamorphosed sandstone. The grains of quartz are recrystallised into a matrix of quartz, producing a durable and very hard-wearing stone used mainly as a flooring material. The presence of mica allows the material to be split along smooth cleavage planes, producing a riven finish. Quartzite is mainly imported from Norway and South Africa, and is available in white, grey, grey-green, blue-grey and ochre colours.

ALABASTER

Alabaster is naturally occurring gypsum or calcium sulphate. Historically, it has been used for building as in the Palace of Knossos, Crete, but in the UK its use has been mainly restricted to carved monuments and ornaments. The purest form is white and translucent, but traces of iron oxide impart light-brown, orange or red colorations.

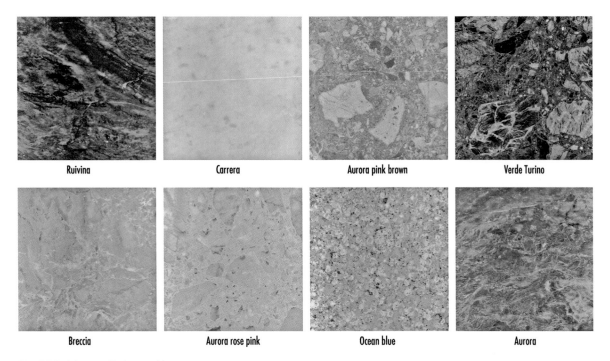

Ruivina	Carrera	Aurora pink brown	Verde Turino
Breccia	Aurora rose pink	Ocean blue	Aurora

Fig. 9.13 Selection of Italian marbles

Stonework

TRADITIONAL WALLING

Dressed stone may be used as an alternative to brick or block in the external leaf of standard cavity construction. Limestone and sandstone are the most frequently used for walling, but slate is also used where it is available locally. Although random rubble and hand-dressed stone can be supplied by stone suppliers, sawn-bedded (top and bottom) stones are generally the most freely available. These are normally finished split faced, pitch faced, fine rubbed or sawn. The standard sizes are 100 or 105 mm on bed, with course heights typically 50, 75, 100, 110, 125, 150, 170, 225 and 300 mm (Fig. 9.14). Stones may be to a particular course length (e.g. 300 mm or 450 mm), although they are frequently to random lengths. Quoin blocks, window and door surrounds, cills and other components are often available as standard. In ashlar masonry, the stones are carefully worked and finely jointed. Stones within horizontal courses are of the same height and are perfectly rectangular in elevation. Joints are generally under 6 mm in width.

The mortar for stone masonry should be weaker than the stone selected. For porous limestones and sandstones, crushed stone aggregate is frequently used as the aggregate in the mortar, typically in a 1 : 3 : 12 mix of Portland cement, lime putty and crushed stone. For ashlar Bath stone a typical mix would be 1 : 2 : 8 cement, lime and stone dust. Dense sandstones may be bonded with a stronger 1 : 1 : 6 mix, and granite a 1 : 2 or 1 : 3 Portland cement to fine aggregate mix. Jointing should generally be to a similar texture and colour as that of the dressed stone itself, and should be slightly recessed to emphasise the stones rather than the joints. In ashlar work, in which accurately cut squared stones are used, a matching 5 mm flush joint is appropriate.

The David Mellor Cutlery Factory, Hathersage, Derbyshire (Fig. 9.15) illustrates the use of traditionally detailed Derbyshire stone as load-bearing masonry worked in conjunction with precast concrete quoins and padstones. The building takes its form from the base of an old gasholder which provides its foundations.

Gabions

Gabions are wire cages filled with crushed rocks or recycled concrete rubble. They are frequently used in civil engineering applications as retaining walls, and are simply stacked to the required height either vertically or to an appropriate incline. Compressive loads are

Roughly squared split faced random rubble

Polygonal random rubble

Sawn bedded pitched face random walling

Sawn bedded pitched face coursed walling

Fig. 9.14 Traditional stone walling

transmitted through the stones or concrete rubble, and any spreading movement is restrained by the tensile forces within the wire cage. Normally, the cages are of heavy gauge woven or welded steel mesh, which may be zinc, aluminium/zinc alloy, or PVC coated, but for use in load-bearing building applications, such as walls, stainless steel should be used. Gabions are now being used as significant components in building construction, where the particular rugged aesthetic is required (Fig. 9.16). Gabions may be delivered on site filled or flat packed for filling and fastening, usually with a helical binder in alloy-coated or stainless steel. A range of sizes is available based mainly on a metre module.

As an alternative, softer aesthetic, landscape-retaining structures may be constructed from a combination of stones, steel mesh and timber (Fig. 9.17).

STONE CLADDING

For the majority of large commercial buildings, stone is used as a cladding material mechanically fixed to the structural system. The standard BS 8298–1: 2010 gives a broad outline of the structural requirements, including fixings, back-up material and joints. The strength of the stone largely determines the appropriate cladding panel thickness. For granites, marbles and slate 40 mm slabs are usual for external elevations above ground-floor level, but for the softer limestones and sandstones a minimum thickness of 80 mm is frequently recommended. The standard BS 8298–2: 2010 relates flexural strength, span between fixings and external stone cladding thickness to high, medium or low wind exposure.

Fixings (Fig. 9.18) must be manufactured from stainless steel or non-ferrous metal and sized to sustain the dead load of the cladding together with applied

Fig. 9.15 Load-bearing stone masonry — David Mellor Cutlery Factory, Hathersage, Derbyshire. *Architects*: Hopkins Architects.
Photographs: Arthur Lyons

Fig. 9.16 Rock-filled gabions – London Regatta Centre. *Architects*: Ian Ritchie Architects. *Photographs*: Arthur Lyons

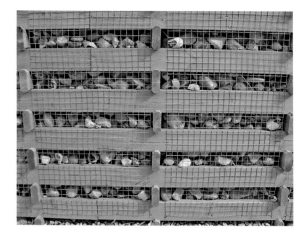

Fig. 9.17 Timber, mesh and stone earth retaining system

loads from wind and maintenance equipment. Movement joints are required to accept the differential structural movements of the frame and the thermal and moisture movements of the cladding. Horizontal compression joints of 15 mm minimum should be located at each floor level; vertical movement joints of 10 mm should be at approximately 6 m centres. Polysulphides, polyurethanes and silicones are used as joint sealants, although non-staining silicones should be used on stones which darken by absorption of silicone fluid. Stone-cladding systems should ideally be protected from impact damage at ground level by the design detailing.

Stone-faced precast concrete cladding systems

An alternative approach to traditional stone cladding is the use of an integral stone veneer on concrete cladding panels. Stone is fixed to the concrete with a series of non-corroding dowels inclined in opposite directions, creating a mechanical fixing, not dependent on the bond between stone and cast concrete. With limestone and sandstone a stone veneer of 50 mm is required, but for granite, slate, quartzite and marble 30 mm is usually appropriate. The concrete should be cast with appropriate reinforcement and fixings for attachment to the building structure. Technical requirements are detailed in the standard BS 8298–4: 2010.

Stone-faced masonry blocks

Stone-faced concrete blocks are manufactured to the standard size of 440 × 215 mm, and faced with a

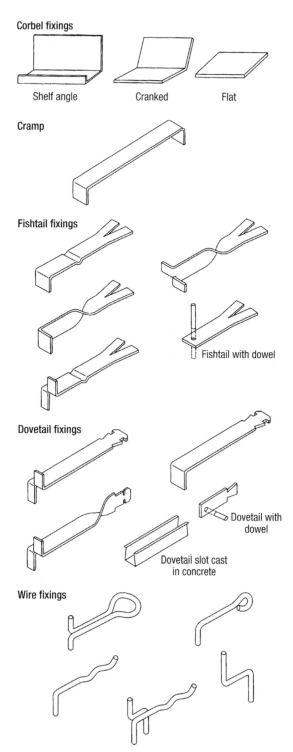

Fig. 9.18 Typical fixings for stone cladding

10 mm veneer of polished marble or granite, fixed with a frost-resistant adhesive. A range of matching special shapes including quoins, end blocks and lintels is available, formed with mitred-stone corner joints.

Lightweight stone cladding

Thin-section stone (approximately 6 mm) may be bonded to lightweight backing materials to reduce the dead weight of stone cladding (Fig. 9.19). The reduction in dead load is significant compared to thick-stone sections which would require traditional stone-cladding techniques. One such material, originally used in the aerospace industry, is a sandwich panel consisting of a core of honeycomb aluminium faced with glassfibre-reinforced epoxy resin skins. The polished stone facing is bonded to one face with epoxy resin, to create a lightweight stone-finished panel, which, if detailed appropriately, has all the visual qualities associated with solid stone masonry.

Rainscreen stone cladding

Rainscreen cladding systems incorporate stone panels fixed to the backing wall by a durable metal framing system, leaving a drained void space which may be partially filled with thermal insulation. The stone panels may be solid or a laminate of stone bonded to lightweight concrete. External and internal corner special units are prefabricated to match the façade. The standard BS 8298–4: 2010 describes the basis of both drained and ventilated rainscreens and pressure-equalised rainscreen cladding systems.

Deterioration of stone

The main agencies causing the deterioration of stone are soluble salt action, atmospheric pollution, frost, the corrosion of metal components and poor design or workmanship.

SOLUBLE SALT ACTION

If moisture containing soluble salts evaporates from the surface of stonework, then the salts will be left either on the surface as white efflorescence or as crystals within the porous surface layer. If the wetting and drying cycles continue, the crystalline material builds up within the pores to the point at which the pressure produced may exceed the tensile strength of the stone, causing it to crumble. The actual pore size significantly influences the durability of individual stones, but generally the more porous stones, such as limestone and sandstone, are susceptible to soluble salt action.

ATMOSPHERIC POLLUTION

Stones based on calcium carbonate are particularly vulnerable to attack by acid atmospheric pollutants.

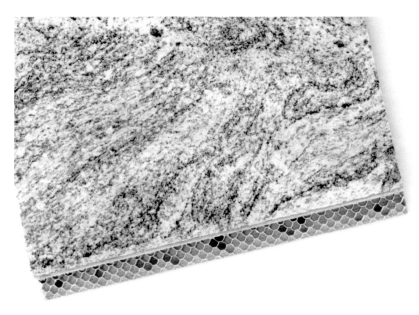

Fig. 9.19 Lightweight veneer stone cladding panels. *Photograph*: Reproduced by permission of IHS

Sulphur dioxide in the presence of water and oxygen from the air produces sulphuric acid, which attacks calcium carbonate to produce calcium sulphate. Limestones and calcareous sandstones are vulnerable to attack. In the case of limestone, the gypsum (calcium sulphate) produced at the surface is slightly soluble and on exposed surfaces it gradually washes away leaving the eroded limestone clean. In unwashed areas, the surface becomes blackened with soot producing a hard crust which eventually blisters, exposing powdered limestone. Magnesian limestones react similarly, except that in some cases the recrystallisation of magnesium sulphate under the blackened crust causes a more serious cavernous decay of the stone. Calcareous sandstones, when rain-washed, gradually decay to powder; however, in unwashed areas they produce a hard crust in which the pores are blocked with gypsum. The crust eventually fails due to differential thermal expansion. Dolomitic sandstones are less vulnerable to acid attack, unless they contain a significant proportion of vulnerable calcite. Silicious sandstones, which are not attacked directly by atmospheric acids, may be damaged by the calcium sulphate washings from limestone, which then cause crystallisation damage to the sandstone surface. Marble, which is essentially calcium carbonate, is also affected by atmospheric acids. Any polished surface is gradually eroded; however, as marble is generally non-porous, crystallisation damage is unusual, and is limited to *sugaring* in some cases.

FROST ACTION

Frost damage occurs in the parts of a building which become frozen when very wet, such as copings, cornices, string courses, window hoods and cills. Frost causes the separation of pieces of stone, but it does not produce powder as in crystallisation attack. Generally, limestones and magnesian limestones are more vulnerable to frost damage than sandstones. Marble, slate and granite used in building are normally unaffected by frost due to their low porosities.

CORROSION OF METALS

Rainwater run-off from copper and its alloys can cause green colour staining on limestones. Iron and steel produce rust staining which is difficult to remove from porous stones. Considerable damage is caused by the expansion of iron and steel in stonework caused by corrosion. All new and replacement fixings should normally be manufactured from stainless steel or non-ferrous metals.

FIRE

Fire rarely causes the complete destruction of stonework. In the case of granite, marble and most sandstones, the surfaces may be blackened or spall. Limestones are generally unaffected by fire, although the paler colours may turn permanently pink due to the oxidation of iron oxides within the stone. Reigate stone, a calcareous sandstone, is also resistant to heat, but it is not a durable stone for exterior use.

PLANTS

Generally, large plants including ivy should be removed from old stonework; however, Virginia creeper and similar species are not considered harmful. Lichens may contribute to deterioration of limestones, and affected stonework should be treated. Damp, north-facing walls and sloping sandstone surfaces are vulnerable to developing algae and lichen growth.

Maintenance of stonework

CLEANING

External granite, marble and slate claddings require regular washing with a mild detergent solution. In particular, highly polished external marble should be washed at least twice a year to prevent permanent dulling of the surface. Limestone, where it is not self-cleaned by rainwater, should be cleaned with a fine water spray and brushing, removing only deposit and not the gypsum-encrusted surface. However, the washing of limestone may cause a *ginger* staining or efflorescence as the stone dries out and risk the possible corrosion of embedded ferrous cramps, so water quantities should be adequately controlled. Sandstone is usually cleaned mechanically by abrasive blasting or chemical cleaning. Abrasive blasting with sand or grit is satisfactory for hard stones but can seriously damage soft stone and moulded surfaces. Hydrofluoric acid and sodium hydroxide (caustic soda) are used in the chemical cleaning of sandstones, but both are hazardous materials which need handling with extreme care by specialist contractors. Unpolished granite may be cleaned by abrasive or chemical processes, but polished granite should only be treated with alkaline solution and then neutralised.

The standard BS 8221–1: 2012 advises careful inspection and the testing of trial cleaning areas of stonework before the full cleaning process is undertaken by water washing, air or abrasive cleaning or chemical agents. In the case of buildings of historic or architectural interest specialist advice or permissions may be required.

STONE PRESERVATION

Generally, coatings such as silicone water repellents should only be applied to stonework following expert advice and testing. Silicone treatment may in certain cases cause a buildup of salt deposits behind the treated layer, eventually causing failure. Silicone treatment should not be applied to already decayed stone surfaces. Polymeric silanes may be used to consolidate decaying stone. The silane is absorbed up to 50 mm into the stone where it polymerises, stabilising the stone but without significantly changing its external appearance. Different silane compositions offer a range of solidification and water-repellent properties which must be matched to the substrate. Generally, such treatment is appropriate for small artefacts which are in immediate danger of loss if left untreated. The standard pr EN 16581: 2013 describes the appropriate testing of water repellents for use on stonework.

Cast stone

The appearance of natural stones, such as Bath, Cotswold, Portland and York, can be recreated using a mixture of stone dust and natural aggregates with cement. In certain cases, iron oxide pigments may also be added to match existing stonework as required. Many architectural components such as classical columns, capitals, balustrades and porticoes are standard items (Figs 9.20 and 9.21), but custom-made products may be cast to designers' specifications as illustrated in the façade of the Thames Water building at Reading (Fig. 9.22). High-quality finishes are achieved by the specialist manufacturers, and cast stone often surpasses natural stone in terms of strength and resistance to moisture penetration. Cast stone may be homogeneous, or, for reasons of economy, may have the facing material intimately bonded to a backing of concrete, in which case the facing material should be at least 20 mm thick. The standard BS 1217: 2008

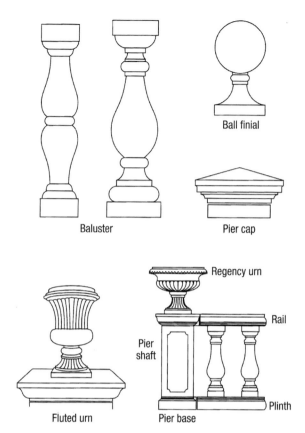

Fig. 9.20 Typical cast stone units

requires that untreated steel reinforcement should have at least 40 mm cover from any visual faces or a minimum of 30 mm if galvanised. Corrosion-resistant metals (e.g. stainless steel, bronze or copper) require at least 10 mm cover from any exposed face. Most masonry units are designed to be installed with 5 or 6 mm joints, and locating holes for dowel joints should be completely filled. Mortars containing lime are recommended rather than standard sand and cement (Table 9.5). Careful workmanship is required to prevent staining of the cast stone surfaces with mortar as it is difficult to remove. Cast stone should weather in a similar manner to the equivalent natural stone.

Table 9.5 Recommended grades of mortar for cast stonework

Exposure	Masonry cement : sand	Plasticised cement : sand	Cement : lime : sand
Severe	1 : 4½	1 : 6	1 : 1 : 6
Moderate	1 : 6	1 : 8	1 : 2 : 9

Fig. 9.21 Cast stone classical Ionic column

DRY AND WET CAST STONE

Cast stone is manufactured by either the dry cast or wet cast process. Dry cast stone is formed from zero-slump concrete, which is densely compacted by vibration. The process is used for the repetitive casting of smaller components, which can be removed from the mould immediately after compaction, allowing many units to be made each day. The wet cast stone system is used for the manufacture of larger units, which remain in the mould for 24 hours, and may incorporate anchor fixings and more complex reinforcement.

AGGLOMERATED STONE

Agglomerated stone, defined in BS EN 14618: 2009, is manufactured from a mixture of aggregates bound with a binder which is either based on cement or thermosetting resin. The aggregate is frequently natural stone, but other recycled materials including crushed ceramics, concrete or glass may be included. The product is cast into blocks or slabs for machining into tiles or worktops. The surface finish may be polished, honed, sand-blasted, flamed or sawn according to the application.

Fig. 9.22 Reconstructed stone cladding – Thames Water, Reading. *Photograph*: Courtesy of Trent Concrete Ltd

Terrazzo tiles (BS EN 13748: 2004) are manufactured with appropriate aggregate (usually crushed natural stone) in a cement binder either as a single layer, or as a double layer with a facing layer and a base concrete layer.

References

FURTHER READING

Allbury, K., Franklin, E. and Anderson, J. 2013: *Environmental impact of brick, stone and concrete.* Bracknell: IHS BRE Press.

Chacon, M.A. 1999: *Architectural stone: fabrication, installation and selection.* New York: John Wiley.

CIRIA. 2009: *Dry stone retaining walls and their modifications – condition appraisal and remedial treatment.* C676. London: Construction Industry Research and Information Association.

Dernie, D. 2003: *New stone architecture.* London: Laurence King.

Dernie, D. 2009: *Stone architecture.* London: Laurence King.

Environment & Heritage Service 2006: *Cleaning masonry buildings (Brick, stone and external renders).* Technical Note 52. Belfast: Environment & Heritage Service.

Garner, L. 2007: *Dry stone walls*, 2nd edn. Princes Risborough: Shire Publications.

Hugues, T., Steiger, L. and Weber, J. 2005: *Dressed stone. Types of stone, details, examples.* Basel: Birkhäuser.

Kicklighter, C.E. 2003: *Modern masonry, brick, block, stone.* Illinois: Goodheart Willcox.

Mäckler, C. (ed.). 2003: *Material stone. Constructions and technologies for contemporary architecture.* Basel: Birkhäuser.

National Federation of Roofing Contractors. 2012: *Selecting natural slates for roof covering.* London: NFRC.

Pavan, V. 2004: *New stone architecture in Italy.* Basel: Birkhäuser.

Pavan, V. 2005: *New stone architecture in Germany.* Basel: Birkhäuser.

Price, M. 2007: *The sourcebook of decorative stone. An illustrated identification guide.* Ontario: Firefly.

Price, M. 2007: *Decorative stone. The complete source book.* London: Thames and Hudson.

Scottish Executive. 2007: *Igneous rock.* Edinburgh: Scottish Executive.

Scottish Executive. 2007: *Building, paving and roofing stone.* Edinburgh: Scottish Executive.

Studio Marmo. 2008: *Stone sampler.* New York: W.W. Norton.

UKCSA. 2010: *Specifying cast stone with confidence.* Crowthorne: UK Cast Stone Association.

UKCSA. 2011: *Technical manual for cast stone.* Crowthorne: UK Cast Stone Association.

UKCSA. 2011: *Specification for cast stone.* Crowthorne: UK Cast Stone Association.

STANDARDS

BS 845–2: 2003
Specification for ancillary components for masonry. Lintels.

BS 1217: 2008
Cast stone. Specification.

BS 5080
Structural fixings in concrete masonry:

Part 1: 1993	Method of test for tensile loading.
Part 2: 1986	Method for determination of resistance to loading in shear.

BS 5385
Wall and floor tiling:

Part 1: 2009	Design and installation of ceramic and natural stone and mosaic wall tiling and in normal internal conditions. Code of practice.
Part 5: 2009	Design and installation of terrazzo, natural stone and agglomerated stone tile and slab flooring. Code of practice.

pr BS 5534: 2013
Slating and tiling for pitched roofs and vertical cladding. Code of practice.

BS 5642
Sills and copings:

Part 1: 1978	Specification for window sills of precast concrete, cast stone, clayware, slate and natural stone.
Part 2: 1983	Specification for copings of precast concrete, cast stone, clayware, slate and natural stone.

BS 6093: 2006
Design of joints and jointing in building construction. Guide.

BS 6100
Building and civil engineering. Vocabulary:
Part 6: 2008   Construction parts.

BS 7533
Pavements constructed with clay, natural stone or concrete pavers:

Part 1: 2001   Guide for the structural design of heavy duty pavements constructed of clay pavers or precast concrete blocks.

Part 2: 2001   Guide to the structural design of lightly trafficked pavements constructed of clay pavers or precast concrete blocks.

Part 3: 2005   Code of practice for laying precast concrete paving blocks and clay pavers.

Part 4: 2006   Code of practice for the construction of pavements of precast concrete flags or natural stone slabs.

Part 6: 1999   Code of practice for laying natural stone, precast concrete and clay kerb units.

Part 7: 2010   Code of practice for the construction of pavements of natural stone paving units and cobbles and rigid construction with concrete block paving.

Part 8: 2003   Guide for the structural design of lightly trafficked pavements of precast concrete and natural stone flags.

Part 9: 2010   Code of practice for the construction of rigid pavements of clay pavers.

Part 10: 2010   Guide for the structural design of trafficked pavements constructed of natural stone setts.

Part 11: 2003   Code of practice for the opening, maintenance and reinstatement of pavements of concrete, clay and natural stone.

Part 12: 2006   Guide to the structural design of trafficked pavements using concrete paving flags and natural stone slabs.

Part 13: 2009   Guide for the design of permeable pavements with concrete paving blocks and flags, natural stone slabs and setts and clay pavers.

BS 8000
Workmanship on building sites:

Part 6: 2013   Code of practice for slating and tiling of roofs and walls.

Part 11: 2011   Internal and external wall and floor tiling. Ceramic and agglomerated stone tiles, natural stone and terrazzo tiles and slabs and mosaics.

BS 8221
Code of practice for cleaning and surface repair of buildings:

Part 1: 2012   Cleaning of natural stone, brick, terracotta and concrete.

Part 2: 2000   Surface repair of natural stones, brick and terracotta.

BS 8297: 2000
Code of practice for design and installation of non-loadbearing precast concrete cladding.

BS 8298
Code of practice for design and installation of natural stone cladding and lining:

Part 1: 2010   General.

Part 2: 2010   Traditional handset external cladding.

Part 3: 2010   Stone-faced pre-cast concrete cladding systems.

Part 4: 2010   Rainscreen and stone on metal frame cladding systems.

BS EN 771
Specification for masonry units:

Part 5: 2011   Manufactured stone masonry units.

Part 6: 2011   Natural stone masonry units.

BS EN 772
Methods of test for masonry units:

Part 4: 1998   Determination of bulk density and porosity.

Part 11: 2011   Determination of water absorption.

Part 14: 2002   Determination of moisture movement.

Part 20: 2000   Determination of flatness of faces.

BS EN 1341: 2012
Slabs of natural stone for external paving. Requirements and test methods.

BS EN 1342: 2012
    Setts of natural stone for external paving.
    Requirements and test methods.
BS EN 1343: 2012
    Kerbs of natural stone for external paving.
    Requirements and test methods.
BS EN 1468: 2012
    Natural stone. Rough slabs. Requirements.
pr EN 1469: 2012
    Natural stone products. Slabs for cladding.
    Requirements.
BS EN 1925: 1999
    Natural stone test methods. Determination of
    water absorption coefficient.
BS EN 1926: 2006
    Natural stone test methods. Determination of
    uniaxial compressive strength.
BS EN 1936: 2006
    Natural stone test methods. Determination of real
    and apparent density.
pr EN 12057: 2012
    Natural stone products. Modular tiles.
    Requirements.
pr EN 12058: 2012
    Natural stone products. Slabs for floors and stairs.
    Requirements.
BS EN 12059: 2008
    Natural stone products. Dimensional stonework.
    Requirements.
BS EN 12326
    Slate and stone products for discontinuous
    roofing and cladding:
        pr Part 1: 2014   Specifications for slate and
                          carbonate slate.
        Part 2: 2011   Methods of test for slate and
                          carbonate slate.
BS EN 12370: 1999
    Natural stone test methods. Determination of
    resistance to salt crystallization.
BS EN 12371: 2010
    Natural stone test methods. Determination of
    frost resistance.
BS EN 12372: 2006
    Natural stone test methods. Determination of
    flexural strength.
BS EN 12407: 2007
    Natural stone test methods. Petrographic
    examination.
BS EN 12440: 2008
    Natural stone. Denomination criteria.

BS EN 12670: 2002
    Natural stone. Terminology.
BS EN 13161: 2008
    Natural stone test methods. Determination of
    flexural strength.
BS EN 13364: 2002
    Natural stone test methods. Determination of
    breaking load at dowel hole.
BS EN 13748
    Terrazzo tiles:
        Part 1: 2004   Terrazzo tiles for internal use.
        Part 2: 2004   Terrazzo tiles for external use.
BS EN 13755: 2008
    Natural stone test methods. Determination of
    water absorption.
BS EN 14157: 2004
    Natural stones. Determination of abrasion
    resistance.
BS EN 14579: 2004
    Natural stone test methods. Determination of
    sound speed propagation.
BS EN 14580: 2005
    Natural stone test methods. Determination of
    static elastic modulus.
BS EN 14581: 2004
    Natural stone test methods. Determination of
    linear thermal expansion coefficient.
BS EN 14617
    Agglomerated stone. Test methods:
        Part 1: 2013   Determination of apparent
                        density and water absorption.
BS EN 14618: 2009
    Agglomerated stone. Terminology and
    classification.
DD CEN/TS 15209: 2008
    Tactile paving surface indicators produced from
    concrete, clay and stone.
BS EN 15286: 2013
    Agglomerated stone. Slabs and tiles for wall
    finishes (internal and external).
BS EN 15388: 2008
    Agglomerated stone. Slabs and cut-to-size
    products for vanity and kitchen tops.
BS EN 15886: 2010
    Conservation of cultural property. Test methods.
    Colour measurement of surfaces.
BS EN 16140: 2011
    Natural stone test methods. Determination of
    sensitivity to changes in appearance produced by
    thermal cycles.

BS EN 16301: 2013
Natural stone test methods. Determination of sensitivity to accidental staining.

BS EN 16306: 2013
Natural stone test methods. Determination of resistance of marble to thermal and moisture cycles.

pr EN 16515: 2013
Conservation of cultural heritage. Guidelines to characterise natural stone used in cultural heritage.

pr EN 16572: 2013
Conservation of cultural heritage. Glossary of technical terms concerning mortars for masonry, renders and plasters.

pr EN 16581: 2013
Conservation of cultural heritage. Surface protection for porous materials.

BUILDING RESEARCH ESTABLISHMENT PUBLICATIONS

BRE Digests

BRE Digest 420: 1997
Selecting natural building stones.

BRE Digest 448: 2000
Cleaning buildings: legislation and good practice.

BRE Digest 449: 2000
Cleaning exterior masonry (Parts 1 and 2).

BRE Digest 467: 2002
Slate and tile roofs: avoiding damage from aircraft wake vortices.

BRE Digest 502: 2007
Principles of masonry conservation management.

BRE Digest 508: 2008
Conservation and cleaning of masonry (Part 1 Stonework).

BRE Good building guide

BRE GBG 64: 2005
Tiling and slating pitched roofs (Parts 1–3).

BRE Information papers

BRE IP 10/00
Flooring, paving and setts.

BRE IP 10/01
Lightweight veneer stone cladding panels.

BRE Reports

SO 36: 1989
The building limestones of the British Isles, E. Leary.

BR 84: 1986
The building sandstones of the British Isles, E. Leary.

BR 134: 1988
The building magnesian limestones of the British Isles, D. Hart.

BR 195: 1991
The building slates of the British Isles, D. Hart.

ADVISORY ORGANISATIONS

Men of the Stones, The Old School House, Clipsham, Oakham, Rutland LE15 7SE (01780 410382).

National Federation of Terrazzo, Marble & Mosaic Specialists, PO Box 1218, RH10 0HE (0845 609 0050).

Stone Federation Great Britain, Channel Business Centre, Ingles Manor, Castle Hill Avenue, Folkestone, Kent CT20 2RD (01303 856123).

UK Cast Stone Association, 15 Stone Hill Court, The Arbours, Northampton NN3 3RA (01604 405666).

PLASTICS

CONTENTS

Introduction

The plastics used in the construction industry are generally low-density non-load-bearing materials. Unlike metals they are not subject to corrosion, but they may be degraded by the action of direct sunlight, with a corresponding reduction in mechanical strength. Many plastics are flammable unless treated; the majority emit noxious fumes in fires. Approximately 20% of plastics produced within the UK are used in the building industry. Polyvinyl chloride (PVC), which has a high embodied energy content, accounts for 40% of this market share, predominantly in pipes, but also in cladding, electrical cable insulation, windows, doors and flooring applications. Foamed plastics for thermal and acoustic insulation are formulated as either open- or closed-cell materials, the latter being resistant to the passage of air and water.

In terms of their chemical composition, plastics form a diverse group of materials which have chain-like molecular structures composed of a large number of small repeat units. While some materials such as rubber

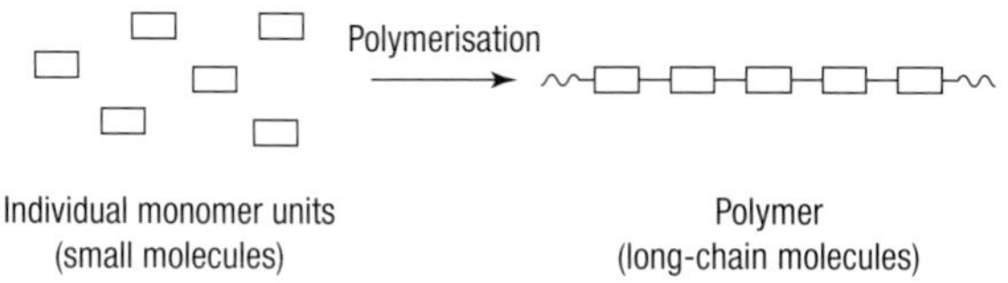

Individual monomer units
(small molecules)

Polymerisation

Polymer
(long-chain molecules)

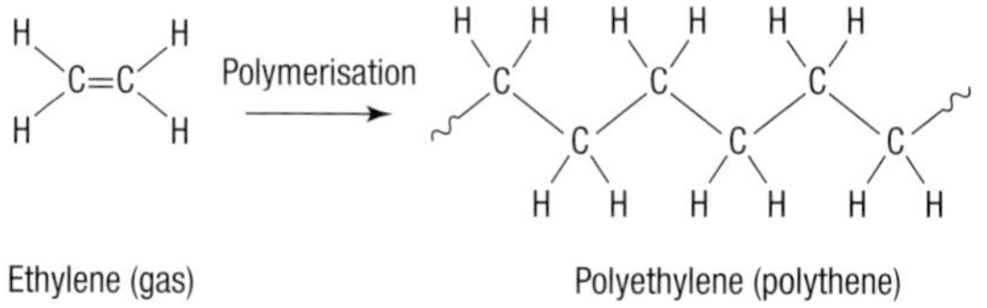

Ethylene (gas)

Polymerisation

Polyethylene (polythene)

Fig. 10.1 Polymerisation of ethylene to polyethylene (polythene)

Vinyl chloride
(gas)

Addition
polymerisation

Polyvinyl chloride
(PVC)

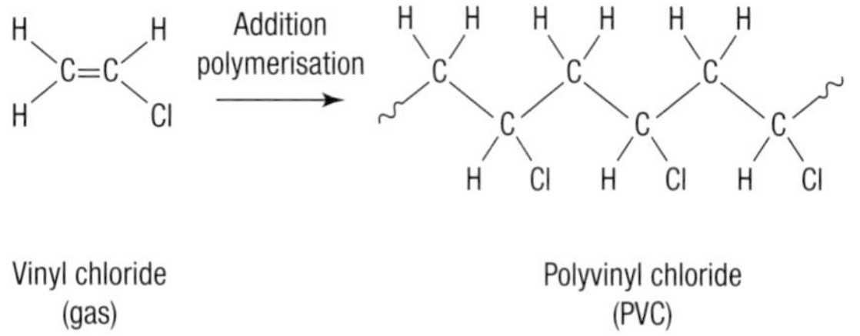

Fig. 10.2 Polymerisation of vinyl chloride to PVC

and cellulose derivatives are based on natural products, the majority of plastics are produced from petrochemical products. The manufacture of polythene, which dates back to 1933, involves the polymerisation of ethylene monomer, a colourless gas, which under high pressure at 200°C is converted into the clear polymer polyethylene or polythene (Fig. 10.1).

Polymerisation

In the production of polythene the small molecular units of ethylene are joined end to end by an addition polymerisation process to produce the long-chain macromolecules. A similar process converts vinyl chloride into polyvinyl chloride (PVC) (Fig. 10.2), styrene monomer into polystyrene and tetrafluoroethylene into polytetrafluoroethylene (PTFE).

While the molecular backbones of plastics are predominantly composed of chains of carbon atoms, variations occur, particularly when the polymerisation process involves the elimination of water between adjacent monomer units. Thus, in the case of condensation polymerisation (Fig. 10.3), oxygen or nitrogen atoms are incorporated into the backbone of the

macromolecular chains as in the polyesters (resins) and polyamides (nylons).

BRANCHED CHAINS

Depending on the conditions during the polymerisation process, the polymer chains produced may be linear or branched. In the case of polythene, this affects the closeness of packing of the chains and therefore the bulk density of the material. Thus high-density polythene, (HDPE) (s.g. 0.97) which is relatively stiff, has few branched chains compared to low-density polythene (LDPE) (s.g. 0.92) which is softer and waxy (Fig. 10.4).

COPOLYMERS

Where two or more different monomers are polymerised together, the product is a copolymer. The properties of the copolymer are significantly dependent upon whether the two components have joined together in alternating, random or block sequences (Fig. 10.5).

More complex plastics can be produced for their specific physical properties by combining several

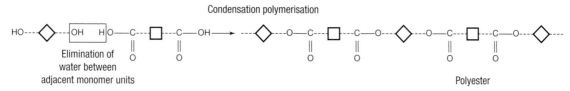

Condensation polymerisation

Elimination of
water between
adjacent monomer units

Polyester

Fig. 10.3 Condensation polymerisation

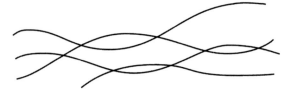

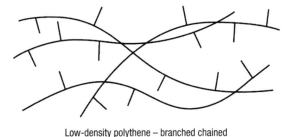

High-density polythene – straight chained

Low-density polythene – branched chained

Fig. 10.4 *Straight- and branched-chain plastics*

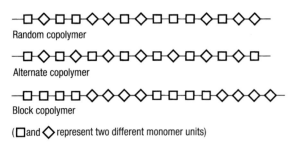

Random copolymer

Alternate copolymer

Block copolymer

(☐ and ◇ represent two different monomer units)

Fig. 10.5 *Random, alternate and block copolymers*

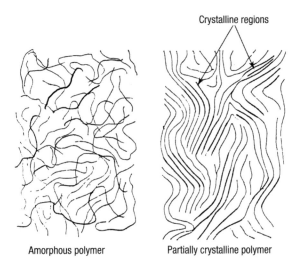

Amorphous polymer   Partially crystalline polymer

Fig. 10.6 *Crystallinity in polymers*

components. Thus, acrylonitrile butadiene styrene (ABS) is produced by grafting styrene-acrylonitrile onto a preformed butadiene latex of a carefully controlled particle size.

CRYSTALLINITY

In their initial manufactured state, most polymers consist of amorphous randomly oriented molecular chains. However, if the plastic material is stretched in one direction, such as during the drawing of spun fibres, it causes an alignment of the molecular chains, leading to partial formation of crystalline regions and an associated anisotropy (Fig. 10.6). Crystalline regions may also be produced during the solidification of simple polymers such as polyethylene, but they will be limited in their extents due to the general entanglement of the molecular chains.

GLASS TRANSITION TEMPERATURE

In their molten state, the individual molecular chains of a plastic material move freely relative to each other, allowing the material to be moulded within the various forming processes used for the manufacture of components. As the temperature of melted plastic material is lowered, the freedom of movement of the molecular chains is reduced; gradually the plastic becomes more viscous, until eventually it solidifies at its characteristic melting-point temperature. However, even when solid,

most plastics remain rubbery or flexible, due to rotations within the individual molecular chains. As the temperature is lowered further, the material will eventually become rigid and brittle, as movement can no longer take place within the individual molecular units. The temperature at which a particular plastic changes from flexible to rigid is defined as its characteristic *glass transition temperature*. Depending on the nature of the particular plastic material this may be above or below normal ambient temperatures. Further, the glass transition temperature for a particular plastic can be significantly changed by, for example, the addition of plasticisers, characterised by the differences in physical properties between PVC-U (unplasticised) and PVC (plasticised polyvinyl chloride).

Polymer types

Polymers are normally categorised in respect of their physical properties as either thermoplastic, thermosetting or elastomeric.

THERMOPLASTICS

Thermoplastics soften on heating and reset on cooling. The process is reversible and the material is unaffected by repeating the cycle, providing that excessive temperatures, which would cause polymer degradation, are not applied. Many thermoplastics are soluble in organic solvents, while others swell by solvent absorption. Thermoplastics are usually produced initially in the form of small granules for subsequent fabrication into components.

THERMOSETTING PLASTICS

Thermosetting plastics have a three-dimensional cross-linked structure, formed by the linkage of adjacent macromolecular chains (Fig. 10.7). Thermosets are not softened by heating, and will only char and degrade if heated to high temperatures. Thermosets are usually produced from a partially polymerised powder or by mixing two components, such as a resin and a hardener. The resin is essentially the macromolecular component

and the hardener cross-links the liquid resin into the thermoset plastic. Curing for epoxy resin adhesives and polyesters as in glassfibre-reinforced polyester (GRP) occurs at room temperature, while for phenolic and formaldehyde-based resins, a raised temperature and pressure are required. Thermosets, because of their three-dimensional structure, are usually solvent resistant and harder than thermoplastics.

ELASTOMERS

Elastomers are long-chain polymers in which the naturally helical or zigzag molecular chains are free to straighten when the material is stretched, and recover when the load is removed. The degree of elasticity depends on the extensibility of the polymeric chains. Thus natural rubber is highly extensible, but when sulphur is added, the vulcanisation process increasingly restricts movement by locking together adjacent polymer chains (Fig.10.8). For most uses some cross-linking is required to ensure that an elastomeric material returns to its original form when the applied stress is removed.

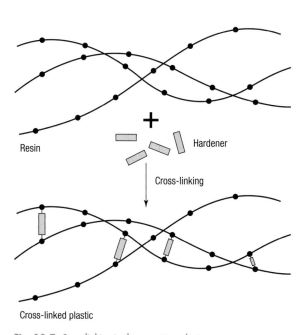

Fig. 10.7 Cross-linking in thermosetting plastics

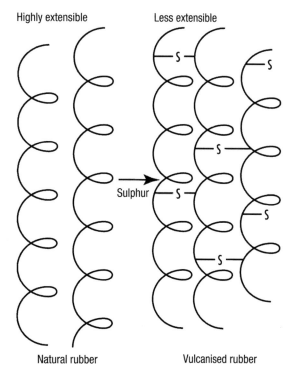

Fig. 10.8 Elastomers and the effect of cross-linking

Additives

PLASTICISERS

Plasticisers are frequently incorporated into plastics to increase their flexibility. The addition of the plasticiser separates the molecular chains, decreasing their mutual attraction. Thus unplasticised PVC (PVC-U) is suitable for the manufacture of rainwater goods, window units and glazing, whereas plasticised PVC is used for flexible single-layer roof membranes, tile and sheet floor coverings, and electrical cable insulation. Loss of plasticiser by migration can cause eventual embrittlement of plasticised PVC components.

FILLERS

Chalk, sand, china clay or carbon black are often added to plastics to reduce costs, improve fire resistance or opacity. Titanium dioxide is added to PVC-U to produce a good shiny surface. Glass fibres are added to polyester resins to give strength to the composite material, glassfibre-reinforced polyester (GRP), as described in Chapter 11.

PIGMENTS AND STABILISERS

Dyes and pigments may be added to the monomer or polymer. Stabilisers are added to absorb ultraviolet light which would otherwise cause degradation. For example, organotin compounds are used in clear PVC sheet to preferentially absorb incident ultraviolet light, in order to prevent degradation by the elimination of hydrogen chloride.

Degradation of plastics

The degradation of plastics is most frequently attributed to the breakdown of the long molecular chains (Fig. 10.9), or, in the case of PVC, the loss of plasticiser. Polymeric molecular chains may be broken by the effect of either heat, ultraviolet light or ozone, or by a combination of any of these factors, thus reducing their average molecular chain length. Discoloration occurs through the production of molecular units with double bonds, usually causing a yellowing of the plastic. Surface crazing and stress cracks may develop where degradation has caused cross-linking, resulting in embrittlement of the surface.

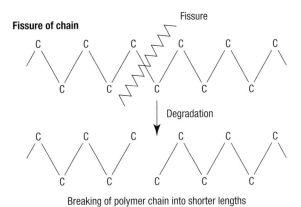

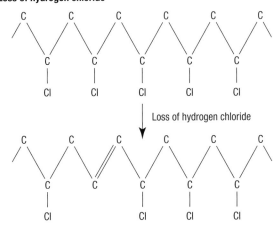

Fig. 10.9 Degradation of plastics

Where plasticiser is lost by migration from PVC, the glass transition temperature is gradually raised, so eventually the material becomes brittle at ambient temperatures. Typically, high-boiling-point oils such as dibutyl phthalate and dioctyl phthalate are incorporated into the original PVC, but these gradually evaporate. leaving the surface vulnerable to cracking and shrinkage.

Properties of plastics

FIRE

All plastics are combustible, producing noxious fumes and smoke (Table 10.1). Carbon monoxide is produced by most organic materials, but in addition, plastics containing nitrogen, such as polyurethane foam, generate hydrogen cyanide and PVC produces hydrochloric acid. Some plastics, particularly acrylics and expanded polystyrene, have a high surface spread of flame and produce burning droplets; however, others when treated with fire retardant are difficult to ignite and some are self-extinguishing.

STRENGTH

Although plastics have a good tensile strength-to-weight ratio, they also have a low modulus of elasticity which renders them unsuitable for most load-bearing situations, the only exception being glassfibre-reinforced polyester (GRP) which has been used for some limited load-bearing applications. Generally, thermoplastics soften at moderate temperatures and are subject to creep under ambient conditions.

Table 10.1 Behaviour of common building plastics in fire

Material	Behaviour in fire
Thermoplastics	
Polythene, Polypropylene	Melts and burns readily
Polyvinyl chloride	Melts, does not burn easily, but emits smoke and hydrogen chloride
PTFE/ETFE	Does not burn, but at high temperatures evolves toxic fumes
Polymethyl methacrylate	Melts and burns rapidly, producing droplets of flaming material
Polystyrene	Melts and burns readily, producing dense black smoke and droplets of flaming material
ABS copolymer	Burns readily
Polyurethane	The foam burns readily producing highly toxic fumes including cyanides and isocyanates
Thermosetting plastics	
Phenol formaldehyde, Melamine formaldehyde, Urea formaldehyde	Resistant to ignition, but produce noxious fumes including ammonia
Glass-reinforced polyester (GRP)	Burns producing smoke, but flame-retarded grades are available
Elastomers	
Rubber	Burns readily, producing black smoke and sulphur dioxide
Neoprene	Better fire resistance than natural rubber

THERMAL AND MOISTURE MOVEMENT

The thermal expansion of most plastics is high. The expansion of GRP is similar to that of aluminium, but most other plastics have larger coefficients of linear expansion. For this reason, attention must be paid to careful detailing to allow for adequate thermal movement, particularly where weather exclusion is involved. Most plastics are resistant to water absorption, and therefore do not exhibit moisture movement. (Typical coefficients of linear expansion are polythene (HD) $(110-130) \times 10^{-6}°C^{-1}$, polypropylene $110 \times 10^{-6}°C^{-1}$, ABS $(83-95) \times 10^{-6}°C^{-1}$, PVC $(40-80) \times 10^{-6}°C^{-1}$, GRP $(20-35) \times 10^{-6}°C^{-1}$.)

Plastics forming processes

Depending on the nature of the product, plastics may be formed by either continuous or batch processes. With thermoplastics, frequently a two-stage process is most appropriate in which the raw materials, supplied by the primary manufacturer as powder or granules, are formed into an extrusion or sheet which is then reformed into the finished product. However, thermosetting plastics must be produced either from a partially polymerised material or directly from the resin and hardener mix in a single-stage process. Most foamed plastics are either blown with internally generated gas, or produced by a vacuum process, which reduces reliance on the previously used environmentally damaging CFCs and HCFCs.

CONTINUOUS PROCESSES

Extrusion

Plastic granules are fed continuously into the heated barrel of a screw extruder, which forces the molten thermoplastic through an appropriately shaped die to produce a rod, tube or the required section (Fig. 10.10). Products include pipes, rainwater goods and fibres.

Film blowing

As a molten thermoplastic tube is produced in the extrusion process, air is blown in to form a continuous cylindrical plastic sheet, which is then rolled flat and trimmed to produce a folded sheet. Adjustment of the applied air pressure controls the sheet thickness.

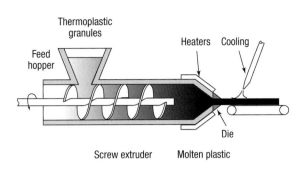

Fig. 10.10 Formation of plastics by extrusion

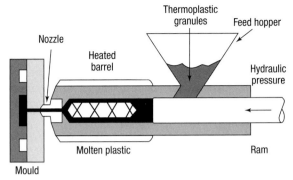

Fig. 10.11 Formation of plastics by injection moulding

Calendering

Sheet thermoplastic materials may be produced from plastics granules by compression and fusion between a series of heated rollers. Laminates may be produced by heating together two or more thermoplastic sheets, and during this process, sheet reinforcement material may be incorporated.

BATCH PROCESSES

Injection moulding

Thermoplastic granules are melted in a screw extruder to fill a ram which injects the plastic into an appropriate mould. After cooling, the components are removed from the mould and trimmed as necessary. The process is low cost and rapid. A series of moulds may be attached to the injection moulding machine to ensure continuity of production (Fig. 10.11). Thermosetting polymers may be injection moulded by initial forming at a low temperature followed by heating of the mould to cross-link the liquid plastic.

Compression moulding

In the compression moulding process for thermosetting resins, the appropriate quantity of uncross-linked resin powder is subjected to pressure and heat within the mould. When the polymer has melted and cross-linked, the mould may be opened and the component removed.

Pressing

Pressing is used to form thermoplastic sheet into components. The sheet plastic is initially heated to softening point and then pressed between an appropriately shaped pair of dies.

Vacuum forming and blow moulding

During vacuum forming, thermoplastic sheet is heated over a mould which is then evacuated through a series of fine holes, drawing the soft plastic into the appropriate form. In the similar process, blow moulding, positive air pressure is applied inside a molten polymer tube which is expanded into the shape of the mould.

THREE-DIMENSIONAL PRINTING

Advances in 3-D printing enable prototypes and components to be manufactured to very close tolerances from three-dimensional digital models. This has implications not only for the design of building components but also for the manufacture of detailed architectural models.

The systems are based on the successive buildup of very thin layers of solid material to the exact pattern of layered digital model sections. Various lay-up systems have been developed for the deposition of plastic and other materials. These range from a fine nozzle, to using laser technology to accurately polymerise viscous resin in very thin layers and the use of adhesive-backed paper cut by laser to the required section shapes. Each system produces a highly accurate three-dimensional solid, over a period of several hours depending upon the product size. Where any part of the buildup of the solid object needs support during manufacture, the system automatically produces additional material in a weak form. This can be broken away easily after the whole object is complete, and in the case of laser/resin production finally cured. In all these manufacturing processes the buildup layers are extremely thin, so smooth and accurate surfaces are achieved.

The reverse of this process allows prototype complex-shape components or small-scale architectural models to be turned into accurate three-dimensional CAD files, using a delicate probe mechanism which senses all over the object's surfaces. This allows the designer to generate CAD files for highly complex three-dimensional forms which would be virtually impossible to draw directly into a CAD system.

Plastics in construction

The broad ranges of thermoplastic, thermosetting and elastomeric plastics are collated into families in Figure 10.12. Typical uses in construction are listed in Table 10.2.

(Glassfibre-reinforced polyester is described in Chapter 11, foamed plastics as insulation materials in Chapter 13 and plastics used primarily as adhesives in Chapter 14.)

THERMOPLASTICS

Polythene (polyethylene)

Polyethylene (PE) is one of the cheapest plastics and is available in a range of grades of differing densities and physical properties. Low-density polyethylene (LDPE) has a softening point of 90°C and high-density polyethylene (HDPE) softens at 125°C.

Grades of polyethylene:

		Density kg/m³
LLDPE	Linear low-density polyethylene	900–939
VLDPE	Very low-density polyethylene	880–915
LDPE	Low-density polyethylene	916–925
MDPE	Medium-density polyethylene	926–940
HDPE	High-density polyethylene	941–970
HMWPE	High molecular weight polyethylene	947–950
UHMWPE	Ultra-high molecular weight polyethylene	930–935
PEX	Cross-linked polyethylene	926–970

Polyethylene is resistant to chemicals and tough at low temperatures, but it is rapidly embrittled by ultraviolet light unless carbon black is incorporated. Polyethylene burns and has a relatively high coefficient of thermal expansion. Low-density polyethylene is used widely for damp-proof membranes, damp-proof courses and vapour barriers. High-density polyethylene, which is stiffer than the low-density material, is used for tanking membranes to basements. Ultra-high molecular weight polyethylene (melting point 130°C) has chemical inertness and good environmental stress crack resistance to solvents, oils and detergents. Linear low-density polyethylene, used for pipes and cable jacketing, is easily extruded and has good impact resistance.

Polyethylene is used for the production of cold-water cisterns, but is only suitable for cold-water plumbing applications due to its high thermal expansion. Mains water pressure pipes which are manufactured from medium-density (PE80) and ultra-high molecular weight (PE100) polyethylene require a significant wall thickness due to the relatively low tensile strength of polyethylene.

Cross-linked polyethylene (PEX), manufactured by the action of peroxide catalyst on normal polyethylene, is used for domestic hot-water and underfloor heating systems as it can withstand operating temperatures of up to 90°C. In underfloor heating systems, an interlayer of aluminium is incorporated into the PEX pipe to prevent the ingress of oxygen which would cause corrosion of steel components within the system. Certain barrier composite pipes can operate at temperatures of up to 95 °C.

The standard BS EN ISO 11542–1: 2001 describes ultra-high molecular weight polyethylene.

Polypropylene

Polypropylene (PP), with a softening point of 150°C, is slightly stiffer than polyethylene, to which it is closely related chemically. Like polyethylene, it is resistant to chemicals and susceptible to ultraviolet light, but unlike polyethylene, it becomes brittle below 0°C. However, the block copolymer with ethylene does have improved low temperature impact resistance. Polypropylene is used for pipes, drainage systems, water tanks, DPCs, connecting sleeves for clay pipes and WC cisterns. Polypropylene fibres are used in fibre-reinforced concrete to produce an increase in impact resistance over the equivalent unreinforced material. Polypropylene permanent shuttering for concrete is frequently a double-layer system incorporating polystyrene insulation and radon protection where necessary. Certain breather membranes used for tile underlay and timber

Table 10.2 Typical uses of plastics in construction

Material		Examples of plastics in construction
Thermoplastics		
Polythene	(Low-density LD)	DPC, DPM, vapour checks, roof sarking
	(High-density HD)	Cold water tanks, cold water plumbing
	(Cross-linked PEX)	Hot and cold water plumbing
Polypropylene	(PP)	Pipework and fittings, cold water plumbing, drainage systems, water tanks, WC cisterns, DPCs, fibres in fibre-reinforced concrete
Polybutylene	(PB)	Hot and cold water pipework and fittings
Polyvinyl chloride	(PVC-U)	Rainwater goods, drainage systems, cold water, underground services, window and door frames, conservatories, garage doors, translucent roofing sheets, cold water tanks
	(PVC-UE)	Claddings, barge boards, soffits, fascias, window boards
	(PVC)	Tile and sheet floor coverings, single-ply roofing, cable insulation, electrical trunking systems, sarking, tensile membrane structures, glazing to flexible doors, door seals, handrail coatings, vinyl-film finishes to timber products
	(CPVC)	Hot and cold water systems, window and door frames
ETFE		Inflated systems for translucent wall and roof membranes
PTFE		Sealing tape for plumbing, tensile membrane structures, low-friction movement joints
Polymethyl methacrylate		Baths, shower trays, kitchen sinks, glazing, roof lights, luminaires
Polycarbonate		Vandal-resistant glazing, spa baths, kitchen sinks
Polystyrene		Bath and shower panels, decorative expanded polystyrene tiles
ABS copolymer		Pipes and fittings, rainwater goods, drainage systems, shower trays
Nylons		Electrical conduit and trunking, low-friction components – hinges, brush strips for sealing doors and windows, carpet tiles and carpets, shower curtains
Thermosetting plastics		
Phenol formaldehyde		Decorative laminates
Melamine formaldehyde		Laminates for working surfaces and doors, moulded electrical components, WC seats
Urea formaldehyde		Decorative laminates
Glass-reinforced polyester (GRP)		Cladding and roofing panels, simulated cast-iron rainwater goods, cold water tanks, spa baths, garage doors, decorative tiles and panels
Elastomers		
Rubber		Flooring, door seals, anti-vibration bearings
Neoprene		Glazing seals, gaskets
EPDM		Glazing seals, gaskets, single-ply roofing systems
Butyl rubber		Sheet liners to water features and landfill sites
Nitrile rubber		Tile and sheet flooring

frame construction are manufactured from multilayer systems incorporating polypropylene with polyethylene and glassfibre reinforcement. Such products are wind- and watertight, but vapour-permeable. Many geotextiles for soil stabilisation are manufactured as a mat material from non-woven heat-bonded polypropylene continuous fibres. The material may be reinforced by woven polyester fibres. Polypropylene is highly fatigue resistant and therefore used for integral hinges in lightweight components.

The standard BS EN ISO 1873–1: 1995 describes the designation system for polypropylene.

Polybutylene (Polybutene-1)

Polybutylene (polybutene-1 PB-1) is used for pipework as an alternative to copper. It has the advantage of flexibility and the very smooth internal surface is resistant to the buildup of scale and deposits. It can withstand continuous operating temperatures of up to

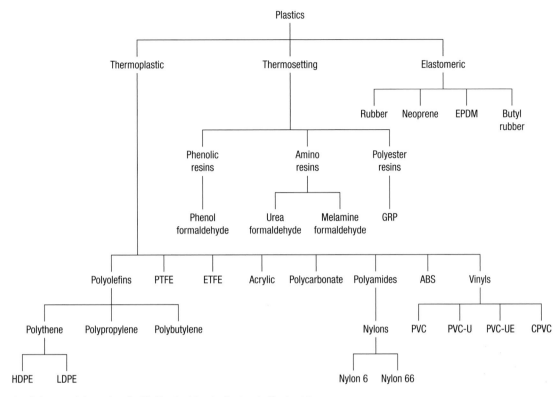

Fig. 10.12 Plastics used in construction

82°C. As polybutylene is slightly permeable to oxygen, some pipe manufacturers incorporate an ethyl vinyl alcohol copolymer (EVOH) barrier as an interlayer. The standard BS EN ISO 8986–1: 2009 describes the designation system for polybutene-1. Polybutylene pipes and fittings are described in BS 7291–2: 2010.

Polyvinyl chloride

Polyvinyl chloride is the most widely used plastic material in the construction industry. It is available both in the unplasticised form (PVC-U) and as the plasticised product (PVC). In both forms polyvinyl chloride is combustible, giving off noxious hydrogen chloride fumes; however, the unplasticised form tends to burn only with difficulty. PVC begins to soften at 75°C, and therefore cannot be used for hot-water systems, although chlorinated PVC (CPVC) may be used at higher temperatures. PVC is soluble in certain organic solvents which, therefore, may be used for the solvent welding of joints, but PVC is unaffected by acids and alkalis.

Plasticised PVC is extensively used in the manufacture of floor coverings, either as individual tile units or as continuously jointed sheet. It is also the most widely used material for single-layer roofing systems, due to its durability, colour range and ease of application. It also offers an alternative to bitumen sheet for sarking. Plasticised PVC is the standard for electrical cable insulation, and many small building components are made from injection-moulded PVC.

PVC-U

PVC-U is widely used for rainwater goods, usually in white, grey, black or brown, and similarly for soil- and waste-pipes. It is also used colour-coded for underground water, gas, electrical and telecommunications systems. PVC-U is used extensively for the manufacture of extruded window frames, door frames and conservatories, usually incorporating sealed double-glazing units. Where insufficient rigidity is achieved with the PVC-U alone, steel inserts within the extruded sections give strength and provide additional protection against forced entry. PVC-U is used in the manufacture of translucent, transparent and coloured profiled sheeting for domestic structures such as carports and conservatories, where an economical product is required, although eventually the products discolour and craze due to the effects of direct ultraviolet light. PVC-U may be recycled several times without loss of performance. The recycled material is generally ground into granular form and reused in the co-extrusion of profile sections or in the production of 100% recycled PVC-U cavity closers.

PVC-UE

Extruded cellular unplasticised PVC (PVC-UE) is used for cladding, fascias, soffits, window boards, barge boards and other components of uniform section. It is manufactured by the co-extrusion of a high-impact PVC-U surface material over a core of closed-cell PVC foam. The foaming agent is usually sodium bicarbonate. The high stiffness-to-weight ratio arises from the combination of a dense outer skin and a cellular core. Both the cellular core and the wearing surface are stabilised with metallic additives to prevent degradation and discoloration. Contact with bitumen should be avoided. The material will char and melt in fire, but with a limited surface spread of flame. The material is described within the standard BS 7619: 2010.

Tensile membrane structures
PVC-coated polyester is the standard material used for tensile membrane structures and canopies. The durability depends directly on the degree of translucency; and at 15% transmission, a period of 15 years may be reasonably expected. At greater levels of translucency the expected serviceable lifetime is considerably reduced; however, fluoropolymer lacquers to the top surface of the fabric enhance durability. While white fabric is the standard, coloured and patterned membranes are available according to client requirements. PVC-coated polyester membranes are a cheaper alternative to PTFE-coated fabrics, but are not noncombustible. PVC-coated polyester is more flexible than PTFE-coated fibreglass and is therefore the preferred material for temporary structures which may be folded for transport and storage. The thermal insulation afforded by single-layer tensile membrane roofs is negligible.

Tensile membrane structures are manufactured from a set of tailored panels stitched or welded together. They are usually tensioned by wires or rods running through edge pockets, or by fixing directly to structural elements. Accurate tensioning is required to generate the correct form and resistance to wind and snow loads. The use of double curvature within the panel elements imparts structural rigidity to the overall membrane structure. Damage by accident or vandalism can usually be repaired on site.

Polytetrafluoroethylene

Polytetrafluoroethylene (PTFE) coated glassfibre woven fabrics are used for permanent tensile membrane structures. In a fire, PTFE gives off toxic combustion products, but only at temperatures above which any fabric would have already failed and vented the heat and smoke. With a fire rating of Class 0, PTFE-coated glassfibre tensile membranes are more expensive than the Class 1 rated PVC-coated polyesters, but are generally more durable with an anticipated lifespan in excess of 20–25 years. The low-friction PTFE surface has good self-cleaning properties.

The Lawn Tennis Association's National Tennis Centre, Roehampton (Fig. 10.13) and the Millennium Dome at Greenwich are roofed with PTFE (*Teflon*)-coated glassfibre tensile membranes. The translucent fabric gives well-lit internal spaces during the daytime, and striking glowing surfaces at night. The 321 m-high Burj Al Arab seven-star hotel in Dubai (Fig. 10.14) is clad on one face with a PTFE-coated glassfibre tensile membrane. It was the first building to use this material orientated vertically.

PTFE tape has a very low coefficient of friction and a high melting point. It is therefore ideal for use as a sealing tape for threaded joints in water and gas pipes. It is also used to form sliding joints in large structures.

The standard BS EN ISO 12086–1: 2006 describes the designation system for fluoropolymers.

Ethylene tetrafluoro ethylene

Ethylene tetrafluoro ethylene copolymer (ETFE) is used as a translucent foil for low-pressure pneumatic metal-framed building envelope cushions. The fluorocopolymer has the advantages over glass that when used to form two- to five-layer air cushion systems, it offers higher thermal insulation with greater transparency to UV light. ETFE is strong, shatter proof, half the cost and only 100th the weight of the equivalent glass, thus offering significant economies to the required structural supporting system. ETFE, with an anticipated life span of 25 years, can withstand maintenance loads, be easily

Fig. 10.13 Tensile Membrane – National Tennis Centre, Roehampton. *Architects*: Hopkins Architects. *Photograph*: Arthur Lyons

repaired and is recyclable. It has been used very effectively on the galvanised-tubular steel space-frame envelope for the biomes (domes sheltering plants from around the world) at the Eden Project, Cornwall (Fig. 10.15). The structure is formed from an icosahedral geodesic outer layer, with a combination of hexagons, pentagons and triangles as the inner layer of the three-dimensional space frame. Only a small pumping system, powered by photovoltaic cells, is required to maintain the air-fill of the ETFE cushions. At the National Space Centre in Leicester, also designed by Grimshaw Architects, the ETFE cushion-clad tower houses the main space rocket exhibits (Fig. 10.16).

If automatic smoke venting of an ETFE atrium is required, electrical wiring may be incorporated into the cushion frames, which release the cushions, except at one point, in case of fire. An enclosed atrium is thus turned into a fully open light well.

Transmitted light levels through ETFE cushions may be constantly adjusted by the use of partially printed internal layers within the cushions, which may be

Fig. 10.14 Teflon-coated glass fibre membrane – Burj Al Arab Hotel, Dubai. *Architects*: WS Atkins. *Photograph*: Arthur Lyons

Fig. 10.15 ETFE – Eden Project, Cornwall. *Architects*: Grimshaw Architects. *Photographs*: Arthur Lyons and courtesy of Perry Hooper (interior)

moved closer or further apart by changing the pumped air pressure, thus modifying the shadowing effect. Interesting patterns can be created by the use of coloured cushions, while aluminium-coated foils will give a highly reflective effect with reduced sunlight penetration.

Polymethyl methacrylate

Acrylic or polymethyl methacrylate (PMMA) is available in a wide variety of translucent or transparent,

clear or brightly coloured sheets. It softens at 90°C, and burns rapidly with falling droplets of burning material. Stress crazing may occur where acrylic has been shaped in manufacture and not fully annealed, but generally the material is resistant to degradation by ultraviolet light. Acrylic is frequently used for decorative signs, roof lights and light fittings. Baths and shower trays are manufactured from acrylic as a lighter alternative to cast iron and ceramics. Although not resistant to abrasion, scratches can usually be polished out with proprietary metal polish. Polymethyl methacrylate

from industrial and commercial sources can be fully recycled when clean and separated from other waste materials.

The standard BS EN ISO 8257–1: 2006 describes the designation system for polymethyl methacrylate.

Polycarbonate

Polycarbonates (PCs) are used as vandal-resistant glazing, due to their high impact resistance, good optical transparency and low ignitability. Polycarbonate glazing is used extensively for roof lights, roof domes, smoking shelters, carports, covered walkways and road barriers. Polycarbonate blocks offer a lightweight alternative to traditional cast glass blocks. Proprietary extruded cellular systems of double-, triple-, five- and seven-walled polycarbonate offer combined thermal insulation and vandal-resistant properties. The protective outer surface prevents ultraviolet degradation for 10 years, and sections may be curved on site within the limits of the manufacturers' specifications.

The standard BS EN ISO 7391–1: 2006 describes the designation system for polycarbonate.

Acrylonitrile butadiene styrene

Acrylonitrile butadiene styrene (ABS) plastics are a range of complex terpolymers manufactured by combining together the two copolymers, styrene-acrylonitrile and butadiene-styrene. ABS plastics are relatively expensive but tough and retain their strength at low temperatures. They are used to manufacture moulded components, rainwater and drainage goods. A special ABS solvent cement is required for solvent welding.

Nylons

Nylons, usually nylon 66 or nylon 6, are used for the manufacture of small components where low friction is required. Nylons are tough and strong but tend to be embrittled and become powdery upon prolonged exposure to sunlight. Carpet tiles in nylon 66 are durable and hard-wearing.

Kevlar

Kevlar (polyparabenzamide) fibres are produced by extrusion of a cold solution of the polymer into a cylinder at 200°C, which causes the solvent to evaporate. The resulting fibres are stretched by a drawing process, which aligns the polymer molecules along the fibres to produce a very high modulus material, used in ropes and composite plastics.

THERMOSETTING PLASTICS

Phenol formaldehyde

Phenol formaldehyde (PF) was the original, and remains the cheapest thermosetting resin. Currently, its main use is in the production of laminates by the hot pressing of layers of resin-impregnated paper, fabric or glassfibre. The cured resin is brown, but heat-resistant laminates for working surfaces and wallboards are laminated with a decorative printed paper film and coated with a clear melamine formaldehyde finish. Phenol formaldehyde is resistant to ignition, but produces a phenolic smell on burning.

Urea formaldehyde

Urea formaldehyde (UF) is similar to phenol formaldehyde except that because it is clear it can be produced in a range of colours including white. It is used in the manufacture of electrical components and other moulded components such as WC seats. Urea formaldehyde is resistant to ignition, but produces a fishy smell on burning. Urea formaldehyde foam is no longer used for cavity wall insulation.

Melamine formaldehyde

Melamine formaldehyde (MF) is available clear and in a wide range of colours. When heat cured, it is hard-wearing, durable and resistant to heat, and is therefore used as the surface laminate over the cheaper brown phenol formaldehyde layers in the production of working surface and wallboard laminates. Melamine formaldehyde is resistant to ignition, but produces a fishy smell on burning.

ELASTOMERS

Natural rubber

Natural rubber is harvested from the species *Hevea brasiliensis*, in Africa, South America and Malaysia. The white latex is predominantly *cis*-polyisoprene, a macromolecule containing some double bonds within the carbon chain. It is these double bonds which permit cross-linking with sulphur when natural rubber is heated under pressure in the vulcanisation process. Natural rubber is usually reinforced with carbon and treated with antioxidants to prevent degradation. It is used for flooring and in anti-vibration bearings for buildings and large structures. Natural rubber has the advantage that it is a renewable resource and also the rubber trees absorb carbon dioxide during growth. Uncontaminated natural rubber may be recycled several times without severe degradation.

Neoprene

Neoprene (polychloroprene), unlike natural rubber, is resistant to chemical attack, and is therefore used for glazing seals and gasket systems. It is available only in black.

EPDM

Unlike neoprene, ethylene propylene diene mon-omer (EPDM) may be obtained in any colour, and is characterised by high elongation and good weathering resistance to ultraviolet light and ozone. It is therefore taking over from neoprene as the key material for gaskets and weather strips. EPDM is extensively used in single-ply roofing systems.

Butyl rubber

A copolymer of isobutylene and isoprene, this material has good chemical and weathering resistance. It is used in liners to landfill sites and decorative water features.

Nitrile rubber

Nitrile rubber is formed by the copolymerisation of acrylonitrile and butadiene. It is water and oil resistant, so it is frequently used in structural movement joints which may be subject to surface oil.

COMPOSITE PLASTICS

Composite plastic materials, such as glassfibre-reinforced polyester (Chapter 11), have physical properties which differ significantly from the individual component materials. An increasing variety of composite plastics are reaching the construction industry, driven by the demand for product diversity and in some cases recycling.

Wood plastic composites

Wood plastic composites (WPC) encompass a range of materials incorporating polymers such as polyethylene, polypropylene and polyvinyl chloride blended with natural fibres and sawdust. The plastics materials may be virgin or recycled. The natural fibres are usually wood chips, but hemp, sisal, jute or rice may also be used. The materials are defined in the publications BS EN 15534 Parts 1, 2 and 4–6 and BRE Digest 480: 2004.

For the standard production of wood plastic composites, sawmill wood chips are dried to 2% or 3% moisture and ground down through a hammer mill to wood fibres (<5 mm). Wood flour may be incorporated as a filler giving bulk to the product. The prepared wood fibre material and any filler are mixed with lubricants, UV stabilisers and pigments into the molten polymer, either in a batch or continuous process. The components are then formed by injection moulding for items such as architectural mouldings. Extrusion or pultrusion, a combination of extrusion and pulling, are both used for continuous sections such as window profiles and decking.

Wood plastic composites for outdoor products such as decking, fencing and garden furniture can be manufactured using a significant proportion of recycled polymer and scrap wood, which reduces the quantities of these materials in the waste stream.

Wood plastic composites are generally less stiff than wood and have lower strengths, but they are normally resistant to rot and insect attack. Fire resistance is similar to that of wood of the same density, but is improved by the incorporation of flame and smoke retardants during the manufacturing process. The materials are generally coloured grey or brown to reduce the effects of fading. Thermal movement is large and appropriate expansion joints are necessary.

High-pressure decorative laminates

High-pressure decorative laminates (HPL), usually referred to as laminates, are described in the standards series BS EN 438 Parts 1–9. Products range from thin laminates of less than 2 mm thick for bonding to substrates such as furniture or flooring to thicker materials for internal and external wall and ceiling finishes. Materials are manufactured from decorative surface layers and various core laminates bonded under high pressure at 120°C with thermosetting phenolic resins to produce dense non-porous products. Thin laminates for furniture usually have a cellulosic core and a melamine decorative finish. Thicker exterior grade compact laminates are manufactured with a wood fibre-reinforced core and decor paper coated with acrylic urethane for weathering and UV protection. Fire-resistant grades are available.

Corian

Corian® is a composite of natural minerals, pigments and acrylic polymer, which combine together to produce a highly durable and tough material, available in a wide range of colours. A typical composition is 33% acrylic polymer and 67% aluminium trihydrate. The proprietary product is frequently used for kitchen and other counter tops as it may be moulded into complex forms and inconspicuously joined into single units with the base plastic composite. Thicknesses range from 6 mm to 19 mm in sheets up to 930 × 3650 mm.

LIGHT-CONDUCTING PLASTICS

A range of light- and colour-sensitive materials has been developed by embedding a matrix of light-conducting plastic channels into a substrate of either concrete or acrylic polymer. Each of the light-conducting channels, which operate like fibre optics, gathers light or shadow from one end and transfers it to the other end, creating scintillation or darkening,

respectively. Overall this creates an optical rippling effect as an object or light passes over the surface. The material, according to the substrate, may be used as floor tiles, walls, partitions, façades or table top surfaces. In each case the surface is seen to respond to object movement or changes in light intensity and colour. Units may be individual tiles or larger panels and are available in a range of standard or custom colours.

VARIABLE-COLOUR PLASTIC FILMS

Dichroic plastic films cause the observed colour and opacity to change depending on the viewpoint as well as on the direction and intensity of the light source. For example, one dichroic sheet changes between green, gold and orange, and another between purple and blue. These films may be used to make the external envelope or internal environment of buildings appear active.

Thermochromic pigments in fibre-reinforced plastic sheet cladding change colour with the temperature, allowing the building to visually respond to outside temperatures. The thermochromic pigment is incorporated into the core fibres and the gel coat to gain maximum effect.

Recycling of plastics

The increasing use of plastics in our everyday lives has led to a concern regarding large accumulations of waste, which can only be resolved by extensive recycling. Many plastics are slow to degrade in landfill sites, and, as most are based on products from the petrochemical industry, this finite resource should not be wasted.

The standard BS EN 15343: 2007 addresses the issue of recycling plastics. In addition, the publication PD CEN/TS 14541: 2013 describes the use of non-virgin PVC-U, polypropylene and polyethylene materials in the manufacture of pipes. The total use of plastics within the UK is approximately 5 million tonnes per year, of which 22% is consumed within the building industry, although construction accounts for over half of the consumption of PVC. Much of this material is for profile sections, mainly for the window industry. Currently, in Europe over half of the window profile waste is recycled and ground down into granules for re-extrusion. Most PVC-U can be recycled at least 10 times without significant loss of physical properties, giving it potentially a long lifespan. Recycling of PVC within the UK has increased ten-fold in the period from 2005 to 2012 and is now at about 80,000 tonnes per year.

Expanded polystyrene waste can be recycled by solvent extraction into a material, which has the appearance and many characteristics of wood. PVC bottles can be recycled into plastic pipes by co-extruding new PVC as the inner and outer skins over a recycled PVC core.

However, many recycled plastics have a reduced resistance to degradation as stabilisers are lost in reprocessing, and the products would therefore fail to reach the technical standards, which are normally related to the quality achieved by new materials rather than to fitness for purpose.

Much domestic plastics waste can be identified by the recycling coding (Fig. 10.17) separated into seven categories: polyethylene terephthalate (PET-1), high-density polyethylene (HDPE-2), polyvinyl chloride (PVC-3), low-density polyethylene (LDPE-4), poly-propylene (PP-5), polystyrene (PS-6) and others (7). With appropriate separation there is now the potential for enhanced levels of recycling rather than landfill disposal. Currently some domestic plastic bottle waste is recycled into loft insulation.

Fig. 10.17 Domestic plastic recycling icons

RECYCLED PLASTIC LUMBER

Mixed domestic plastic waste is cut up into small flakes, melted at 200°C into a grey viscous liquid and cast into moulds to produce structural components. The product, *polywood*, has been used to create a 7.5 m-span lightweight bridge with a capacity of 30 tonnes in America. Recycled plastic was used to construct the l-beams and other structural sections. Recycled plastic lumber has the advantage over timber in that it requires no maintenance or treatment with noxious chemicals and its use reduces the demand on landfill sites. Polywood is light, although more dense than timber; however, it suffers from creep. In addition, it has greater thermal movement and lower stiffness (modulus of elasticity) than timber. Polywood has been used for fencing, garden furniture and various marine applications. One UK product incorporating 90% thermoplastic recycled material is extruded into wood-coloured profiles for cladding, decking, windows, doors, conservatories, sheds and similar products normally associated with timber.

BLENDED RECYCLED PLASTIC

Crushed and ground plastic waste, when mixed with resin and reagents causing an exothermic reaction, produces a liquid plastic which can be poured and cast into moulds. The material sets quickly and cures after several hours to a strong, waterproof and durable product, which has the potential to be an alternative to both timber and concrete. The product is based largely on recycled domestic plastic which would otherwise have been sent to landfill sites.

References

FURTHER READING

Armijos, S.J. 2008: *Fabric architecture. Creative resources for shade, signage and shelter.* London: W.W. Norton.

ASFP. 2010: *Guide to class 0 and class 1 as defined in the UK Building Regulations.* Technical Guidance Document 5. Aldershot: Association for Specialist Fire Protection.

Bechthold, M. 2008: *Innovative surface structures technologies and applications.* Abingdon: Taylor & Francis.

Bell, M. and Buckley, C. 2014: *Permanent change. Plastics in architecture and engineering.* New York: Princetown Architectural Press.

Engelsmann, S., Spalding, V. and Peters, S. 2010: *Plastics in architecture and construction.* Basel: Birkhäuser.

Halliwell, S. 2003: *Polymers in building and construction.* Shrewsbury: Smithers RAPRA Press.

HVCA. 2006: *Guide to the use of plastic pipework*. TR/11. London: Heating and Ventilating Contractors' Association.

Jeska, S. 2007: *Transparent plastics. Design and technology*. Basel: Birkhäuser.

Kaltenbach, F. 2004: *Translucent materials: Glass, plastics, metals*. Basel: Birkhäuser.

Kim, J.K. and Pal, K. 2011: *Recent advances in the processing of wood-plastic composites*. Berlin: Springer-Verlag.

Knippers, J., Cremers, J. and Gabler, M. 2011: *Construction manual for polymers and membranes*. Basel: Birkhäuser.

Koch, K-M. (ed.) 2004: *Membrane structures. The fifth building material*. Munich: Prestel.

Ritter, S. 2006: *Smart materials in architecture, interior architecture and design*. Basel: Birkhäuser.

STANDARDS

BS 476
  Fire tests on building materials:

Part 4: 1970	Non-combustibility test for materials.
Part 6: 1989	Method of test for fire propagation for products.
Part 7: 1997	Classification of the surface spread of flame of products.

BS 743: 1970
  Materials for damp-proof courses.

BS 1254: 1981
  Specification for WC seats (plastics).

BS ISO 1382: 2012
  Rubber. Vocabulary.

BS 3012: 1970
  Specification for low and intermediate density polythene sheet for general purposes.

BS 3757: 1978
  Specification for rigid PVC sheet.

BS 3837–1: 2004
  Expanded polystyrene boards. Boards and blocks manufactured from expandable beads.

BS 3953: 1990
  Synthetic resin bonded woven glass fabric laminated sheet.

BS 4023: 1975
  Flexible cellular PVC sheeting.

pr BS ISO 4097: 2014
  Rubber, ethylene-propylene-diene (EPDM). Evaluation procedure.

BS 4213: 2004
  Cisterns for domestic use.

BS 4514: 2001
  Unplasticized PVC soil and ventilating pipes, fittings and accessories.

BS 4592–6: 2008
  Industrial flooring and stair treads. Glass reinforced plastics (GRP).

BS 4607
  Non-metallic conduit fittings for electrical installations:

Part 1: 1984	Specification for fittings and components.
Part 5: 1982	Specification for rigid conduits.

BS 4660: 2000
  Thermoplastics ancilliary fittings for below ground drainage.

BS 4840
  Rigid polyurethane (PUR) foam in slab form:

Part 1: 1985	Specification for PUR foam for use in transport containers.
Part 2: 1994	Specification for PUR foam for use in cold rooms and stores.

BS 4841
  Rigid polyisocyanurate (PIR) and polyurethane (PUR) products for building end-use applications:

Part 1: 2006	Laminated insulation boards.
Part 2: 2006	Thermal insulation for internal wall linings and ceilings.
Part 3: 2006	Roofboard thermal insulation under bituminous built up roofing membranes.
Part 4: 2006	Roofboard thermal insulation under non-bituminous single-ply roofing membranes.
Part 5: 2006	Thermal insulation for pitched roofs.
Part 6: 2006	Thermal insulation for floors.

BS 4901: 1976
  Plastics colours for building purposes.

BS 4962: 1989
  Specification for plastic pipes and fittings for use as subsoil field drains.

BS 4965: 1999
  Specification for decorative laminated plastics sheet veneered boards and panels.

BS 4991: 1974
  Specification for polypropylene copolymer pressure pipe.

BS 5051: 1988
Bullet-resistant glazing. Specification for glazing for interior use.

BS 5241
Rigid polyurethane (PUR) and polyisocyanurate (PIR) foam when dispensed or sprayed on a construction site:
Part 1: 1994   Specification for sprayed foam thermal insulation applied externally.
Part 2: 1991   Specification for dispensed foam for thermal insulation or buoyancy applications.

BS 5255: 1989
Thermoplastics waste pipe and fittings.

BS 5391
Acrylonitrile-butadiene-styrene (ABS) pressure pipe:
Part 1: 2006   Specification.

BS 5608: 1993
Specification for preformed rigid urethane (PUR) and polyisocyanurate (PIR) foams for thermal insulation of pipework and equipment.

BS 5617: 1985
Specification for urea-formaldehyde (UF) foam systems suitable for thermal insulation of cavity walls with masonry or concrete inner and outer leaves.

BS 5618: 1985
Code of practice for thermal insulation of cavity walls by filling with urea-formaldehyde (UF) foam systems.

BS 5955
Plastics pipework (thermoplastics materials):
Part 8: 2001   Specification for the installation of thermoplastics pipes and associated fittings for use in domestic hot and cold water services and heating systems.

BS 6203: 2003
Guide to the fire characteristics and fire performance of expanded polystyrene materials (EPS and XPS) used in building applications.

BS 6206: 1981
Specification for impact performance requirements for flat safety glass and safety plastics for use in building.

BS 6515: 1984
Specification for polyethylene damp-proof courses for masonry.

BS 7291
Thermoplastic pipes and associated fittings for hot and cold water:
Part 1: 2010   General requirements.
Part 2: 2010   Polybutylene (PB) pipes and fittings.
Part 3: 2010   Cross-linked polyethylene (PE-X) pipes and fittings.

BS 7412: 2007
Specification for windows and doorsets made from unplasticized polyvinyl chloride (PVC-U) extruded hollow profiles.

BS 7414: 1991
White PVC-U extruded hollow profiles with heat welded corner joints for plastics windows: materials type B.

BS 7619: 2010
Extruded cellular unplasticized white PVC (PVC-UE) profiles. Specification.

BS 7722: 2010
Surface covered PVC-U profiles for windows and doors.

BS 8203: 2001
Code of practice for installation of resilient floor coverings.

BS 8204
Screeds, bases and in situ floorings:
Part 6: 2008   Synthetic resin floorings. Code of practice.

BS 8215: 1991
Code of practice for the installation of damp proof courses in masonry construction.

BS ISO 10840: 2008
Plastics. Guidance for the use of standard fire tests.

BS ISO 12230: 2012
Polybutene-1 (PB-1) pipes. Effect of time and temperature on the expected strength.

BS ISO 18225: 2012
Plastics piping systems. Multilayer piping systems for outdoor gas installations. Specifications for systems.

BS ISO 22621: 2010
Plastics piping systems for the supply of gaseous fuels.

BS ISO 23559: 2011
Plastics. Film and sheeting. Guidance on the testing of thermoplastic films.

BS EN 198: 2008
Sanitary appliances. Baths made from cross-linked cast acrylic sheets.

BS EN 438

High-pressure decorative laminates. Sheets based on thermosetting resins:

Part 1: 2005	Introduction and general information.
Part 2: 2005	Determination of properties.
Part 3: 2005	Classification and specifications. Less than 2 mm thick.
Part 4: 2005	Classification and specifications. Thickness 2 mm and greater.
Part 5: 2005	Flooring grade less than 2 mm thick.
Part 6: 2005	Exterior grade of thickness 2 mm or greater.
Part 7: 2005	Composite panels for internal and external wall and ceiling finishes.
Part 8: 2009	Classification and specifications for design laminates.
Part 9: 2010	Classification and specifications for alternative core laminates.

BS EN 607: 2004

Eaves gutters and fittings made of PVC-U. Definitions, requirements and testing.

BS EN 1013: 2012

Light transmitting single skin profiled plastic sheeting for internal and external roofs, walls and ceilings.

BS EN ISO 1043–2: 2011

Symbols and abbreviated terms. Fillers and reinforcing materials.

BS EN 1329

Plastics piping systems for soil and waste discharge – unplasticized PVC-U:

Part 1: 2014	Specification for pipes, fittings and the system.
PD CEN/TS	Part 2: 2012 Guidance for the assessment of conformity.

BS EN 1401

Plastics piping systems for non-pressure underground drainage and sewerage. Unplasticized polyvinyl chloride (PVC-U):

Part 1: 2009	Specification for pipes, fittings and the system.
PD CEN/TS	Part 2: 2012 Guidance for the assessment of conformity.
Part 3: 2001	Guidance for installation.

BS EN 1451

Plastics piping systems for soil and waste discharge. Polypropylene (PP):

Part 1: 2000	Specification for pipes, fittings and the system.
PD CEN/TS	Part 2: 2012 Guidance for the assessment of conformity.

BS EN ISO 1452

Plastics piping systems for water supply. Unplasticized polyvinyl chlorid (PVC-U):

Part 1: 2009	General.
Part 2: 2009	Pipes.
Part 3: 2010	Fittings.
Part 4: 2009	Valves.
pr Part 5: 2010	Fitness for purpose of the system.
PD CEN/TS	Part 7: 2014 Guidance for the assessment of conformity.

BS EN 1455

Plastics piping systems for soil and waste discharge. ABS:

Part 1: 2000	Specification for pipes, fittings and the system.
PD CEN/TS	Part 2: 2012 Guidance for the assessment of conformity.

BS EN 1462: 2004

Brackets and eaves gutters.

BS EN 1519

Plastics piping systems for soil and waste discharge. Polyethylene (PE):

Part 1: 2000	Specification for pipes, fittings and the system.
PD CEN/TS	Part 2: 2012 Guidance for the assessment of conformity.

BS EN 1565

Plastics piping systems for soil and waste discharge. Styrene copolymer blends:

Part 1: 2000	Specification for pipes, fittings and the system.
PD CEN/TS	Part 2: 2012 Guidance for the assessment of conformity.

BS EN 1566

Plastics piping systems for soil and waste discharge. Chlorinated polyvinyl chloride (PVC-C):

Part 1: 2000	Specification for pipes, fittings and the system.
PD CEN/TS	Part 2: 2012 Guidance for the assessment of conformity.

BS EN 1796: 2013

Plastics piping systems for water supply. Glass reinforced thermo-setting plastics (GRP) based on unsaturated polyester resin (UP).

BS EN 1873: 2005
Prefabricated accessories for roofing. Individual
roof lights of plastics.

BS EN ISO 1873
Plastics. Polypropylene moulding and extrusion
materials:
    Part 1: 1995   Designation system and basis for
                        specifications.
    Part 2: 2007   Preparation of test specimens
                        and determination of properties.

BS EN ISO 5999: 2007
Flexible cellular polymeric materials. Polyurethane
for load-bearing applications. Specifications.

BS EN ISO 7391
Plastics. Polycarbonate (PC) moulding and
extrusion materials:
    Part 1: 2006   Designation system and basis for
                        specifications.
    Part 2: 2006   Preparation of test specimens.

BS EN ISO 7792
Plastics. Thermoplastic polyester (TP) moulding
and extrusion materials:
    Part 1: 2012   Designation system and basis for
                        specifications.
    Part 2: 2012   Preparation of test specimens
                        and determination of
                        properties.

BS EN ISO 8257
Plastics. Polymethyl methacrylate (PMMA)
moulding and extrusion materials:
    Part 1: 2006   Designation system and basis for
                        specifications.
    Part 2: 2006   Preparation of test specimens
                        and determination of properties.

BS EN ISO 8986
Plastics. Polybutene-1 (PB-1) moulding and
extrusion materials:
    Part 1: 2009   Designation system and basis for
                        specification.
    Part 2: 2009   Preparation of test specimens
                        and determination of
                        properties.

BS EN ISO 11295: 2010
Classification and information on design of
plastics piping systems used for renovation.

BS EN ISO 11296
Plastics piping systems for the renovation of
underground non-pressure drainage networks:
    Part 1: 2011   General.
    Part 3: 2011   Lining with close-fit pipes.
    Part 4: 2011   Lining with cured-in-place

pipes.
    Part 7: 2013   Lining with spirally-wound
                        pipes.

BS EN ISO 11298
Plastics piping systems for the renovation of
underground water supply networks:
    Part 1: 2011   General.
    Part 3: 2011   Lining with close-fit pipes.

BS EN ISO 11299
Plastic piping systems for renovation of
underground gas supply networks:
    Part 1: 2013   General.
    Part 3: 2013   Lining with close-fit pipes.

BS EN ISO 11542
Plastics. Ultra high molecular weight
polyethylene (PE-UHMW) moulding and
extrusion materials:
    Part 1: 2001   Designation system and basis for
                        specification.
    Part 2: 1998   Preparation of test specimens
                        and determination of
                        properties.

BS EN ISO 12086
Plastics. Fluoropolymer dispersions and moulding
and extrusion materials:
    Part 1: 2006   Designation system and basis for
                        specifications.
    Part 2: 2006   Preparation of test specimens
                        and determination of
                        properties.

BS EN 12200
Plastics rainwater piping systems. Unplasticized
polyvinyl chloride (PVC-U):
    Part 1: 2000   Specification for pipes, fittings
                        and the system.
    DD CEN/TS   Part 2: 2003 Guidance for the
                        assessment of conformity.

BS EN 12201
Plastics piping systems for water supply.
Polyethylene (PE):
    Part 1: 2011   General.
    Part 2: 2011   Pipes.
    Part 3: 2011   Fittings.
    Part 4: 2012   Valves.
    Part 5: 2011   Fitness for purpose of the
                        system.
    PD CEN/TS   Part 7: 2014 Guidance for the
                        assessment of conformity.

BS EN 12224: 2000
Geotextiles and geotextiles-related products.
Resistance to weathering.

BS EN 12225: 2000
Geotextiles and geotextiles-related products. Microbiological resistance.

BS EN 12608: 2003
Unplasticized polyvinyl chloride (PVC-U) profiles for the fabrication of windows and doors.

BS EN 12666
Plastics piping systems for non-pressure underground drainage. Polyethylene (PE):

Part 1: 2005	Specification for pipes, fittings and the system.
PD CEN/TS	Part 2: 2012 Guidance for the assessment of conformity.

BS EN 13162: 2012
Thermal insulation products for buildings. Factory made mineral wool (MW) products. Specification.

BS EN 13163: 2012
Thermal insulation products for buildings. Factory made products of expanded polystyrene (EPS). Specification.

BS EN 13164: 2012
Thermal insulation products for buildings. Factory made products of extruded polystyrene foam (XPS). Specification.

BS EN 13165: 2012
Thermal insulation products for buildings. Factory made rigid polyurethane foam (PUR) products. Specification.

BS EN 13166: 2012
Thermal insulation products for buildings. Factory made products of phenolic foam (PF). Specification.

BS EN 13167: 2012
Thermal insulation products for buildings. Factory made cellular glass (CG) products. Specification.

BS EN 13168: 2012
Thermal insulation products for buildings. Factory made wood wool (WW) products. Specification.

BS EN 13169: 2012
Thermal insulation products for buildings. Factory made products of expanded perlite (EPB). Specification.

BS EN 13170: 2012
Thermal insulation products for buildings. Factory made products of expanded cork (ICB). Specification.

BS EN 13171: 2012
Thermal insulating products for buildings. Factory made wood fibre (WF) products. Specification.

BS EN 13172: 2012
Thermal insulating products. Evaluation of conformity.

BS EN 13245
Plastics. Unplasticised polyvinyl chloride (PVC-U) profiles for building applications:

Part 1: 2010	Designation of PVC-U profiles.
Part 2: 2008	PVC-U and PVC-UE profiles for internal and external wall and ceiling finishes.
Part 3: 2010	Designation of PVC-UE profiles.

BS EN 13476
Plastic piping systems for non-pressure underground drainage and sewerage systems. Structured wall piping systems:

Part 1: 2007	General requirements.
Part 2: 2007	Specifications for pipes and fittings. Type A.
Part 3: 2007	Specifications for pipes and fittings. Type B.
PD CEN/TS	Part 4: 2013 Guidance for the assessment of conformity.

BS EN 13566
Plastics piping systems for renovation of underground drainage networks:

Part 2: 2005	Lining with continuous pipes.

BS EN 13598
Plastics piping systems for non-pressure underground drainage and sewerage. Unplasticized polyvinyl chloride (PVC-U), polypropylene (PP) and olyethylene (PE):

Part 1: 2010	Specification for ancillary fittings.
Part 2: 2009	Specification for manholes and inspection chambers.
PD CEN/TS	Part 3: 2012 Guidance for assessment of conformity.

BS EN 13859
Flexible sheets for waterproofing. Definitions and characteristics of underlays:

Part 1: 2010	Underlays for discontinuous roofing.
Part 2: 2010	Underlays for walls.

BS EN 13956: 2012
Flexible sheets for waterproofing. Plastic and rubber sheets for roof waterproofing. Definitions and characteristics.

BS EN 13967: 2012
Flexible sheets for waterproofing. Plastic and rubber damp proof sheets including plastic and

rubber basement tanking sheet. Definitions and characteristics.

BS EN 14364: 2013

Plastics piping systems for drainage and sewerage. Glass-reinforced thermoplastics (GRP) based on unsaturated polyester (UP) resin. Specification for pipes, fittings and joints.

PD CEN/TS 14541: 2013

Plastics pipes and fittings for non-pressure applications. Utilisation of non-virgin PVC-U, PP and PE materials.

PD CEN/TS 14632: 2012

Plastics piping systems for drainage. GRP based on polyester resin (UP).

BS EN 14693: 2006

Flexible sheets for waterproofing on concrete surfaces.

BS EN 14758

Plastics piping systems for non-pressure underground drainage. Polypropylene with mineral modifiers (PP-MD):

Part 1: 2012	Specification for pipes, fittings and the system.
DD CEN/TS	Part 2: 2007 Guidance for the assessment of conformity.
DD CEN/TS	Part 3: 2006 Guidance for installation.

BS EN 14909: 2012

Flexible sheets for waterproofing. Plastic and rubber damp proof courses. Definitions and characteristics.

BS EN 15012: 2007

Plastics piping systems. Soil and waste discharge systems within the building structure.

BS EN 15014: 2007

Plastics piping systems. Buried and above ground systems for water.

BS EN 15342: 2007

Plastics. Recycled plastics. Characterisation of polystyrene (PS) recyclates.

BS EN 15343: 2007

Plastics. Recycled plastics. Plastics recycling traceability.

BS EN 15344: 2007

Plastics. Recycled plastics. Characterisation of polyethylene (PE) recyclates.

BS EN 15345: 2007

Plastics. Recycled plastics. Characterisation of polypropylene (PP) recyclates.

BS EN 15346: 2007

Plastics. Recycled plastics. Characterisation of polyvinyl chloride (PVC) recyclates.

BS EN 15347: 2007

Plastics. Recycled plastics. Characterisation of plastics waste.

PD CEN/TR 15438: 2007

Plastics piping systems. Guidance for coding of products and their intended uses.

BS EN 15534

Composites made from cellulose-based materials (usually called wood polymer composites (WPC) or natural fibre composites (NFC)):

Part 1: 2014	Test methods for characterisation of compounds and products.
DD CEN/TS	Part 2: 2007 Characterisation of WPC materials.
Part 4: 2014	Specifications for decking profiles and tiles.
Part 5: 2014	Specifications for cladding profiles and tiles.
pr Part 6: 2014	Specification for fencing profiles and systems.

BS EN ISO 15877

Plastics piping systems for hot and cold water installations. Chlorinated polyvinyl chloride (CPVC):

Part 1: 2009	General.
Part 2: 2009	Pipes.
Part 3: 2009	Fittings.
Part 5: 2009	Fitness for purpose of the system.
DD CEN ISO/TS Part 7: 2009 Guidance for the assessment of conformity.	

PD CEN/TS 16011: 2013

Plastics. Recycled plastics. Sample preparation.

BS EN 16153: 2013

Light transmitting flat multiwall polycarbonate (PC) sheets for internal and external use in roofs, walls and ceilings. Requirements.

BS EN ISO 21003

Multilayer piping systems for hot and cold water installations inside buildings:

Part 1: 2008	General.
Part 2: 2008	Pipes.
Part 3: 2008	Fittings.
Part 5: 2008	Fitness for purpose of the system.

DD CEN ISO/TS Part 7: 2008 Guidance for the assessment of conformity.

PD CEN ISO/TR 27165: 2012
Thermoplastics piping systems. Guidance for definitions of wall constructions for pipes.

BUILDING RESEARCH ESTABLISHMENT PUBLICATIONS

BRE Digests

BRE Digest 440: 1999
Weathering of white external PVC-U.

BRE Digest 442: 1999
Architectural use of polymer composites.

BRE Digest 480: 2004
Wood plastic composites and plastic lumber.

BRE Good building guides

GBG 73: 2008
Radon protection for new domestic extensions and conservatories with solid concrete ground floors.

GBG 74: 2008
Radon protection for new dwellings. Avoiding problems and getting it right.

GBG 75: 2009
Radon protection for new large buildings.

BRE Information papers

BRE IP 8/01
Weathering of plastics pipes and fittings.

BRE IP 12/01
Hot air repair of PVC-U profiles.

BRE IP 2/04
Wood plastic composites: market drivers and opportunities in Europe.

BRE IP 8/08
Determining the minimum thermal resistance of cavity closers.

BRE Report

BR 405: 2000
Polymer composites in construction.

ADVISORY ORGANISATIONS

British Laminate Fabricators Association, PO Box 775, Broseley Wood, Telford TF7 9FG (0845 0568496).

British Plastics Federation, 6 Bath Place, Rivington Street, London EC2A 3JE (0207 457 5000).

British Rubber Manufacturers' Association Ltd, 6 Bath Place, Rivington Street, London EC2A 3JE (0207 457 5040).

GLASSFIBRE-REINFORCED PLASTICS, CONCRETE AND GYPSUM

CONTENTS

Introduction

Composite materials, such as the glassfibre-reinforced materials glassfibre-reinforced polyester (GRP), glassfibre-reinforced concrete (GRC) and glassfibre-reinforced gypsum (GRG), rely for their utility on the advantageous combination of the disparate physical properties associated with the individual component materials. This is possible when a strong bond between the glass fibres and the matrix material ensures that the two materials within the composite act in unison. Thus polyester, which alone has a very low modulus of elasticity, when reinforced with glass fibres produces a material which is rigid enough for use as a cladding material. Concrete, which alone would be brittle, when reinforced with glass fibres, may be manufactured into thin, impact-resistant sheets. Similarly, glassfibre reinforcement in gypsum considerably increases its impact and fire resistance.

Glass fibres

The glass fibres for GRP and GRG are manufactured from standard E-glass as shown in Figure 11.1. Molten glass runs from the furnace at 1200°C into a forehearth, and through a spinneret of fine holes from which it is drawn at high speed down to approximately 9 μm in thickness. The glass fibres are coated in size and bundled before winding up on a collet. Subsequently, the glassfibre 'cake' is either used as continuous rovings or cut to 20–50 mm loose chopped strand. Glassfibre rovings may be manufactured into woven mats; chopped strand mats are formed with organic binder.

Glassfibre-reinforced plastics

The standard matrix material for glassfibre-reinforced plastics is polyester resin, although other thermosetting

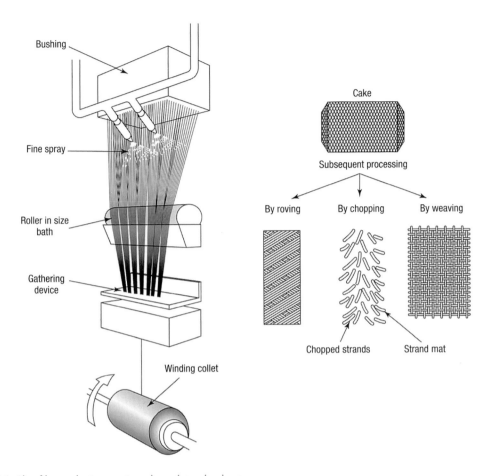

Fig. 11.1 Glass fibre production – rovings, chopped strand and mat

resins including phenolic, epoxy and polyurethane may be used. Glass fibres as continuous rovings or chopped strand are used for most purposes; however, the highest-strength products are obtained with woven glass fabrics and unidirectionally aligned fibres. The proportion of glass fibres ranges widely from 20–80% by weight depending on the strength required. Enhanced performance may be achieved by using the more expensive S-grade high-strength and high-modulus glass fibres used mainly in the aerospace industry. Alternative higher tensile strength fibres include the polyaramids such as *Kevlar* and carbon fibres, but these are considerably more expensive than glass.

FABRICATION PROCESS

A major investment in the manufacture of GRP cladding panels lies within the production of the high-quality moulds. These are usually made from timber, but steel or GRP itself may also be used. Moulds are reused, sometimes with minor variations (e.g. the insertion of a window void within a wall unit), as many times as possible to minimise production costs. The number of different mould designs for any one building is therefore kept to a minimum, and this may be reflected in the repetitiveness of the design.

In the fabrication process, the mould is coated with a release agent to prevent bonding and associated damage to the finished panel surface. A gel coat, which will ultimately be the weathering surface, is applied to a finished thickness of 0.25–0.4 mm. Early examples of GRP without sufficient gel coat have weathered to a rough surface with consequent exposure of the glass fibres; however, modern gel coats when applied to the correct thickness are durable. The subsequent fabrication involves the *laying up* of layers of glass fibres and polyester resin to the required thickness, usually with either sprayed rovings or chopped strand mat. Reinforcement and fixings, normally in aluminium due to similarities in coefficients of expansion, may be incorporated and areas requiring additional strength thickened as appropriate by the laying-up process. Plastic foam insulation may be encapsulated to give the required thermal properties. Curing may take up to two weeks, after which the unit is stripped from the mould, trimmed around the edges and fitted out.

PHYSICAL PROPERTIES AND DESIGN CONSIDERATIONS

The choice of GRP (for example, as a cladding panel) imparts its own aesthetic on a building design. The high strength-to-weight ratio of GRP allows for the use of large panel units, but cost constraints in the mould-making reduce the number of panel variations to a minimum. Curved edges to panels and openings are preferred to reduce stress-raising points at very sharp corners. The high thermal expansion coefficient of GRP demands careful detailing of movement joints and their appropriate sealing where necessary with components that retain their flexibility. In some cases the high expansion may be resolved by the use of profiled forms, which also impart strength. Colour fading and yellowing of GRP panels have been problems; however, recent products with ultraviolet light protection are more colour-fast. Slightly textured finishes are generally more durable than smooth ones for exposure to full direct sunlight. GRP may be manufactured with fire-resistant additives; the phenolic resins have the advantage of lower flammability and smoke emissions. Long-term creep precludes the use of GRP as a significant load-bearing material, although single-storey structures, two-storey mobile units and structural overhead walkways are frequently constructed from the material. GRP is vandal resistant and may be laminated sufficiently to be bullet resistant. Where both surfaces are to be exposed, the material may be pressed between the two halves of a die.

USES OF GLASSFIBRE-REINFORCED PLASTICS

The lightweight properties of GRP make it an eminently suitable material for the manufacture of large cladding panels and custom-moulded structures, as illustrated in the belfry and spire of St James, Piccadilly, London (Fig. 11.2). Finishes may be self-coloured or incorporate a natural stone aggregate finish. In addition, GRP is frequently used for the production of architectural

Fig. 11.2 GRP replacement spire – Church of St James, London. *Photograph:* Courtesy of Smith of Derby Ltd

features such as barge boards, dormer windows, classical columns and entrance canopies (Fig. 11.3). GRP may be pigmented to simulate various timbers, slate, Portland or Cotswold stone and lead or copper. It is also used to produce a wide range of small building components including baths, valley troughs, flat roof edge trim and water drainage systems. In addition, a wide range of composite cladding panels are manufactured from glass fibre-reinforced resins incorporating stone granules within the core of the material. These products, which are impact and fire resistant, are available with a granular stone, painted or gel-coat finish.

CARBON FIBRE AND ARAMID COMPOSITES

Carbon fibres, which were originally developed for the aerospace industry, combine strength and stiffness with low weight, but have poor impact resistance. They are produced from polyacrylonitrile fibres by

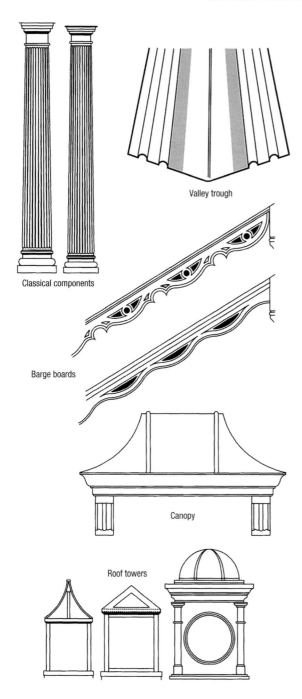

Valley trough

Classical components

Barge boards

Canopy

Roof towers

Fig. 11.3 Typical GRP components

controlled oxidation at 250°C followed by carbon-isation at 2600°C in an inert atmosphere. Three grades – high strength (HS), high modulus (HM) and intermediate modulus (IM) – are produced. Carbon fibres, like glass fibres, are available as woven material, chopped strand or continuous filament. Carbon fibres have a small negative coefficient of expansion along the fibre axis; thus, composite materials of zero thermal expansion may be produced.

Aramids are aromatic polyamide liquid crystalline polymers, with high strength-to-weight ratio in tension, but poorer properties under compression or bending. Impact resistance is greater than that for carbon fibres. Aramid fibres, typically *Kevlar*, are produced by spinning the continuous fibre from solution. A variety of products is available with a range of modulus, elongation and impact-resistance properties. Aramid composites exhibit good abrasion resistance. Carbon and aramid fibres may be combined into a composite material where strength, stiffness and impact resistance are all required.

Although polyester resins may be used as the matrix material for either carbon or aramid fibres, usually these more expensive fibres are incorporated into higher performance epoxy resins. In addition to the standard GRP laying-up production process, pultrusion and preimpregnation are used for manufacturing carbon fibre-reinforced components. Pultrusion, a combination of extruding and pulling (Fig. 11.4), is used for making continuous profiles which may be either solid or hollow. Preimpregnation involves coating the continuous fibre or woven fibre fabric with a mixture of resin and curing agent, which may be stored frozen at this stage, then thawed and moulded into shape when required. By using low-temperature moulding processes, large and complex structures may be fabricated for the construction industry.

For externally bonded reinforcement to concrete structures, pultruded carbon fibre reinforced plates may be bonded to the concrete with a thixotropic epoxy resin. Alternatively, woven carbon fibre mat is wrapped around the concrete and pasted on with epoxy resin. The required level of reinforcement can be achieved by building up the appropriate thickness of epoxy resin saturated carbon fibre mat. The technique may also be applied to reinforcing steel, masonry, timber or cast iron. Although usually used for remedial work, this type of reinforcement could also be considered for new-build elements.

RECYCLING FIBRE-REINFORCED POLYMERS

According to 2012 data, the UK/Ireland fibre-reinforced polymer (FRP) industry produces 134,000 tonnes per year, of which 15% is used within the construction industry. Currently, the majority of waste

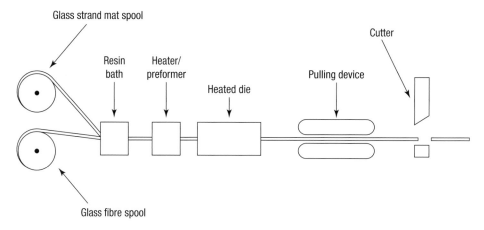

Fig. 11.4 Pultrusion

fibre-reinforced polymers are disposed of into landfill sites. While melting and remoulding is possible for thermoplastic FRPs, this is not applicable to thermo-set GRP.

One solution for GRP is to grind the waste into powder and use this material in conjunction with other binders; however, this process is difficult where embedded metal fixings are incorporated into the original components. The recyclate powder may be blended with other recycled plastics to produce GRP/plastic lumber, which may be used for lightly loaded piles, decking, fencing and similar applications. This material may be cut and worked like the natural timber which it replaces. Alternatively, ground GRP may be incorporated into particle board to make GRP-reinforced wood particleboard, which has enhanced mechanical properties compared to the standard grade (P5) of particleboard used for domestic flooring. However, when energy costs, transportation and other factors are considered, the ecological balance towards recycling fibre-reinforced polymers may be dependent upon future considerations of recycling at the initial design stage.

Alternative potential processes for recycling GRP include thermal or chemical processes to recover the fibres, although the quality of the fibres is reduced in the process. Incineration with energy recovery is a viable option for GRP waste, but currently most incineration sites burn material to reduce volume rather than to generate electricity. GRP has a high calorific content compared to most domestic refuse, which on balance means that less domestic waste is incinerated and therefore more is sent to landfill.

Glassfibre-reinforced concrete

Glassfibre-reinforced concrete (GRC) is a material that was developed in the early 1970s by the Building Research Establishment. The standard material is produced from a mixture of alkali-resistant glass fibres with Portland cement, sand aggregate and water. Admixtures such as pozzolanas, superplasticisers and polymers are usually incorporated into the mix to give the required fabrication or casting properties. The breakthrough in the development of the material was the production of the alkali-resistant (AR) glass fibres, as the standard E-glass fibres, which are used in GRP and GRG, corrode rapidly in the highly alkaline environment of hydrated cement. Alkali-resistant glass, in addition to the sodium, silicon and calcium oxide components of standard E-glass, contains zirconium oxide. Alkali-resistant glass fibres, which have been improved by progressive development, are manufactured under the trade name *Cem-FIL*. The addition to GRC mixes of metakaolin, a pozzolanic material produced by calcining china clay at 750–800°C, prevents the development of lime crystals around the glass fibres. In the unblended GRC this leads to some gradual loss of strength. Standard grey GRC has the appearance of sheet cement and is non-combustible.

MANUFACTURE OF ALKALI-RESISTANT GLASS FIBRES

Silica, limestone and zircon are melted in a furnace; the alkali-resistant glass produced is drawn into fibres of 14 or 20 µm diameter and rolled into cakes for subsequent use as continuous rovings or for conversion into chopped strand. The process is comparable to that for standard E-glass fibres. *Cem-FIL* glass tissue with a fine texture is also available.

CEMENT MATRIX

Portland cement, strength class 42.5 or 42.5R (rapid early strength), is normally used. Portland cement will produce a grey finish, but white Portland cement or added pigments may be used to give different effects. However, with the use of pigments care must be taken to ensure uniformity of colour. Washed sand and fly ash (pulverised fuel ash) are the usual aggregates, but crushed marble, limestone or granite may be used when a particular exposed aggregate finish is required.

FABRICATION PROCESSES

Fibre-reinforced concrete components may be formed either by using a spray-gun, which mixes the glass fibres with a slurry of cement as it sprays directly into the mould, or by premixing a blend of cement, sand, water, admixtures and glass fibres before casting. Moulds similar to those required for the production of GRP components are used. Extrusion and injection moulding techniques are applicable for linear or small components, and bagged pre-blended mixes may be used for on-site applications.

Sprayed glassfibre-reinforced concrete

Spray techniques, which may be manual (Fig. 11.5) or robotic, are used to build up the required thickness, usually between 10 and 20 mm. During spraying, the gun chops the fibres into 25–40 mm lengths, depositing a uniform felt of fibres and mortar into the mould. A typical sprayed mix would contain 5% glass fibres, 36% Portland cement, 36% washed sand, 11% additives/polymer and 12% water. The curing of GRC is relatively slow, with 95% strength developed after seven days.

Premixed glassfibre-reinforced concrete

It is normal to mix the cement, sand, water and admixtures before adding the chopped fibres. A typical mix would contain up to 3.5% of 12 mm fibres in a sand : cement mix of 0.5 : 1 with a water/cement ratio of 0.35. The mix is then cast and vibrated, or pressed into form for smaller components. For renderings, a glass fibre content of between 1% and 2% is appropriate.

Fig. 11.5 Spraying glassfibre-reinforced cement.
Photograph: Courtesy of Trent Concrete Ltd

An alternative system involves the direct spraying of the premixed material using a specialist spray-gun.

PROPERTIES OF GLASSFIBRE-REINFORCED CONCRETE

Appearance

While standard GRC has the appearance of concrete, a wide diversity of colours, textures and simulated materials can be manufactured. A gloss finish should be avoided, as it tends to craze and show any defects or variations. The use of specific aggregates followed by grinding can simulate marble, granite, terracotta, etc., while reconstructed stone with either a smooth or tooled effect can be produced by the action of acid etching. An exposed aggregate finish is achieved by the use of retardants within the mould, followed by washing and brushing. Applied finishes, which are usually water-based synthetic latex emulsions, may be applied to clean, dust-free surfaces.

Moisture and thermal movement

GRC exhibits an initial irreversible shrinkage followed by a reversible moisture movement of approximately 0.2%. The coefficient of thermal expansion is within the range $(7–20) \times 10^{-6}\,°C^{-1}$, typical for cementitious materials.

Thermal conductivity

The thermal conductivity of GRC is within the range 0.21–1.0 W/m K. Double-skin GRC cladding panel units usually incorporate expanded polystyrene, mineral wool or foamed plastic insulation. Cold bridging should be avoided where it may cause shadowing effects.

Durability

GRC is less permeable to moisture than normal concrete, so it has good resistance to chemical attack; however, unless manufactured from sulphate-resisting cement, it is attacked by soluble sulphates. GRC is unaffected by freeze/thaw cycling.

Impact resistance

GRC exhibits a high-impact resistance but toughness and strength does decrease over long periods of time. However, the incorporation of metakaolin $(2SiO_2 \cdot Al_2O_3)$ into the mix appears to improve the long-term performance of the material.

USES OF GLASSFIBRE-REINFORCED CONCRETE

GRC is used extensively for the manufacture of cladding and soffit panels because it is lightweight and easily moulded (Fig. 11.6). It is used in conservation work as a replacement for natural stone and in architectural mouldings, including sophisticated decorative screens within countries of the Middle East. It is used as permanent formwork for concrete, fire-resistant partitioning and in the manufacture of small components including slates, tiles and decorative

Fig. 11.6 GRC components. *Photograph*: Courtesy of Trent Concrete Ltd

ridge tiles. Fibre-reinforced cement slates are manufactured to simulate the texture and colour of natural slate. Some manufacturers incorporate blends of other non-asbestos natural and synthetic fibres together with pigments and fillers to produce a range of coloured products with glossy, matt or simulated riven finishes.

Glassfibre-reinforced gypsum

Glassfibre-reinforced gypsum (GRG) combines the non-combustibility of gypsum plaster with the reinforcing strength of glass fibres. Products contain typically 5% of the standard E-glass fibres, which considerably improve impact as well as fire resistance. Commercial GRG products are available as standard panels, encasement systems for the fire protection of steel and decorative wall panels. As with all gypsum products, standard GRG should not be used in damp conditions or at temperatures regularly over 50°C.

GLASSFIBRE-REINFORCED GYPSUM BOARDS

The standard boards, available in a range of thicknesses from 6–12.5 mm, are manufactured with a glassfibre-reinforced gypsum core and glassfibre tissue immediately below the gypsum faces. The material is suitable for a wide range of applications including wall linings, ceilings and protected external positions such as roof soffits. The material can be easily cut on site and fixed with nails or screws; in addition, owing to the effect of the glassfibre reinforcement, it can be curved to fit, for example, barrel-vault ceilings. The minimum radius of curvature depends on the board thickness. The material has a smooth off-white finish; joints should be taped before finishing board plaster is applied.

For the fire protection of steelwork, high-performance, Class 0 board thicknesses of 15, 20, 25 and 30 mm are available. Depending on the steel section factor, A/V (surface area per unit length/volume per unit length), with double layers and staggered joints, up to 120 minutes' fire resistance may be achieved (Fig. 11.7). With specialist steel fixings to the structural columns, up to 180 minutes' fire protection is achieved with glassfibre-reinforced gypsum boards.

Decorative glassfibre-reinforced gypsum boards and components

Decorative boards manufactured with a range of motifs may be used as dado or wall panels, false ceilings or

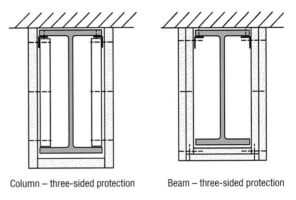

Column – three-sided protection   Beam – three-sided protection

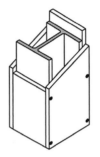

Column – four-sided protection

Fig. 11.7 Fire protection with GRG panels

organic forms (Fig. 11.8). Dabs are used to fix panels to the substrate allowing for adjustment to a flush joint. Finally panels may be painted.

Ceiling tiles manufactured from GRG are available in a wide range of designs, including plain, textured, patterned, open- or closed-cell surface and with square, tapered or bevelled edges. The standard size is usually 600 × 600 mm, although some manufacturers produce a wide range of units. GRG has good fire-resistant properties; it is non-combustible to BS 476 Part 4: 1970, Class 1 surface spread of flame to BS 476 Part 6: 1989, up to Class 0 to Building Regulations Section E15 and Euroclass A2-s1,d0 (no significant contribution to the fire load, no evolution of smoke or flaming droplets). It does not emit noxious fumes in fire. Acoustic tiles with enhanced sound absorption and attenuation properties are normally part of the standard range, which may also include Imperial sizes for refurbishment work.

A wide range of small decorative components is manufactured in GRG, including cornices, coving, ceiling centres, corbels, columns, arches and niches.

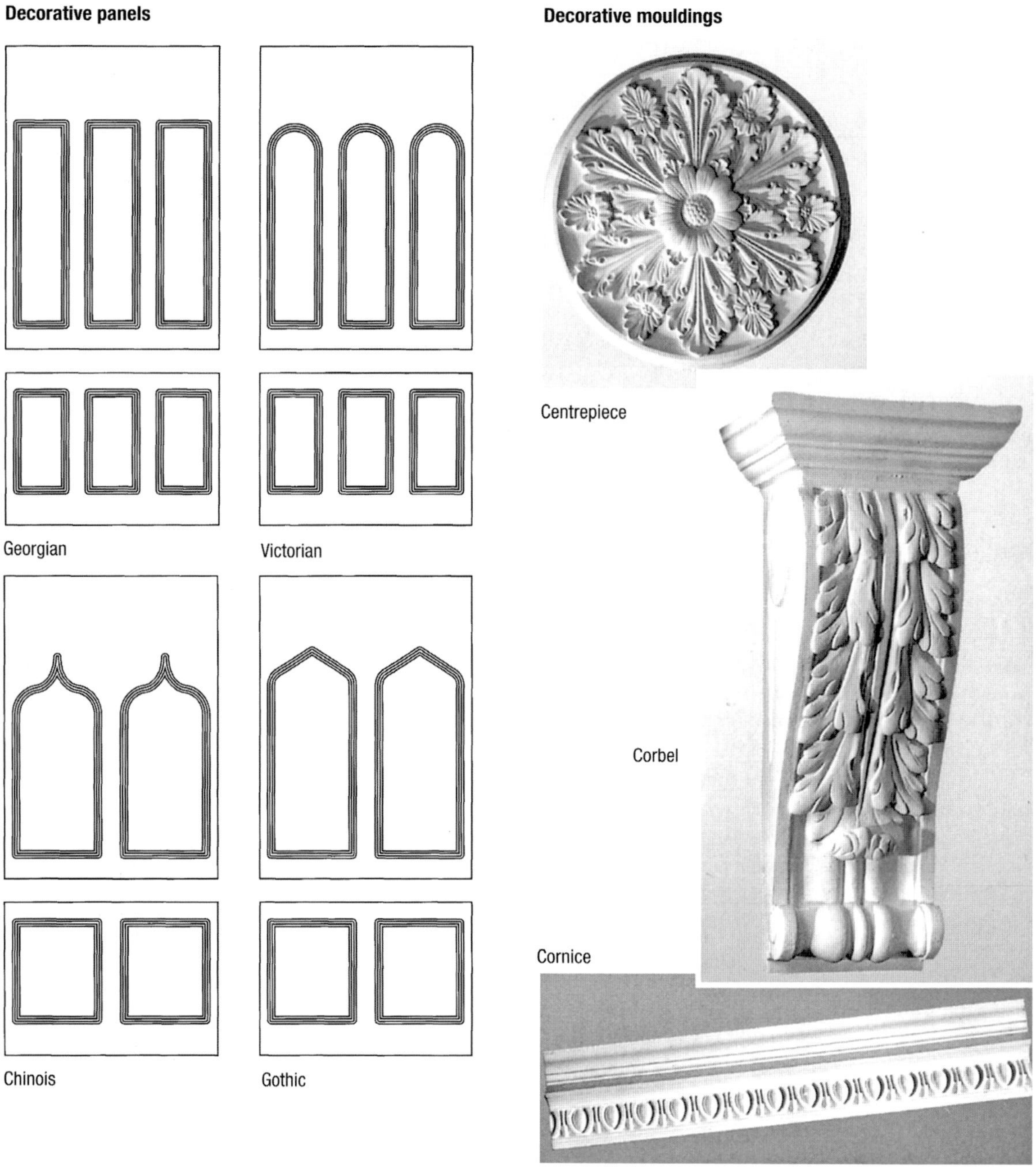

Decorative panels

Georgian

Victorian

Chinois

Gothic

Decorative mouldings

Centrepiece

Corbel

Cornice

Fig. 11.8 Decorative GRG plaster components

References

FURTHER READING

Bai, J. 2013: *Advanced fibre-reinforced polymer composites for structural applications*. Cambridge: Woodhead Publishing Ltd.

British Gypsum. 2009: *The white book*. Loughborough: British Gypsum Ltd.

British Gypsum. 2011: *The fire book*. Loughborough: British Gypsum Ltd.

Concrete Repair Association. 2009: *Structural strengthening with fibre reinforced polymers*. Bordon: Concrete Repair Association.

Concrete Society. 2012: *Design guidance for strengthening concrete structures using fibre composite materials*, 3rd edn. Technical Report 55. Camberley: Concrete Society.

Cripps, A. 2002: *Fibre-reinforced polymer composites in construction*. Publication C 564. London: CIRIA.

Glass and Glazing Federation. 2000: *Plastics glazing sheet materials – basic products*. London: GGF.

Glassfibre Reinforced Cement Association. 2012: *Specifiers guide to glass reinforced concrete*. Camberley: Glassfibre Reinforced Cement Association.

IGRCA. 2007: *Practical design guide for glass reinforced concrete using limit state theory*. Camberley: International Glassfibre Reinforced Cement Association.

Shenoi, R., Moy, S. and Holloway, L. (eds) 2002: *Advanced polymer composites for structural applications in construction*. London: Thomas Telford.

Transport Research Laboratory. 2008: *Fibre reinforced concrete update (2005–2008)*. CT125–3. Wokingham: Transport Research Laboratory.

STANDARDS

BS PL 4: 2005
Properties of unsaturated polyester resins for low pressure laminating of high strength fibre reinforced composites. Specification.

BS PL 5: 2005
Unsaturated polyester resins for low pressure laminating of high strength fibre reinforced composites. Test methods.

BS PL 6: 2005
Properties of unsaturated polyester resins for low pressure laminating of glass fibre reinforced composites. Specification.

BS PL 7: 2005
Unsaturated polyester resins for low pressure laminating of glass fibre reinforced laminates. Test methods.

BS 476
Fire tests on building materials and structures:
Part 4: 1970   Non-combustibility test for materials.
Part 6: 1989   Methods of test for fire propagation for products.
Part 7: 1997   Method of test to determine the classification of the surface spread of flame of products.

BS ISO 3598: 2011
Textile glass. Yarns. Basis for specification.

BS 4592
Industrial type flooring and stair treads:
Part 4: 2006   GRP open bar gratings. Specification.
Part 5: 2006   Solid plates and metal in GRP. Specification.
Part 6: 2008   GRP moulded open mesh gratings and protective barriers. Specification.

BS 6206: 1981
Specification for impact requirements for flat safety glass and safety plastics for use in buildings.

BS ISO 14127: 2008
Carbon-fibre reinforced composites. Determination of the resin, fibre and void contents.

BS ISO 22314: 2006
Plastics. Glass-fibre reinforced products. Determination of fibre length.

BS EN 492: 2012
Fibre-cement slates and fittings. Product specification and test methods.

BS EN 494: 2012
Fibre-cement profiled sheets and fittings. Product specification and test methods.

BS EN 1013: 2012
Light transmitting single skin profiled sheets for single skin for internal and external roofs, walls and ceilings.

BS EN 1169: 1999
Precast concrete products. Factory production control of glass-fibre reinforced cement.

BS EN 1170
Precast concrete products. Test method for GRC:
Part 1: 1998   Measuring the consistency. Slump test method.

Part 2: 1998   Measuring the fibre content in fresh GRC.

Part 3: 1998   Measuring the fibre content in sprayed GRC.

Part 4: 1998   Measuring bending strength. Simplified bending test.

Part 5: 1998   Measuring bending strength. Complete bending test.

Part 6: 1998   Determination of the absorption of water.

Part 7: 1998   Measurement of extremes of dimensional variations.

Part 8: 2008   Cyclic weathering type test.

BS EN 1447: 2009
Plastics piping systems. Glass reinforced thermoplastics (GRP) pipes.

BS EN 1796: 2013
Plastics piping systems for water supply with or without pressure. Glass-reinforced thermosetting plastics (GRP) based on unsaturated polyester resin (UP).

BS EN 12467: 2012
Fibre-cement flat sheets. Product specification and test method.

BS EN 13121–3: 2008
GRP tanks and vessels for use above ground. Design and workmanship.

BS EN 13280: 2001
Specification for glass fibre reinforced cisterns.

BS EN 14364: 2013
Plastics plumbing systems for drainage and sewerage with or without pressure. Glass-fibre reinforced thermosetting plastics (GRP).

BS EN 14649: 2005
Precast concrete panels. Test method for strength retention of glass fibres in cement and concrete.

BS EN 14845
Test methods for fibres in concrete:

Part 1: 2007   Reference concretes.

Part 2: 2006   Effect on concrete.

BS EN 15191: 2009
Precast concrete products. Classification of glass-fibre reinforced concrete performances.

BS EN 15422: 2008
Precast concrete products. Specification for glass fibres for reinforcement of mortars and concretes.

BS EN 16245
Fibre-reinforced plastic composites. Declaration of raw material characteristics:

Part 1: 2013   General requirements.

Part 2: 2013   Specific requirements for resin, curing systems, additives and modifiers.

Part 3: 2013   Specific requirements for fibre.

Part 4: 2013   Specific requirements for fabrics.

Part 5: 2013   Specific requirements for core materials.

BS EN ISO 25762: 2012
Plastics. Guidance on the assessment of fire characteristics and fire-performance of fibre reinforced composites.

BUILDING RESEARCH ESTABLISHMENT PUBLICATIONS

BRE Digests

BRE Digest 442: 1999
Architectural use of polymer composites.

BRE Digest 480: 2004
Wood plastic composites and plastic lumber.

BRE Good repair guide

BRE GRG 34: 2003
Repair and maintenance of FRP structures.

BRE Information papers

BRE IP 7/99
Advanced polymer composites in construction.

BRE IP 19/01
The performance of fibre cement slates.

BRE IP 10/03
Fibre reinforced polymers in construction: durability.

BRE IP 11/03
Fibre reinforced polymers in construction: predicting weathering.

BRE IP 2/04
Wood plastic composites: market drivers and opportunities in Europe.

BRE IP 4/04
Recycling fibre reinforced polymers in the construction industry.

BRE IP 5/04
Fibre reinforced polymers in construction.

BRE Reports

BR 405: 2000
    Polymer composites in construction.
BR 461: 2003
    Fibre reinforced polymers in construction: long-term performance in service.
BR 467: 2004
    Recycling fibre reinforced polymers in construction: A guide to best practicable environmental option.
FB 8: 2004
    Effective use of fibre reinforced polymer materials in construction.

ADVISORY ORGANISATIONS

British Plastics Federation, 6 Bath Place, Rivington Street, London EC2A 3JE (0207 457 5000).

Glassfibre Reinforced Concrete Association, Concrete Society, 4 Meadows Business Park, Station Approach, Blackwater, Camberley, Surrey GU17 9AB (01276 607140).

PLASTER AND BOARD MATERIALS

Introduction

Plastering, based on lime, was brought to Britain by the Romans. In Britain, it was originally used to strengthen and seal surfaces and, in the case of combustible materials, to afford some fire protection, but by the eighteenth century its value as a decorative finish had been appreciated. The use of gypsum plaster both as a sealant and as a decorative material by the Minoan civilisation is well documented, and current UK practice is now based on gypsum (hydrated calcium sulphate), rather than lime. Gypsum is mined from geological deposits produced by the gradual evaporation of lakes containing the mineral; there are extensive reserves within the UK, mainly in the North of England, but also in the East Midlands.

Historically, fibrous materials have been used to reinforce plaster and particularly to control shrinkage in lime plaster. Traditionally, ox, horse and goat hair were the standard materials; however, straw, hemp and jute have also been used. The earliest lightweight support for plasters was interwoven hazel twigs, but by the fifteenth century split timber laths were common. The modern equivalent is the use of galvanised and stainless steel expanded metal.

Gypsum plaster

MANUFACTURE OF GYPSUM PLASTER

Rock gypsum is mined, crushed and ground to a fine powder. The natural mineral may be white or discoloured pale pink, grey or brown due to small quantities of impurities, which do not otherwise affect the product. On heating to temperatures in the range 130–170°C, water is driven off the hydrated gypsum; the type of plaster produced is largely dependent upon the extent of this dehydration process.

$$CaSO_4 \cdot 2H_2O \xrightarrow{130°C} CaSO_4 \cdot \tfrac{1}{2}H_2O \xrightarrow{170°C} CaSO_4$$

hydrated   hemihydrate   anhydrous

gypsum   gypsum

Gypsum plasters including those for special purposes are defined in the standard BS EN 13279–1: 2008. Gypsum plasterboards are defined in BS EN 520: 2004.

TYPES OF PLASTER

Plaster of Paris

Plaster of Paris is produced by driving off three-quarters of the water content from natural hydrated gypsum. Plaster of Paris sets very quickly on the addition of water, and is therefore often used as a moulding material.

Retarded hemi-hydrate gypsum plaster

The majority of plasters in current use within construction are based on retarded hemi-hydrate gypsum. The addition of different quantities of a retarding agent, usually keratin, is used to adjust the setting time (usually between 1.5 and 2 hours) for different products. Other additives and admixtures include fillers, fibres, lime, lightweight aggregates, pigments, retarders and plasticisers.

Types of gypsum binders and plasters (BS EN 13279–1: 2008):

Notation	Designation
A	Gypsum binder for further processing
A1	Gypsum binders for direct use
A2	Gypsum binders for direct use on site
A3	Gypsum binders for further processing
B	Gypsum plaster
B1	Gypsum building plaster
B2	Gypsum-based building plaster (minimum 50% gypsum)
B3	Gypsum-lime building plaster (>5% lime)
B4	Lightweight gypsum building plaster (inorganic or organic aggregates)
B5	Lightweight gypsum-based building plaster
B6	Lightweight gypsum-lime building plaster
B7	Enhanced surface hardness gypsum plaster
C	Gypsum plaster for special purposes
C1	Gypsum plaster for fibrous plasterwork
C2	Gypsum mortar
C3	Acoustic plaster
C4	Thermal insulation plaster
C5	Fire protection plaster
C6	Thin-coat plaster, finishing product
C7	Finishing product

Undercoat and one-coat plasters

The main constituents of undercoat and one-coat plasters are retarded hemi-hydrate gypsum, with lightweight aggregates for the lightweight products, together with small quantities of limestone, anhydrite (anhydrous gypsum), clay and sand. In addition, other materials are incorporated to adjust the product specification and setting time, which normally ranges between one and two hours. Thus, lime is added to undercoat plaster, and for backgrounds of high suction, a water-retention agent is also required. For example, *browning* is suitable for use on backgrounds with moderate or high-suction and a good mechanical key (e.g. brickwork and blockwork), but on smooth surfaces or low-suction backgrounds such as concrete, an initial bonding coat is required. For higher impact resistance, cement and granulated blast furnace slag are incorporated, and for a one-coat plaster, limestone is added. Typical applications for walls would be 11 mm for undercoats with a finish coat of 2 mm, or a single one-coat application of 13 mm. The undercoat for ceilings would normally be 8 mm with a 2 mm-finish coat. A maximum total plaster thickness of 25 mm is recommended.

Finish-coat plasters

For finish-coat plasters, like undercoat plasters, the main constituent is retarded hemi-hydrate gypsum, but with a small addition of lime to accelerate the set. The lightweight products contain exfoliated vermiculite. Finish coats on masonry substrates are usually 2 mm in thickness, and board finish plaster is normally applied to 2–3 mm.

Lightweight plasters

Lightweight plasters usually contain inorganic expanded perlite or exfoliated vermiculite, but organic lightweight aggregates may also be used. Premixed expanded polystyrene dry plaster/mortar for thermal or impact noise insulation is described in BS EN 16025: 2013. The range of densities and associated thermal conductivities is listed in Table 12.1.

Table 12.1 Typical relationship between density and thermal conductivity for gypsum plasters

Density (kg/m³)	Thermal conductivity (W/m K)
600	0.18
700	0.22
800	0.26
900	0.30
1000	0.34
1100	0.39
1200	0.43
1300	0.47
1400	0.51
1500	0.56

BACKGROUNDS FOR PLASTER

Plaster bonds to the background by a combination of mechanical key and adhesion. Backgrounds should be clean, dry and free from other contamination, and the specification of the plaster should be appropriate to the suction of the background surface. Where possible, as in the case of brickwork, a good mechanical key should be obtained by raking out the joints. On hard, low-suction materials such as smooth concrete and ceramic tiles, a polyvinyl acetate (PVA) or proprietary bonding agent should be applied. Similarly, to control the high suction in substrates such as aerated concrete blocks, a bonding agent may be applied or the substrate wetted prior to the application of plaster. Plaster may, however, be applied directly to dense aggregate concrete blocks without prior wetting. Where two or more coats of plaster are applied, the undercoats should be scratched to ensure good subsequent bonding. Gypsum plasters, if applied correctly, do not shrink or crack on drying out and subsequent coats may be applied in quick succession.

PLASTERBOARD

Plasterboard consists of a gypsum core bonded to strong paper liners. Most wallboards have one light surface for direct decoration or plaster skim and one grey surface. The decorative surface may be either tapered or square. The standard board sizes are 1200 and 900 mm wide to coordinate with timber or metal stud partitioning systems. Plasterboard may be cut with a saw or scored and snapped. Nail fixings should be driven in straight, leaving a shallow depression but without fracturing the paper surface. Alternatively boards may be screwed. Standard thicknesses are 12.5, 15 and 19 mm, although 9.5 mm board may also be obtained.

Only the moisture-resistant grades of plasterboard (Type H) normally require the application of a bonding agent before plastering. These have a water-resistant core and treated liners, so may be used in moist and humid conditions such as kitchens or bathrooms and behind external finishes such as vertical tiling and weatherboarding, or in external sheltered positions protected from direct rain.

Fire-resistant boards (Type F), available in 12.5 and 15 mm thicknesses and reinforced with glass fibres, offer increased fire resistance over standard gypsum boards. Fire-resistant boards are colour-coded pink. Impact-resistant boards are also reinforced with glass fibres and have a high-strength paper liner. Glassfibre-reinforced gypsum is described in Chapter 11.

Sound insulation boards, colour-coded blue, have a modified gypsum core, making them heavier than standard wallboards. The extra weight enhances the sound attenuation by up to 5 dB R_w compared to standard gypsum board. When used in conjunction with *Robust Details*, sound insulation boards provide enhanced acoustic performance. The Approved Document – Building Regulations Part E – Resistance to the Passage of Sound requires a surface mass of 10 kg/m² for plasterboard separating and internal walls as well as floors.

The heavy 19 mm gypsum planks, produced to a width of 600 mm, are used in walls, ceilings and floors to comply with the requirements of the Building Regulations when constructed according to *Robust Details*.

Boards are available finished with PVC, backed with aluminium foil or laminated to insulation (expanded polystyrene, extruded polystyrene, phenolic foam or mineral wool) for increased thermal properties. (The thermal conductivity of standard plasterboard is 0.19 W/m K.)

Types of plasterboards (BS EN 520: 2004):

Type	Designation
A	Gypsum plasterboard with a face suitable for a gypsum finish coat or decoration
H1, H2, H3	Gypsum plasterboard with reduced water absorption rates

E   Gypsum sheathing board for external walls but not permanently exposed to weather conditions

F   Gypsum plasterboard with improved core cohesion at high temperatures

P   Gypsum baseboard to receive gypsum plaster

D   Gypsum plasterboard with controlled density

R   Gypsum plasterboard with enhanced strength

I   Gypsum plasterboard with enhanced surface hardness

Note: Some boards may combine more than one designation.

Plasterboard systems

Plasterboard non-load-bearing internal walls may be constructed using proprietary metal stud systems or as traditional timber stud walls. Where appropriate, acoustic insulation should be inserted within the void spaces. Dry-lining to masonry may be fixed with dabs of adhesive, or alternatively with metal or timber framing. Plasterboard suspended ceiling systems are usually supported on a lightweight steel framework fixed directly to either concrete or timber. Convex and concave surfaces can be achieved. Sound transmission through existing upper-storey timber-joist floors can be reduced by a combination of resiliently mounted plasterboard and mineral wool insulation. Compliance with the acoustic requirements of the Building Regulations Approved Document E for domestic building can be achieved using *Robust Details.* Components for plasterboard metal-framing systems are defined in BS EN 14195: 2005.

Plasterboard ceiling tiles
Plasterboard ceiling tiles are available in a range of smooth, textured and perforated finishes to produce various levels of sound-insulating and sound-absorbing properties. The standard tiles are 600 × 600 mm, for fixing to metal sub-framing. Fire classification is Class 0 and Euroclass A2-s1, d0.

Fibre-reinforced gypsum boards

Fibre-reinforced gypsum boards are manufactured with either natural or glass fibres. Glassfibre-reinforced gypsum (GRG) is described in Chapter 11.

Natural fibre-reinforced gypsum boards are manufactured from cellulose fibres, frequently from recycled paper, or wood fibres within a matrix of gypsum. The boards may be uniform with dispersed fibres or laminated with woven or non-woven gypsum-reinforced sheets encasing a perlite and gypsum core. Boards are impact and fire resistant and easily fixed by nails, screws, staples or adhesive as a dry-lining system to timber, metal framing or masonry. Standard boards are 1200 × 2400 mm with thicknesses normally in the range 12.5–18 mm. Joints are filled or taped and corners beaded as for standard plasterboard products. A composite board of fibre-reinforced gypsum and expanded polystyrene offers enhanced insulation properties. Gypsum boards with fibrous reinforcement are defined in BS EN 15283: 2008. (The thermal conductivity of gypsum board containing 13% wood fibres is 0.24 W/m K.)

Types of gypsum boards with fibrous reinforcement (BS EN 15283: 2008):

Type	Designation
GM	Gypsum board with mat reinforcement
GM-H1 ⎱ GM-H2 ⎰	Gypsum board with mat reinforcement and reduced water absorption rate
GM-I	Gypsum board with mat reinforcement and enhanced surface hardness
GM-R	Gypsum board with mat reinforcement and enhanced strength
GM-F	Gypsum board with mat reinforcement and enhanced core cohesion at high temperatures
GF	Gypsum fibre board
GF-H	Gypsum fibre board with reduced water absorption rate
GF-W1 ⎱ GF-W2 ⎰	Gypsum fibre board with reduced surface water absorption rate
GF-D	Gypsum fibre board with enhanced density
GF-I	Gypsum fibre board with enhanced surface hardness
GF-R1 ⎱ GF-R2 ⎰	Gypsum fibre board with enhanced strength

SPECIAL PLASTERS

Renovating plaster

Renovating plaster is used where walls have been stripped of existing plaster during the successful installation of a new damp-proof course. Renovating plasters contain aggregates which promote surface drying when applied to structures with residual moisture, but they should not be used in permanently damp locations below ground level. Renovating plaster should also not be used where masonry is heavily contaminated with salts, such as in buildings not originally constructed with damp-proof courses, and on the brickwork of chimney breasts. Renovating plasters contain a fungicide to inhibit mould growth during the drying-out process.

Projection plaster

Projection plaster is sprayed onto the background from a projection machine as a continuous ribbon which flows sufficiently for the ribbons to coalesce. The plaster should be built up to the required thickness, ruled to an even surface, then flattened and trowelled to a flat surface. As with all plastering the process should not be carried out under freezing, excessively hot or dry conditions. A typical application to masonry would be 13 mm and should not exceed 25 mm.

Acoustic plaster

Acoustic plaster has a higher level of sound absorption than standard gypsum plasters owing to its porosity and surface texture. Aluminium powder is added to the wet plaster mix to produce fine bubbles of hydrogen gas, which remain trapped as the plaster sets giving it a honeycomb structure. One form of acoustic plasterboard consists of perforated gypsum plasterboard which may be backed with a 100 mm glass wool sound-absorbing felt.

X-ray plaster

X-ray plaster is retarded hemi-hydrate plaster containing barium sulphate (barytes) aggregate. It is used as an undercoat plaster in hospitals, etc., where protection from X-rays is required. Typically, a 20 mm layer of X-ray plaster affords the same level of protection as a 2 mm sheet of lead, providing that it is free of cracks.

Textured plaster

Textured plaster is frequently applied to plasterboard ceilings. A variety of different patterns and textures can be achieved. The textured surface may be left as a natural white finish or painted as required.

Polished plaster

Polished plaster is usually applied in three coats over a primer. The first coat is not tinted, but the subsequent coats are pigmented to the required colour. Coats are applied with stainless steel tools allowing 10 hours' drying time between coats. The final coat is polished within 5–10 minutes of its application to give the required smooth and hard surface. After 48 hours a wax finish is applied and polished to produce the required highly reflective surface finish.

Fibrous plaster

Fibrous plaster is Plaster of Paris reinforced with jute, sisal, hessian, glass fibres, wire mesh or wood laths. It is used for casting in moulds, ornate plasterwork, such as fire surrounds, decorative cornices, dados, friezes, panel mouldings, corbels and centre-pieces for ceilings in both restoration and new work. The reinforcement material may be elementary in the form of random fibres or sheet material, or complementary as softwood laths or lightweight steel sections. Fibrous plaster is described in the standards BS EN 13815: 2006 and BS EN 15319: 2007.

Phase change material plaster

An alternative method of stabilising internal room temperatures, rather than using thermal mass, is to incorporate phase change materials (PCMs) into the building fabric. One approach is to use a layer of gypsum plaster containing 26–30% by volume of phase change material. A commercial system uses wax encapsulated within 3 μm particles of PVA, which are formed as dispersion, and then dried to a 0.1–0.3 mm powder for mixing into gypsum. The wax undergoes a phase change at 23°C or 26°C, absorbing heat as it melts and releasing it as it solidifies. A 30 mm layer of phase change gypsum plaster has an equivalent thermal stabilising effect to a mass of 180 mm of concrete or 230 mm of brickwork, and a 15 mm PCM gypsum board is equivalent to 100 mm of concrete. The boards may be used for ceilings or wall linings as appropriate, and the phase change temperature should be specified. Current costs are of the order of 10 times that of standard plasterboard.

VOC absorbing plasterboard

Formaldehyde and other VOCs are absorbed into gypsum boards or tiles and converted into inert compounds which remain trapped, preventing re-emission. Internal air quality is therefore maintained at a higher level by the removal of approximately 70% of the organic pollutants.

ACCESSORIES FOR PLASTERING

Beads

Angle and stop beads are manufactured from galvanised or stainless perforated steel strip or expanded metal. They provide a protected, true straight arris or edge with traditional plastering to masonry or for

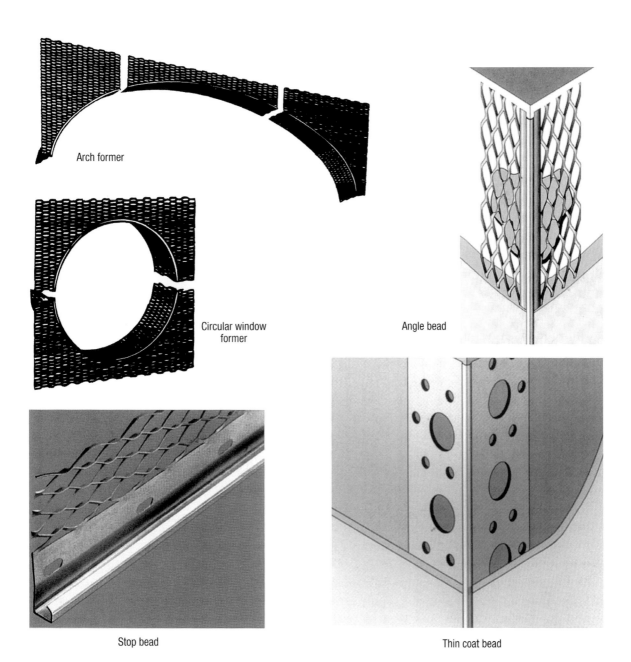

Arch former

Circular window former

Angle bead

Stop bead

Thin coat bead

Fig. 12.1 Plastering beads and arch formers

thin-coat plasterboard. Proprietary systems are manufactured similarly from perforated galvanised or stainless steel to form movement joints in dry-lining systems (Fig. 12.1). These components are defined in BS EN 14353: 2007 and BS EN 13658: 2005.

Scrim and jointing tape

Scrim, an open-weave material, is used across joints between plasterboards and in junctions between plaster and plasterboard. Both self-adhesive glass fibre mesh and traditional jute scrim are available. For the prevention of thermal movement cracking at plasterboard butt joints, 50 mm paper tape bedded into the plaster skim is often more effective than self-adhesive scrim. The paper-based tape and bedding compounds for plasterboards are defined in BS EN 14353: 2007.

Coves and cornices

Decorative coves and cornices are manufactured from gypsum plaster encased in a paper liner. In some cases the gypsum is reinforced with glass fibres. The components (Fig. 12.2) can be cut to size with a saw, and are normally fixed with proprietary adhesives. The standard sizes of plain cove are 100 mm and 127 mm, although a wide variety of decorative forms are also produced.

GYPSUM FLOOR SCREED

Gypsum interior floor screed, manufactured from a mixture of hemi-hydrate gypsum, limestone and less than 2% cement, may be used as an alternative to a traditional sand and cement screed, provided that a floor covering is to be used. The material is self-smoothing and may be pumped. It is laid on a polythene membrane to a minimum thickness of 35 mm for floating screeds, and may be used over underfloor heating systems. When set, the hard plaster has a minimum 28-day compressive strength of 30 MPa.

FIRE RESISTANCE OF PLASTER MATERIALS

Gypsum products afford good fire protection within buildings due to their basic chemical composition. Gypsum, hydrated calcium sulphate ($CaSO_4 \cdot 2H_2O$) as present in plaster and plasterboard, contains nearly 21% water of crystallisation. When exposed to a fire this chemically combined water is gradually expelled in the form of vapour. It is this process which absorbs the incident heat energy from the fire, considerably reducing the transmission of heat through the plaster, thus protecting the underlying materials. The process of dehydrating the gypsum commences on the face adjacent to the fire, and immediately the dehydrated material, because it adheres to the unaffected gypsum, acts as an insulating layer slowing down further dehydration. Even when all the water of crystallisation has been expelled, the remaining anhydrous gypsum continues to act as an insulating layer while it retains its integrity. The inclusion of glass fibres into gypsum plasterboards increases the cohesiveness of the material within fires. Gypsum binders and gypsum plasters are classified as Class A1 (no contribution to fire) if they contain less than 1% of organic material.

RECYCLING GYPSUM PRODUCTS

A large proportion of gypsum for plasterboard (over 1 million tonnes per annum) is produced as a by-product of the removal of sulphur dioxide from the flue gasses of coal-fired electricity-generating stations. A slurry of pulverised limestone absorbs the sulphur dioxide, producing high-purity calcium sulphate. However, the supply of this synthetic gypsum (known as desulphogypsum DSG) is dependent upon the continuing generation of electricity by coal-fired stations as well as the sulphur content of the coal used. Low-sulphur coal imported from Australia may reduce the production of desulphogypsum.

Waste of plasterboard on site is high, ranging from 10–25% of the delivered material. Disposal in landfill sites is very limited, as gypsum, in association

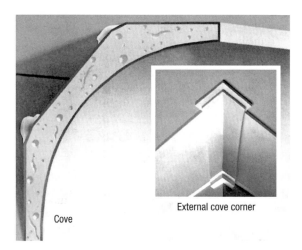

External cove corner

Cove

Fig. 12.2 Preformed plaster coves

with biodegradable materials, can produce toxic hydrogen sulphide. Gypsum products must therefore be deposited in biodegradable-free locations. Currently, the industry maximum recycled gypsum content for plasterboard is 18%. The paper liners for plasterboard are made from 97% recycled paper and cardboard. One manufacturer will collect all of its own surplus gypsum board, ceiling tiles, coving and GRG from site for recycling into new products.

The publication PAS 109: 2008 gives the specification for the production of recycled gypsum from waste plasterboard.

Lime plaster

Hydraulic lime plaster is suitable for interior application, particularly on earth structures and unfired clay walls. It is usually applied in two or three coats – the best-quality work requiring the three-coat system. In this case a 13 mm coat of *coarse stuff* containing 5 mm sand (lime : sand, 1 : 2.5) is followed, when dry, with a similar thickness of a 1 : 3 mix and a thin final coat of between 1 : 1 and 1 : 2 lime to sand. Other additions, including horsehair and cow dung, may be added to improve the setting properties of the lime plaster.

Calcium silicate boards

Calcium silicate boards are manufactured from silica with lime and/or cement, usually incorporating cellulose fibres or softwood pulp and mica or exfoliated vermiculite filler, to produce a range of densities. The high-density material is laminated under steam and pressure, while the lower-density material is produced by rolling followed by curing in an autoclave. Calcium silicate boards, like gypsum boards, are non-combustible. The material is grey or off-white in colour, easily worked and nailed. Calcium silicate boards are durable, moisture-, chemical- and impact-resistant with dimensional stability and a good strength-to-weight ratio. They are available with a range of smooth or textured factory finishes for interior or exterior use and also laminated to extruded polystyrene for enhanced insulation properties. Standard thicknesses range from 4.5–20 mm, although thicknesses up to 100 mm are available in the vermiculite lightweight boards used for fire protection giving up to 240 minutes' resistance. The standard fire-resistance class is A2-s1, d0. Typical applications include wall, roof and partition linings, suspended ceilings, fascias, soffits, weatherboarding and fire protection to structural steelwork. External cladding boards may be finished with a sprayed or trowelled render to produce a seamless finish. (The thermal conductivities of calcium silicate boards are usually within the range 0.13–0.29 W/m K depending upon their composition.)

References

FURTHER READING

British Geological Survey. 2006: *Gypsum*. London: Office of the Deputy Prime Minister.

British Gypsum. 2009: *The white book*. Loughborough: British Gypsum Ltd.

British Gypsum. 2011: *The fire book*. Loughborough: British Gypsum Ltd.

Millar, W. and Bankart, G. 2009: *Plastering plain and decorative*, 4th edn. Shaftesbury: Donhead Publishing.

STANDARDS

BS 476
Fire tests on building materials and structures:
Part 4: 1970   Non-combustibility test for materials.
Part 6: 1989   Methods of test for fire propagation for products.
Part 7: 1997   Method of test to determine the classification of the surface spread of flame of products.
BS 5270
Bonding agents for use with gypsum plasters and cement:
Part 1: 1989   Specification for polyvinyl acetate (PVAC) emulsion bonding agents for indoor use with gypsum building plasters.
BS 6100
Building and civil engineering. Vocabulary:
Part 9: 2007   Work with concrete and plaster.
BS 7364: 1990
Galvanized steel studs and channels for stud and sheet partitions and linings using screw fixed gypsum wallboards.
BS 8000
Workmanship on building sites:

Part 8: 1994   Code of practice for
plasterboard partitions and dry
linings.

BS 8212: 1995
Code of practice for dry lining and partitioning
using gypsum plasterboard.

BS 8481: 2006
Design, preparation and application of internal
gypsum, cement, cement and lime plastering
systems. Specification.

BS 9250: 2007
Code of practice for the design of airtightness of
ceilings in pitched roofs.

BS EN 520: 2004
Gypsum plasterboards. Definitions, requirements
and test methods.

BS EN 998
Specification for mortar for masonry:
Part 1: 2010   Rendering and plastering
mortar.

BS EN ISO 1182: 2010
Reaction to fire tests for products. Non-
combustibility test.

BS EN ISO 1716: 2010
Reaction to fire tests for products. Determination
of the gross heat of combustion.

BS EN ISO 11925–2: 2010
Reaction to fire tests. Ignitability of products.
Single flame source.

BS EN 12859: 2011
Gypsum blocks. Definitions, requirements and
test methods.

BS EN 12860: 2001
Gypsum based adhesives for gypsum blocks.
Definitions.

BS EN 13055–1: 2002
Lightweight aggregates for concrete, mortar and
grout.

BS EN 13139: 2013
Aggregates for mortar.

BS EN 13279
Gypsum binders and gypsum plasters:
Part 1: 2008   Definitions and requirements.
Part 2: 2014   Test methods.

BS EN 13501
Fire classification of construction products and
building elements:
Part 1: 2007   Classification using data from
reaction to fire tests.
Part 2: 2007   Classification using data from
fire resistance tests.

BS EN 13658
Metal laths and beads. Definitions, requirements
and test methods:
Part 1: 2005   Internal plastering.
Part 2: 2005   External rendering.

BS EN 13815: 2006
Fibrous gypsum plaster casts. Definitions,
requirements and test methods.

BS EN ISO 13823: 2010
Reaction to fire tests for building products.
Building products excluding floorings.

BS EN 13914
Design, preparation and application of external
rendering and internal plastering:
pr Part 1: 2013 External rendering.
pr Part 2: 2013 Design considerations and
essential principles for internal
plastering.

BS EN 13915: 2007
Prefabricated gypsum plasterboard panels with a
cellular paperboard core.

pr EN 13950: 2011
Gypsum plasterboard thermal/acoustic insulation
composite panels. Definitions, requirements and
test methods.

pr EN 13963: 2011
Jointing materials for gypsum plasterboards.
Definitions, requirements and test methods.

pr EN 13964: 2014
Suspended ceilings. Requirements and test
methods.

pr EN 14190: 2011
Gypsum, plasterboard products from processing.
Definitions, requirements and test methods.

BS EN 14195: 2005
Metal framing components for gypsum
plasterboard systems. Definitions, requirements
and test methods.

BS EN 14209: 2005
Preformed plasterboard cornices. Definitions,
requirements and test methods.

BS EN 14246: 2006
Gypsum elements for suspended ceilings.
Definitions, requirements and test methods.

BS EN 14353: 2007
Metal beads and feature profiles for use with
gypsum plasterboards. Definitions, requirements
and test methods.

BS EN 14496: 2005
Gypsum-based adhesives for thermal/acoustic
insulation composite panels and plasterboards.

BS EN 14566: 2008
   Mechanical fasteners for gypsum plasterboard systems. Definitions, requirements and test methods.
PD CEN/TR 15123: 2005
   Design, preparation and application of internal polymer plastering systems.
BS EN 15254
   Extended application of results from fire-resistance tests:
   | Part 2: 2009 | Non-loadbearing walls. Masonry and gypsum blocks. |
   | Part 7: 2012 | Non-loadbearing ceilings. Metal sandwich panels construction. |
BS EN 15283
   Gypsum boards with fibrous reinforcement:
   | Part 1: 2008 | Gypsum boards with mat reinforcement. |
   | Part 2: 2008 | Gypsum fibre boards. |
BS EN 15318: 2007
   Design and application of gypsum blocks.
BS EN 15319: 2007
   General principles of design of fibrous (gypsum) plaster works.
BS EN 15824: 2009
   Specifications for external renders and internal plasters based on organic binders.
BS EN 16025: 2013
   Thermal and/or sound insulating products in building construction. Bound EPS ballasting:

   | Part 1: 2013 | Requirements for factory premixed EPS dry plaster. |
   | Part 2: 2013 | Processing of the factory premixed EPS dry plaster. |
PD CEN/TR 16239: 2011
   Installation rules of fibrous gypsum plaster works.
NS/PAS 109: 2013
   Specification for the production of reprocessed gypsum from waste plasterboard.

BUILDING RESEARCH ESTABLISHMENT PUBLICATIONS

BRE Good building guides

BRE GBG 65: 2005
   Plastering and internal rendering.
BRE GBG 70: 2007
   Plasterboard; types and their applications (Parts 1, 2 and 3).

BRE Good repair guide

BRE GRG 18: 1998
   Replacing plasterwork.

ADVISORY ORGANISATION

Gypsum Products Development Association, PO Box 35084, London NW1 4XE (0207 935 8532).

13

INSULATION MATERIALS

Introduction

With the increasing emphasis on energy-conscious design and the broader environmental impact of buildings, greater attention is being focused on the appropriate use of thermal and sound insulation materials for both new-build and refurbishment work. The UK government has set the target of reducing overall greenhouse gas emissions from the 1990 baseline by 80% by 2050, with an interim 50% cut by 2030. Currently, 27% of UK carbon emissions are from domestic buildings and 17% from non-domestic buildings. It is a UK government requirement that all new homes will be net zero carbon from 2016. To achieve this, thermal insulation specifications need to support the Fabric Energy Efficiency Standards which are the maximum space heating and cooling demands for zero carbon homes in relation to their house types. Even higher building fabric specifications are associated with the Passivhaus standards (Chapter 18, page 475). The equivalent target date of net zero carbon for new non-domestic buildings is 2019. The broad energy conservation criteria within the Approved Documents of the Building Regulations Part L – Conservation of Fuel and Power (2013 edition) are outlined in Chapter 2 (page 44) and Chapter 7 (page 286), although there is some variation of requirements among the countries of the UK.

Thermal and sound insulation materials

To consider the relative efficiency of insulating materials, the thermal conductivities (λ W/m K) are quoted at the standard 10°C to allow direct comparisons. U-values would not illustrate direct comparability owing to the varying thicknesses used and the wide variety of combinations of materials typically used in construction.

In considering acoustic control, distinction is made between the reduction of sound transmitted directly through the building components and the attenuation of sound reflected by the surfaces within a particular enclosure. Furthermore, transmitted sound is considered in terms of both impact and airborne sound. Impact sound is caused by direct impact onto the building fabric which then vibrates, transmitting the sound through the structure; it is particularly significant in the case of intermediate floors. Airborne soundwaves, from the human voice and sound-generating equipment, cause the building fabric to vibrate, thus transmitting the sound. Airborne sound is particularly critical in relation to separating walls and is significantly increased by leakage at discontinuities within the building fabric, particularly around unsealed openings. Flanking sound is transmitted between rooms via the adjoining elements rather than directly through the separating wall. The reduction in sound energy passing through a building element is expressed in decibels (dB). The doubling of the mass of a building component reduces the sound transmission by approximately 5 dB; thus, sound-insulating materials are generally heavy structural elements. However, the judicious use of dissipative absorbers within walls can reduce the reliance for sound absorption on mass alone. Noise may be transmitted through services installations, so consideration should be given to the use of acoustic sleeves and linings as appropriate.

For dwellings, the Approved Documents of the UK Building Regulations Part E (2010) 'Resistance to the Passage of Sound' requires protection against sound from adjacent buildings and within buildings including internal walls and floors. The use of *Robust Details* in housing can ensure adherence to the regulations and eliminate the requirement for pre-completion sound testing to demonstrate compliance. The Building Regulations list minimum criteria for the attenuation of airborne and impact sound.

The absorption of sound at surfaces is related to the porosity of the material. Generally, light materials with fibrous or open surfaces are good absorbers, reducing ambient noise levels and reverberation times, whereas smooth, hard surfaces are highly reflective to sound (Table 13.1). Sound absorption is measured on a 0–1 scale, with 1 representing total absorption of the sound.

Table 13.1 Typical sound absorption coefficients at 125, 500 and 2000 Hz for various building materials

Material	Absorption coefficient		
	125 Hz	500 Hz	2000 Hz
Concrete	0.02	0.02	0.05
Brickwork	0.05	0.02	0.05
Plastered solid wall	0.03	0.02	0.04
Glass 6 mm	0.1	0.04	0.02
Timber boarding, 19 mm over air space against solid backing	0.3	0.1	0.1
Wood wool slabs, 25 mm, on solid backing, unplastered	0.1	0.4	0.6
Fibreboard, 12 mm on solid backing	0.05	0.15	0.3
Fibreboard, 12 mm over 25mm air space	0.3	0.3	0.3
Mineral wool, 25 mm with 5% perforated hardboard over	0.1	0.85	0.35
Expanded polystyrene board, 25 mm over 50 mm airspace	0.1	0.55	0.1
Flexible polyurethane foam, 50 mm on solid backing	0.25	0.85	0.9

Table 13.2 Typical thermal conductivity values for various building materials

Material	Thermal conductivity (W/m K)
Aerogel blanket	0.013
Phenolic foam	0.018–0.031
Polyurethane foam (rigid)	0.019–0.023
Foil-faced polyisocyanurate foam	0.021
Polyisocyanurate foam	0.023–0.025
Sprayed polyurethane foam	0.024
Extruded polystyrene	0.028–0.036
Expanded PVC	0.030
Mineral wool	0.031–0.040
Glass wool	0.031–0.040
Expanded polystyrene	0.033–0.040
Cellulose (recycled paper)	0.035–0.040
Sheep's wool	0.037–0.039
Rigid foamed glass	0.037–0.055
Flax	0.038
Urea-formaldehyde foam	0.038
Wood fibre	0.038
Hemp wool	0.040
Corkboard	0.042
Coconut fibre boards	0.045
Perlite board	0.045–0.050
Fibre insulation board	0.050
Straw bales	0.060
Exfoliated vermiculite	0.062
Hempcrete	0.060–0.090
Thatch	0.072
Wood wool slabs	0.077
Medium-density fibreboard (MDF)	0.10
Foamed concrete (low density)	0.10
Lightweight to dense concrete	0.10–1.7
Compressed straw slabs	0.10
Softwood	0.13
Oriented strand board (OSB)	0.13
Hardboard	0.13
Particleboard/plywood	0.14
Gypsum plasterboard	0.19
Bituminous roofing sheet	0.19
Cement bonded particleboard	0.23
Unfired clay blocks	0.24
Calcium silicate boards	0.29
GRC – lightweight	0.21–0.5
GRC – standard density	0.5–1.0
Mastic asphalt	0.5
Calcium silicate brickwork	0.67–1.24
Clay brickwork	0.65–1.95
Glass – sheet	1.05

Notes:
Individual manufacturers' products may differ from these typical figures. Additional data is available in BS 5250: 2011, BS EN 12524: 2000 and BS EN ISO 10456: 2007.

FORMS OF INSULATION MATERIALS

Thermal and sound insulation materials may be categorised variously according to their appropriate uses in construction, their physical forms or their material origin. Many insulating materials are available in different physical forms, each with their appropriate use in building. Broadly, the key forms of materials may be divided into:

- structural insulation materials;
- rigid and semi-rigid sheets and slabs;
- loose fill, blanket materials and applied finishes;
- aluminium foil;
- vacuum- and gas-filled panels.

However, within this grouping it is clear that certain materials span two or three categories. Insulation materials are therefore categorised according to their composition, with descriptions of their various forms, typical uses in construction and, where appropriate, fire protection properties. Materials are initially divided into those of inorganic and organic origin respectively.

The broad range of non-combustible insulating materials is manufactured from ceramics and inorganic minerals including natural rock, glass, calcium silicate and cements. Some organic products are manufactured from natural cork or wood fibres but materials developed by the plastics industry predominate. In some cases these organic materials offer the higher thermal insulation properties but many are either inflammable or decompose within fire. Cellular plastics include open- and closed-cell materials. Generally, the closed-cell products are more rigid and have better thermal insulation properties and resistance to moisture, whereas the open-cell materials are more flexible and permeable. Aluminium foil is considered as a particular case, as its thermal insulation properties relate to the transmission of radiant rather than conducted heat. Typical thermal conductivity values are indicated in Table 13.2.

Inorganic insulation materials

FOAMED CONCRETE

The manufacture of foamed concrete is described in Chapter 3.

Foamed concrete with an air content in the range 30–80% is a fire- and frost-resistant material. Foamed concrete can be easily placed without the need for compaction but it does exhibit a higher drying shrinkage than dense concrete. It is suitable for insulating under floors and on flat roofs where it may be laid to a fall of up to 1 in 100. (Thermal conductivity ranges from 0.10 W/m K at a density of 400 kg/m^3 to 0.63 W/m K at a density of 1600 kg/m^3.)

LIGHTWEIGHT AGGREGATE CONCRETE

Lightweight concrete blocks and in situ concrete are discussed in Chapters 2 and 3, respectively. Lightweight concrete materials offer a range of insulating and load-bearing properties, starting from 0.10 W/m K at a crushing strength of 2.8 MPa. Resistance to airborne sound in masonry walls is closely related to the mass of the wall. However, any unfilled mortar joints which create air paths will allow significant leakage of sound. In cavity walls, again mass is significant, but, in addition, to reduce sound transmissions the two leaves should be physically isolated, with the exception of any necessary wall ties, to comply with the Building Regulations.

GYPSUM PLASTER

Plasterboard thermal linings will increase the thermal response in infrequently heated accommodation; the effect may be enhanced with metallised polyester-backed boards, which reduce radiant, as well as transmitted, heat loss. The addition of such linings for either new or upgrading existing buildings reduces the risk of thermal bridging at lintels, etc. (The thermal conductivity of gypsum plaster is typically 0.16 W/m K.)

Sound transmission through lightweight walls can be reduced by the use of two layers of differing thicknesses of gypsum plasterboard (e.g. 12.5 and 19 mm), as these resonate at different frequencies. The addition of an extra layer of plasterboard attached to existing ceilings with resilient fixings can reduce sound transmission from upper floors, particularly if an acoustic quilt is also incorporated.

Gypsum board thermal/acoustic composite and sandwich panels are described in pr EN 13950: 2011. Class 1 composites comprise gypsum board bonded to expanded polystyrene, extruded polystyrene, rigid polyurethane foam or phenolic foam. Class 2 composites incorporate mineral wool.

WOOD WOOL PRODUCTS

Wood wool products manufactured from wood fibres and cement (Chapter 4) are both fire and rot resistant. Wood wool boards, slabs (WW) and composite boards (WW-C) are described in BS EN 13168: 2012. With their combined load-bearing and insulating properties, wood wool slabs are suitable as a roof decking material, which may be exposed, painted or plastered to the exposed lower face. Wood wool slabs offer good sound absorption properties due to their open-textured surface, and this is largely unaffected by the application of sprayed emulsion paint. Acoustic insulation for a pre-screeded 50 mm slab is typically 30 dB. (The thermal conductivity of wood wool is typically 0.077 W/m K.)

MINERAL WOOL

Mineral wool (MW) is manufactured from volcanic rock (predominantly silica, with alumina and magnesium oxide), which is blended with coke and limestone and fused at 1500°C in a furnace. The melt runs onto a series of rotating wheels which spin the droplets into fibres; they are then coated with resin binder and water-repellent mineral oil. The fibres fall onto a conveyor belt, where the loose mat is compressed to the required thickness and density, then passed into an oven where the binder is cured; finally, the product is cut into rolls or slabs. New binders are bio-polymers rather than phenolic resins, thereby reducing the embodied energy. Mineral wool is non-combustible, water repellent and rot proof, and contains no CFCs or HCFCs.

Mineral wool is available in a range of forms dependent upon its degree of compression during manufacture and its required use:

- loose for blown cavity insulation;
- mats for insulating lofts, lightweight structures and within timber framed construction;
- batts (slabs) for complete cavity fill of new masonry;
- semi-rigid slabs for partial cavity fill of new masonry;
- rigid slabs for warm pitched roof and flat roof insulation;
- rigid resin-bonded slabs for floor insulation;
- weather-resistant boards for inverted roofing systems;
- dense pre-painted boards for exterior cladding;
- ceiling tiles.

The mats and board materials may be faced with aluminium foil to enhance their thermal properties. Roof slabs may be factory cut to falls or bitumen faced for torch-on bitumen membrane roofing systems. Floor units are coated with paper when they are to be directly screeded. A resilient floor can be constructed with floor units manufactured from mineral wool slabs, with the fibres orientated vertically rather than horizontally, bonded directly to tongued and grooved flooring-grade chipboard.

The thermal conductivity of mineral wool products for internal use varies typically between 0.031 and 0.039 W/m K at 10°C, although products for external use have higher conductivities ranging to 0.045 W/m K.

Mineral wool may be used effectively to attenuate transmitted sound. In lightweight construction, acoustic absorbent quilts are effective in reducing transmitted sound through separating walls when combined with double plasterboard surfaces and a wide airspace, as well as in traditional timber joist floors when combined with a resilient layer between joists and floor finish. Pelletised mineral wool may be used for *pugging* between floor joists to reduce sound transmission, and is particularly appropriate for upgrading acoustic insulation during refurbishment.

Mineral wool, due to its non-combustibility, is used for the manufacture of fire stops to prevent fire spread through voids and cavities, giving fire resistance ratings of between 30 and 120 minutes. Mineral wool slabs give typically between 60 and 240 minutes' fire protection to steel. Similar levels of protection may be achieved with sprayed-on mineral wool, which may then be coated with a decorative finish.

Ceiling tiles for suspended ceilings manufactured from mineral wool typically provide Class 1 Spread of Flame to BS 476 Part 7 (1997) and Class 0 to Part 6 (1989) on both their decorative and back surfaces. The thermal conductivity of mineral wool suspended ceiling tiles is typically within the range 0.052–0.057 W/m K. Sound attenuation of mineral wool ceiling tiles usually lies within the range 34–36 dB, but, depending upon the openness of the tile surface, the sound absorption coefficient may range from 0.1 for smooth tiles, through 0.5 for fissured finishes to 0.95 for open-cell tiles overlaid with 20 mm mineral wool.

The standard BS EN 13162: 2012 covers both mineral wool and glass wool boards, but in situ insulation products are specified in BS EN 14064: 2010 and products for industrial purposes are described in BS EN 14303: 2009. For in situ insulation, settlement should be limited to 1% in walls and not more than 10% in loft applications.

GLASS WOOL

Glass wool is made by the Crown process (Fig. 13.1), which is similar to the process used for mineral wool. A thick stream of glass flows from a furnace into a forehearth and by gravity into a rapidly rotating steel alloy dish, punctured by hundreds of fine holes around its perimeter. The centrifugal force expels the filaments which are further extended into fine fibres by a blast of hot air. The fibres are sprayed with a bonding agent and then sucked onto a conveyor to produce a mat of the appropriate thickness. This is cured in an oven to set the bonding agent, then finally cut, trimmed and packaged.

Glass wool is non-combustible, water repellent and rot proof, contains no CFCs or HCFCs and is available in a range of product forms:

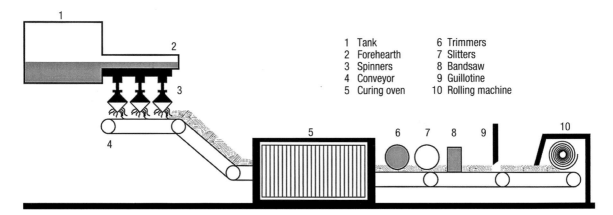

1 Tank	6 Trimmers
2 Forehearth	7 Slitters
3 Spinners	8 Bandsaw
4 Conveyor	9 Guillotine
5 Curing oven	10 Rolling machine

Fig. 13.1 Crown process for the manufacture of glass wool

- loose for blown cavity wall insulation;
- rolls, either unfaced or laminated between kraft paper and polythene, for roofs, within timber frame construction, internal walls and within floors;
- semi-rigid batts with water-repellent silicone for complete cavity fill of new masonry;
- rigid batts for partial cavity fill within new masonry;
- compression-resistant slabs for solid concrete or beam and slab floors;
- a laminate of rigid glass wool and plasterboard for dry linings;
- PVC-coated rigid panels for exposed factory roof linings.

(The thermal conductivity of glass wool products ranges typically between 0.031 and 0.040 W/m K at 10°C.)

The sound- and fire-resistant properties of glass wool are similar to those of mineral wool. Glass wool sound-deadening quilts, which have overlaps to seal between adjacent units, are used to reduce impact sound in concrete and timber floating floors. Standard quilts are appropriate for use in lightweight partitions and over suspended ceilings.

Resin-bonded glass wool treated with water repellent is used to manufacture some ceiling tiles which meet the Class 0 fire-spread requirements of the Building Regulations (BS 476: Parts 6 and 7) and also offer sound absorption to reduce reverberant noise levels.

CELLULAR OR FOAMED GLASS BLOCKS

Cellular or foamed glass (CG) is manufactured from a mixture of crushed glass and fine carbon powder which, upon heating to 1000°C, causes the carbon to oxidise, creating bubbles within the molten glass. The glass is annealed, cooled and finally cut to size. The black material is durable, non-combustible, easily worked and has a high compressive strength. It is water resistant due to its closed-cell structure, impervious to water vapour and contains no CFCs. Cellular glass products are described in BS EN 13167: 2012.

Cellular glass slabs are appropriate for roof insulation, including green roofs and roof-top car parks owing to their high compressive strength. Tapered slabs are available to create roof falls. The slabs are usually bonded in hot bitumen to concrete screeds, profile metal decking or reinforced bitumen membrane on timber roofing. Foamed glass is suitable for floor insulation under the screed, and may be used internally, externally or within the cavity of external walls. Externally, it may be rendered or tile hung and internally finished with plasterboard or expanded metal and conventional plaster. (The thermal conductivity of cellular glass is within the range 0.037–0.055 W/m K at 10°C, depending upon the grade.)

EXFOLIATED VERMICULITE

Exfoliated vermiculite (EV) is manufactured by heating the natural micaceous mineral. The material, containing up to 90% air by volume, is used as a loose fill for loft insulation and within a cementitious spray produces a hard fire-protection coating for exposed structural steelwork. Where thicknesses over 30 mm are required, application should be in two coats. The product has a textured surface finish which may be exposed internally or painted in external applications. Depending upon the thickness of application and the A/V ratio (section factor) between exposed surface area and steel volume (Chapter 5), up to 240 minutes' fire protection may be obtained. Vermiculite is used for certain demountable fire stop seals where services penetrate through fire compartment walls. The standards BS EN 14317 Parts 1 and 2 describe coated and hydrophobic vermiculite as well as premixed vermiculite with binders. (The thermal conductivity of exfoliated vermiculite is 0.062 W/m K. Within lightweight aggregate concrete a thermal conductivity of typically 0.11 W/m K may be achieved.)

EXPANDED PERLITE

Expanded perlite is manufactured by heating natural volcanic rock minerals. It is used for loose and bonded in situ insulation for roofs, ceilings, walls and floors as well as predominantly as preformed boards which may be fibre reinforced. Expanded perlite boards (EPB) are described in BS EN 13169: 2012 and the loose fill material in BS EN 14316 Parts 1 and 2. (The thermal conductivity of expanded perlite boards is 0.05 W/m K.)

CALCIUM SILICATE

Calcium silicate (CS), which is described in Chapter 12, has the advantage of good impact resistance and is very durable. Various wallboards are manufactured with calcium silicate boards laminated to extruded polystyrene or reinforced with fibres. The standard

BS EN 14306: 2009 specifies calcium silicate products. (Calcium silicate typically has a thermal conductivity of 0.29 W/m K.)

GLASS AND MULTIPLE GLAZING

The thermal and sound insulation effects of double and triple glazing and the use of low-emissivity glass are described in Chapter 7.

AEROGEL

Aerogels are extremely lightweight, hydrophobic amorphous silica materials with densities as low as 2 kg/m^3 (ρ_{air} = 1.2 kg/m^3). They are manufactured by solvent evaporation from silica gel under reduced pressure. Aerogels are highly porous with typically 95–97% and even 99.8% air space, but significantly the pore size of 20 nm is so small that it is less than the mean free path of nitrogen and oxygen in the air. This prevents the air particles from moving and colliding with each other, which would normally give rise to gas phase heat conduction. With only 3–5% solid material, heat conduction in the solid phase is very limited. When used to fully fill the cavity in glazing units, 0.5–4 mm aerogel granules prevent the movement of air, thus reducing heat transfer by convection currents. Limited heat transfer can therefore only occur across the glazing unit by radiation.

Light transmission through aerogel is approximately 80% per 10 mm thickness, giving a diffuse light and eliminating the transmission of ultraviolet light. Airborne sound transmission is reduced, particularly for lower frequencies of less than 500 Hz. The material is hydrophobic and therefore resistant to mould growth.

Polycarbonate glazing units filled with aerogel are available as 10 and 16 mm panels, over a range of sizes to fit profiled glass trough sections (Chapter 7) or as translucent roof lights or wall panels (e.g. 1220 × 3660 mm). One proprietary cladding and roofing system uses aluminium framing to support a sandwich panel of fibreglass sheets separated by aerogel. The panels, 1200 × 3600 mm or 1500 × 3000 mm (maxima), have U-values of 0.28W/m^2 K. (The thermal conductivity of aerogel blanket is typically 0.013 W/m K.)

Recently aerographite and graphene aerogel have become the lightest manufactured solid materials at densities of 0.18 and 0.16 kg/m^3 respectively.

Organic insulation materials

The use of straw bales is described in Chapter 17 with other recycled products.

CORK PRODUCTS

Cork is harvested from the cork oak (*Quercus suber L.*) on a nine-year (or more) cycle and is therefore considered to be an environmentally friendly material (Fig. 13.2). For conversion into boards, typically used for roof insulation, cork granules are expanded, then formed under heat and pressure into blocks using the natural resin within the cork. The blocks are trimmed to standard thicknesses or to a taper to produce falls for flat roofs (Fig. 13.3). For increased thermal insulation properties, the cork may be bonded to closed-cell polyurethane or polyisocyanurate foam. In this case the laminate should be laid with the cork uppermost. Cork products are unaffected by the application of hot bitumen in flat roofing systems. Insulation cork board (ICB) is described in BS EN 13170: 2012. (The thermal conductivity of corkboard is typically 0.042 W/m K but some denser products have higher thermal conductivity.)

SHEEP'S WOOL

Sheep's wool is a very efficient renewable resource insulation material, with a low conductivity that compares favourably to other fibrous insulants. It is available in grey batts ranging in thickness from 50 and 75 to 100 mm thick. Wool is a hygroscopic material; that is, it reversibly absorbs and releases water vapour, and this effect is advantageous when it is used for thermal insulation. When the building temperature rises, wool releases its moisture causing a cooling effect in the fibre and thus a reduced flow of heat into the building. In winter the absorption of moisture warms the material. This evolution of heat helps prevent interstitial condensation in construction cavities by maintaining the temperature of the fibres above the dew point, and also effectively reduces the heat loss from the building.

Wool is safe to handle, only requiring gloves and a dust mask as minimal protection. It causes no irritation except in the rare cases of people with a specific wool allergy. Wool batts, which contain 85% wool and 15% polyester to maintain their form, can easily be cut with a sharp knife or torn to size. Wool is potentially susceptible to rodents which may use it as a nesting

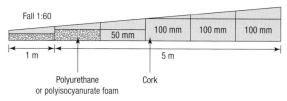

Fig. 13.3 Cork insulation to falls for flat roofs – additional insulation dependent on the structure

material if it is accessible, but the batts are treated with an insecticide to prevent moth or beetle attack and with an inorganic fire retardant.

Wool batt insulation is suitable for ventilated loft applications between joists or rafters and for timber-frame construction. It should be installed with a vapour-permeable breather membrane on the cold side, and kept clear of any metal chimneys or flues. Wool also acts as an effective acoustic insulating material. (The thermal conductivity of wool batts is 0.039 W/m K.)

Sheep's wool has also been used experimentally as loose fill insulation for lofts, sloping ceilings, timber-frame walls and timber floors. Natural wool from sheep that have not been dipped is washed several times to remove the natural oil lanolin, then opened out to the required density. It is sprayed with borax as a fire retardant and insect repellent. Supplied loose as hanks, wool is only suitable for locations where it will not get wet, which would cause it to sag, thus reducing its thermal efficiency. Wool insulation is a renewable source with low embodied energy, but it is currently more expensive than the standard mineral wool alternative. (The thermal conductivity of loose wool is 0.037 W/m K.)

CELLULOSE INSULATION

Cellulose insulation is manufactured from shredded recycled paper and other organic waste. It is treated with borax for flammability and smouldering resistance; this also makes it unattractive to vermin, and resistant to insects, fungus and dry rot. Unlike mineral fibre and glass fibre insulation it does not cause skin irritation during installation. Recycled cellulose has a low embodied energy compared to mineral and glass fibre insulation, and when removed from a building it may be recycled again or disposed of safely without creating toxic waste. (Treatment with the inorganic salt, borax, ensures that cellulose insulation conforms

Fig. 13.2 Production of cork. *Photographs*: Arthur Lyons

to BS 5803 Part 4: 1985 – Fire Test Class 1 and Smoulder Test Class B2.)

Cellulose insulation may be used directly from bags for internal floors and also lofts where the required eaves ventilation gap must be maintained. For other cavities, including sloping roof voids, the material is dry injected under pressure, completely filling all spaces to prevent air circulation. In breathing walls, cellulose insulation is filled inside a breathing membrane, which allows the passage of water vapour through to the outer leaf of the construction. Cellulose may be damp-sprayed in between wall studs before the wall is closed. Cellulose is a hygroscopic material, which under conditions of high humidity absorbs water vapour and then releases it again under dry conditions. Cellulose is an effective absorber of airborne sound.

Loose-fill cellulose insulation (LFCI) is detailed in the standards BS EN 15101 Parts 1 and 2: 2013. Classification with respect to settlement, resistance to mould and fire (BS EN 13501–1: 2007, BS EN 13823: 2010 and BS EN ISO 11925–2: 2010) is required. (The thermal conductivity of cellulose insulation is 0.035 W/m K in horizontal applications and 0.038–0.040 W/m K in walls.)

RECYCLED PLASTIC

Loft insulation material is manufactured from re-cycled plastic bottles which are converted into a non-woven, clean material (Fig. 13.4). The product contains approximately 85–90% of recycled material, and unlike

Fig. 13.4 Recycled plastic insulation

the equivalent glass fibre or mineral wool, has no loose fibres which may cause irritation. The material is available in rolls of thicknesses from 100–200 mm and has a thermal conductivity of 0.043 W/m K.

FLAX, HEMP AND COCONUT FIBRE

As the demand for sustainable insulating materials increases, products derived from renewable flax, hemp and coconut fibres are becoming available. Flax insulation is suitable for ventilated or breathing con-structions. The batts may be used in ceilings and walls, and rolls in lofts, suspended floors and walls. Flax is treated with borax for fire and insect resistance and bonded with potato starch, giving a moisture-absorbing, non-toxic product with good thermal and acoustic insulation properties. (The thermal conduct-ivity of flax is 0.038 W/m K.)

Chopped hemp fibres, treated with borax for fire resistance, are used to produce insulation batts, and also as loose fill for floors and roofs. Hemp is also blended with either wood fibres or waste cotton and 15% polyester fibres to form insulation batts. Hemp, a very tough material, is used in the manufacture of certain particleboards in Germany and generally for paper production. *Hempcrete* is described in Chapter 3. (The thermal conductivities of hemp insulation products are in the range 0.038–0.040 W/m K.)

Coconut fibre thermal and acoustic insulation boards have the advantage of natural rot resistance. They are available in a range of thicknesses from 10–25 mm, and typical uses include ceiling and floor insulation including under-screed applications. (The thermal conductivity of coconut fibre is 0.045 W/m K. The sound reduction for a typical 18 mm under-screed application is 26 dB.)

WOOD FIBRE PRODUCTS

Wood fibre (WF) insulation to BS EN 13171: 2012 is produced to a range of forms including mat, batt, felt, roll and board. The manufacture of softboard, which is a low-density wood fibre building board, is described in Chapter 4. Standard grades of insulating board should only be used in situations where they are not in contact with moisture, or at risk from the effects of condensation. Insulating board is used for wall linings and may be backed with aluminium foil for increased thermal insulation.

Insulating board may be impregnated with inorganic fire retardants to give a Class 1 Surface Spread of Flame to BS 476 Part 7 or finished with plaster-board to give a smooth Class 0 fire rated surface. The Euroclass fire performance rating under the conditions specified in BS EN 13986: 2004 for 9 mm untreated high-density medium board of 600 kg/m^3 is Class D-s2, d0 for non-floor use. For untreated low-density medium board of 400 kg/m^3 the equivalent rating is Class E, pass for non-floor use, and for 9 mm untreated softboard of 250 kg/m^3 the rating is Class E, pass for non-floor use. Manufacturers are required to specify fire resistance to BS EN 13501–1: 2007.

Exposed insulating board has good sound-absorbing properties due to its surface characteristics. Standard 12 mm lining softboard has a noise-reduction coefficient of 0.42, although this is increased to 0.60 for the 24 mm board.

Bitumen-impregnated insulating board, with its enhanced water-resistant properties, is used as a thermal insulation layer on concrete floors. The concrete floor slab is overlaid with polythene, followed by bitumen-impregnated insulation board and the required floor finish such as particleboard. In the upgrading of existing suspended timber floors, a loosely laid layer of bitumen-impregnated insulating board under a new floor finish can typically reduce both impact and airborne sound transmission by 10 dB. Bitumen-impregnated insulating board is frequently used in flat-roofing systems as a heat protective layer to polyurethane, polystyrene or phenolic foams prior to the application of the hot bitumen waterproof membrane. It is also used for sarking in pitched roofs. (The thermal conductivity of insulating board is typically 0.050 W/m K.)

Rigid and semi-rigid wood fibre insulation panels with thicknesses ranging from 20–240 mm are used for wall, roof and intermediate floor applications. Boards are available with either a butt or tongue and groove edge profile. Fire classification to BS EN 13501–1: 2007 is usually Class E. Thermal conductivity is typically within the range 0.036–0.042 W/m K. The equivalent blown material may be used to fill void spaces giving a thermal conductivity of 0.038–0.043 W/m K, depending upon the density of the fill.

EXPANDED POLYSTYRENE

Expanded polystyrene (EPS) is a combustible material, which, in fire, produces large quantities of noxious black smoke, although Type A, with a flame-retardant additive, is not easily ignitable. Expanded polystyrene, a closed-cell product, is unaffected by water, dilute acids and alkalis but is readily dissolved by most organic solvents. It is rot and vermin proof. Expanded polystyrene is described in BS EN 13163: 2012.

Polystyrene beads

Expanded polystyrene beads are used as loose fill for cavity insulation. To prevent subsequent slippage and escape through voids, one system bonds the polystyrene beads by spraying them with atomised PVA adhesive during the injection process, although other processes leave the material loose. Walls up to 12 m in height can be insulated by this type of system. Polystyrene bead insulation should not be used where electrical wiring is present in the cavity, as the polystyrene gradually leaches the plasticiser out from plastic cables, causing their embrittlement, which could lead to problems later if the cables are subsequently moved. Polystyrene bead aggregate cement is used to form an insulating sandwich core in concrete cladding panel systems.

Expanded polystyrene boards

Expanded polystyrene rigid lightweight boards are used for thermal insulation and four standard grades are available (Table 13.3). The standard material is classified as Euroclass F in relation to fire, but certain flame-retardant modified boards are classified as Euroclass E. (Grades range from A1 and A2 through to F.) Non-load-bearing expanded polystyrene for impact sound insulation properties is designated type EPS SD to BS EN 13163: 2012.

The boards, which are manufactured by fusing together pre-foamed beads under heat and pressure, can easily be cut, sawn or melted with a hot wire. Polystyrene boards provide thermal insulation for walls, roofs and floors. In addition, polystyrene may be cast into reinforced concrete, from which it is easily removed to create voids for fixings.

In cavity wall insulation, a 50 mm cavity may be retained to prevent the risk of water penetration, with proprietary wall ties fixing the boards against the inner leaf. Alternatively, with a fully filled cavity system, the boards may be slightly moulded on the outer surface to shed any water back onto the inside of the external masonry leaf. Interlocking joints prevent cold bridging, air leakage and water penetration at the board joints. In upgrading existing walls (Fig. 13.5), external expanded polystyrene insulation should be protected by suitably supported rendering, tile hanging or brick slips. Clay, concrete or acrylic brick slips may be

Table 13.3 Standard grades of expanded polystyrene (BS 3837–1: 2004 and BS EN 13163: 2012)

Grade BS 3837	Type BS EN 13163	Description	Typical density (kg/m³)	Thermal conductivity (W/m K)
SD	EPS 70	Standard duty	15	0.038
HD	EPS 100	High duty	20	0.036
EHD	EPS 150	Extra-high duty	25	0.035
UHD	EPS 200	Ultra-high duty	30	0.034

Notes:
– BS EN 13163: 2012 lists the range of types from EPS 30 to EPS 500 (Compressive strengths from 30 to 500 kPa respectively).
– BS EN 13163: 2012 refers to four types of expanded polystyrene:
  EPS – load-bearing applications
  EPS S – non-load-bearing applications
  EPS SD – non-load-bearing applications with acoustic properties
  EPS T – floating floor applications.

mounted on steel panels or mesh and may be pointed or mortarless. Weatherproof rendering should tolerate thermal and moisture movement and be frost resistant to minimise maintenance. For internal wall insulation, expanded polystyrene may be used in conjunction with 12.5 mm plasterboard either separately or as a laminate. Expanded polystyrene (Type EPS T) is used to give thermal insulation in ground floors. It may be laid below or above the oversite slab; in the latter case, it may be screeded or finished with chipboard. Composite floor panels manufactured from expanded polystyrene and oriented strand board are suitable for beam and block floors, while proprietary systems offer thermal insulation to prestressed concrete beam and reinforced concrete screed floors. Expanded polystyrene boards reduce impact and airborne sound transmission through intermediate floors.

Expanded polystyrene is suitable for thermal insulation in flat and pitched roofs. For flat roofs it may be cut to falls. Where hot bitumen products are to be applied, the expanded polystyrene boards must be protected by an appropriate layer of bitumen-impregnated fibreboard, perlite board or corkboard. In metal deck applications the insulating layer may be above or below the purlins, whereas in traditional pitched roofs expanded polystyrene panels are normally installed over the rafters. Expanded polystyrene, although a closed-cell material, acts as a sound absorber, provided that it is installed with an air gap between it and the backing surface. It particularly absorbs sound at low frequencies and may be used in floors and ceilings. It is, however, less effective than the open-cell materials, such as flexible polyurethane foam.

Expanded polystyrene incorporating graphite is grey in colour and has enhanced thermal insulation properties, as it absorbs infrared radiation and reflects heat. (The thermal conductivity of expanded polystyrene is in the range 0.033–0.040 W/m K depending upon the grade. The graphite product has a thermal conductivity of 0.033 W/m K.)

EXTRUDED POLYSTYRENE

Extruded polystyrene (XPS) is manufactured by the extrusion of a heated mixture of polystyrene and expansion agent through a die. The continuous ribbon of expanded material is cooled and trimmed. It is slightly denser and therefore slightly stronger in compression than expanded polystyrene but has a lower thermal conductivity. It has a closed-cell structure with very low water-absorption and vapour-transmission properties. It is produced either with or without a polystyrene skin. Extruded polystyrene is described in BS EN 13164: 2012. It is available with densities ranging from 20–40 kg/m³. Extruded polystyrene is widely used for cavity wall and pitched roof insulation. Because of its high resistance to water absorption, extruded polystyrene may be used for floor insulation below the concrete slab and on inverted roofs where its resistance to mechanical damage from foot traffic is advantageous. Extruded polystyrene is also available laminated to tongued and grooved moisture-resistant flooring grade particleboard for direct application to concrete floor slabs, and laminated to plasterboard as a wallboard. (The thermal conductivity of extruded polystyrene is typically 0.033–0.035 W/m K.)

Fig. 13.5 Retrofit external polystyrene insulation

EXPANDED PVC

Plasticised PVC open-, partially open- and closed-cell foams are manufactured as flexible or rigid products within the density range of 24–72 kg/m³. The rigid closed-cell products provide low water permeability and are self-extinguishing in fire. Expanded PVC boards are used in sandwich panels and for wall linings. The low-density open-cell material has particularly good acoustic absorbency and may be used to reduce sound transmission through unbridged cavities and floating floors. (The thermal conductivity of expanded PVC is typically 0.030 W/m K.)

POLYISOCYANURATE FOAM

Polyisocyanurate foam (PIR) is usually blown with pentane. It is used as a roof insulation material since it is more heat resistant than other organic insulation foams, which cannot be directly hot-bitumen bonded. Polyisocyanurate is also appropriate for use in wall and floor insulation. PIR is combustible (BS 476 Part 4) with a Class 1 Surface Spread of Flame (BS 476 Part 7) but is more fire resistant than polyurethane foam and can be treated to achieve Class 0 rating. Polyisocyanurate foam boards have a fire classification of Class F, but the Building Regulations permit their use in wall cavities within masonry leaves of at least 75 mm thickness (BS 4841: 2006). Polyisocyanurate tends to be rather friable and brittle. Certain proprietary systems for insulated cavity closers use PVC-U-coated polyisocyanurate insulation. Such systems offer a damp-proof barrier and can assist in the elimination of cold bridging, which sometimes causes condensation and mould growth around door and window openings. In situ formed polyisocyanurate foam, described in BS EN 14315: 2013, is delivered as a two-component sprayed mix which subsequently expands to fill the void space. (The thermal conductivity of polyisocyanurate foam is usually in the range 0.023–0.025 W/m K.)

POLYURETHANE FOAM

Rigid polyurethane (PUR) is a closed-cell foam manufactured using pentane or carbon dioxide. Pentane remains trapped within the closed cells, enhancing the thermal performance, but carbon dioxide diffuses out. Certain polyurethanes are modified with polyisocyanurates. Polyurethane foam products are defined in BS EN 13165: 2012.

Rigid polyurethane is a combustible material producing copious noxious fumes and smoke in fire, although a flame-resistant material is available. Polyurethane boards have a fire classification of Class F, but the Building Regulations permit their use in wall cavities within masonry leaves of at least 75 mm thickness (BS 4841: 2006). Polyurethane is used to enhance the thermal insulation properties of concrete blocks either by filling the void spaces in hollow blocks or by direct bonding onto the cavity face. Roofboards, in certain systems pre-bonded to bitumen roofing sheet, are suitable for mastic asphalt and reinforced bitumen membrane roofing systems. Owing to the temperature stability of polyurethane no additional protection from the effects of hot bitumen application is required; the durability of the material also makes it suitable for use in inverted roofs. Laminates with foil or kraft paper are available. Factory-manufactured double-layer profiled-metal sheeting units are frequently filled with rigid polyurethane foam due to its good adhesive and thermal insulation properties. Polyurethane laminated to plasterboard is used as a wallboard. When injected as a premixed two-component system into cavity walls polyurethane adheres well to the masonry, foaming and expanding in situ to completely fill the void space (BS EN 14315: 2013). It has been used in situations where the cavity ties have suffered serious corrosion, and where additional bonding between the two leaves of masonry is required, but it is not now widely used as a cavity insulation material. However, two-component polyurethane foam spray is effective at closing gaps around service voids, eliminating cold bridges, under-tile insulation and for filling inaccessible locations.

Flexible polyurethane foam is an open-cell material offering good noise absorption properties. It is therefore used in unbridged timber frame partitions, floating floors and duct linings to reduce noise transmission. Polyurethane foams are resistant to fungal growth, aqueous solutions and oils, but not to organic solvents. (The thermal conductivity of rigid polyurethane foam is usually in the range 0.019–0.023 W/m K at a nominal density of 32 kg/m³. Flexible polyurethane foam typically has a thermal conductivity of 0.048 W/m K.)

UREA-FORMALDEHYDE FOAM

Urea-formaldehyde foam (UF) was used extensively in the 1980s for cavity wall insulation, but it can shrink after installation, creating fissures which link the outer and inner leaves. Occasionally, in conditions of high exposure, this had led to rain water penetration. After installation the urea-formaldehyde foam emits formaldehyde fumes, which have in certain cases entered buildings, causing occupants to suffer from eye and nose irritation. The problem normally arises only if the inner leaf is permeable and a cavity greater than 100 mm is being filled. Recent advances claim to have reduced formaldehyde emissions but all installations must be undertaken to the stringent British Standard BS 5618: 1985. The Health & Safety Executive advises against the use of urea formaldehyde where the inner leaf of the cavity wall is porous or has unsealed connections with the interior. (The thermal conductivity of urea-formaldehyde foam is typically 0.038 W/m K.)

PHENOLIC FOAM

Phenolic foams (PF), which have very low thermal conductivities, are used as alternatives to rigid polyurethane and polyisocyanurate foams, where a self-extinguishing, low smoke emission material is required. Phenolic foams are produced with densities in the range 35–200 kg/m³. The closed-cell material is usually expanded with pentane. Phenolic foam is described in BS EN 13166: 2012. Wallboard laminates with plasterboard offer good thermal insulation properties due to the very low thermal conductivity of phenolic foam, compared to polyurethane or extruded polystyrene. Phenolic foams are stable up to a continuous temperature of 120°C. (The thermal conductivity of phenolic foam in the density range 35–60 kg/m³ is typically 0.021 W/m K, although the open-cell material has a thermal conductivity of 0.031 W/m K.)

Aluminium foil

Aluminium foil is frequently used as an insulation material in conjunction with organic foam or insulating gypsum products. It acts by a combination of two physical effects. First, it reflects back incident heat due to its highly reflecting surface. Second, owing to its low emissivity, the re-radiation of any heat that is absorbed is reduced. Thin aluminium reflective foil insulation can be inserted between studs, joists or rafters, leaving a 25 mm air gap on either side. In addition to insulation it acts as an air infiltration and vapour barrier.

THERMO-REFLECTIVE INSULATION PRODUCTS

Proprietary quilt systems incorporating multi-layers of aluminium foil, fibrous materials and cellular plastics act as insulation by reducing conduction, convection and radiation (Fig. 13.6). A range of these thermo-reflective insulation products is manufactured using different combinations of thin plastic foam, plastic bubble sheet, and non-woven fibrous wadding with plain and reinforced aluminium foil. For example, the combination of a nine-layer and 19-layer of thermo-reflective insulation within standard timber studding produces a typical wall U-value of 0.22 W/m^2 K. The similar combination in conjunction with 25 mm insulated plasterboard gives a typical roof U-value of 0.18 W/m^2 K. The standard BS EN 16012: 2012 describes four types of product ranging from aluminium-faced rigid foam to flexible multi-layer bubble-foil systems. (The thermal conductivity of foil-faced foam is typically 0.020 W/m K.)

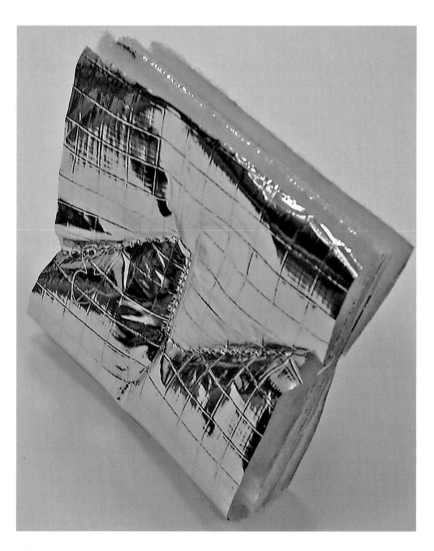

Fig. 13.6 Multi-layer aluminium foil insulation system

Panel systems

Panel systems based on the insulating effect of vacuum or inert gas fill are under development, but not yet commercially viable.

VACUUM INSULATION PANELS

Vacuum insulation panels (VIPs) are evacuated materials such as polyurethane foam or porous silica powder, held within an aluminium-based multi-layered envelope. Desiccants are usually incorporated to maintain the vacuum. Care is required during transportation and installation to avoid damage to the vacuum.

GAS-FILLED PANELS

Gas-filled panels are similar to vacuum insulation panels but are filled with argon, krypton or xenon. The envelope therefore does not need to resist atmospheric pressure, but the insulating effect is less than with a vacuum. The incorporation of low-emissivity reflective layers reduces heat transfer by radiation.

References

FURTHER READING

Anderson, J. and Allbury, K. 2011: *Environmental impact of insulation*. Garston: IHS BRE Press.

Bynum, R. and Rubino, D. 2000: *Insulation handbook*. Maidenhead: McGraw.

Colwell, S. and Baker, T. 2013: *Fire performance of external thermal insulation of walls of multi-storey buildings*. Garston: IHS BRE Press.

Communities and Local Government. 2007: *Accredited construction details*. Wetherby: Communities and Local Government Publications.

Doran, D. and Anderson, J. 2011: *Environmental impact of vertical cladding*. Garston: IHS BRE Press.

Pearson C. 2009: *The complete guide to external wall insulation*. York: Wellgarth Publishing.

Peters, R.J., Smith, B.J. and Hollins, M. 2011: *Acoustics and noise control*. Harlow: Prentice Hall.

Pfundstein, M., Gellert, R., Spitzner, M. and Rudolphi, A. 2008: *Insulating materials. Principles materials and applications*. Basel: Birkhäuser.

Robust Details. 2013: *Robust details handbook*, 3rd edn. Milton Keynes: Robust Details Ltd.

TRADA. 2008: *England and Wales Building Regulations Part E – Resistance to the passage of sound*. High Wycombe: TRADA Technology Ltd.

Zero Carbon Hub. 2012: *Fabric energy efficiency for Part L 2013. Worked examples and fabric specifications*. Milton Keynes: NHBC.

Zero Carbon Hub. 2012: *Fabric energy efficiency for Part L 2013. Classification methodology for different dwelling types*. Milton Keynes: NHBC.

STANDARDS

BS 476
    Fire tests on building materials and structures:

Part 4: 1970	Non-combustibility test for materials.
Part 6: 1989	Method of test for fire propagation of products.
Part 7: 1997	Method of test to determine the classification of the surface spread of flame of products.

BS ISO 633: 2007
    Cork. Vocabulary.

BS ISO 2219: 2010
    Thermal insulation products for building. Factory made products of expanded cork (ICB). Specification.

BS 3379: 2005
    Combustion modified flexible polyurethane cellular materials for loadbearing applications. Specification.

BS 3533: 1981
    Glossary of thermal insulation terms.

BS 3837
    Expanded polystyrene boards:

Part 1: 2004	Boards and blocks manufactured from expandable beads.

BS 4023: 1975
    Flexible cellular PVC sheeting.

BS 4841
    Rigid polyisocyanurate (PIR) and polyurethane (PUR) for building end-use applications:

Part 1: 2006	Specification for laminated insulation boards with auto-adhesively or separately bonded facings.
Part 2: 2006	Specification for laminated boards for use as thermal

insulation for internal wall linings and ceilings.

Part 3: 2006 Specification for laminated boards for use as roofboard thermal insulation under built up bituminous roofing membranes.

Part 4: 2006 Specification for laminated boards for use as roofboard thermal insulation under non-bituminous single-ply roofing membranes.

Part 5: 2006 Specification for laminated boards for use as thermal insulation boards for pitched roofs.

Part 6: 2006 Specification for laminated boards for use as thermal insulation for floors.

BS 5241
Rigid polyurethane (PUR) and polyisocyanurate (PIR) foam when dispensed or sprayed on a construction site:

Part 1: 1994 Specification for sprayed foam thermal insulation applied externally.

Part 2: 1991 Specification for dispensed foam for thermal insulation or buoyancy applications.

BS 5250: 2011
Code of practice for control of condensation in buildings.

BS 5422: 2009
Method for specifying thermal insulating materials for pipes, tanks, vessels ductwork and equipment.

BS 5608: 1993
Specification for preformed rigid polyurethane (PUR) and polyisocyanurate (PIR) foams for thermal insulation of pipework and equipment.

BS 5617: 1985
Specification for urea-formaldehyde (UF) foam systems suitable for thermal insulation of cavity walls with masonry or concrete inner and outer leaves.

BS 5618: 1985
Code of practice for thermal insulation of cavity walls by filling with urea-formaldehyde (UF) foam systems.

BS 5803
Thermal insulation for use in pitched roof spaces in dwellings:

Part 2: 1985 Specification for man-made mineral fibre thermal insulation in pelleted or granular form for application by blowing.

Part 3: 1985 Specification for cellulose fibre thermal insulation for application by blowing.

Part 4: 1985 Methods for determining flammability and resistance to smouldering.

Part 5: 1985 Specifications for installations of man-made mineral fibre and cellulose fibre insulation.

BS 5821
Methods for rating the sound insulation in buildings and of building elements:

Part 3: 1984 Airborne sound insulation of façade elements and façades.

BS 6203: 2003
Guide to the fire characteristics and fire performance of expanded polystyrene materials (EPS and XPS) used in building applications.

BS 7021: 1989
Code of practice for thermal insulation of roofs externally by means of sprayed rigid polyurethane (PUR) or polyisocyanurate (PIR) foam.

BS 7456: 1991
Code of practice for stabilization and thermal insulation of cavity walls by filling with polyurethane (PUR) foam systems.

BS 7457: 1994
Specification for polyurethane (PUR) foam systems suitable for stabilization and thermal insulation of cavity walls with masonry or concrete inner and outer leaves.

BS 8233: 2014
Guidance on sound insulation and noise reduction for buildings.

BS ISO 12575–1: 2012
Thermal insulation. Exterior insulating systems for foundations. Material specification.

BS ISO 16346: 2013
Energy performance of buildings. Assessment of overall energy performance.

BS ISO 16818: 2008
Building environment design. Energy efficiency. Terminology.

BS EN ISO 717
   Acoustics. Rating of sound insulation in buildings
   and of building elements:
      Part 1: 1997   Airborne sound insulation.
      Part 2: 1997   Impact sound insulation.
BS EN ISO 5999: 2007
   Flexible cellular polymeric materials. Polyurethane
   foam for load-bearing applications excluding
   carpet underlay. Specification.
BS EN ISO 6946: 2007
   Building components and building elements.
   Thermal resistance and thermal transmittance.
   Calculation method.
BS EN ISO 7345: 1996
   Thermal Insulation. Physical quantities and
   definitions.
BS EN ISO 8990: 1996
   Thermal insulation. Determination of steady-state
   thermal transmission properties.
BS EN ISO 9229: 2007
   Thermal insulation. Vocabulary.
BS EN ISO 10077
   Thermal performance of windows, doors and
   shutters. Calculation of transmittance:
      Part 1: 2006   General.
      Part 2: 2012   Numerical method for frames.
BS EN ISO 10456: 2007
   Building materials and products. Hygrothermal
   properties. Tabulated design values.
BS EN ISO 10848
   Acoustics. Laboratory measurement of the flank-
   ing transmission of airborne and impact sound:
      Part 1: 2006   Frame document.
      Part 2: 2006   Application to light elements.
                      Junction small influence.
      Part 3: 2006   Application to light elements.
                      Junction substantial influence.
      Part 4: 2010   Applications to junctions with
                      at least one heavy element.
BS EN ISO 11654: 1997
   Acoustics. Sound absorbers for use in buildings.
BS EN ISO 11925–2: 2010
   Reaction to fire tests. Ignitability of products
   subjected to direct impingement of flame. Single-
   flame source test.
BS EN ISO 12241: 2008
   Thermal insulation for building equipment and
   industrial installations. Calculation rules.
BS EN 12354
   Building acoustics. Estimation of acoustic
   performance of buildings from the performance
   of elements:

      Part 1: 2000   Airborne sound insulation
                      between rooms.
      Part 2: 2000   Impact sound insulation
                      between rooms.
      Part 3: 2000   Airborne sound insulation
                      against outdoor sound.
      Part 4: 2000   Transmission of indoor sound
                      to the outside.
      Part 5: 2009   Sound levels due to the service
                      equipment.
      Part 6: 2003   Sound absorption in enclosed
                      spaces.
BS EN 12524: 2000
   Building materials and products. Hygrothermal
   properties. Tabulated design values.
BS EN ISO 12567
   Thermal performance of windows and doors:
      Part 1: 2010   Complete windows and doors.
      Part 2: 2005   Roof windows and other
                      projecting windows.
BS EN ISO 12631: 2012
   Thermal performance of curtain walling.
BS EN ISO 12655: 2013
   Energy performance of buildings. Presentation of
   measured energy use of buildings.
BS EN 12758: 2011
   Glass in building. Glazing and airborne sound
   insulation.
BS EN ISO 12999–1: 2014
   Acoustics. Determination and application of
   measurement uncertainties in building acoustics.
   Sound insulation.
BS EN 13162: 2012
   Thermal insulation products for building. Factory
   made mineral wool (MW) products. Specification.
BS EN 13163: 2012
   Thermal insulation products for building.
   Factory made products of expanded polystyrene
   (EPS). Specification.
BS EN 13164: 2012
   Thermal insulation products for building.
   Factory made products of extruded polystyrene
   foam (XPS). Specification.
BS EN 13165: 2012
   Thermal insulation products for building.
   Factory made rigid polyurethane foam (PUR)
   products. Specification.
BS EN 13166: 2012
   Thermal insulation products for building.
   Factory made products of phenolic foam (PF).
   Specification.

BS EN 13167: 2012
Thermal insulation products for building.
Factory made cellular glass (CG) products.
Specification.

BS EN 13168: 2012
Thermal insulation products for building.
Factory made wood wool (WW) products.
Specification.

BS EN 13169: 2012
Thermal insulation products for building.
Factory made products of expanded perlite (EPB).
Specification.

BS EN 13170: 2012
Thermal insulation products for building.
Factory made products of expanded cork (ICB).
Specification.

BS EN 13171: 2012
Thermal insulation products for building. Factory
made wood fibre (WF) products. Specification.

BS EN 13172: 2012
Thermal insulating products. Evaluation of
conformity.

BS EN ISO 13370: 2007
Thermal performance of buildings. Heat transfer
via the ground. Calculation methods.

BS EN 13467: 2001
Thermal insulation for building equipment and
industrial installations. Preformed pipe insulation.

BS EN 13469: 2012
Thermal insulation for building equipment and
industrial installations. Preformed pipe insulation.
Water vapour transmission properties.

BS EN 13470: 2012
Thermal insulation for building equipment and
industrial installations. Preformed pipe insulation.
Apparent density.

BS EN 13471: 2001
Thermal insulation for building equipment and
industrial installations. Coefficient of thermal
expansion.

BS EN 13472: 2012
Thermal insulation for building equipment and
industrial installations. Preformed pipe insulation.
Water absorption.

BS EN 13494: 2002
Thermal insulation products for building
applications. Bond strength to thermal insulation
material.

BS EN 13495: 2002
Thermal insulation products for building
applications. Pull off resistance.

BS EN 13496: 2013
Thermal insulation products for building
applications. Determination of the mechanical
properties of glass fibre meshes as reinforcement
for external thermal insulation composite systems
(ETICS).

BS EN 13497: 2002
Thermal insulation products for building
applications. External thermal insulation.
Resistance to impact.

BS EN 13498: 2002
Thermal insulation products for building
applications. External thermal insulation.
Resistance to penetration.

BS EN 13499: 2003
Thermal insulation products for building
applications. External thermal insulation
composite systems (ETICS) based on polystyrene.
Specification.

BS EN 13501
Fire classification of construction products and
building elements:
    Part 1: 2007   Classification using test data
                from reaction to fire tests.
    Part 2: 2007   Classification using data from
                fire resistance tests.

BS EN ISO 13788: 2012
Hygrothermal performance of building
components and building elements. Calculation
methods.

BS EN ISO 13789: 2007
Thermal performance of buildings. Transmission
and ventilation heat transfer coefficients.
Calculation method.

BS EN ISO 13790: 2008
Thermal performance of buildings. Calculation of
energy use for space heating and cooling.

BS EN ISO 13791: 2012
Thermal performance of buildings. Calculation of
internal temperatures of a room in summer.
General criteria and validation procedures.

BS EN ISO 13792: 2012
Thermal performance of buildings. Calculation of
internal temperatures. Simplified methods.

BS EN 13823: 2010
Reaction to fire tests for building products.
Building products excluding floorings exposed to
the thermal attack by a single burning item.

pr EN 13950: 2011
Gypsum plasterboard thermal/acoustic insulation
composite panels.

BS EN 13986: 2004
Wood-based panels for use in construction. Characteristics, evaluation of conformity and marking.

BS EN 14063
Thermal insulation products for building. In-situ formed expanded clay lightweight aggregate products:

Part 1: 2004	Specification for the loose-fill products before installation.
Part 2: 2013	Specification for the installed products.

BS EN 14064
Thermal insulation in buildings. In-situ formed loose-fill mineral wool (MW) products:

Part 1: 2010	Specification for the loose-fill products before installation.
Part 2: 2010	Specification for the installed insulation products.

BS EN 14303: 2009
Thermal insulation products for building equipment and industrial installations. MW.

BS EN 14304: 2009
Thermal insulation products for building equipment and industrial installations. Flexible elastomeric foam (FEF).

BS EN 14305: 2009
Thermal insulation products for building equipment and industrial installations. CG.

BS EN 14306: 2009
Thermal insulation products for building equipment and industrial installations. Calcium silicate CS.

BS EN 14307: 2009
Thermal insulation products for building equipment and industrial installations. XPS.

BS EN 14308: 2009
Thermal insulation products for building equipment and industrial installations. Polyurethane foam (PUR) and polyisocyanurate foam (PIR).

BS EN 14309: 2009
Thermal insulation products for building equipment and industrial installations. EPS.

BS EN 14313: 2009
Thermal insulation products for building equipment and industrial installations. Polyethylene foam (PEF).

BS EN 14314: 2009
Thermal insulation products for building equipment and industrial installations. PF.

BS EN 14315
Thermal insulation products for buildings. In-situ formed sprayed rigid polyurethane (PUR) and polyisocyanurate (PIR) products:

Part 1: 2013	Specification for the rigid foam before installation.
Part 2: 2013	Specification for the installed insulation products.

BS EN 14316
Thermal insulation products for buildings. Expanded perlite (EP) products:

Part 1: 2004	Specification for bonded and loose-fill products before installation.
Part 2: 2007	Specification for the installed products.

BS EN 14317
Thermal insulation products for buildings. Exfoliated vermiculite (EV) products:

Part 1: 2004	Specification for bonded and loose-fill products before installation.
Part 2: 2007	Specification for the installed products.

BS EN 14318
Thermal insulating products for buildings. Polyurethane and polyisocyanurate:

Part 1: 2013	Specification for the rigid foam before installation.
Part 2: 2013	Specification for the installed insulation products.

BS EN 14319
Thermal insulation products for building equipment and industrial installations. In-situ formed dispensed rigid polyurethane (PUR) and polyisocyanurate foam (PIR):

Part 1: 2013	Specification for the rigid foam before installation.
Part 2: 2013	Specification for the installed insulation products.

BS EN 14320
Thermal insulation products for building equipment and industrial installations. In-situ formed sprayed rigid polyurethane and polyisocyanarate foam:

Part 1: 2013	Specification for the rigid foam before installation.
Part 2: 2013	Specification for the installed insulation products.

BS EN 14933: 2007
  Thermal insulation and lightweight fill products
  for civil engineering applications. Expanded
  polystyrene (EPS).
BS EN 14934: 2007
  Thermal insulation and lightweight fill products
  for civil engineering applications. Extruded
  polystyrene foam (XPS).
BS EN 15101
  Thermal insulation products for building. In-situ
  formed loose fill cellulose (LFCI) products:
    Part 1: 2013   Specification for the products
                   before installation.
    Part 2: 2013   Specification for the installed
                   insulation products.
PD CEN/TR 15131: 2006
  Thermal performance of building materials.
BS EN 15265: 2007
  Energy performance of buildings. Calculation of
  energy needs for space heating and cooling using
  dynamic methods.
BS EN 15501: 2013
  Thermal insulation products for building
  equipment and industrial installations. Factory
  made perlite (EP) and exfoliated vermiculite (EV)
  products.
BS EN 15599
  Thermal insulation products for building
  equipment and industrial installations:
    Part 1: 2010   In situ expanded perlite (EP)
                   products before installation.
    Part 2: 2010   Specification for the installed
                   products.
BS EN 15600
  Thermal insulation products for building
  equipment and industrial installations:
    Part 1: 2010   In situ exfoliated vermiculite
                   (EV) products.
    Part 2: 2010   Specification for the installed
                   products.
pr EN 15603: 2013
  Energy performance of buildings. Overarching
  standard EPBD.
BS EN 16012: 2012
  Thermal insulation for buildings. Reflective
  insulation products.
BS EN 16025
  Thermal and/or sound insulating products in
  building construction. Bound EPS ballasting:
    Part 1: 2013   Requirements for factory
                   premixed EPS dry plaster.

    Part 2: 2013   Processing of the factory
                   premixed EPS dry plaster.
BS EN 16069: 2012
  Thermal insulation products for building. Factory
  made products of polyethylene foam (PEF).
  Specification.
BS EN ISO 16283
  Acoustics. Field measurement of sound insulation
  in buildings and of building elements:
    Part 1: 2014   Airborne sound insulation.
    pr Part 2: 2013 Impact sound insulation.
    pr Part 3: 2014 Façade sound insulation.
PD ISO/TR 16344: 2012
  Energy performance of buildings. Common terms,
  definitions and symbols for the overall energy
  performance rating and certification.
pr BS EN 16491: 2012
  Thermal insulation products for building. Factory
  made composite products. Specification.
BS EN ISO 23993: 2010
  Thermal insulation products for building
  equipment and industrial installations.
  Determination of design thermal conductivity.
PAS 2030: 2012 Ed. 2
  Improving the energy efficiency of existing
  buildings. Specification.
PD 6680: 2002
  Guidance on the new European Standards for
  thermal insulation materials.

BUILDING RESEARCH ESTABLISHMENT PUBLICATIONS

BRE Digest

BRE Digest 453: 2000
  Insulating glazing units.

BRE Information papers

BRE IP 4/01
  Reducing impact and structure-borne sound in
  buildings.
BRE IP 14/02
  Dealing with poor sound insulation between new
  dwellings.
BRE IP 3/03
  Dynamic insulation for energy saving and
  comfort.
BRE IP 2/05
  Modelling and controlling interstitial
  condensation in buildings.

BRE IP 15/05
  The scope for reducing carbon emissions from housing.
BRE IP 1/06
  Assessing the effects of thermal bridging at junctions and around openings.
BRE IP 18/11
  Natural fibre insulation. An introduction to low-impact building materials.
BRE IP 4/13
  Advanced thermal insulation technologies in the built environment.

BRE Good building guides

BRE GBG 37: 2000
  Insulating roofs at rafter level: sarking insulation.
BRE GBG 43: 2000
  Insulating profiled metal roofs.
BRE GBG 44: 2001
  Insulating masonry cavity walls (Parts 1 and 2).
BRE GBG 45: 2001
  Insulating ground floors.
BRE GBG 50: 2002
  Insulating solid masonry walls.
BRE GBG 68: 2006
  Installing thermal insulation: Good site practice (Parts 1 and 2).
BRE GBG 83: 2014
  Sound insulation in dwellings. An introduction (Part 1).

BRE Good repair guide

BRE GRG 30: 2001
  Remedying condensation in domestic pitched tiled roofs.

BRE Reports

BR 262: 2002
  Thermal insulation: Avoiding risks.
BR 406: 2000
  Specifying dwellings with enhanced sound insulation: A guide.
FB 61: 2013
  Reducing thermal bridging at junctions when designing and installing solid wall insulation.

ADVISORY ORGANISATIONS

British Rigid Urethane Foam Manufacturers Association Ltd, 12a High Street East, Glossop, Derbyshire SK13 8DA (01457 855884).

Cork Industry Federation, 13 Felton Lea, Sidcup, Kent DA14 6BA (0208 302 4801).

Eurisol-UK Mineral Wool Association, PO Box 35084, London NW1 4XE (0207 935 8532).

European Phenolic Foam Association, Kingsley House, Ganders Business Park, Kingsley, Bordon, Hampshire GU35 9LU (01420 471617).

Insulated Render and Cladding Association Ltd, 6–8 Bonhill Street, London EC2A 4BX (0844 2490040).

National Insulation Association, 2 Vimy Court, Vimy Road, Leighton Buzzard, Bedfordshire LU7 1FG (01525 383313).

Thermal Insulation Manufacturers & Suppliers Association, Kingsley House, Ganders Business Park, Kingsley, Bordon, Hampshire GU35 9LU (01420 471624).

14

SEALANTS, GASKETS
AND ADHESIVES

CONTENTS

Introduction

Although used in relatively small quantities compared with the load-bearing construction materials, sealants, gaskets and adhesives play a significant role in the perceived success or failure of buildings. A combination of correct detailing and appropriate use of these materials is necessary to prevent the need for expensive remedial work. Many manufacturers are producing solvent-free sealants and adhesives as environmentally friendly alternatives to the traditional solvent-based systems.

Sealants

Sealants are designed to seal the joints between adjacent building components while remaining sufficiently flexible to accommodate any relative movement. They may be required to exclude wind, rain and airborne sound. A wide range of products is available matching the performance characteristics of the sealant to the requirements of the joint. Incorrect specification or application, poor joint design or preparation is likely to lead to premature failure of the sealant. The standard BS EN ISO 11600: 2003 + A1: 2011 classifies sealants into type G for glazing applications and type F (façade) for other construction joints. For both types, classes are defined by movement capability, modulus and elastic recovery (Fig. 14.1). The standard BS 8000–16: 1997 + A1: 2010 is the code of practice for sealing joints in buildings with sealants.

Key factors in specifying the appropriate sealant are:

- understand the cause and nature of the relative movement;
- match the nature and extent of movement to the appropriate sealant;
- match the sealant to the substrate;

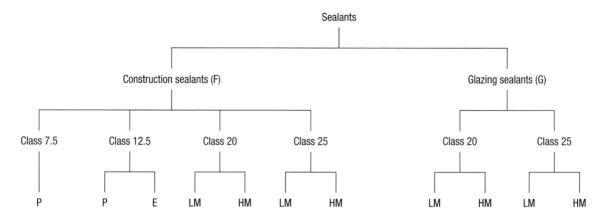

Notes: F refers to façade, G to glazing.
Class number indicates the movement accommodation as a percentage.
P refers to plastic, E to elastic, LM to low modulus and HM to high modulus.

Fig. 14.1 Classification of sealants in construction

- ensure appropriate joint design, surface preparation and sealant application;
- anticipate the exposure and temperature range;
- determine the required service life of the sealant.

RELATIVE MOVEMENT WITHIN BUILDINGS

The most common causes of movement in buildings are associated with settlement, dead and live load effects including wind loading, fluctuations in temperature, changes in moisture content and, in some cases, the deteriorative effects of chemical or electrolytic action. Depending on the prevailing conditions, the various effects may be additive or compensatory.

Settlement

Settlement is primarily associated with changes in loadings on the foundations during the construction process, although it may continue for some time, frequently up to five years after the construction is complete. Subsequent modifications to a building or its contents may cause further relative movement. Settlement is usually slow and in one direction, creating a shearing effect on sealants used across the boundaries.

Thermal movement

All building materials expand and contract to some degree with changes in temperature. For timber the movement is low, but moderate for glass, steel, brick, stone and concrete, and relatively high for plastics and aluminium. Such thermal movements are accentuated by the effects of colour, insulation and thickness of the material. Dark materials absorb solar radiation and heat more quickly than light-reflective materials. In addition, well-insulated claddings respond quickly to changes in solar radiation, producing rapid cyclical expansion movements, whereas heavy construction materials respond more slowly but will still exhibit considerable movements over an annual cycle. Typical thermal movements are shown in Table 14.1.

Moisture movement

Moisture movement falls into two categories: irreversible movements as new materials acclimatise to the environment, and reversible cyclical movements due to climatic variations. Many building materials, especially concrete and mortars, exhibit an initial contraction during the drying-out process. Incorrectly seasoned timber will also shrink but new bricks used too quickly after manufacture will expand. After these initial effects, all materials which absorb moisture will expand and contract to varying degrees in response to changes in their moisture content. Depending on climatic conditions, moisture and thermal movements may oppose or reinforce each other. Typical irreversible and reversible moisture movements are shown in Table 14.2.

Table 14.1 Thermal movements of building materials

Material	Typical thermal movement (mm/m for 85°C change)	Coefficient of linear expansion per °C $\times$ 10^{-6}
Masonry		
Concrete – standard aggregates	1.2	10–14
Calcium silicate brickwork	1.2	8–14
Concrete blockwork	1.0	6–12
Concrete – aerated	0.7	8
Concrete – limestone aggregate	0.6	7–8
Clay brickwork	0.5–0.7	5–8
GRC	1.0	7–12
Plaster		
Dense plaster	1.5	18–21
Lightweight plaster	1.4	16–18
Metals		
Zinc (along roll)	2.7	32 (22 across roll)
Lead	2.5	29
Aluminium	2.0	24
Titanium zinc	1.8	20–22
Copper	1.4	17
Stainless steel (austenitic)	1.4	18
Stainless steel (ferritic)	0.8	10
Terne coated stainless steel	1.4	17
Structural steel	1.0	12
Stone and glass		
Glass	0.9	9–11
Slate	0.9	9–11
Granite	0.8	8–10
Sandstone	0.8	7–12
Marble	0.4	4–6
Limestone	0.3	3–4
Plastics		
ABS	8.0	83–95
PVC (rigid)	6.0	40–80
GRP	3.0	18–35
Timber		
Wood (along grain)	0.5	4–6

Note:
Typical thermal movements of building materials in use calculated for a temperature variation of 85°C (e.g. −15 to +70°C) (measured in mm/m).

Table 14.2 Moisture movements of typical building materials

Material	Reversible (mm/m)	Irreversible (mm/m)	
Concrete	0.2–0.6	0.3–0.8	(shrinkage)
Aerated concrete	0.2–0.3	0.7–0.9	(shrinkage)
Brickwork – clay	0.2	0.2–1.0	(expansion)
Brickwork – calcium silicate	0.1–0.5	0.1–0.4	(shrinkage)
Blockwork – dense	0.2–0.4	0.2–0.6	(shrinkage)
Blockwork – aerated	0.2–0.3	0.5–0.9	(shrinkage)
Glass-fibre reinforced cement	1.5	0.7	(shrinkage)
Softwood	5–25	(60–90% relative humidity)	
Hardwood	7–32	(60–90% relative humidity)	
Plywood	2–3	(60–90% relative humidity)	

(timber has no irreversible movement)

Note:
Typical reversible and irreversible moisture movements of building materials in use (measured in mm/m).

Loading and deterioration

Movements associated with live loads such as machinery, traffic and wind can cause rapid cyclical movements within building components. The deterioration of materials, such as the corrosion of steel or sulphate attack on concrete, is often associated with irreversible expansion, causing movement of adjacent components. Concrete structures may exhibit creep, which is gradual permanent deformation under load, over many years.

Types of sealant

There are three distinct types of sealant – plastic, elastoplastic and elastic – each of which exhibits significantly different properties which must be matched to the appropriate application (BS 6213: 2000).

PLASTIC SEALANTS

Plastic sealants, which include general-purpose mastics, allow only a limited amount of movement, but when held in a deformed state they stress-relax. Elastic recovery is limited to a maximum of 40%. Plastic sealants dry by the formation of a surface skin, leaving liquid material encased to retain flexibility. However, with time, the plastic core continues to harden; thus durability is related to the thickness of the material used. Plastic sealants (Code P to BS EN ISO 11600: 2003 + A1: 2011) are more suitable for locations in which the majority of movement is irreversible (BS 6213: 2000 + A1: 2010).

Oil-based mastics

For oil-based mastics a 10 mm depth is required for optimum durability with a typical life expectancy of two to ten years. The effects of ultraviolet degradation are reduced by painting. Typical uses include sealing around window and door frames in traditional low-rise buildings. Oil-based mastics are not suitable for use with PVC-U windows. (The typical movement accommodation for oil-based mastics is 10%.)

Butyl sealants

Butyl sealants are plastic but with a slightly rubbery texture. They are used in small joints as a gap filler and general-purpose sealant where oil-based mastics would dry too rapidly. Life expectancy is between 10 and 20 years if they are protected from sunlight by painting, but only up to five years in exposed situations. (The typical movement accommodation for butyl sealants is 10%.)

Acrylic sealants

Water-based acrylic sealants are frequently used for internal sealing such as between plaster and new windows. The solvent-based acrylic sealants are durable for up to 20 years, with good adhesion to slightly contaminated surfaces. They accommodate only limited movement but produce a good external seal around windows, both for new and remedial work. (The typical movement accommodation for water-based and solvent-based acrylic sealants is 15% and 20% respectively.)

Polymer/bitumen sealants

Solvent-based bitumen sealants are generally suitable for low-movement joints in gutters and flashings. Hot-poured bitumen is used for sealing movement joints in asphalt and concrete floor slabs, although compatibility with any subsequent floor coverings should be verified. Non-hardening bitumen sealant is used for blending in new and old bitumen roof membranes.

Linseed oil putty

Traditional putty contains a mixture of linseed oil and inorganic fillers, which sets by a combination of aerial oxidation of the oil and some absorption into the timber. A skin is produced initially, but the mass ultimately sets to a semi-rigid material. Application is with a putty knife onto primed timber. For application to steel window frames, non-absorbent hardwoods and water-repellent preservative-treated softwoods, non-linseed oil putty is appropriate. Linseed oil putty should be painted within two weeks, whereas metal casement putty may be left for three months before painting.

ELASTOPLASTIC SEALANTS

Elastoplastic sealants will accommodate both slow cyclical movements and permanent deformations. A range of products offer appropriately balanced strength, plastic flow and elastic properties for various applications.

Polysulphide sealants

Polysulphide sealants are available as one- or two-component systems. The one-component systems have the advantage that they are ready for immediate use. They cure relatively slowly by absorption of moisture from the atmosphere, initially forming a skin and fully curing within two to five weeks. One-component systems are limited in their application to joints up to 25 mm in width, but their ultimate performance is comparable to that of the two-component materials. Typical uses include structural movement joints in masonry, joints between precast concrete or stone cladding panels and sealing around windows. The two-component polysulphide sealants require mixing immediately before use and fully cure within 24–48 hours. They are more suitable than one-component systems for sealing joints which are wider than 25 mm, have large movements, or are subject to vandalism during setting. Uses include sealing joints within concrete and brickwork cladding systems and also within poorly insulated lightweight cladding panels. Polysulphides have a life expectancy of 20–25 years. (The typical movement accommodation for polysulphide sealants is up to 25% for one-part systems and up to 30% for two-part systems.)

ELASTIC SEALANTS

Elastic sealants are appropriate for sealing dynamic joints where rapid cyclic movement occurs. They are often sub-classified as low or high modulus depending upon their stiffness. High-modulus (HM) sealants are stiffer than low-modulus (LM) sealants. Low-modulus sealants should be used where joints are exposed to long periods of compression or extension and where the substrate material is weak. Elastic sealants are categorised as Code E to BS EN ISO 11600: 2003 + A1: 2011.

Polyurethane sealants

Polyurethane sealants are available as one- or two-component systems. The products are highly elastic but surfaces should be carefully prepared and usually primed to ensure good adhesion. Polyurethane sealants have good abrasion resistance and durability is good, ranging from 20–25 years. Typical applications are joints within glazing, curtain walling, lightweight cladding panels and floors. Two-component thixotropic sealants may be used in industrial locations subject to hydrocarbon spillage. (The typical movement accommodation for polyurethane sealants is between 10% and 30% depending on the modulus.)

Silicone sealants

Silicone sealants are usually one-component systems which cure relatively quickly in air, frequently with the evolution of characteristic smells such as acetic acid. Generally, silicone sealants adhere well to metals and glass, but primers may be necessary on friable or porous surfaces, such as concrete or stone. Silicone sealants may cause staining on stonework which is difficult to remove. High-modulus silicone sealants are resilient. Typical applications include glazing and curtain-wall systems, movement joints in ceramic tiling and around sanitary ware. Low-modulus silicone sealants are very extensible and are appropriate for use in joints subject to substantial thermal or moisture movement. Typical applications are the perimeter sealing of PVC-U and aluminium windows, and also cladding systems. Specialist silicone sealants will withstand a working temperature of 300°C. Silicone sealants are durable with life expectancies within the range of 25–30 years. (The typical movement accommodation for silicone sealants ranges from 20%–70% depending on the modulus.)

Epoxy sealants

Epoxy sealants are appropriate for stress-relieving joints where larger movements in compression than tension are anticipated. Typical applications include floor joints and the water sealing of tiling joints within swimming pools. Two-component systems may be used for car parks subject to heavy low-speed traffic. (Epoxy sealants have a life expectancy of 10–20 years. The typical movement accommodation of epoxy sealants is within the range 5–15%.)

JOINT DESIGN

There are three forms of joint: butt, lap and fillet (Fig. 14.2). However, only butt and lap joints will accommodate movement. Generally, lap joints, in which the sealant is stressed in shear, will accommodate double the movement of butt joints in which the sealant is under tension or compression. Furthermore, lap joints tend to be more durable, as the sealant is partially protected from the effects of weathering. However, lap joints are generally more difficult to seal than butt joints. Frequently, joints are made too narrow, either for aesthetic reasons or due to miscalculation of component tolerances. The effect is that extent of movement is excessive in proportion to the width of sealant, causing rapid failure.

To correctly control the depth of the sealant and to prevent it from adhering to the back of the joint, a compressible back-up material, usually rectangular or round closed-cell polyethylene, is inserted (Fig. 14.3). The polyethylene acts as a bond breaker by not adhering to the sealant. Where the joint is filled with a filler board, such as impregnated fibreboard or corkboard, a plastic bond-breaker tape or closed-cell polyethylene strip should be inserted. Normally, the depth of the sealant should be half the width of the joint for elastic and elastoplastic sealants and equal to the width of the joint for plastic sealants. A minimum depth of 6 mm is normally required. The minimum width of the joint is calculated from the maximum total relative movement (TRM) to be accommodated and the movement accommodation factor (MAF); that is, the extensibility of the sealant to be used. Where insufficient depth is available to insert a polyethylene foam strip, a tape bond breaker should be inserted at the back of the joint.

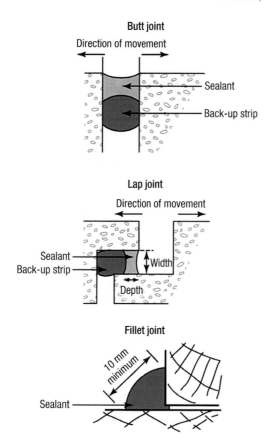

Fig. 14.2 Butt, lap and fillet joints

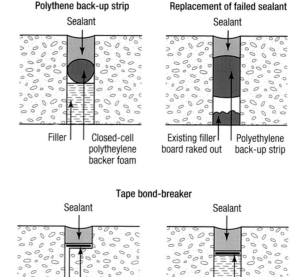

Fig. 14.3 Typical sealant systems

Minimum joint width calculation (BS 6093: 2006 + A1: 2013):

Total relevant movement (TRM)	=	3 mm
Movement accommodation factor (MAF)	=	25%
Width of sealant to accommodate movement	=	3/0.25 + 3
	=	15 mm.

In order to obtain good adhesion, the joint surfaces should be prepared by the removal of contaminants, loose material or grease and by the application of a primer if specified by the sealant manufacturer. Most sealants are applied directly by gun application, although tooled, poured and tape/strip sealants are also used. Tooling helps to remove air bubbles entrained in two-component mixes; if left, air bubbles would reduce the durability of the seal. Externally, recessed cladding joints show less staining than flush joints, although the usual finish is a slightly concave surface. Where stonework is being sealed, non-staining silicone sealants must be used to prevent the migration of plasticiser into the stone, which could cause discoloration. Sealants to floor joints need to be tough, and therefore wider to accommodate the necessary movements and recessed to prevent mechanical damage. Alternatively, proprietary mechanical jointing systems should be used.

Colour matching

While most sealants, except for the black bituminous products, are available in white, translucent, greys and browns, the silicone sealants appropriate for use around kitchen and bathroom units are available in a wide range of colours. For these purposes, fungicides are often included within the formulation.

FIRE-RESISTANT SEALANTS

Many fire-resistant sealants are based on the use of intumescent materials which expand copiously in fire. The intumescent components commonly used are either ammonium phosphate, hydrated sodium silicate or intercalated graphite (layers of water and carbon), and these are incorporated into the appropriate sealant. Intumescent oil-based mastics and acrylic sealants are suitable for sealing low-movement joints around fire check doors. For the fire-resistant sealing of structural movement joints, fire-resistant grades of low-modulus silicone, two-part polysulphide and acrylic sealants are available. Maximum fire resistance is obtained if the sealant is applied to both faces of the joint, with mineral wool or glass fibre insulation in the void space. Four hours' fire resistance with respect to both integrity and insulation may be achieved for a 20 mm-wide movement joint within 150 mm concrete (BS 476–20: 1987). The low-modulus silicone is appropriate for sealing fire-resisting screens, curtain walls, claddings and masonry subject to movement. The two-part polysulphide is designed for use in concrete and masonry fire-resisting joints. Acrylic sealants are appropriate over a wide range of materials but where timber is involved an allowance must be made for its loss by charring.

Intumescent fillers manufactured from acrylic emulsions with inert fillers and fire-retardant additives may be applied either by gun or trowel to fill voids created around service ducts within fire-resistant walls. Four hours of fire resistance may be achieved with these materials. Intumescent tapes are appropriate for application within structural movement joints. Most intumescent sealants are now *low-smoke* and evolve no halogenated products of combustion in fire situations. (The typical movement accommodation for intumescent acrylic sealants is 15%.)

FOAM SEALANTS

Compressible strips of closed-cell PVC and polyethylene, or open-cell polyurethane foams, coated on one or both edges with pressure-sensitive adhesive, are used to seal thermal movement and differential settlement joints, gaps around window and door frames, and in air-conditioning ductwork. Strips may be uniform in section or profiled for particular applications. Aerosol-dispensed polyurethane foam is widely used as an all-purpose filler. It is available either as foam or as expanding foam, and acts as an adhesive, sealant, filler and insulator.

CONCRETE JOINT FILLERS AND SEALANTS

Concrete joint fillers for use in pavements are specified by the standards BS EN 14188: 2004, Parts 1 and 2 for hot and cold application sealants, respectively. Sealants for cold application are classified as single-component systems (S) or multi-component systems (M), and subdivided into self-levelling (sl) or non-sag (ns) types. An additional classification A, B, C or D relates to increasing level of resistance to chemicals. Standard hot-applied joint sealants are classified as elastic (high extension) Type N1, and normal (low extension) Type N2. Where fuel resistance is also required, the higher specification grades F1 and F2 are necessary.

Gaskets

Gaskets are lengths of flexible components of various profiles, which may be solid or hollow and manufactured from either cellular or non-cellular materials. They are held in place either by compression or encapsulation into the adjacent building materials, and maintain a seal by pushing against the two surfaces (Fig. 14.4). Typical applications include the weather sealing of precast cladding units and façade systems. Within precast concrete, glassfibre-reinforced polyester (GRP) or glassfibre-reinforced concrete (GRC) cladding units, the gaskets are typically inserted into recessed open-drained joints. The gaskets therefore act as a rain barrier, but because they do not necessarily fit tightly along their full length, they may be backed up by a compressed cellular foam wind penetration seal. Gaskets should not be either stretched or crammed in during insertion as they will subsequently shrink, leaving gaps, or pop out, causing failure.

In glazing and related curtain-walling systems, gaskets may be applied as capping seals, retained by appropriate profiles within the mullions and transoms; alternatively, the gaskets may be recessed within the joints of the glazing system to give narrower visual effect to the joint. Some glazing gaskets of H- or U-sections are sealed with a zipper or filler strip which is inserted into the profile, compressing the material into an air- and watertight seal. Gaskets and weatherstripping for use on doors, windows and curtain walling are classified by a letter and digit code which defines the use and key physical properties of the particular product, enabling appropriate specification (Table 14.3).

The standard materials for gaskets used in construction are neoprene which is highly elastic, ethylene propylene diene monomer (EPDM) which has better weathering characteristics than neoprene, and silicone rubbers which are highly resistant to ultraviolet light, operate over a wide range of temperatures and are available in almost any colour.

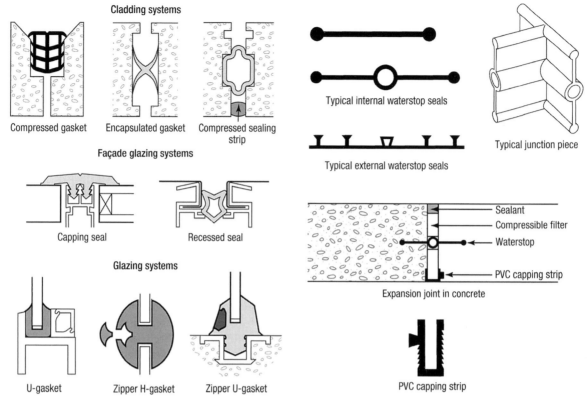

Fig. 14.4 Typical gaskets for cladding and glazing systems

Fig. 14.5 Concrete waterstop seals

Cruciform section gaskets of polychloroprene rubber are suitable for vertical joints between precast concrete panels.

Proprietary systems offer watertight expansion jointing for horizontal surfaces, such as roof car parks and pedestrian areas. Systems usually combine complex aluminium or stainless steel profiles with extruded synthetic rubber inserts. Materials can withstand high loads, with good resistance to bitumen and salt water.

Dry glazing strips are based on elastomeric polymers, typically EPDM or butyl rubber. Usually, the synthetic rubber strip has a self-adhesive backing which adheres to the rebate upstand. With external beading, the dry glazing strip may also be applied to each bead, which is then fixed with suitable compression to ensure a good seal to the glass. The performance requirements and classification for gaskets and weatherstripping for doors, windows and curtain walling are described in the standard BS EN 12365–1: 2003.

Table 14.3 Classification of gaskets and weatherstripping to BS EN 12365–1: 2003 (letter and five-number code)

Letter (G or W)	Digit 2	Digit 3	Digit 4	Digit 5	Digit 6
Category	Working range (mm)	Compression force (KPa)	Working temperature range (°C)	Deflection recovery (%)	Recovery after ageing (%)
Gasket (G) Weatherstripping (W)	9 grades identified (1–9)	9 grades identified (1–9)	6 grades identified (1–6)	8 grades identified (0–7)	8 grades identified (1–8)

WATERSTOPS

Waterstops for embedding into in situ concrete for sealing movement and construction joints are manufactured in PVC or rubber according to the required movement (Fig. 14.5). Sections are available in long extruded lengths and factory-produced intersections. Applications include water-containing structures and water exclusion from basements. Waterstops placed centrally within concrete will resist water pressure from either side, but externally positioned waterstops, not encased below the concrete slab or within permanent concrete shuttering, will only resist water pressure from the outer face. Hydrophilic waterstops, including modified chloroprene rubber or sodium bentonite/butyl rubber, expand in contact with water to form the seal. Their typical application is within concrete construction joints. PVC waterstops incorporating hydrophilic elements are also available.

Adhesives

TYPES OF ADHESIVE

The traditional adhesives based on animal and vegetable products have largely been superseded by synthetic products manufactured by the polymer industry, except for casein, manufactured from skimmed milk, which is currently used as a timber adhesive (BS EN 12436: 2002). The range of adhesives is under constant development and particular applications should always be matched to manufacturers' specifications. Special notice should be taken of exclusions where materials and adhesives are incompatible; also to safety warnings relating to handling and the evolution of noxious fumes or flammable vapours. Adhesives are more efficient when bonding components subject to shear forces rather than direct tension. They are least efficient against the peeling stresses. Most adhesives have a *shelf life* of 12 months when stored unopened under appropriate conditions. The *pot life* after mixing the two-component systems ranges from a few minutes to several hours.

Tile adhesives

The standard BS EN 12004: 2007 + A1: 2012 classifies adhesives for tiles into three types: cementitious (C), dispersion (D) and reaction resin (R). Each of these types may have further characteristics defined by classes

Table 14.4 Classification of tile adhesives by composition and properties (BS EN 12004: 2007 + A1: 2012)

Classification	Composition and properties
Type C	cementitious adhesive – hydraulic binding resin
Type D	dispersion adhesive – aqueous organic polymer resin
Type R	reaction resin adhesive – one- or two-component synthetic resin
Class 1	normal adhesive
Class 2	improved adhesive
Class F	fast-setting adhesive
Class T	reduced slip adhesive
Class E	extended open time adhesive
Class S1	deformable adhesive
Class S2	highly deformable adhesive

relating to enhanced adhesive properties, faster setting, reduced slip or extended open time (the time between spreading the adhesive and applying the tiles) (Table 14.4). Dispersion adhesives are the ready-for-use aqueous polymer dispersions, while the reaction resin adhesives are one- or two-component systems which set by chemical reaction. Tile adhesives with no more than 1% organic material are classified for fire resistance as Class A1 or $A1_{fl}$.

Ceramic wall tile adhesives

Wall tile adhesives are usually polyvinyl acetate (PVA), acrylic- or cement-based compositions. The standard PVA thin-bed adhesives, typically to 3 mm, will only tolerate moisture, whereas the thin bed water-resistant acrylic-based adhesives are suitable for fixing wall tiles and mosaics in damp and wet conditions associated, for example, with domestic showers. Some acrylic-based products evolve ammonia on setting. The water-resistant cements and polymer-modified cement products are appropriate both for internal and external use and can usually be applied with either thin or thick bedding. The polymer-modified cement adhesives are also suitable for fixing marble, granite and slate tiles up to 15 mm thick. For chemical resistance thin bed epoxy resin-based adhesives are available. In all cases the substrate must be sound, with new plaster, brickwork and concrete fully dried out for two to six weeks. Plasterboard and timber products must be adequately fixed at 300 mm centres horizontally and vertically to ensure rigidity. In refurbishment work, flaking or multi-layered paint should be removed and glazed surfaces made good. Where the tile adhesive is classified as waterproof either acrylic- or cement-based, it may be used as the grouting medium. Alternatively, equivalent waterproof grouting is available in a wide range of colours to blend or contrast with the wall tiles. Epoxy-resin tile grout is available for very wet conditions. Two-component thixotropic polyurethane non-slump tile adhesives prevent slippage during wall tiling.

Ceramic floor tile adhesives

The majority of ceramic floor tile adhesives are cement-based, used either as thick bed (up to 25 mm) or thin bed according to the quality of the substrate. Standard products are suitable for fixing ceramic tiles, quarries, brick slips, stone and terrazzo to well-dried-out concrete or cement/sand screed. Where suspended timber floors are to be tiled, they must be well ventilated and strong enough to support the additional dead load. An overlay of 12 mm exterior grade plywood, primed with bonding agent and screwed at 200 mm centres, may be necessary. In refurbishment work it is better to remove all old floor finishes, but ceramic floor tiles may be fixed over cleaned ceramic or possibly primed vinyl tiles, provided that all loose material is first removed.

Cement-based grouting can be pigmented to the required colour, but care must be taken to ensure that excess grout is removed from the surface of the tiles before staining occurs. Thin bed two-component epoxy-based adhesives are more water and chemical resistant than the standard cement-based products and are appropriate for use where repeated spillage is likely from industrial processes. Where there is likely movement of the substrate, two-component rubber-based adhesives are generally appropriate. Two-component polyurethane systems are also appropriate for stone, tile and mosaic flooring.

Contact adhesives

Contact adhesives based on polychloroprene rubber, either in organic solvents or aqueous emulsions, are normally suitable for bonding decorative laminates, and other rigid plastics such as PVC and ABS to timber, timber products and metals. The adhesive is usually applied to both surfaces; the solvent or emulsion is allowed to become touch dry, prior to bringing the two surfaces into contact, when an immediate strong bond is produced. The aqueous emulsion products can also be suitable for fixing sealed cork and expanded polystyrene, and have the advantage that no fumes are evolved. Expanded polystyrene tiles may be adversely affected by solvent-based formulations.

Vinyl floor tile and wood block adhesives

Most vinyl floor tile and wood block adhesives are based on either rubber/bitumen, rubber/resin or modified bitumen emulsions. Alternative materials include polyurethane and epoxy-polyurethane adhesives. In all cases it is essential that the sub-floor is dry, sound, smooth and free from any contamination which would affect the adhesion. Where necessary, cement/acrylic or cement/latex floor levelling compound should be applied to concrete, asphalt or old ceramic tiled floors. Some cement/latex materials evolve ammonia during application.

Wood adhesives

Wood joints should generally be in close contact with a gap of less than 0.15 mm, but so-called *gap-filling* adhesives satisfactorily bond up to 1.3 mm. Poly vinyl acetate (PVA) wood glues are widely used for most on-site work and in the factory assembly of mortice and tenon joints for doors, windows and furniture. The white emulsion sets to a colourless translucent thermoplastic film, giving a bond of similar strength to the timber itself, but is insufficient for bonding load-bearing structural members. Components should be clamped into position for up to 12 hours to ensure maximum bonding, although this may be reduced by increasing the temperature. Waterproof PVA adhesives, which partially cross-link on curing, are suitable for protected external use but not immersion in water. PVA adhesives generally retain their strength up to 60°C and do not discolour the timber, except by contact with ferrous metals.

The thermosetting wood resins are mainly two-component systems based on phenolic compounds, such as urea, melamine, resorcinol or phenol, which cure with formaldehyde to produce load-bearing adhesives (BS EN 301: 2013). Most formulations require the mixing of the resin and hardener, but a premixed dry powder to which water is added is also available. Structural resin-based adhesives are designated for exterior (Service Class 1, 2 or 3 to BS EN 1995–1-1: 2004) or high-temperature (70/90°C) exposure (Type I), or internal Service Class 1 use (Type II). Melamine formaldehyde adhesives will not resist prolonged exposure to weathering. Urea formaldehyde adhesives are generally moisture resistant or for interior use only. Certain timber fire-retardant and preservative treatments reduce the efficiency of timber adhesives, although generally those based on phenol formaldehyde/resorcinol formaldehyde are unaffected.

Wallpaper adhesives

Standard wallpaper adhesives are based on methyl cellulose, a white powder which is water soluble giving a colourless solution. For fixing the heavier papers and decorative dado strips, poly vinyl acetate is an added component. Cold water starch is also available as both a wall-sizing agent and wallpaper adhesive. Most wallpaper pastes contain fungicide to inhibit mould growth. The standard BS 3046: 1981 describes five types of adhesive ranging from low solids to high wet and dry strength with added fungicide.

Epoxy resin adhesives

Epoxy resins are two-component, cold-curing adhesives which produce high-strength, durable bonds. Most require equal quantities of the resin and hardener to be mixed and various formulations are available giving curing times ranging from minutes to hours. Strong bonds can be obtained to timber, metal, glass, concrete, ceramics and rigid plastics. Epoxy resins may be used internally or externally and are resistant to oils, water, dilute acids, alkalis, and most solvents except for chlorinated hydrocarbons. Epoxy resins are frequently used for attaching stainless steel fixings into stone and brick slips prior to their casting into concrete cladding panels. Epoxy flooring adhesives may be used for bonding vinyl floor finishes in wet service areas and to metal surfaces. Thixotropic two-part epoxy adhesives are designed to inject into cracks and holes in timber and masonry.

Cyanoacrylate adhesives

Cyanoacrylates are single-component adhesives which bond components held in tight contact within seconds. A high-tensile bond is produced between metals, ceramics, most plastics and rubber. The curing is activated by adsorbed moisture on the material surfaces, and only small quantities of the clear adhesive are required. The bond is resistant to oil, water, solvents, acid and alkalis but does not exhibit high impact resistance. A range of adhesive viscosities is manufactured to match to particular applications. Generally the low-viscosity material is appropriate for

close-fitting joints and higher viscosities for the larger gaps.

Hot-melt adhesives

Hot-melt adhesives for application by glue gun are usually based on the thermoplastic copolymer, ethylene vinyl acetate (EVA). Formulations are available for joining materials to either flexible or rigid substrates. Generally, the adhesive should be applied to the less easily bonded surface first (e.g. the harder or smoother surface) and then the two components should be pressed together for at least one minute. Where metals are to be bonded they should be pre-warmed to prevent rapid dissipation of the heat. Similar adhesives are used in iron-on edging veneers for plastic- and wood-faced particleboard.

Bitumen sheet roofing adhesives

Bitumen adhesives are available for hot application, emulsion or in hydrocarbon solvent for the cold bonding of bituminous sheet roofing. The adhesives should be poured and spread by trowel to avoid air pockets, which may cause premature delamination of the sheet from the substrate. Excess bitumen should be removed, as it may stain adjacent materials.

Plastic pipe adhesives

Solvent-based vinyl resin adhesives are used for bonding PVC-U and ABS pipes and fittings. The adhesive is brush-applied to both components which are then united and slightly rotated to complete the seal. Curing is rapid but in cold-water supply systems, water pressure should not be applied for several hours.

Gap-filling adhesive

Gun grade gap-filling adhesives, either solvent-borne rubber/synthetic rubber resins with filler reinforcement or solvent-free polymer emulsion systems, are versatile in their applications. They are generally formulated to bond timber, timber products, decorative laminates, sheet metals, PVC-U and rigid insulating materials (except for polystyrene) to themselves and also to brickwork, blockwork, concrete, plaster and GRP. Typical applications include the fixing of decorative wall panels, dado rails, architraves and skirting boards without nailing or screwing. Surfaces to be bonded must be sound and clean, but the gap-filling properties of the products can allow fixing to uneven surfaces. The materials have good immediate adhesion, and allow the components to be adjusted into position.

PVA bonding agent and sealant

Polyvinyl acetate (PVA) is a versatile material which will act not only as an adhesive as described above but also as a bonding agent or surface sealant. As a bonding agent it will bond cement screeds, rendering and plaster to suitable sound surfaces without the requirement for a good mechanical key. PVA will seal porous concrete surfaces to prevent dusting.

References

FURTHER READING

Cognard, P. 2005: *Handbook of adhesives and sealants: Basic concepts and high tech bonding.* Netherlands: Elsevier.

Dunn, D.J. 2004: *Handbook of adhesives and sealants: Applications and markets.* Shrewsbury: RAPRA Technology.

Intumescent Fire Seals Association. 1999: *Sealing apertures and service penetrations to maintain fire resistance.* Princes Risborough: IFSA.

Klosowski, J. 2014: *Sealants in construction,* 2nd edn. Abingdon: CRC Press.

Mittal, K. and Pizzi, A. 2009: *Handbook of sealant technology.* Abingdon: CRC Press.

Petrie, E.M. 2007: *Handbook of adhesives and sealants,* 2nd edn. USA: McGraw-Hill Professional.

SPRA. 2009: *The use of sealants.* London: Single Ply Roofing Association.

STANDARDS

BS 476
    Fire tests on building materials and structures:
        Part 20: 1987 Method for determination of the fire resistance of elements of construction.

BS 1203: 2001
    Hot-setting phenolic and aminoplastic wood adhesives.

BS 2499
    Hot-applied joint sealant systems for concrete pavements:

        Part 2: 1992 Code of practice for application and use of joint sealants.
        Part 3: 1993 Methods of test.

BS 3046: 1981
    Specification for adhesives for hanging flexible wall coverings.

BS 3712
    Building and construction sealants:
        Part 1: 1991 Method of test of homogeneity, relative density and penetration.
        Part 2: 1973 Methods of test for seepage, staining, shrinkage, shelf-life and paintability.
        Part 3: 1974 Methods of test for application life, skinning properties and tack-free time.
        Part 4: 1991 Methods of test for adhesion in peel.

BS 4255
    Rubber used in preformed gaskets for weather exclusion from buildings:
        Part 1: 1986 Specification for non-cellular gaskets.

BS 5212
    Cold applied joint sealants for concrete pavements:
        Part 1: 1990 Specification for joint sealants.
        Part 2: 1990 Code of practice for application and use of joint sealants.
        Part 3: 1990 Methods of test.

BS 5270
    Bonding agents for use with gypsum plaster and cement:
        Part 1: 1989 Specification for polyvinyl acetate (PVAC) emulsion bonding agents for indoor use with gypsum building plasters.

BS 5385
    Wall and floor tiling:
        Part 1: 2009 Design and installation of internal ceramic, natural stone and mosaic wall tiling in normal internal conditions. Code of practice.
        Part 2: 2006 Design and installation of external ceramic and mosaic wall tiling in normal conditions. Code of practice.
        Part 3: 2007 Design and installation of internal and external ceramic floor tiles and mosaics in

normal conditions. Code of
practice.

Part 4: 2009   Design and installation of
ceramic and mosaic tiling in
special conditions. Code of
practice.

Part 5: 2009   Design and installation of
terrazzo, natural stone and
agglomerated stone tile and slab
flooring. Code of practice.

BS ISO 5892: 2013
Rubber building gaskets. Materials for preformed
solid vulcanized structural gaskets. Specification.

BS 6093: 2006
Design of joints and jointing in building
construction. Guide.

BS 6213: 2000
Selection of constructional sealants. Guide.

BS 6446: 1997
Specification for manufacture of glued structural
components of timber and wood based panel
products.

BS 6576: 2005
Code of practice for diagnosis of rising damp
within walls of buildings and installation of
chemical damp-proof courses.

BS 8000
Workmanship on building sites:

Part 11: 2011   Internal and external wall and
floor tiling. Ceramic and
agglomerated stone tiles, natural
stone and terrazzo tiles, slabs
and mosaics. Code of practice.

Part 12: 1989   Code of practice for decorative
wall coverings and painting.

Part 16: 1997   Code of practice for sealing
joints in buildings using
sealants.

BS 8203: 2001
Code of practice for installation of resilient floor
coverings.

BS 8449: 2005
Building and construction sealants with
movement accommodation factors greater than
25%. Method of test.

pr BS ISO 11617: 2013
Buildings and civil engineering works. Sealants.

BS ISO 13007
Ceramic tiles. Grouts and adhesives:

Part 1: 2010   Terms, definitions and
specifications for adhesives.

Part 3: 2010   Terms, definitions and
specifications for grouts.

BS ISO 13640: 1999
Building construction. Jointing products.
Specifications for test substrates.

BS ISO 16938
Building construction. Determination of staining
of porous substrates by sealants used in joints:

Part 1: 2008   Test with compression.

Part 2: 2008   Test without compression.

BS ISO 17087: 2006
Specification for adhesives used for finger joints in
non-structural lumber products.

BS ISO 18280: 2010
Plastics. Epoxy resins. Test methods.

BS EN 204: 2001
Classification of thermoplastic wood adhesives for
non-structural applications.

BS EN 205: 2003
Adhesives. Wood adhesives for non-structural
applications. Determination of tensile shear
strength of lap joints.

BS EN 301: 2013
Adhesives, phenolic and aminoplastic, for load-
bearing timber structures. Classification and
performance requirements.

BS EN 302
Adhesives for load-bearing timber structures:

Part 1: 2013   Determination of longitudinal
tensile shear strength.

Part 2: 2013   Determination of resistance to
delamination.

Part 3: 2013   Determination of effect of acid
damage to wood fibres.

Part 4: 2013   Determination of the effects of
wood shrinkage on shear
strength.

Part 5: 2013   Determination of maximum
assembly time.

Part 6: 2013   Determination of minimum
pressing time.

Part 7: 2013   Determination of the working
life under referenced
conditions.

BS EN 923: 2005
Adhesives. Terms and definitions.

BS EN 1323: 2007
Adhesives for tiles.

BS EN 1903: 2008
Adhesives. Test method for adhesive for plastic or
rubber floor coverings or wall coverings.

BS EN 1965
Structural adhesives. Corrosion:
Part 1: 2011   Determination and classification of corrosion to a copper substrate.
Part 2: 2011   Determination and classification of corrosion to a brass substrate.

BS EN 1966: 2009
Structural adhesives. Characterisation of a surface by measuring adhesion.

BS EN ISO 6927: 2012
Building and civil engineering works. Sealants. Vocabulary.

BS EN ISO 7389: 2003
Building construction. Jointing products. Determination of elastic recovery of sealants.

BS EN ISO 7390: 2003
Building construction. Jointing products. Determination of resistance to flow of sealants.

BS EN ISO 8339: 2005
Building construction. Sealants. Determination of tensile properties.

BS EN ISO 8340: 2005
Building construction. Sealants. Determination of tensile properties at maintained extension.

BS EN ISO 8394: 2010
Building construction. Jointing products. Determination of extrudability of sealants.

BS EN ISO 9046: 2004
Building construction. Jointing products. Determination of adhesion/cohesion properties of sealants at variable temperatures.

BS EN ISO 9047: 2003
Building construction. Jointing products. Determination of adhesion/cohesion properties of sealants at constant temperature.

BS EN ISO 9664: 1995
Adhesives. Test methods for fatigue properties of structural adhesives in tensile shear.

BS EN ISO 10590: 2005
Building construction. Sealants. Determination of tensile properties of sealants at maintained extension.

BS EN ISO 10591: 2005
Building construction. Sealants. Determination of adhesion/cohesion properties.

BS EN ISO 11431: 2002
Building construction. Jointing products. Determination of adhesion/cohesion properties of sealants after exposure to heat, water and artificial light through glass.

BS EN ISO 11432: 2005
Building construction. Sealants. Determination of resistance to compression.

BS EN ISO 11600: 2003
Building construction. Jointing products. Classification and requirements for sealants.

BS EN 12002: 2008
Adhesives for tiles. Determination of transverse deformation for cementitious adhesives and grouts.

BS EN 12004: 2007
Adhesives for tiles. Requirements, evaluation of conformity, classification and designation.

BS EN 12152: 2002
Curtain walling. Air permeability. Performance requirements and classification.

BS EN 12154: 2000
Curtain walling. Watertightness. Performance requirements and classification.

BS EN 12365
Building hardware. Gaskets and weather stripping for doors, windows shutters and curtain walling:
Part 1: 2003   Performance requirements and classification.
Part 2: 2003   Linear compression force test methods.
Part 3: 2003   Deflection recovery test method.
Part 4: 2003   Recovery after accelerated ageing test method.

BS EN 12436: 2002
Adhesives for load-bearing timber structures. Casein adhesives. Classification and performance requirements.

BS EN 12765: 2001
Classification of thermosetting wood adhesives for non-structural applications.

BS EN 12860: 2001
Gypsum based adhesives for gypsum blocks. Definitions, requirements and test methods.

BS EN 13022
Glass in building. Structural sealant glazing:
Part 1: 2014   Glass products for structural sealant glazing systems.
Part 2: 2014   Assembly rules.

BS EN 13119: 2007
Curtain walling. Terminology.

BS EN 13415: 2010
Test of adhesives of floor coverings.

BS EN 13501–1: 2007
  Classification using test data from reaction to fire tests.
BS EN 13880: 2004
  Hot applied sealants. Tests.
BS EN 13888: 2009
  Grouts for tiles. Requirements, evaluation of conformity, classification and designation.
BS EN 14080: 2013
  Timber structures. Glued laminated timber and glued solid timber. Requirements.
BS EN 14187
  Cold applied joint sealants. Test methods:
    Part 9: 2006   Function testing of joint sealants.
BS EN 14188
  Joint fillers and sealants:
    Part 1: 2004   Specification for hot applied sealants.
    Part 2: 2004   Specification for cold applied sealants.
    Part 3: 2006   Specification for preformed joint sealants.
    Part 4: 2009   Specification for primers to be used with joint sealants.
BS EN 14496: 2005
  Gypsum based adhesives for thermal/acoustic insulation composite panels and plasterboards.
BS EN 14680: 2006
  Adhesives for non-pressure plastic piping systems. Specifications.
BS EN 14840: 2005
  Joint fillers and sealants. Test methods for preformed joint seals.
BS EN 15274: 2007
  General purpose adhesives for structural assembly. Requirements and test methods.
BS EN 15275: 2007
  Structural adhesives. Characterisation of anaerobic adhesives.
BS EN 15416: 2006
  Adhesives for load bearing timber structures other than phenolic and aminoplastic:
    Part 2: 2007   Test methods. Static load test.
    Part 3: 2007   Test methods. Creep deformation.
    Part 4: 2006   Test methods. Open assembly time.
    Part 5: 2006   Test methods. Conventional pressing time.
BS EN 15425: 2008
  Adhesives, one component polyurethane, for load

bearing timber structures. Classification and performance requirements.
BS EN 15434: 2006
  Glass in building. Product standard for structural and/or UV resistant sealant.
BS EN 15466
  Primers for cold and hot applied joint sealants:
    Part 1: 2009   Determination of homogeneity.
    Part 2: 2009   Determination of resistance to alkali.
BS EN 15497: 2014
  Structural finger jointed solid timber. Performance requirements and minimum production requirements.
BS EN 15651
  Sealants for non-structural use in joints in buildings and pedestrian walkways:
    Part 1: 2012   Sealants for façade elements.
    Part 2: 2012   Sealants for glazing.
    Part 3: 2013   Sealants for sanitary joints.
    Part 4: 2012   Sealants for pedestrian walkways.
    Part 5: 2012   Evaluation of conformity and marking.
BS EN 15870: 2009
  Adhesives. Determination of tensile strength of butt joints.
BS EN 16254: 2013
  Adhesives. Emulsion polymerized isocyanate (EPI) for load-bearing timber structures. Classification and performance requirements.
pr EN 16759: 2014
  Structural sealant glazing systems (SSGS).
BS EN 28394: 1991
  Building construction. Jointing products. Determination of extrudability of one-component sealants.
BS EN 29048: 1991
  Building construction. Jointing products. Determination of extrudability of sealants under standardized apparatus.

BUILDING RESEARCH ESTABLISHMENT PUBLICATIONS

BRE Digests

BRE Digest 463: 2002
  Selecting building sealants with ISO 11600.

BRE Digest 469: 2002
> Selecting gaskets for construction joints.

BRE Information papers

BRE IP 12/03
> VOC emissions from flooring adhesives.

BRE IP 4/12
> Bio-resins in construction.

ADVISORY ORGANISATION

British Adhesives and Sealants Association, 5 Alderson Road, Worksop, Nottinghamshire S80 1UZ (01909 480888).

15

PAINTS, WOOD STAINS, VARNISHES AND COLOUR

CONTENTS

Introduction

As colour is an important factor in the description of surface finishes including paints, wood stains and varnishes, the key elements of the British Standards, *Natural Colour System, RAL, Colour Palette, Pantone* and *Munsell* are described. Colour is a key feature of architectural design, as illustrated in Figure 15.1 by the vivid spectrum colour sequence in Barajas Airport, Madrid by Rogers Stirk Harbour + Partners and Estudio Lamela. Large electronic panels (Fig. 15.2) operated by programmed LED systems can create dynamic colour patterns or text sequences, creating eye-catching façades.

Colour

BRITISH STANDARDS SYSTEM

The British Standards BS 5252: 1976 (237 colours with approximate Munsell references) and BS 4800: 2011 (122 colours) define colour for building purposes and paints, respectively. A specific colour is defined by the BS 4800 framework with a three-part code consisting of hue (two digits, 00–24), greyness (letter A–E) and weight (two further digits) (Fig. 15.3). Hue is the attribute of redness, yellowness, blueness, etc., and the framework consists of 12 rows of hue in spectral sequence plus one neutral row. Greyness is a measure of the grey content of the colour at five levels from the maximum greyness Group A, to clear Group E. The third attribute, weight, is a subjective term which incorporates both lightness (reflectivity to incident light) and greyness. Within a given column, colours

Fig. 15.1 Colour feature – Barajas Airport, Madrid. *Architects*: Rogers Stirk Harbour + Partners. *Photograph*: AENA / Manuel Renau

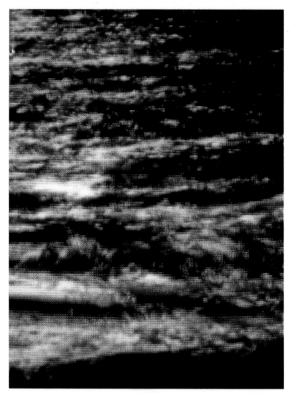

Panel

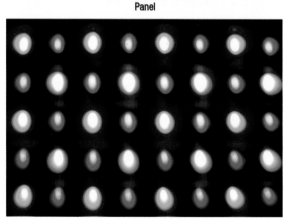

Enlarged detail showing individual LEDs

Fig. 15.2 LED colour panel producing dynamic colour-changing façade. *Photographs*: Arthur Lyons

have the same weight, but comparisons between columns in different greyness groups should only be made in respect of lightness. The framework has up to eight columns of equal lightness in each greyness group commencing with the highest lightness. Thus any colour is defined through the system by its three-part code; for example, Magnolia is yellow-red 08, nearly grey B and low-weight 15 (i.e. 08 B 15), Midnight 20 C 40 and Plum 02 C 39. The standard BS 4800: 2011 gives the approximate Munsell reference and detailed reflectance data for each colour code.

NATURAL COLOUR SYSTEM

The *Natural Colour System*®© (NCS) was developed by the Scandinavian Colour Institute, launched in 1978 and modified in 1995 with a second edition in which extra colours were incorporated and some removed, leaving a total of 1950 colours. It is a colour language system which can describe any colour by a notation, communicable in words without the need for visual matching. It has been used by architects, builders and designers who need to coordinate colour specification across a broad range of building products. A range of materials may be colour-referenced using the system; these include wall, floor and ceiling tiles, carpets, fabrics, wall coverings, flexible floor finishes, paints, architectural ironmongery and metalwork, sanitary fittings, laminates and furniture.

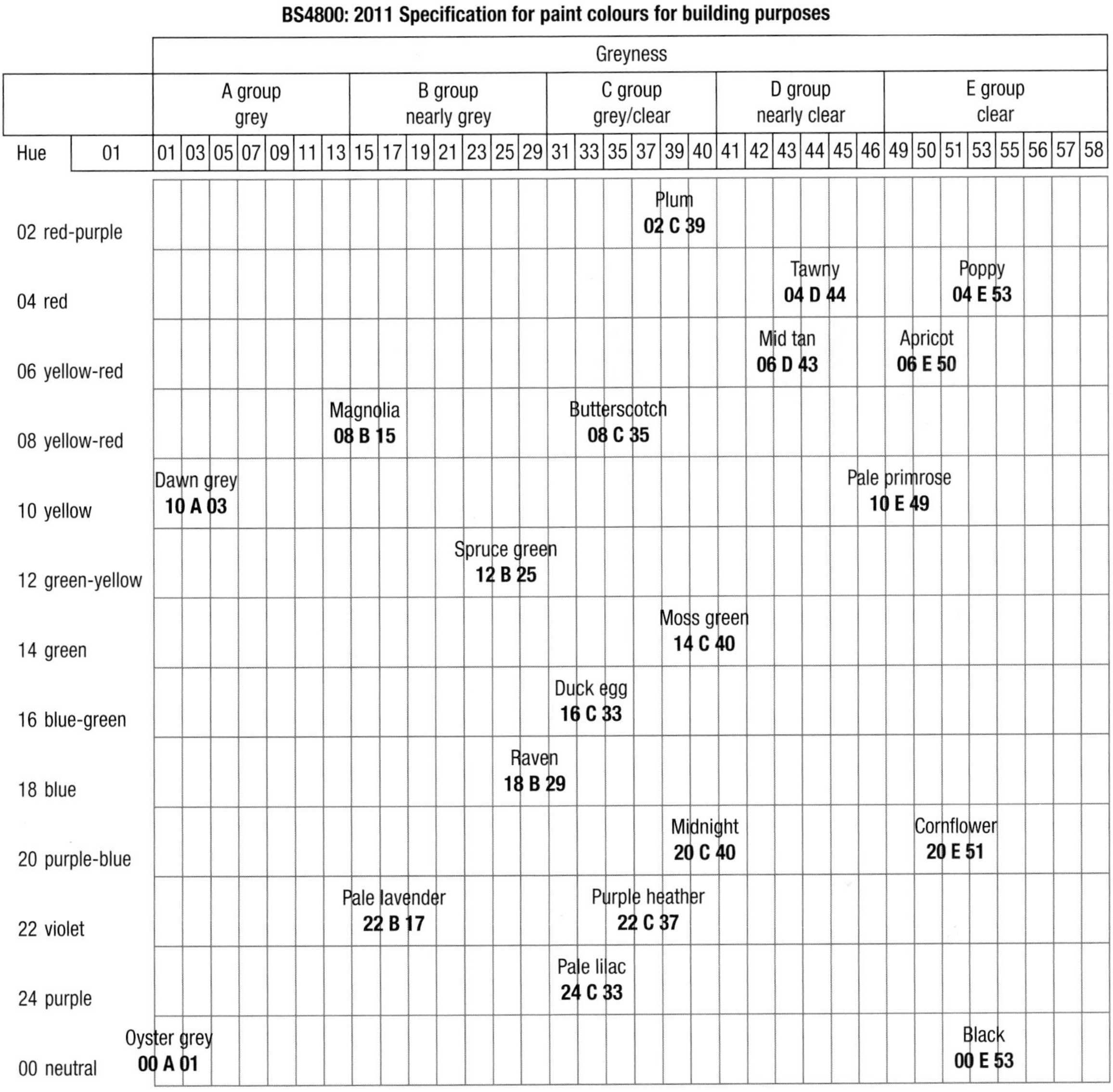

BS4800: 2011 Specification for paint colours for building purposes

Hue	01	A group grey							B group nearly grey							C group grey/clear						D group nearly clear						E group clear							
	01	01	03	05	07	09	11	13	15	17	19	21	23	25	29	31	33	35	37	39	40	41	42	43	44	45	46	49	50	51	53	55	56	57	58
02 red-purple																				Plum 02 C 39															
04 red																									Tawny 04 D 44						Poppy 04 E 53				
06 yellow-red																								Mid tan 06 D 43					Apricot 06 E 50						
08 yellow-red									Magnolia 08 B 15									Butterscotch 08 C 35																	
10 yellow		Dawn grey 10 A 03																										Pale primrose 10 E 49							
12 green-yellow														Spruce green 12 B 25																					
14 green																					Moss green 14 C 40													
16 blue-green																	Duck egg 16 C 33																	
18 blue															Raven 18 B 29																			
20 purple-blue																					Midnight 20 C 40									Cornflower 20 E 51					
22 violet										Pale lavender 22 B 17									Purple heather 22 C 37															
24 purple																	Pale lilac 24 C 33																	
00 neutral		Oyster grey 00 A 01																													Black 00 E 53				

Fig. 15.3 British Standards Colour System with some illustrative examples

The *Natural Colour System* is based on the assumption that for people with normal vision there are six pure colours: yellow, red, blue, green, white and black. The four colours yellow, red, blue and green are arranged around the *colour circle*, which is then subdivided into 10% steps. For example, yellow changes to red through orange, which could be described as Y50R (yellow with 50% red) (Fig. 15.4). In order to superimpose the black/white variation and also intensity of colour, each of the 40 10% steps around the colour circle may be represented by colour triangles, with the pure colour at the perimeter apex and the

vertical axis illustrating blackness/whiteness. A colour may therefore be described as having 10% blackness and 80% chromatic intensity. The full colour specification thus reads NCS S 1080-Y50R for an orange with 10% blackness, 80% chromatic intensity at yellow with 50% red. The S refers to the standardised NCS 1950 original colours.

The system allows for a finer subdivision of the colour circle, and this is necessary to define any colour and to make direct comparisons with colours defined within the British Standards system. Thus Magnolia (BS 08 B 15) is NCS 0606-Y41R (6% blackness,

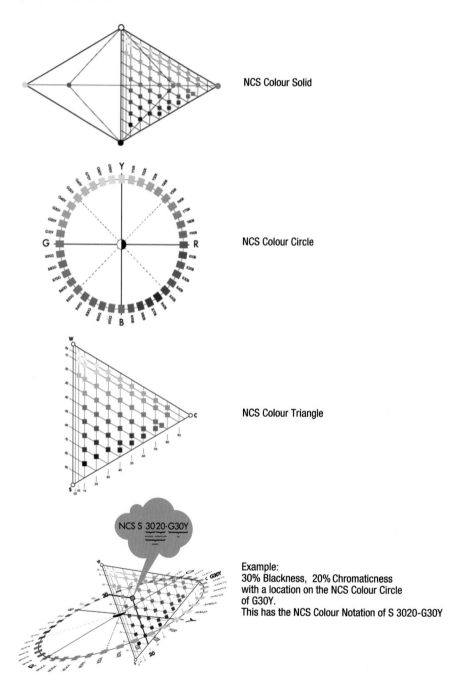

NCS Colour Solid

NCS Colour Circle

NCS Colour Triangle

NCS S 3020-G30Y

Example:
30% Blackness, 20% Chromaticness
with a location on the NCS Colour Circle
of G30Y.
This has the NCS Colour Notation of S 3020-G30Y

Fig. 15.4 NCS – Natural Colour System® © property of and used on licence from NCS Colour AB, Stockholm 2014. References to NCS® © in this publication are used with permission from the NCS Colour AB

6% chromatic intensity on a yellow with 41% red); Plum (BS 02 C 39) is NCS 5331-R21B, and Midnight (BS 20 C 40) is NCS 7415-R82B in the *Natural Colour System®©*. However, for specification purposes the closest standardised NCS references are generally used, i.e. S 0505-Y40R, S 5030-R20B, and S 7020-R80B, respectively.

RAL CLASSIC SYSTEM

The *RAL Classic System®©* is used significantly within the building industry for defining the colours of finishes, particularly to plastics and metals, but also to materials such as glazed bricks. Typical applied finishes include acrylics, polyesters and polyurethane as well as some paints and lacquers. The RAL system, established in Germany in 1925, has developed through several phases. It commenced with 40 colours; subsequently, many were added and others removed, leaving 170 standard colour shades. Because of its development, the RAL system (designated RAL 840-HR) does not have a systematic order of colours with equal steps between shades. Colours are defined by four digits, the first being the colour class (1 yellow, 2 orange, 3 red, 4 violet, 5 blue, 6 green, 7 grey, 8 brown and 9 black/white) and the further three digits relate only to the sequence in which the colours were filed. An official name is also applied to each standard RAL colour (e.g. RAL 1017 Saffron Yellow, RAL 5010 Gentian Blue, RAL 6003 Olive Green). Some additional colours have been added to the RAL classic collection, giving 213 colours. The collection for matt shades is designated RAL 840-HR and that for glossy shades is RAL 841-GL. Computer applications relate the RAL classic colours to RGB (red/green/blue), HLC (hue/lightness/chroma) and offset printing format colours.

RAL DESIGN SYSTEM

Unlike the *RAL Colour Collection* which only has a limited selection of standard colours, the *RAL Design System®©* has 1625 colours arranged in a colour atlas based on a three-dimensional colour space defined by the coordinates of hue, lightness and chroma. Hue is the attribute of colour; for example, red, blue or yellow. Lightness ranges from black to white, and chroma is the saturation or intensity of the colour. The system is equivalent to the HLS (hue, lightness, saturation) system which is used alongside RGB (red, green, blue) in many computer colour systems. The *RAL Design System* is similar to the *Natural Colour System*, except

that it is based on a mathematical division of the whole visible wavelength spectrum, rather than the visually assessed four standard colours: yellow, red, blue and green.

The colour spectrum is therefore divided into 10° steps around a 360° circle (red 0°, yellow 90°, green 180° and blue 270°) Each step, illustrated on a page of the associated colour atlas, represents a particular hue. For each hue on the colour atlas page, samples illustrate lightness decreasing from top to bottom and intensity or saturation increasing from the inside to the outside. Any colour is therefore coded with the three numbers relating to hue, lightness and chroma; for example, 070 70 60. The standard RAL colour collection numbers do not fit neatly into the RAL design system coding but any colour may be defined; thus Saffron Yellow (RAL 1017) becomes 069.9 75.6 56.5. However, as the number defining the hue is not exactly 070, the colour Saffron Yellow will not appear on the atlas page. Computer programs generating colour through the attributes of hue, lightness and chroma can immediately formulate colours according to this system. The current *RAL Design System* offers 1625 shades through its website but 2328 digital colours are available on the RAL APP iColours.

In addition to the established *RAL Classic* and *RAL Design* systems, RAL offers the new *RAL Effect®©* system for water-borne paints and lacquers with 420 solid colours including 70 metallic colours as well as *RAL Plastics®©* (300 colours) and *RAL Feeling®©* – the latter relating to new architectural colour trends.

COLOUR PALETTE NOTATION

The *Colour Palette®© (Dulux)* notation is based on three factors: hue, light reflectance value (LRV) and chroma (Fig. 15.5). The hue or colour family is derived from eight divisions of the spectrum, each of which is subdivided into a further 100 (0–99) divisions to give a precise colour within a particular hue.

Hue families:

RR	magenta to red
YR	red to orange
YY	orange to yellow to lime
GY	lime to green
GG	green to turquoise
BG	turquoise to blue
BB	blue to violet
RB	violet to magenta

Example

30BB 08/263

| HUE | LRV | CHROMA |

Hue
The colour family.

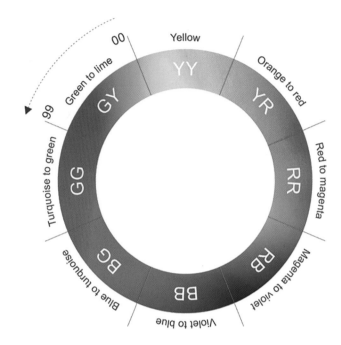

Light Reflectance Value (LRV)
The lightness or darkness of the colour.

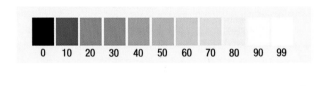

0 10 20 30 40 50 60 70 80 90 99

The higher the number the lighter the colour.

Chroma
The intensity of the colour.

Neutral Grey   Full Chroma

The higher the number the more intense the colour.

Fig. 15.5 Colour Palette Notation. *Images:* Courtesy of AkzoNobel group. (*Dulux Master Palette* and *Colour Palette* are trademarks of the AkzoNobel Group.)

Light reflectance value (LRV) is a measure of lightness or darkness on a 00–99 scale, with light colours having a high two-digit number. Thus, most pastel shades have a light reflectance value between 75 and 83 and the majority of colours fall within the range 04 (very dark) to 83 (very light).

Chroma is a measure of the saturation or strength of the colour measured in 1000 steps from 000–999, with high numbers indicating high saturation or intensity.

A strong bright-lemon colour would be specified as:

37YY　　61 /　　877
hue　　　LRV　　chroma
(where LRV is light reflectance value).

The *Dulux Trade Colour Palette*®© is based on the *Colour Palette* system, specifying the three factors hue, light reflectance value and chroma. Colours are divided into six sections across the spectrum from red to violet. Hues are then subdivided by toning shades.

A deep emerald-green colour would be described as:

10GG　　15 /　　346
hue　　　LRV　　chroma.

PANTONE

The *Pantone Matching System*®© is frequently used by architects and interior designers to specify colours for plastics, fabrics and paint. Colours are normally sampled on paper swatches in either coated (C) or uncoated (U) finish. Within *Pantone* there are various subsets, including one specifically for paints and interiors. This set is defined by two digits followed by a dash, four digits and a letter suffix. The letters TCX and TPX differentiate between editions printed on cotton or paper, respectively. Each colour also has a name (e.g. Pantone 15–1247 TCX Tangerine).

The *Pantone Goe*®© system has 2058 colours arranged chromatically, defined by a three-part hyphenated numbering system. The first number (1–165) defines the colour family, the second number (1–5) defines the page within the family, and the final number defines the position on the page (1–7). A suffix C or U defines coated or uncoated paper (e.g. Pantone 105–5-3 C).

Pantone®© colours may be approximately matched on a RGB screen or in standard CMYK printing, but by using hexachrome printing, which includes orange and green, the vibrancy of the *Pantone* colours can normally be reproduced. The *Pantone* colour selector gives the relevant HTML and RGB values.

MUNSELL

The *Munsell*®© system defines colour by a combination of hue, value (lightness or darkness) and chroma (saturation or brilliance). The three attributes of colour are depicted in a three-dimensional space model, with basic colour around the perimeter, the neutral vertical axis at the centre (black at the bottom) and chroma saturation increasing outward from the central axis. Colours are defined by a colour swatch book in matte or gloss finish. The ten major hues are red, yellow-red, yellow, green-yellow, green, blue-green, blue, purple-blue, purple, red-purple as well as neutral.

VISUAL COMPARISON OF PAINT COLOURS

The European Standard for the comparison of paint colours (BS EN ISO 3668: 2001) is based on observation under specified illumination and viewing conditions (either natural diffuse daylight or scientifically specified artificial light). The colour-matching process is based on an assessment of the differences in hue, chroma and lightness between test panels and reference colour standards.

Cross-referencing between colour systems

The NCS system can supply digital cross-references to the 237 colours in the British Standard BS 5252, to 189 of the 194 colours in RAL and also to the 1150 Colour Matte Finish and 1500 Glossy Finish Collection colours of the Munsell system. A printed cross-reference is available between the NCS system and the European system CMYK (cyan, magenta, yellow, black) used for colour printing. Translation tables from NCS through RGB (computer screen) to CMYK (printing) are available for working with CAD systems on standard operating systems. A NCS colour scanner will identify the closest NCS reference colour to any sample. Commercial paint outlets will scan colour samples to give an accurate computer-generated colour match.

Paints

The British Standard BS 6150: 2006 gives a detailed description of paint systems and their appropriate applications to the range of building materials used within construction. The standard describes pre-treatments for substrates; also the maintenance and deterioration of paint finishes. The choice of a paint system is based on a combination of factors including exposure conditions, function, substrate, constraints, initial cost and maintenance.

COMPONENTS OF PAINTS

Paints consist of a blend of components, each with their specific function. Commonly, these include the binder (or medium), solvent, base, extenders, pigments and driers, although other additives may be incorporated into specialist paints.

The binder solidifies to produce the paint film. Traditionally, the binder was natural linseed oil, which set by gradual oxidation upon exposure to air. However, linseed oil has now been largely replaced by alkyd resins which oxidise in air, or vinyl and acrylic resins which solidify by drying. To ensure adequate fluidity of the paint during application by brushing or spraying, either water or organic solvents (hydrocarbons, ketones or esters) are incorporated; paint thinners have the same effect. The base material, usually white titanium dioxide, produces the required opacity, although the *body* of the paint may be increased by the incorporation of inert extenders such as silica, calcium carbonate, china clay or barytes. Colouring materials are frequently a mixture of organic and inorganic dyes and pigments. Driers which induce the polymerisation of the binder ensure a rapid drying process.

Changes in legislation and environmental concerns have led to the development of paints with reduced levels of volatile organic compounds (VOCs). Mainly this has been through the increased use of water-borne rather than solvent-borne paints. In some respects water-borne paints have the advantage. They have low odour emissions, brushes can be cleaned in water and they will tolerate damp surfaces. However, they are not ideal for external use in cold and wet conditions. Other developments have been towards *high-solids* paints, which have low solvent content and therefore very low VOC emissions.

A further trend is towards the use of *natural paints* based on plant oils, casein, cellulose and mineral compositions. These are free from the high embodied energy materials such as petrochemical products and titanium dioxide. Natural paints consume considerably less energy in manufacture, are environmentally friendly in application and less problematic in waste management; however, they are not suitable for all applications. Some casein emulsions, available in a range of colours, are suitable for internal application only. However, newer *eco-paints* including water-based gloss paints are suitable for exterior use. Many products now carry eco-labels – some indicating the VOC content.

PAINT SYSTEMS

Coats within a paint system perform specific tasks. Usually, a complete system would require primer, undercoat and finishing coat, although in the case of new external materials, four coats may be appropriate.

Primers

The primer must adhere well to the substrate, offer protection from deterioration or corrosion and provide a good base for the undercoat. To ensure adhesion, the substrate surface must be free of loose or degraded material. Appropriate systems are indicated in Table 15.1. For use on timber, primers may be oils, alkyd resins or acrylic emulsions, frequently with titanium oxide. Aluminium wood primer is recommended for resinous woods and to seal aged creosoted and bitumen-coated surfaces. For the corrosion protection of ferrous metals, primers incorporate zinc-rich compounds within oils or alkyd resins. The use of lead-based paints in the UK is subject to the Control of Lead at Work Regulations, 2002. The newly developed low-VOC coatings offer temporary protection against the corrosion of structural steelwork as either pre-fabrication or post-fabrication primers. Alternatively, acrylated rubber paints which form a physical barrier over steel may be used as primers. For non-ferrous metals, zinc phosphate primers are frequently used. The application of primers suitable to ferrous metals may cause increased corrosion on non-ferrous substrates, particularly aluminium. Masonry paints are usually based on alkyd or acrylic resins with titanium oxide; where surfaces are likely to be alkaline, such as new plaster, brickwork or concrete, alkali-resisting primer should be used. Two-component primers are formulated to give good adhesion on glazed tiles, glass, anodised aluminium, powder-coated steel and stove enamel surfaces.

Undercoats

Undercoats provide cover and a good base for the finishing coat. Most undercoats are based on alkyd resins or acrylic emulsions.

Table 15.1 Recommended primers for various substrates

Primer	Suitable substrates and conditions
Timber	
Alkyd primer (solvent-borne)	Softwoods and hardwoods, interior and exterior
Acrylic primer (water-borne)	Softwoods and hardwoods
Aluminium wood primer	Resinous softwoods and hardwoods
Preservative primer	Exterior use, contains fungicides
Plaster and masonry	
Acrylic primer sealer	Loose, friable surfaces
Alkali resisting primer	Plaster, cement and concrete
Ferrous metals	
Pre-treatments	Improve adhesion of paint systems
Zinc phosphate	Steel, iron, galvanised steel. Good rust inhibitor
Zinc-rich primer	Steel. Two-component system. Must be a blast-cleaned surface
Chromate primer	Ferrous and non-ferrous metals
Epoxy primer	Steel. Two-component system containing aluminium powder
Acrylated rubber	Steel, iron, galvanised steel. Must be a full acrylated rubber system
Non-ferrous metals	
Zinc phosphate	Aluminium
Acrylated rubber	Aluminium. Must be a full acrylated rubber system
Acrylic metal primer	Aluminium, copper, lead, brass. Quick-drying water-based primer

Finishing coats

Finishing coats provide a durable and decorative surface. Some gloss, eggshell and satin finishes are still based on oils and alkyd resins, although water-borne products are increasingly becoming predominant. Some water-borne gloss finishes tend to be visually softer and are more moisture permeable than the traditional solvent-borne hard glosses. However, they have the advantage of quick drying without the evolution of solvent odour; generally they are more durable and do not yellow upon ageing. Matt and silk finishes are usually vinyl or acrylic emulsions.

SPECIAL PAINTS

Light-reflecting paint

Paints with lower light absorbency and greater reflectivity can enhance the distribution of light within room enclosures, thus reducing the energy input required to achieve appropriate lighting levels. Products are available in matt and eggshell finishes.

Multicolour paints

Multicolour paints incorporating flecks give a hard-wearing surface which may be glazed over to ease the removal of graffiti. Application is with a spray-gun, which may be adjusted to change the pattern and texture of the fleck. This type of paint system may be applied to most dust- and grease-free internal surfaces.

Broken-colour paints

Broken-colour effects, reflecting the traditional processes of graining, marbling, ragging and stippling, are once again popular. Most modern broken-colour effects require a base coat, applied with a brush or roller, which is then overpainted with a clear coloured glaze. The glaze is then patterned or distressed to create the desired effect. One proprietary system uses a special rag-roller, which flails the wet finish coat giving random partial exposure of the darker first coat. Other appropriate tools include stippling brushes, combs and sponges. Alternative finishes include metallic, pearlescent and graining effects. An iridescent finish produces a two-tone shimmer effect by optical interference of the reflected light. Water-based acrylic glazes are virtually odour-free and are touch dry within two hours.

Acrylated rubber paints

Acrylated rubber paints are suitable for internal and external applications exposed to chemical attack or wet and humid atmospheric conditions. Acrylated rubber paints are tending to replace chlorinated rubber coatings which rely on carbon tetrachloride solvent, now considered environmentally damaging. Acrylated rubber paints may be applied to metal or masonry by either brushing or spraying. Usually, a film of dry thickness 100 μm is applied compared to 25–30 μm for most standard paint products.

Heat-resisting paint

Standard aluminium paint, which has a lustrous metallic finish, is resistant to temperatures up to 230–260°C, but special grades with resistance up to 600°C are available. A dry film thickness of 15 μm is typical. Acrylated rubber paints can usually be used satisfactorily to 100°C.

Flame-retardant paints

Flame-retardant paints emit non-combustible gases when subject to fire, the usual active ingredient being antimony oxide. Combustible substrates such as plywood and particleboard can be raised to Class 1 (BS 476 Part 7) surface spread of flame. Products include matt, semi-gloss and gloss finishes, and may be applied by brush, roller or spray.

Intumescent coatings

Thin-film intumescent coatings, typically 1 or 2 mm in thickness, offer fire protection to structural steel without noticeable visual effect. In the event of fire, the thin coating expands up to 50 times to form a layer of insulating foam. The carbonaceous material in the coating, typically starch, is charred, while the heat also causes the release of acids. These act to produce large volumes of non-inflammable gases which blow up the charring starch within the softened binder into an insulating cellular carbon layer. Coatings may be applied to give 30, 60 or 120 minutes' fire protection. Intumescent emulsion paints or clear varnishes are appropriate for use on timber, although, where timber has been factory impregnated with a flame-retardant salt, the compatibility of the intumescent coating and flame retardant must be verified.

Fungicide paints

Fungicide paints, for application in areas where mould growth is a recurrent problem, usually contain a blend of fungicides to give high initial activity and a steady long-term performance. The latter may be achieved with fungicide constituents of low solubility which are gradually released to the surface during the lifetime of the paint. Matt acrylic finishes are available in a range of colours.

Enamel paints

Enamel paints based on polyurethane or alkyd resins give highly durable, impact-resistant, easily cleaned, hard-gloss surfaces. Colours tend to be strong and bright, suitable for machinery and plant in interior and exterior locations.

Micaceous iron oxide paints

Micaceous iron oxide paints have good resistance to moisture on structural steelwork, iron railings, etc., due to the mica plates which reduce permeability to moisture vapour. A dry film thickness of 45–50 μm is typical, thus requiring longer drying times than standard paint products. Micaceous iron oxide paints should be applied over an appropriate metal primer.

Masonry paints

Smooth and sand-textured masonry paints are suitable for application to exterior walls of brick, block, concrete, stone or renderings. Where fine cracks are present, these can often be hidden using the sand-textured material. Usually, masonry paints contain fungicides to prevent discoloration by moulds and algae. Acrylic resin-based products are predominantly water-based; however, fast-drying solvent-based systems are also produced. Mineral silicate paints form a crystalline protective layer over the masonry surface, which tends to be more durable than the organic finishes from synthetic resins.

Water-repellent and waterproof paints

Silicone water-repellent paints may be applied to porous surfaces including brick, concrete, stone and renderings to prevent damp penetration. Such treatment does not prevent rising damp, but will allow the continued evaporation of moisture within the masonry. Two-pack epoxy waterproofing systems may be applied to sound masonry surfaces to provide an impervious coating. Typical applications are to rooms where condensation causes the blistering of normal paint films; also in basements and solid external walls where penetrating water is a problem, provided that a good bond can be achieved between substrate and epoxy resin. Bituminous paints provide a waterproof finish to metals and masonry and may also be used as a top dressing to asphalt or for renovating bitumen membrane roofing. Aqueous bitumen coatings, if fully protected against physical damage, can provide a vertical membrane where the external ground level is higher than the internal floor level.

Self-cleaning paint

Façade paint, incorporating a microstructure similar to that found on the surface of lotus leaves, prevents dirt from attaching itself to the surface, so it becomes self-cleaning whenever it rains. The material is suitable for masonry and rendered surfaces, and is available in a full range of tints.

Epoxy paints

Epoxy ester paint coatings are highly resistant to abrasion and spillages of oils, detergents or dilute aqueous chemicals. They are therefore frequently used as finishes to concrete, stone, metal or wood in heavily trafficked workshops and factories. Many are produced as two-pack systems requiring mixing immediately before application.

Anti-graffiti paints

The main purpose of anti-graffiti paints is to aid the easy removal of graffiti. Paints may be sacrificial or permanent. The sacrificial paints, usually based on wax emulsions, acrylates or biopolymers, are easily removed by jets of hot water. The permanent anti-graffiti paints based on tough fluoro-acrylics or two-component polyurethane systems are resilient to chemical and abrasive cleaning techniques. The permanent anti-graffiti paints are more successful on smooth surfaces. Multicolour finishes help disguise any discoloration.

Anti-microbial coatings

Anti-microbial wall and ceiling coatings, incorporating silver ions, can reduce bacteria levels within hospitals and hygienic environments.

Wood finishes

Wood finishes include wood stains, varnishes and oils. Wood stains are pigmented resin solutions which penetrate into the surface and may then build up a sheen finish. Varnishes are unpigmented resin solutions which are intended to create a surface film. Timber preservatives are described in Chapter 4.

WOOD STAINS

Most wood stain systems for exterior use include a water- or solvent-based preservative base coat which controls rot and mould growth. Typical formulations include zinc or copper naphthenate, dichlorofluanid, tri-(hexylene glycol) biborate and disodium octaborate tetrahydrate. Wood stain finishes are either low-, medium- or high-build systems, according to the particular application. They usually contain iron oxide pigments to absorb the ultraviolet light which otherwise causes the surface degradation of unprotected timber. Generally, for rough-sawn timber, deeply penetrating wood stains are appropriate, whereas for smooth-planed timber a medium- or high-build system gives the best protection from weathering. Products are based on acrylic and/or alkyd resins.

For sawn timber, both organic solvent-based and water-based materials are available, usually in a limited range of colours. Solvent-based low-build products, which are low in solids, penetrate deeply, leaving a water-repellent matt finish, enhance the natural timber grain and are suitable for timber cladding. Deep penetration should eliminate the risk of flaking or blistering on the surface. Medium- and high-build products for exterior joinery offer the choice of semi-transparency to allow the grain to be partially visible, or opaque colours for uniformity. Products are available in a wide range of colours with matt or gloss finishes. The first coat both penetrates and adheres to the surface, while the second coat provides a continuous microporous film which is both permeable to moisture vapour and water repellent, thus reducing the moisture movement of the timber. Additional coats should be applied to end grain. The coating, typically 30–40 μm thick, should remain sufficiently flexible to accept natural timber movements. Low-VOC products based on water-borne emulsions or high-solids solvent-borne resins are now generally available.

VARNISHES

Traditional varnishes are combinations of resins and drying oils, but most products are now based on modified alkyd resins. Polyurethane varnishes are available in matt, satin or gloss finishes, based on either water- or solvent-based systems. The solvent-based systems produce the harder and more durable coatings up to 80 μm thick, suitable for exterior woodwork. Products either retain the natural wood colour, enhance it, or add colour. Screening agents to protect timber from the effects of ultraviolet light are normally included in the formulations. Urethane-modified alkyd resins are suitable for interior use, and have the advantage of high resistance to scuffing and hot liquids. External weathering causes eventual failure by flaking and peeling as light passing through the varnish gradually degrades the underlying wood surface. For example, hardwood doors decorated with polyurethane varnish, protected from rain and direct sunlight by a porch, should have extended periods between maintenance. End grain should be sealed to prevent trapped moisture from encouraging the development of staining fungi.

NATURAL WOOD FINISHES

Wood finishes entirely composed of natural products are also available. These are based on blends of vegetable oils, beeswax and larch resin, and may incorporate minerals and earth pigments such as iron oxide and titanium dioxide for colour and opacity.

OILS

Oils such as teak oil are used mainly for internal applications. Formulations based on natural oils for exterior use are high in solids, producing an ultraviolet-resistant, microporous finish which may be transparent or opaque. The finish, which should not flake or crack, may be renovated by the application of a further coat.

References

FURTHER READING

Architects' Journal. 2010: *Colour and texture. Specification.* London: Architects' Journal.

Edwards, L. and Lawless, J. 2007: *The natural paint decorator.* London: Kyle Cathie.

ICI Paints: *ICI Dulux colour palette.* Slough: Imperial Chemical Industries plc.

Moor, A. 2006: *Colours of architecture.* London: Mitchell Beazley.

Reichel, A., Hochberg, A. and Kopke, C. 2013: *Plaster, render, paint and coatings: Details, products, case studies.* Basel: Birkhäuser.

Smith, M. 2011: *Eco-paints.* Technical Article. Newcastle-upon-Tyne: NBS.

The Plan. 2010: *Building with colour. Plans and details for contemporary architects.* London: Thames and Hudson.

STANDARDS

BS 381C: 1996
  Specification for colours for identification coding and special purposes.
BS 476
  Fire tests on building materials and structures:
    Part 4: 1970   Non-combustibility test for materials.
    Part 6: 1989   Method of test for fire propagation of products.
    Part 7: 1997   Method of test to determine the classification of the surface spread of flame of products.
BS 1070: 1993
  Black paint (tar-based).
BS 2015: 1992
  Glossary of paint and related terms.
BS 2523: 1966
  Specification for lead-based priming paints.
BS 3416: 1991
  Specification for bitumen-based coatings for cold applications, suitable for use in contact with potable water.
BS 3698: 1964
  Calcium plumbate priming paints.
BS 3761: 1995
  Specification for solvent-based paint remover.
BS 3900
  Paints and varnishes:
    Part 0: 2010   Methods of test for paints. Index of test methods.
    Part D1:1998   Visual comparison of the colour of paints.
BS 4652: 1995
  Metallic zinc-rich priming paint (organic media).
BS 4756: 1998
  Specification for ready-mixed aluminium priming paints for woodwork.
BS 4764: 1986
  Specification for powder cement paints.
BS 4800: 2011
  Schedule of paint colours for building purposes.
BS 4800F: 2011
  Colour matching fan.
BS 4900: 1976
  Specification for vitreous enamel colours for building purposes.

BS 4901: 1976
Specification for plastics colours for building purposes.

BS 4904: 1978
Specification for external cladding colours for building purposes.

BS 5252: 1976
Framework for colour co-ordination for building purposes.

BS 5252F: 1976
Colour matching fan.

BS 6150: 2006
Painting of buildings. Code of practice.

BS 6949: 1991
Specification for bitumen-based coatings for cold application, excluding use in contact with potable water.

BS 7079: 2009
General introduction to standards for preparation of steel substrates before application of paints and related products.

BS 7664: 2000
Specification for undercoat and finishing paint.

BS 7719: 1994
Specification for water-borne emulsion paints for interior use.

BS 7956: 2000
Specification for primers for woodwork.

BS 8000
Workmanship on building sites:
    Part 12: 1989 Code of practice for decorative wallcoverings and painting.

BS 8202
Coatings for fire protection of building elements:
    Part 1: 1995 Code of practice for the selection and installation of sprayed mineral coatings.
    Part 2: 1992 Code of practice for the use of intumescent coating systems to metallic substrates for providing fire resistance.

BS EN ISO 150: 2007
Raw, refined and boiled linseed oil for paints and varnishes. Specifications.

BS EN 351–1: 2007
Durability of wood and wood-based products. Preservative-treated solid wood.

BS EN 927
Paints and varnishes. Coating materials and coating systems for exterior wood:

    Part 1: 2013   Classification and selection.
    pr Part 2: 2013 Performance specification.
    Part 3: 2012   Natural weathering test.
    Part 5: 2000   Assessment of the liquid-water permeability.
    Part 6: 2006   Exposure of wood coatings to artificial weathering.

BS EN 1062
Paints and varnishes. Coating materials and coating systems for exterior masonry and concrete:
    Part 1: 2004   Classification.
    Part 3: 2008   Determination of liquid water permeability.
    Part 6: 2002   Determination of carbon dioxide permeability.
    Part 7: 2004   Determination of crack bridging properties.
    Part 11: 2002 Methods of conditioning before testing.

BS EN ISO 3251: 2008
Paints, varnishes and plastics. Determination of non-volatile matter content.

BS EN ISO 3668: 2001
Paints and varnishes. Visual comparison of the colour of paints.

pr EN ISO 4618: 2013
Paints and varnishes. Terms and definitions.

BS EN ISO 4624: 2003
Paints and varnishes. Pull-off test for adhesion.

BS EN ISO 9717: 2013
Metallic and other inorganic coatings. Phosphate conversion coating of metals.

BS EN ISO 10545–16: 2012
Ceramic tiles. Determination of small colour differences.

BS EN ISO 11341: 2004
Paints and varnishes. Artificial weathering and exposure to artificial radiation.

BS EN ISO 11507: 2007
Paints and varnishes. Exposure of coatings to artificial weathering.

BS EN ISO 11997
Paints and varnishes. Determination of resistance to cyclic corrosion conditions.
    Part 1: 2006   Wet/dry/humidity.
    pr Part 2: 2013 Wet/dry/humidity/UV light.

BS EN 12206–1: 2004
Paints and varnishes. Coating of aluminium and aluminium alloys for architectural purposes.

BS EN 12878: 2014
Pigments for the colouring of building materials based on cement and/or lime. Specifications.

BS EN ISO 12944
Paints and varnishes. Corrosion protection of steel structures by protective paint systems:
Part 5: 2007   Protective paint systems.

BS EN 13300: 2001
Paints and varnishes. Waterborne coating materials and coating systems for interior walls and ceilings. Classification.

BS EN 13438: 2013
Paints and varnishes. Powder organic coatings for hot dip galvanized or sheradised steel products for construction purposes.

BS EN 13523–22: 2010
Coil coated metals. Test methods. Colour difference. Visual comparison.

BS EN 14188: 2009
Joint fillers and sealants. Specification for primers to be used with joint sealants.

BS EN ISO 14680: 2006
Paints and varnishes. Determination of pigment content.

BS EN ISO 15528: 2012
Paints, varnishes and raw materials for paints and varnishes. Sampling.

BS EN 16566: 2014
Paints and varnishes. Fillers for internal and/or external works.

pr EN 16623: 2013
Paints and varnishes. Reactive paints for fire protection of metallic substrates. Definitions, requirements, characteristics and marking.

BUILDING RESEARCH ESTABLISHMENT PUBLICATIONS

BRE Digests

BRE Digest 464: 2002
VOC emissions from building products (Parts 1 and 2).

BRE Digest 466: 2002
EN 927: the new European Standard for exterior wood coatings.

BRE Digest 503: 2007
External timber structures. Preservative treatment and durability.

BRE Information papers

BRE IP 7/03
Planned maintenance painting: Improving value for money.

BRE IP 20/12
Measuring the wellbeing benefits of interior materials.

TRADA PUBLICATIONS

Wood information sheets

WIS 2/3–1: 2012
Finishes for external timber.

WIS 2/3–16: 2012
Preservation treatment for timber. A guide to specification.

WIS 2/3–60: 2011
Specifying timber exposed to weathering.

ADVISORY ORGANISATIONS

NCS Colour Centre, 71 Ancastle Green, Henley-on-Thames, Oxfordshire RG9 1TS (01491 411717).

Paint Research Association, 14 Castle Mews, High Street, Hampton TW12 2NP (0208 487 0800).

Property Care Association, Lakeview Court, Ermine Business Park, Huntingdon PE29 6XR (0844 375 4301).

16

ENERGY-SAVING MATERIALS AND COMPONENTS

CONTENTS

Introduction

The trend towards increasingly energy-conscious design has resulted in a greater focus on energy-saving materials and components. These include photovoltaics and solar collectors, which turn the sun's energy into electricity and hot water, respectively. Light tubes and wind catchers are energy-saving devices which can make modest reductions in the energy consumption of buildings in the context of a holistic energy-efficient strategy. Micro-wind turbines, within the appropriate context, can generate modest quantities of electricity. Rainwater-harvesting systems can eliminate or reduce mains potable water consumption for certain less critical domestic uses.

Photovoltaics

Photovoltaics (PVs) are silicon-based devices which, under sunlight, generate a low-voltage direct electric current. The quantity of electricity produced is directly related to the intensity of incident solar radiation or irradiance (W/m^2). Both direct and diffuse sunlight are effective, although the intensity of direct sunlight is typically ten-fold that of an overcast sky, and the efficiency of energy conversion is around 15%. Photovoltaic cells are connected in series to generate a higher voltage. The supply is then passed through an inverter to convert the direct current into more usable alternating current at the standard voltage. The electricity generated may then be used within the building or sold back into the national supply if generation exceeds the demand.

Photovoltaic units are manufactured from a sandwich of at least two variants of mono- or poly-crystalline silicon (Fig. 16.1). These n- and p-type (negative and positive) silicon crystals generate electricity at their interface under solar (photon) radiation. Cells are arranged in rectangular modules ranging from 0.3–1.5 m², with a typical size of 900 × 1600 mm. A unit of 1.5 m² will produce a maximum of 250 W of electrical power under bright sunshine of 1000 W/m². Panels should last for up to 25 years, maintaining an efficiency of 80% of the original performance.

The cells are usually laminated with a protective layer of glass, backed with metal sheeting and mounted on a steel frame. However, translucent systems built into glass double-glazing units or flexible units faced with a plastic cover are available. Mono- and poly-crystalline silicon modules are usually black or blue in colour, respectively. The mono-crystalline modules are a uniform colour, whereas the poly-crystalline units have a sparkling surface. Other colours may be achieved, but with a reduced level of efficiency.

The alternative amorphous thin-film silicon (TFS) cell modules are matt red, orange, yellow, green, blue or black in colour, and can be laminated into glass or mounted on a flexible plastic backing. These systems have significantly lower levels of efficiency than the crystalline cells in good light conditions, but are the more efficient in poor light conditions. Thin-film amorphous silicone photovoltaics are much cheaper to produce than the crystalline systems which ameliorates their low efficiency at 8%. Hybrid units combining mono-crystalline and thin-film technologies give good output over the range of light conditions. It is anticipated that thin-film silicon photovoltaics will increase their share of the market from 14% in 2008 to 30% by 2015.

Developments include the production of cadmium telluride (CdTe) photovoltaics with an efficiency of 10%. The newer materials including copper indium selenide (CIS) and copper indium gallium selenide (CIGS) can produce 13% energy conversion. Multi-layer thin composites can achieve 20% conversion rates and the expensive gallium arsenide (GaAs) multi-function cells can achieve 40% efficiency, by absorbing the majority of the solar spectrum including infrared and ultraviolet light.

Further advances in photovoltaics include spray-on plastics which require less energy in their production than silicon-based products, but are less efficient in converting solar energy into electricity; also they are not so durable.

More total energy may be obtained by tracking the sun to optimise the orientation of the photovoltaic cells or by using mirrors to increase the solar radiation received. However, these systems involving moving components are more expensive and will require maintenance.

Photovoltaic systems are normally supplied as panels, but PV cladding, slates and glazing systems are

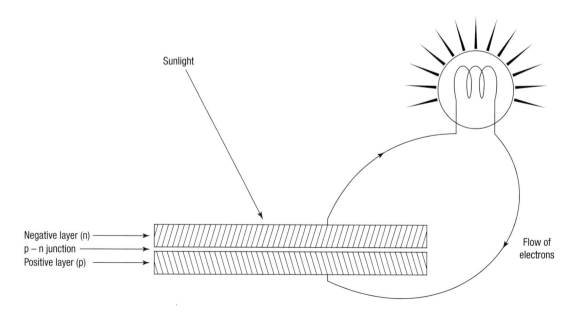

Fig. 16.1 Photovoltaic cell

Fig. 16.2 Photovoltaic array. *Photograph:* Arthur Lyons

also available. The location and tilt angle of PV installations are critical in respect of maximising output. The maximum efficiency in the northern hemisphere is gained from a south orientation with a tilt from the horizontal equal to the geographic latitude minus 20° (Fig. 16.2). Thus, for London, at a latitude of 51°, the optimum tilt is 31° from the horizontal. Generally an elevation of between 30° and 40° with orientation between south-east and south-west gives good results. Even a flat roof system will generate 90% of the potential peak output. However, in urban areas, the effects of the immediate environment must be taken into account when assessing the available solar energy. Shadowing and inter-reflection from adjacent buildings, together with local regular weather patterns, affect the total annual solar energy, which ultimately determines the electrical output.

Photovoltaic units, particularly those manufactured from crystalline silicon, decrease in performance with increased temperature, and any roof, curtain-walling or rainscreen cladding system should be naturally ventilated to maintain efficiency. The use of PVs should be an integral part of the energy strategy for the building.

A 1 m² crystalline photovoltaic unit operating efficiently will generate about 100 kWh per year. The pay-back time for such a unit would be of the order of 10 years including the energy of manufacture, and most installations will last between 20 and 30 years. One appropriately sized system offers a maximum 1.2 kW output, giving 1000 kWh per year, approximately one-fifth of the average UK annual domestic consumption.

PHOTOVOLTAIC ROOFING AND CLADDING MATERIALS

Photovoltaic slates (Fig. 16.3) and tiles, which have the general appearance of fibre cement slates, may be used on suitably orientated roofs as an ecological alternative to standard roofing, subject to appropriate planning consent. Units are compatible with a range of interlocking tiles, plain tiles, natural slates and fibre cement slates. The lower part of the individual slates comprises photovoltaic cells which are connected into a standard photovoltaic system. The tiles are 2.1 m strips of material, marked in units to resemble conventional roof tiles and overlapped to give the required visual effect. To obtain 1 kW of power under optimum conditions, 10 m² of slate or 16 m² of tiling is required depending on the lap. Photovoltaic roof tiles may be coordinated with matching solar panel tiles for hot-water systems, giving a fully integrated appearance to a tile roof.

Integrated photovoltaic materials are available as alternatives to conventional construction materials for façades, windows, roofs and skylights, and may be more appropriate for tall commercial buildings. Thin-film systems are usually based on CIS or CIGS photovoltaic technology.

PHOTOVOLTAIC GLAZING

Thin-film glazing modules can be produced to sizes up to 2.5 × 4.5 m for façade applications with a choice of colours and transparency levels. The electrical connections are made at the perimeter of the units within the frame system. Cell spacing may be tuned to give the

Fig. 16.3 Photovoltaic slates. *Photograph*: Courtesy of Solar Century (solarcentury.com)

optimum balance between electricity generation and daylight transmission.

Combined photovoltaic and solar collectors

Hybrid solar panels which collect both electrical and thermal energy (PV-T) occupy approximately 35% less roof area than separate PV and solar panels. A typical system (Fig. 16.4) is constructed with the PV cells bonded to a copper substrate which conducts the heat into water pipework. The combined system has the advantage that the PV cells maintain their generating efficiency by being kept cooler in bright sunlight. (Above 25°C, photovoltaic cells lose 0.5% efficiency for every degree rise in temperature.) Alternative systems use air cooling, where warmed air is drawn into the space heating system or into a heat exchanger to extract the thermal energy. All different types of photovoltaic cells are applicable to combined PV-T systems. Manufacturers offer systems which are focused on either maximising electrical or thermal output. Collectors of approximately 1.4 m² produce peak outputs of either 190 W electrical/460 W thermal or 170 W electrical/610 W thermal using mono-crystalline PV cells.

Fig. 16.4 Combined photovoltaic and thermal energy collector. *Image*: Courtesy of Caplin Homes Ltd., Newform Energy and Solimpeks – Volther

Solar collectors

The two standard types of solar collectors are the flat-plate and the evacuated-tube systems. Flat-plate collectors consist of a metal heat-absorbing plate, closely bonded to copper water pipes which transport the heated water to a storage system. The maximum efficiency is achieved using a low-emissivity matt black absorbing plate, which limits the loss of energy through re-radiation from the hot surface. A low iron-content double-glazed cover which admits the maximum quantity of short-wave energy protects the absorbing

Fig. 16.5 Evacuated tube solar collectors

plate and retains the entrapped heat. The underside of the pipework is insulated with fibreglass or poly-isocyanurate foam to prevent heat loss to the aluminium casing and the underlying roof structure or support system.

Evacuated-tube collectors (Fig. 16.5) consist of a double-layer glass tube, with a vacuum between the two layers. The outer glass is clear, admitting light and heat with minimal reflection. The inner tube is coated to absorb the maximum quantity of radiation. The heat from the inner tube is transferred in a sealed unit vaporising and condensing system to a heat exchanger within the main liquid flow which is circulating to the heat storage system. Evacuated-tube collectors are substantially more expensive than flat-plate collectors, but are more efficient if angled correctly and will produce higher temperatures.

Flat-plate solar collectors may be located in any unshaded location, at ground level or attached to buildings. The best orientation is directly towards the midday sun, but a variation of up to 15° east or west will have little adverse effect. The optimum tilt from horizontal for solar hot-water collectors for maximum all-year-round efficiency equals the location's latitude.

However, for increased winter efficiency, when solar gains are at a premium, the tilt from horizontal should be increased by 10° to pick up more energy at lower sun altitudes. Generally, in sunny climates flat-plate collectors are more efficient, but in more cloudy conditions evacuated-tube systems perform better. Solar hot-water systems are heavy, and must be fixed securely to suitable substrates. On tiled- or slated-pitched roofs an air gap should allow for the clear passage of rainwater and melting snow.

Solar collector systems with the appearance of standard plain concrete roofing tiles may be integrated into a tiled roof without significant visual effect. Furthermore, solar collector tiles may be matched to photovoltaic tiles giving a fully integrated appearance to a solar-responsive tile roof.

Hot water from the solar collector is usually circulated through an indirect system to a solar storage tank (Fig. 16.6). This acts as a heat store of preheated water to be fed into a standard hot water cylinder system, where the temperature can then be boosted from a boiler to the required level. Circulation may be through either a gravity thermosyphon system operated by hot-water convection with the storage tank located

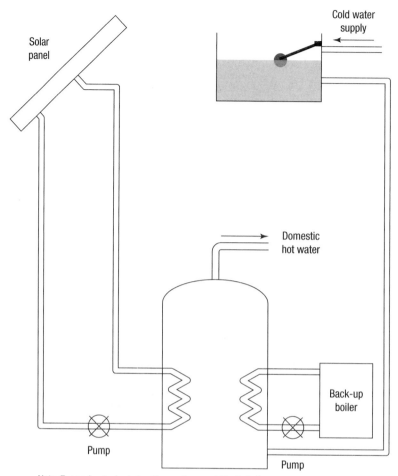

Note: Expansion tanks to heating systems not illustrated

Fig. 16.6 Solar collector and domestic hot water system

above the collector, or a pumped system, in which case the tank may be below the collector. The circulated water must contain antifreeze and a rust inhibitor. An alternative direct system feeds tap water directly into the solar collector, but scaling and corrosion of the pipework can be problematic. A 5 m² solar collector panel will heat 250 litres of hot water per day, which is a typical four-person family demand.

Solar energy district-heating plants in Europe, including Scandinavia, contribute significantly to a direct reduction in the energy requirements for small-town domestic hot-water systems. Water is pre-heated by large arrays of solar collectors before the local conventional-fuel heating system tops up the temperature to the required domestic level. Further-more, solar heating systems, in conjunction with large underground heat storage tanks, can significantly reduce winter energy consumption by preheating the water supplies during periods when direct solar gain is ineffective.

Solar air heating systems

Solar air-heating systems, mainly applicable to commercial buildings, consist of steel wall or roof panels which absorb solar energy. The heat is transmitted to the air within the profile sheeting channels. Low-energy fans draw the warmed air into the building as required, supplementing the main heating and fresh air system. Maximum heating efficiency is gained using darker external colours on the most southerly facing wall. A vertical wall captures the low-level winter sun when heating demands are greatest.

A modified system uses finely perforated steel or aluminium collectors, which are heated by solar energy. The exposed sheet metal warms a fine external boundary layer of air which is drawn through the perforations into the building's mechanical ventilation or heat-recovery system. Collectors are produced in a range of colours and profiles to complement a variety of building façades.

Light pipes

Light pipes or tubes transmit direct sunlight and natural daylight from roof level into the building space below (Fig. 16.7). At roof level a polycarbonate or glass dome or panel admits light into a highly reflective pipe which transmits it down to a diffuser at ceiling level. The mirror-finish reflective tube is available as flexible pipe or in rigid sections.

Standard light pipe diameters are 230 to 450 mm, although larger sizes up to 1000 mm are available for commercial applications. The systems should be free from condensation and not cause winter heat loss or summer solar gain to the building enclosure. Rectangular roof units, similar in appearance to standard or flush-fitting conservation roof lights, are available, and the ceiling unit can be a square diffuser to integrate into suspended ceiling systems.

A 330 mm-diameter system will typically deliver between 100 W from a winter overcast sky to 400 W under full summer sun, through a straight tube not exceeding 3 m in length. Light loss is typically 6% per metre of straight tube and 12% for each 45° bend. Roof units which incorporate prisms or track the sun's path with a moving reflector capture more light when the sun is low in the sky. Light pipe systems can offer energy-saving solutions to existing buildings, and may be considered as one element within a fully integrated lighting strategy for new builds.

A more sophisticated system combines the functions of both a light pipe and a wind catcher to admit natural daylight and ventilation into internal spaces that are poorly served by normal external glazing.

SUNLIGHT TRANSPORTING SYSTEMS

Sunlight received on the roof by collecting panels can be transported to any internal location within 20 m, using 30 mm-diameter fibre optic cable bundles terminated by light-emitting luminaires. The output is a combination of diffuse and parallel light.

Air and ground source heat pumps

Heat pumps operate by withdrawing low-temperature heat from a source and rejecting it at a higher temperature. The standard process uses a refrigeration cycle driven by an electric motor. The refrigerant is compressed and releases heat in a condenser, and then it subsequently expands within the cycle, extracting more heat in the evaporator before being re-compressed. The key environmental aspect of heat-pump systems is that more useful heat is generated from the electricity used in the compressor pump, compared to the fuel consumed by traditional gas or electric heating systems. A typical system may produce 4 kW of useful heat for each 1 kW of electricity consumed.

The standard heat source is the ground, but alternatively air or groundwater flow may be used. Ground temperatures in the UK are usually at around 12°C. For ground -source systems the size of the installation depends upon the anticipated load. The

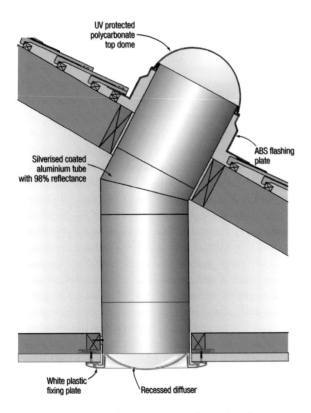

UV protected polycarbonate top dome

ABS flashing plate

Silverised coated aluminium tube with 98% reflectance

White plastic fixing plate

Recessed diffuser

Fig. 16.7 Light pipe. *Illustration:* Courtesy of Monodraught

ground loop may be a horizontal array of pipes to a minimum depth of 1.0 m or a set of 60–200 m-deep vertical bore heat exchangers. Soil conditions significantly affect the efficiency of ground-source systems. Specifically, dry, sandy soils with low thermal conductivity are unsuitable for horizontal systems. Ground-source heat pumps may be used in reverse for summer cooling. Air-source heat pumps can extract heat from the air even when the outside temperature is as low as -15°C.

Small-scale heat pumps are ideally suited to warm air or underfloor heating systems rather than radiators, as the maximum temperature attainable is around 55°C. Large units for non-domestic applications can produce outputs of up to 130 kW and water temperatures of up to 70°C. The design of heat-pump heating systems is described in BS EN 15450: 2007.

Groundwater cooling systems

Open-loop groundwater cooling systems offer an alternative to energy-intensive air-conditioning systems in larger buildings. Groundwater in the UK is typically at 12°C, although in London 14°C is common. Subject to official approval, groundwater may be abstracted for use in cooling systems, provided that appropriate provision is made for its disposal. This may be to lakes, rivers and surface drainage systems or returned to aquifers by injection into the ground. The system in operation for Portcullis House, Westminster uses two 150 m boreholes in the chalk aquifers. These extract water at 14°C, use it within heat exchangers, and then discharge of it at 21°C. Some greywater is used for non-potable functions.

Wind catchers

Wind catchers have been standard architectural features on the roofs of buildings in hot, dry climates for centuries. However, in order to reduce energy costs associated with air-conditioning systems, this additional source of natural ventilation can now be designed into larger temperate-climate buildings to supplement other natural ventilation systems.

A wind catcher operates by capturing the air on the windward side of the shaft and deflecting it down one quadrant by a series of vanes. The force of the wind drives it into the space below. As the entering air

is cooler and denser than that within the building, it displaces the warm vitiated air which rises by natural stack ventilation through the other quadrants of the shaft, leaving through the leeward side of the wind catcher. With a symmetrical system one quadrant will predominantly face the prevailing wind to act as the catcher and the opposite quadrant will provide the majority of the stack-ventilation effect. A glazed top to the wind catcher, which heats up further the stale air, can enhance the stack effect. Dampers may be used to reduce air flow during winter months and permit night-time cooling only in the hot summer months. In densely used locations a combination of temperature and carbon dioxide sensors can activate the systems. Wind catchers should be located near to the ridge on pitched roofs to maximise their efficiency. Solar-powered wind catchers incorporating a photovoltaic panel (Fig. 16.8) will automatically switch on a low-energy extractor fan to increase natural ventilation in sunny conditions.

For functioning in multi-storey buildings, wind catchers require appropriate ducting and damper systems, and may incorporate heat exchangers from the central heating system to admit tempered fresh air in winter operation. Large rectangular ducts may be incorporated into the structure to avoid significant loss of floor space in the upper storeys. Figure 16.9 illustrates wind catchers in a modern school building.

Wind turbine systems

Micro-wind turbines can generate up to 2.5 kW of electrical power depending upon the wind turbine diameter and local wind speed conditions. A 1.75 m-diameter turbine operating at its rated wind speed of 12.5 m/s will generate 1 kW, with a minimum operating wind speed of approximately 4 m/s. The dc current generated is converted into standard voltage at 50Hz ac for use before additional current is taken from the national supply. Micro-wind turbines are only intended to supplement the user's standard mains electricity supply. Planning issues relate to visual impact and noise production.

The exact location of micro-turbines (for example, on house roofs) is important, as the power in the wind is proportional to the cube of wind speed. Urban wind speeds are significantly affected by the immediate topography. Research indicates that the best location for roof-mounted micro-wind turbines is above the ridge at a gable end; however, pole-mounted systems

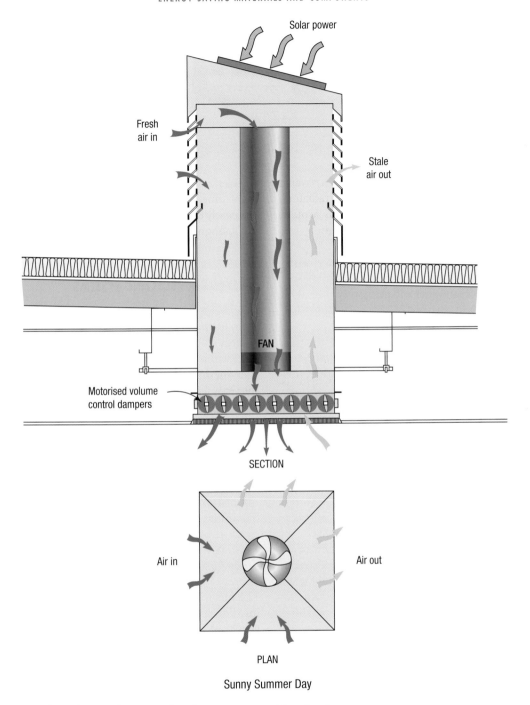

Fig. 16.8 Solar boosted wind catcher – sunny day. *Illustration:* Courtesy of Monodraught

often perform better than those on buildings. Local wind speed data should be obtained from a local meteorological station or the NAOBL wind speed database. Certain environments in large urban conurbations are unsuitable for micro-wind turbines due to the low average wind speeds. This would mean that the environmental cost of the installation, its maintenance, ancillary equipment and decommissioning would never be recovered during the active life of the micro-turbine.

Fig. 16.9 Windcatchers – St Mary's School, Isles of Scilly. *Photograph:* Arthur Lyons

Water management systems

The standard BS 8595: 2013 gives guidance on the selection of water reuse systems including the collection and utilisation of rainwater and previously used water.

RAINWATER-HARVESTING SYSTEMS

Rainwater-harvesting systems allow for the collection, filtering and storage of roof rainwater for use in non-potable functions. The storage capacity depends upon the roof area and the anticipated demand. In certain systems, when the storage tank level is insufficient, the mains water supply automatically switches in and admits a measured quantity of water without completely filling the system.

Three types of rainwater-harvesting systems are described in BS 8515: 2009. Water from the storage tanks may be delivered by gravity or self-activating pump directly to the points of use; alternatively, the stored water is pumped to an elevated cistern which then gravity feeds the points of use. Filtration systems require routine annual maintenance and should retain particles of size >1.25 mm. Rainwater is suitable for flushing toilets, laundry operations, car washing, garden watering and other non-potable functions. Rainwater outlets must be clearly marked as non-potable.

Guidance on calculating the required storage capacity is given in BS 8515: 2009. For the approximate calculation of the required storage capacity, a domestic consumption of 50 litres/day/person is appropriate. To anticipate fluctuations in supply, storage capacity should equal 5% of the annual demand or 5% of the average annual rainwater yield, which ever is the lesser. Storage tanks range upwards in size from 1000 litres. Suitable provision is required for overflow from the storage tank.

Rainwater harvesting may also provide benefits from attenuating surface water run-off.

GREYWATER RECYCLING SYSTEMS

Greywater from domestic showers and baths may be recycled to flush toilets. One electronically controlled system collects the greywater from showers and baths and, without the addition of chemical agents, skims off the floating material, allows the settlement of sand residues and retains up to 100 litres of greywater. This is sufficient for approximately 20 flushes of the directly connected WCs. The system automatically purges itself as appropriate and is topped up with fresh water when necessary. The system conforms to the strict water regulations which prevent contamination of potable water supplies. The system can save up to 30% of normal domestic potable water use, and in new housing contributes to the Code for Sustainable Homes which requires reduced water consumption.

Code for Sustainable Homes water specifications:

Water consumption (litre/person/day)	Mandatory code levels
⩽120	Levels 1 and 2
⩽ 105	Levels 3 and 4
⩽ 80	Levels 5 and 6

Phase change systems

A small unit filled with phase change material (PCM) can act as a large thermal mass by drawing in hot air and cooling it by absorption of the latent heat of fusion of the phase change material during excessively hot periods. The process is reversed as the ambient temperature falls when heat is released by the solidification of the phase change material. Passive cooling systems based on PCM technology may be located on outside walls to draw in either fresh or recycled air. Phase change materials can also act as energy stores in integrated low-energy systems based on photovoltaics or solar thermal panels. Phase change materials are described in Chapter 12 (page 388).

Energy-generating paving slabs

The kinetic energy from walking on specialist rubber paving slabs is converted into electrical power. When installed at entrances to shopping malls, a significant quantity of electrical energy is produced which may be used immediately for local LED lighting systems.

References

FURTHER READING

Addington, M. and Schodek, D. 2004: *Smart materials and technologies*. Oxford: Elsevier.

Berge, B. 2009: *Ecology of building materials*, 2nd edn. Oxford: Architectural Press.

Boyle, G. 2012: *Renewable energy. Power for a sustainable future*. Oxford: Oxford University Press.

Brown, G.Z. and DeKay, M. 2000: *Sun, wind and light: Architectural design strategies*, 2nd edn. New York: John Wiley.

CIBSE. 2005: *Reclaimed water*. London: Chartered Institution of Building Services Engineers.

CIBSE. 2008: *Groundwater cooling systems*. TM45. London: Chartered Institution of Building Services Engineers.

CIBSE. 2009: *Capturing solar energy*. London: Chartered Institution of Building Services Engineers.

Construction Products Association. 2012: *Guide to understanding the embodied impacts of construction products*. London: CPA.

Dickson, M. and Farnelli, M. 2005: *Geothermal energy. Utilization and technology*. London: Earthscan Publications.

Edwards, B. 2014: *Rough guide to sustainability. A Design Primer*, 4th edn. London: RIBA Publishing.

German Solar Energy Society. 2008: *Planning and installing photovoltaic systems: A guide for installers, architects and engineers*. Abingdon: Routledge.

Goetzberger, A. and Hoffmann, V.U. 2005: *Photovoltaic solar energy generation*. Berlin: Springer-Verlag.

Gonzalo, M. and Habermann, K. 2006: *Energy efficient architecture. Basics for planning and construction*. Basel: Birkhäuser.

Haas-Arndt, D. 2008: *Basics. Water cycles*. Basel: Birkhäuser.

Herzog, T. (ed.) 2008: *European charter for solar energy in architecture and urban planning*. Munich: Prestel.

HVAC. 2007: *Guide to good practice. Heat pumps*. TR/30. London: Heating and Ventilating Contractors' Association.

Hyde, R. (ed.) 2007: *Bioclimatic housing. Innovative design for warm climates*. London: Earthscan Publications.

Luque, A. and Hegedus, S. 2003: *Handbook of photovoltaic science and engineering*. London: John Wiley and Sons.

McCrea, A. 2013: *Renewable energy. A user's guide*. Marlborough: Crowood Press.

Mendler, S. and Odell, W. 2005: *The guide book to sustainable design*. New Jersey: John Wiley and Sons.

NBS. 2012: *Thin film and polymer photovoltaics*. Technical Advice. Newcastle-upon-Tyne: National Building Specification.

Ochsner, K. 2007: *Geothermal heat pumps. A guide for planning and installing*. Abingdon: Routledge.

Parker, D. 2008: *Microgeneration. Low energy strategies for larger buildings*. Oxford: Elsevier.

Pennycook, K. 2008: *The illustrated guide to renewable technologies*. Bracknell: Building Services Research and Information Association.

Roaf, S., Fuentes, M. and Thomas, S. 2012: *Ecohouse*, 4th edn. Abingdon: Routledge.

Roberts, S. 2009: *Building integrated photovoltaics*. Basel: Birkhäuser.

Smith, P.F. 2009: *Building for a changing climate*. London: Earthscan Publications.

Ternaux, E. 2011: *Material World 3. Innovative materials for architecture and design*. Amsterdam: Frame.

Thomas, R. and Garnham, T. 2007: *The environments of architecture. Environmental design in context*. Abingdon: Taylor & Francis.

Turrent, D. 2007: *Sustainable architecture*. London: RIBA Enterprises.

Vale, B. and Vale, R. 2002: *The new autonomous house: Design and planning for sustainability*. London: Thames and Hudson.

Vallero, D. and Brasier, C. 2008: *Sustainable design. The science of sustainability and green engineering*. New Jersey: John Wiley and Sons.

STANDARDS

pr BS 5918: 2013
Solar heating systems for domestic hot water. Code of practice.

BS 8515: 2009 + A1: 2013
Rainwater harvesting systems. Code of practice.

BS 8525
Greywater systems:

Part 1: 2010	Code of practice.
Part 2: 2011	Domestic greywater treatment equipment. Requirements and test methods.

BS 8582: 2013
Code of practice for surface water management for development sites.

BS 8595: 2013
Code of practice for the selection of water reuse systems.

BS ISO 9459
Solar heating. Domestic water heating systems:

Part 4: 2013	System performance characterization by means of component tests.
Part 5: 2007	System performance characterization by means of whole-system tests.

pr BS ISO 18178: 2014
Glass in building. Laminated solar PV glass.

BS EN 1085: 2007
Wastewater treatment. Vocabulary.

BS EN 1717: 2001
Protection against pollution of potable water in water installations.

BS EN ISO 9488: 2000
Solar energy. Vocabulary.

BS EN 12975
Thermal solar systems and components. Solar collectors:

pr Part 1: 2011	General requirements.
pr Part 3.1: 2011	Qualification of solar absorber surface durability.

BS EN 12977
Thermal solar systems and components. Custom built systems:

Part 1: 2012	General requirements for solar water heaters and combisystems.
Part 2: 2012	Test methods for solar water heaters and combisystems.

Part 3: 2012	Performance test methods for solar water heater stores.
Part 4: 2012	Performance test methods for solar combistores.
Part 5: 2012	Performance test methods for control equipment.

BS EN 15193: 2007
Energy performance of buildings. Energy requirements for lighting.

BS EN 15217: 2007
Energy performance of buildings. Methods for expressing energy performance and for energy certification of buildings.

BS EN 15232: 2012
Energy performance of buildings. Impact of building automation controls and building management.

BS EN 15255: 2007
Energy performance of buildings. Sensible room cooling load calculation.

BS EN 15265: 2007
Energy performance of buildings. Calculation of energy needs for space heating and cooling using dynamic methods.

BS EN 15450: 2007
Heating systems in buildings. Design of heat pump heating systems.

BS EN 15603: 2008
Energy performance of buildings. Overall energy use and definition of energy ratings.

pr EN 50583: 2012
Photovoltaics in buildings.

BS EN 61400
Wind turbines:

Part 1: 2005	Design requirements.
pr Part 2: 2012	Small wind turbines.

BS EN 61646: 2008
Thin film terrestrial photovoltaic (PV) modules. Design, qualification and type approval.

BS EN 61730
Photovoltaic (PV) module safety qualification:

Part 1: 2007	Requirements for construction.
Part 2: 2007	Requirements for testing.

DD CLC/TS 61836: 2009
Solar photovoltaic energy systems. Terms, definitions and symbols.

pr EN 62548: 2011
Design requirements for photovoltaic (PV) arrays.

BUILDING RESEARCH ESTABLISHMENT PUBLICATIONS

BRE Digests

BRE Digest 438: 1999
  Photovoltaics integration into buildings.
BRE Digest 446: 2000
  Assessing environmental impacts of construction: industry consensus, BREAM and UK ecopoints.
BRE Digest 452: 2000
  Whole life-cycle costing and life-cycle assessment for sustainable building design.
BRE Digest 457: 2001
  The carbon performance rating for offices.
BRE Digest 486: 2004
  Reducing the effects of climate change by roof design.
BRE Digest 489: 2004
  Wind loads on roof-based photovoltaic systems.
BRE Digest 495: 2005
  Mechanical installation of roof-mounted photovoltaic systems.
BRE Digest 498: 2014
  Selecting lighting controls.
BRE Digest 499: 2006
  Designing roofs for climate change. Modifications to good practice guidance.

BRE Information papers

BRE IP 13/00
  Green buildings revisited (Parts 1 and 2).
BRE IP 17/00
  Advanced technologies for 21st century building services.
BRE IP 5/01
  Solar energy in urban areas.
BRE IP 3/03
  Dynamic insulation for energy saving and comfort.
BRE IP 13/03
  Sustainable buildings (Parts 1–4).
BRE IP 10/04
  Whole life value: sustainable design in the built environment.
BRE IP 12/05
  Small scale, building integrated wind power systems.
BRE IP 15/05
  The scope for reducing carbon emissions from housing.

BRE IP 16/05
  Domestic energy use and carbon emissions. Scenarios to 2050.
BRE IP 6/06
  Balanced Value for sustainable procurement.
BRE IP 4/07
  Environmental weightings. Their use in the environmental assessment of construction products.
BRE IP 1/08
  The price of sustainable schools.
BRE IP 2/08
  New build and refurbishment in the Sustainable Communities Plan.
BRE IP 3/08
  Delivering sustainable objectives through planning.
BRE IP 4/08
  Micro-wind turbines on house roofs.
BRE IP 9/08
  Applying the code for sustainable homes on the BRE Innovation Park (Parts 1, 2, 3 and 4).
BRE IP 12/08
  An introduction to PassivHaus.
BRE IP 13/08
  An introduction to intelligent buildings (Parts 1 and 2).
BRE IP 3/09
  Lessons learned from the Barratt green house. Delivering a zero carbon home using innovative concrete systems.
BRE IP 4/09
  Delivering sustainable development in the built environment.
BRE IP 7/09
  Carbon reduction commitment.
BRE IP 1/10
  Micro-wind turbines on tall buildings.
BRE IP 2/10
  Ground source heat pumps.
BRE IP 15/10
  Specifying LED lighting.
BRE IP 7/11
  Biomass systems. Key factors for successful installation.
BRE IP 8/11
  Photovoltaic systems. Key factors for successful installation.
BRE IP 10/11
  Ground source heat pumps. Key factors for successful installation.

BRE IP 11/11
    Solar thermal systems. Key factors for successful installation.
BRE IP 2/12
    Smart meters and smart energy networks for dwellings.
BRE IP 11/12
    Building-integrated photovoltaic systems.
BRE IP 14/12
    Static and dynamic wind loads on building-mounted microwind turbine.
BRE IP 18/12
    Funding renewable energy projects.
BRE IP 22/12
    Site layout planning for sunlight and solar gain.
BRE IP 23/12
    Site layout for daylight.
BRE IP 7/13
    Energy surveys and audits. A guide to best practice.
BRE IP 15/13
    Delivering energy efficiency in commercial buildings. A guide for facilities managers.
BRE IP 17/13
    Planning of domestic air source heat pumps to mitigate noise impacts.

BRE Good building guides

BRE GBG 63: 2004
    Climate change.
BRE GBG 82: 2012
    Rainwater harvesting for domestic purposes.

BRE Reports

FB 17: 2007
    Micro-wind turbines in urban environments.
FB 18: 2008
    Siting micro-wind turbines on house roofs.
FB 22: 2010
    Building mounted microwind turbines on high rise and commercial buildings.
FB 44: 2012
    Energy management in the built environment. A review of best practice.
FB 60: 2013
    Performance of photovoltaic systems on non-domestic buildings.

ADVISORY ORGANISATIONS

Centre for Alternative Technologies, Machynlleth, Powys SY20 9AZ (01654 705950).

Ground Source Heat Pump Association. Davy Avenue, Knowlhill, Milton Keynes MK5 8NG (01908 354545).

Heat Pump Association, 2 Waltham Court, Milley Lane, Hare Hatch, Reading, Berkshire RG10 9TH (0118 940 3416).

Renewable Energy Association, 2nd Floor, 25 Eccleston Place, Victoria, London SW1W 9NF (0207 925 3570).

Solar Trade Association, 2nd Floor, 25 Eccleston Place, Victoria, London SW1W 9NF (0207 925 3575).

UK Rainwater Harvesting Association, Millennium Green Business Centre, Rio Drive, Collingham, Newark, Nottinghamshire NG23 7NB (08450 260240).

17

RECYCLED AND ECOLOGICAL MATERIALS

CONTENTS

Introduction

Realisation of the finite nature of global resources and the greenhouse effect of ever-increasing carbon dioxide emissions has promoted consideration of the potential for recycling into construction of many mass-produced waste products which are currently either burned or buried in landfill sites. Such materials include plastics, cardboard, straw, paper and tyres. While some recycled products are only in the experimental stage, others are now becoming recognised as standard building materials. The re-use of building materials is well illustrated by the former Earth Centre, Doncaster, which is constructed using many recycled and reclaimed products including crushed concrete, telegraph poles, glass and radiators (Fig. 17.1).

Straw bales

Straw bales, a by-product of the mechanical harvesting and threshing of grain, are produced in large quantities.

The traditional rectangular bales, which are cheap and can be manhandled individually, are appropriate for building. The large cylindrical and very large rectangular bales, which require mechanical lifting, are less useful in construction and are not considered here. Standard bales (typically $375 \times 500 \times 990$ mm) are produced within the baler by compressing quantities of straw into flakes about 100 mm thick. These layers are built up along the length of the bale, which is then automatically tied, usually with two polypropylene strings. There is inevitably some variation in length, and the ends are slightly rounded. For construction, the bales should be well compressed in manufacture, dry (maximum 20% moisture) to prevent the growth of moulds and fungi and with the minimum amount of remaining grain, which might attract rodents.

In building construction, bales are stacked, large faces down in stretcher bond, making the orientation of the straw fibres predominantly horizontal. At ground level, straw bales must be protected from rising

damp and from any risk of saturation from surface water. In addition, steel mesh protection from rodents is necessary. Adjacent bales must be firmly packed together to ensure stability and to reduce settling under load both during and after construction. Bales are normally secured with metal spikes or hazel rods from coppiced timber and may be sprayed with insecticide for added protection. Externally, lime/clay or lime render on wire mesh to 35 mm is appropriate, as it is flexible, self-healing, and will breathe to prevent the buildup of trapped moisture. Alternatively, a rainscreen, separated from the external face of the bales, may be used. Internally, straw bales are usually finished with gypsum plaster on wire mesh. Openings in straw bale construction may be formed with timber-framing, but careful detailing is required to prevent water penetration at these locations. Roofs are normally set onto a timber wall plate fixed through the top bales for stability. Two-storey load-bearing straw bale construction with appropriate finishes can achieve a U-value of 0.13 W/m² K.

An alternative approach to using load-bearing straw bales is illustrated in the University of Nottingham Gateway Building on the Sutton Bonington Campus. The building is constructed from panels, each comprising a cross-laminated timber- frame filled with compressed straw and externally rendered with a breathable coating (Fig. 17.2). Manufacturers are producing a variety of structural insulating panels (SIPs) incorporating straw insulation. While fire is a risk during straw bale construction, the non-combustible internal and external finishes and the compact nature of the straw make the completed construction resistant to fire. (The thermal conductivity of baled straw is approximately 0.050 W/m K.)

Cardboard

The structural cardboard of the exhibition shelter in Pouilly-en-Auxois, France (Fig. 17.3) illustrates the potential of this largely recycled product as a useful construction material. A grid of cardboard tubes is connected by steel nodes and protected from the environment with polycarbonate profile sheeting.

Following the disastrous earthquake of 2011 in Christchurch, New Zealand, a cardboard cathedral designed by Shigeru Ban has been opened. The Λ-shaped building (Fig. 17.4) constructed from a series of interlocking cardboard tubes accommodates a congregation of 700 people and has a steeple of cardboard tubes inserted with timber. The 600 mm cardboard tubes, each weighing 120 kg, are coated with polyurethane waterproofing and flame retardants, and the building is expected to last for 50 years. The architect had previously designed a temporary cardboard church for Kobe, following the 1995 earthquake.

In the recycling process, waste-paper and cardboard are broken down and converted into pulp, which is a suspension of cellulose fibres in water. The pulp flows onto a conveyor belt, where it is drained of the excess water and compressed, causing the fibres to felt together, producing a long roll of paper. Flat cardboard sheets are formed by gluing together successive layers of paper. Tubes are manufactured from multiple layers of spirally wound paper plies, starting on a steel tube former of the appropriate size, the adhesive being starch or PVA glue. The final layer of paper gives the required visual finish.

Interior panel
showing straw infill

Fig. 17.2 Straw bales in construction – Gateway Building, University of Nottingham Sutton Bonington Campus. *Architect*: Make Architects. *Photographs*: Arthur Lyons

Fig. 17.3 Cardboard tube structure – Pouilly-en-Auxois, France. *Photograph*: Arthur Lyons

Fig. 17.4 Cardboard tube structure – Christchurch Cathedral, New Zealand. *Architect:* Shigeru Ban.
Photographs: Linda Bartley and Brenda Griffiths

PROPERTIES

Cardboard, like timber, is combustible and can be treated to improve its fire performance, particularly in relation to the surface spread of flame test. However, some fire-retardant materials are environmentally unfriendly and should be avoided if the material is to be subsequently recycled.

The structural strength of cardboard is seriously affected by water. Cardboard, even if specially treated in its manufacture, is a hygroscopic material which will readily absorb moisture. It is therefore necessary to protect it internally from warm, moist air and externally from rain.

Cardboard is potentially vulnerable to rot and insect attack. This could be prevented by treatment with boron products; however, this would adversely affect the potential for ultimate recycling of the cardboard.

Rammed-earth and cob construction

Earth construction is one of the oldest forms of building used by mankind. Rammed-earth buildings may be found in most countries, and many have survived for hundreds of years. The ideal material is a well-graded mixture of gravel, sand, silt and clay fines. The clay content should be sufficient to act as an efficient binder, but not in excess to cause large moisture movement or cracking of the finished construction. In modern rammed-earth construction Portland cement is frequently incorporated as a binder to improve cohesion of the stabilised earth mix.

In rammed-earth construction, the mix is placed in layers, typically 100–150 mm deep, within the rigid formwork and firmly tamped down, thus inevitably giving some variation in density between the top and bottom of each lift. The compaction should be sufficient to ensure good strength and a smooth finish.

Window and door openings should be limited to no more than one-third of the length of any wall to ensure structural stability. Lintels should be sufficiently robust to take static loading and the effects of ramming further lifts of earth. A minimum bearing of 300 mm at each end is required. A wall plate of timber or poured reinforced concrete, which may be hidden within the top lift of the earth wall, is necessary to spread the loading from the roof structure. The eaves should be detailed to ensure appropriate shelter to rammed-earth walling, which is usually protected with several coats of limewash finish. To comply with the thermal requirements of the Building Regulations, a wall thickness of more than 700 mm is normally required. (The thermal

Fig. 17.5 Mud and straw construction – The House for Stories, Bleddfa Centre for the Arts, Powys, Wales. *Architect:* Tono Mirai. *Photograph:* Courtesy of Richard Weston

conductivity of soil is typically in the range 0.8–1.5 W/m K.) Rammed-earth construction has been used by Buro Happold at the Eden Project in Cornwall and the Centre for Alternative Technology at Machynlleth, Powys, where the rammed-earth walls supporting the roof are 7.2 m high.

Cob construction differs from rammed earth in that the clay is mixed with straw or flax fibres. In the traditional process, a fine tilth of clay is spread about 100 mm deep over a thin straw bed; water and a second thicker layer of straw is added. The mixture is well trodden to produce a reasonably uniform mix. Devon clay is ideal for this process as it is well graded with a range of particle size from coarse gravel through fine sand to coarse clay. Devon clay has only a low expansion and contraction which may otherwise cause cracking of the completed structure. In the construction process, the mixture of the straw and clay is tamped together, in lifts of 500 mm at a time, usually commencing from a minimum 450 mm stone plinth. Free-form designs may be achieved without the use of shuttering. This type of construction was common in many parts of the UK and many old cob buildings still exist in Devon. The thermal mass of cob construction stabilises seasonal variations, helping to keep the interior cool in summer and warm in winter. The hygroscopic properties of clay can moderate internal humidity, which in turn reduces the numbers of house-dust mites and their associated allergies. The external finish should be of lime plaster rather than a Portland cement render, which cracks or breaks away allowing rainwater to penetrate the wall. As with rammed-earth construction, the eaves should be deep enough to protect the walls from severe weathering.

The 'House for Stories' designed by Tono Mirai at the Bleddfa Centre for the Arts in Powys, Wales is constructed from mud and straw. The building, which is intended to be a quiet space for contemplation and imagination, is partly subterranean, but spirals out of the ground into the delightful helical form shown in Figure 17.5.

Earth-sheltered buildings

Earth-sheltered buildings, including homes, are defined as those where the roof and some sides are covered by earth. Increased depth of earth cover improves the thermal performance, but this has to be balanced against consequential increased structural strength requirements. Typically 400–450 mm of earth cover is

South elevation

North elevation

Fig. 17.6 Earth-sheltered housing – south and north elevations. The Hockerton Housing Project, Southwell, Nottinghamshire.
Photographs: Copyright of Hockerton Housing Project

appropriate, and the weight of this material is usually supported by concrete or masonry construction. Water exclusion is key to the design, requiring land drainage and the use of reinforced membranes.

One method of construction uses fibrous plaster shells to create the internal organic form. These are sprayed with a layer of lightweight aggregate concrete insulation, followed by 100 mm of structural concrete. After the concrete is structurally sound, the building may be fitted out and covered externally with soil and grass. An alternative construction system uses extruded polystyrene insulation between the structural concrete and the soil backfill. Either approach uses the temperature stabilising effect of the mass concrete and soil cover to significantly reduce energy consumption. In order to introduce sufficient light, at least one elevation is usually glazed, and, in addition, interesting effects can be achieved with roof lights or light pipes. Ventilation may be mechanical but is normally supplied through opening glazing. At the same time, unwanted cold air infiltration is eliminated by the earth enclosure.

The Hockerton Housing Project at Southwell, Nottinghamshire (Fig. 17.6) illustrates an ecological development of earth-sheltered housing in which the residents generate their own energy, harvest their own water and recycle waste materials, eliminating pollution and carbon dioxide emissions. Only the south elevation of the development overlooking the reed pond is visible, as grass covers the majority of the construction.

Clay products

The concern over the intrinsic energy in most manufactured building products has led to the further development of a range of clay-based products including clay boards and plasters. Unfired clay building products are hygroscopic and have the positive effect of controlling internal environments by absorbing odours and stabilising humidity and temperature.

CLAY BOARDS

Clay boards, which are an alternative to gypsum plasterboards, are manufactured from clay and layers of reed along and across the board length. Hessian on both faces acts as reinforcement and a key for a 2–3 mm finish of lime-based or earth plaster. Joints should be scrimmed before skimming, although clay boards if

sealed can be painted directly. The 25 mm boards may be used for drywall construction and ceilings, where they should be screw-fixed at 600 or 400 mm centres, respectively.

CLAY PLASTERS

Clay plasters, also known as earth plasters, are available in a range of self-colours, which do not require paint decoration. Clay plaster, manufactured from a blend of clay and fine aggregate, may be applied in two layers of 10 and 3 mm, respectively, or as a single 10 mm coat. If necessary, an initial 1 mm bonding coat may be applied to the substrate. Clay plasters harden only by drying out without any chemical processes. As with all clay products, these plasters absorb moisture and are responsive to environmental conditions, helping to control the internal relative humidity.

Recycled tyres

Recycled tyres have been used to create the structural walls of homes called *earthships* in Fife and near Brighton, UK. The tyres are laid in courses, filled with rammed earth and finished internally with plaster. Externally, 1 metre of soil is banked up and the walls are then rendered with mud, adobe or cement. Rainwater harvesting and solar collection are key to the low-impact constructions. To date only a few small housing units have been built, but subject to Building Regulations there is no theoretical limitation on size. The UK produces 48 million used tyres per year, enough to create 24,000 low-carbon autonomous earthships annually. Used tyres are now banned from landfill sites and need to be usefully recycled.

Papercrete

Papercrete is made from recycled paper and/or cardboard with sand and Portland cement. Pulverised waste glass from recycled bottles may be used instead of sand, and glossy magazines may be mixed in with standard newsprint. The material is made by dry mixing shredded paper with sand and Portland cement in the approximate ratio of 3 : 1 : 1. Water is added to make a papier mâché slurry that can be cast into block units or into monolithic structures. Where papercrete blocks are used in construction, the same material may be used as the mortar. The material dries to a grey colour. It is, however, very water-absorbent, and must be protected from moisture and weather by appropriate detailing. A stronger mix of 1 : 1 paper and cement may be used as an external stucco layer or for the manufacture of denser blocks. Internally, papercrete plaster may be used to give a textured or patterned finish. Papercrete has the potential to remove up to 20% of the material currently deposited in landfill sites as it has a recycled content of between 50 and 80% waste-paper or cardboard. As a lightweight material it has good insulating properties (λ = 0.074 W/m K) and the cement content significantly increases its fire resistance.

Sandbags

In office/home accommodation adjacent to a noisy main railway line in Islington, London, sandbags have been used to create a sound-absorbing façade (Fig. 17.7). The sandbags, filled with a sand-cement-lime mixture and exposed to the elements, will eventually set hard. Over a further period of time, possibly 30 years, the hessian bags will disintegrate, leaving the undulating concrete exposed, imprinted with the texture of the hessian fabric. The wall is experimental in that the weathering effects cannot be predicted, as with other more standard forms of construction.

Fig. 17.7 Sand bags in construction. *Architects and Photograph:* Sarah Wigglesworth Architects

References

FURTHER READING

Addis, B. 2006: *Building with reclaimed components and materials.* London: Earthscan Ltd.

Berge, B. 2009: *The ecology of building materials.* Abingdon: Routledge.

Bingham, W. and Smith, C. 2007: *Strawbale home plans.* Utah: Gibbs M. Smith Inc.

Design Center Stuttgart. 2008: *Focus green.* Stuttgart: AVEdition.

Easton, D. and Wright, C. 2007: *The rammed earth house.* Vermont: Chelsea Green Publishing.

Fernandez, J. 2006: *Material architecture: emergent materials and issues for innovative buildings and ecological construction.* Oxford: Elsevier.

Guelberth, C.R. and Chiras, D. 2002: *The natural plaster book, earth, lime and gypsum plasters for natural homes.* Canada: New Society Publishers.

Hall, K. 2008: *The green building bible*, 4th edn. Vol 1. Llandysul: Green Building Press.

Halliday, S. 2008: *Sustainable construction.* Oxford: Butterworth-Heinemann.

Hewitt, M. and Telfer, K. 2012: *Earthships in Europe.* Watford: IHS BRE.

Hren, S. and Hren, R. 2008: *The carbon-free home.* Vermont: Chelsea Green Publishing.

Hyde, R., Watson, S., Cheshire, W. and Thomson, M. 2007: *The environmental brief. Pathways for green design.* Abingdon: Taylor & Francis.

Jones, B. 2010: *Building with straw bales: A practical guide for the UK and Ireland*, 2nd edn. Dartington: Green Books.

King, B. 2007: *Design of straw bale buildings.* Dartington: Green Building Press.

Lacinski, P. and Bergeron, M. 2000: *Serious straw bale: A home construction guide for all climates.* Totnes: Chelsea Green Publishing.

Magwood, C. and Mack, P. 2005: *More straw bale building: A complete guide to designing and building with straw.* USA: New Society Publishers.

McCann, J. 2004: *Clay and cob buildings.* Princes Risborough: Shire Publications.

Minke, G. 2012: *Building with earth. Design and technology of a sustainable architecture.* Basel: Birkhäuser.

Minke, G. and Friedemann, M. 2005: *Building with straw: Design and technology of a sustainable architecture.* Basel: Birkhäuser.

NBS. 2007: *Rammed earth. The hardened soil of architecture.* NBS Shortcut 33. Newcastle-upon-Tyne: National Building Specification.

Pauwels, W. 2012: *The 100 best projects with reclaimed materials.* Belgium: Beta-plus.

Schofield, J. and Smallcombe, J. 2004: *Cob buildings: A practical guide.* South Carolina: Black Dog Press.

Spiegel, R. and Meadows, D. 2010: *Green building materials.* New Jersey: John Wiley and Sons.

Transport Research Laboratory. 2009: *Reinforced earth update. CT65.5.* Wokingham: TRL.

Walker, P., Keable, R., Marton, J. and Maniatidis, V. 2005: *Rammed earth. Design and construction guidelines.* Englewood: IHS BRE.

Weismann, A. and Bryce, K. 2006: *Building with cob. A step-by-step guide.* Totnes: Greenbooks.

Wells, M. 2009: *Earth sheltered house. An architect's sketchbook.* Totnes: Chelsea Green Publishing.

Williams-Ellis, C. 2010: *Cottage building with cob, pisé and stabilised earth.* London: FQ Books.

Woolley, T. 2013: *Low impact building using renewable materials.* Chichester: Wiley Blackwell.

Woolley, T. and Kimmins, S. 2000: *Green building handbook. v2. A guide to building products and their impact on the environment.* London: E. & F.N. Spon.

Yates, T., Ferguson, A., Binns, B. and Hartless, R. 2013: *Cellulose-based building materials. Use, performance and risk.* Publication NF55. Milton Keynes: NHBC Foundation.

BUILDING RESEARCH ESTABLISHMENT PUBLICATIONS

BRE Digests

BRE Digest 446: 2000
   Assessing environmental impacts of construction.
BRE Digest 452: 2000
   Whole life costing and life-cycle assessment for sustainable building design.
BRE Digest 470: 2002
   Life cycle impacts of timber. A review of the environmental impacts of wood products in construction.

BRE Good repair guide

BRE GRG 35: 2006
   Earth, clay and chalk walls: inspection and repair methods.

BRE Information papers

BRE IP 7/00
    Reclamation and recycling of building materials.
BRE IP 13/03
    Sustainable buildings (Parts 1–4).
BRE IP 6/06
    Balanced Value for sustainable procurement.
BRE IP 8/06
    Non-ferrous metal wastes as aggregates in
    highway construction.
BRE IP 15/11
    Straw bale. An introduction to low-impact
    building materials.
BRE IP 14/13
    Critical materials and materials security for the
    construction industry and its supply chain.

BRE Reports

BR 501: 2009
    The Green Guide to Specification (4th edn).
EP 62: 2005
    Rammed earth design and construction.
EP 80: 2008
    Earth masonry. Design and construction
    guidelines.
EP 101: 2011
    Earth building . History, science and
    conservation.
FB 54: 2013
    Dealing with difficult demolition wastes.
    A guide.

18

SUSTAINABILITY

CONTENTS

Introduction

In 2008, the UK government published a strategy for sustainable construction which reflected the industry's commitment to reduce its carbon footprint and consumption of natural resources, while maintaining a strong construction sector. The initiative was within the wider context of concern for global warming and climate change, which has led to UK national targets for reduction in carbon dioxide emissions by 34% from 1990 levels by 2020 and by 80% by 2050. An indicator of the problem is that the middle-range emissions scenario for London predicts by 2020 a 1.5°C increase in mean summer temperature (+1.3°C in winter), with 6% more rainfall in the winter and 6% less rainfall in the summer.

As buildings account either directly or indirectly for approximately 44% of the UK's carbon emissions (27% from homes and 17% from non-domestic buildings), it is appropriate that the targets placed on the construction industry reflect this significant factor. The aim is therefore to move to zero carbon for homes by 2016 and for other buildings by 2019. In addition, the production of materials and transportation adds another 10% to the UK's carbon emissions.

While technological solutions can ameliorate carbon emissions from domestic, commercial and industrial buildings, the cornerstone for sustainable construction is holistic design, taking into consideration site, local climate, orientation, external form, building envelope and openings.

UK government initiatives have included significant changes to the Building Regulations and the development of the Code for Sustainable Homes. The BRE Green Guide Calculator enables BREEAM assessors to determine Green Guide ratings for new buildings.

Methods for expressing energy performance and for energy certification of buildings are described in the standard BS ISO 16343: 2013, while the standard BS ISO 16346: 2013 describes the calculated and measured energy ratings for buildings. Guidance on energy management and audits of constructed buildings is available from IHS BRE Press (IP 7/13, IP 1/14, IP 2/14 and BRE Report FB44: 2012).

Zero carbon targets

CODE FOR SUSTAINABLE HOMES

The Code for Sustainable Homes is an environmental assessment method for rating and certifying the performance of new homes. Code assessment procedure is normally a two-stage process which includes a Design Stage (DS) assessment and a Post-Construction Stage (PCS) assessment. Only on completion of both stages is the property awarded its final Code-level certificate. An issue to be addressed is the gap between design and as-built performance, which requires reconsideration of the existing SAP procedures and an enhancement of construction joint details.

The Code for Sustainable Homes covers nine categories, of which six have mandatory performance requirements denoted by M. The other three criteria are non-mandatory.

Code for Sustainable Homes Design categories:

Energy and carbon dioxide emissions (M)
Water (M)
Materials (M)
Surface water run-off (M)
Waste (M)
Health and well-being (M)
Pollution
Management
Ecology

The overall Code level rating depends upon a combination of reaching or exceeding the six mandatory standards together with achievements in the three non-mandatory criteria.

All nine categories carry a certain number of potential credits which are awarded according to the level of achievement in that particular category (e.g. indoor water consumption). The credits within each category are weighted with an appropriate factor. For each category the number of credits achieved is divided by the number available and then multiplied by the category weighting factor to give a final points score. The total points score is calculated by a summation of the point scores in each of the nine categories, giving final totals which range from Code level 1 (36 points) and Code level 3 (57 points) to Code level 6 (90 points). However, there are minimum mandatory standards in certain categories, including the environmental impact of materials, which must be achieved before additional points may be applied and the final Code level determined.

The property will receive a Final Certificate with an overall one to six Code level rating, and a breakdown into the nine design categories indicating the percentage score achieved in each. Initially, the main focus of the Code is on the measureable issues including energy, carbon, water and waste.

All new homes given planning approval before 2014 must be at Code level 3 corresponding to the Building Regulations 2010, and from 2014 a further 6% fabric energy-efficiency enhancement from the 2010 standard will be required. The target for 2016 of zero carbon (Code level 6) is mandatory. In Wales, all new dwellings must now be at Code level 5. In addition, to achieve a Code level 3 rating, a new house must be designed to use only 105 litres of water per day.

A Code level 6 house must be completely zero carbon with respect to space and water heating, cooking and use of all appliances. The anticipated water use should be only 80 litres per day, which means that approximately 30% of the water requirement will be taken from rainwater harvesting or recycling. To achieve Code level 6, high levels of insulation and airtightness are required. In addition, the design must provide good daylighting levels while incorporating appropriate solar control to prevent summertime overheating. A range of renewable energy technologies will include solar water heating and combinations of other systems, such as photovoltaics, heat recovery, passive cooling and ventilation, a biomass boiler, and highly efficient water and waste management systems. A minimum number of construction materials must meet the D grade of the BRE Green Guide at least.

It is government policy that the energy efficiency rules within the Code for Sustainable Homes will be incorporated into future UK Building Regulations.

PROTOTYPICAL ZERO CARBON HOMES

A number of experimental homes have been built to demonstrate the alternative approaches to achieving various Code level homes. These include several houses on the BRE Innovation Park at Watford, although only the Sheppard Robson 'Lighthouse' (Fig. 18.1) and the Barratt 'Green House' (Fig. 18.2), both of which feature some off-site construction, achieve a Code level 6 rating. The 'Cub House' and the 'Sigma Home' (both Code level 5) illustrate modern methods of construction and sustainable construction features. The 'Prince's Natural House' illustrates an energy-efficient house constructed from a range of eco-friendly materials. The University of Nottingham has several experimental homes including the 'Tarmac Homes Project' (Fig. 18.3), a pair of semi-detached properties built to Code levels 4 and 6 using traditional masonry building techniques. The Caplin Homes 'Zero-Carbon Solar House' in Leicestershire (Fig. 18.4) illustrates a skilful combination of innovative technologies which achieve carbon neutrality and is an exemplar for general building development.

The 'Lighthouse', built with high-performance structural insulated panels (SIPs), features a 40° pitched roof with a photovoltaic array and incorporates an integrated wind catcher for passive cooling and ventilation with a sun pipe for bringing additional daylight into the core of the home. West and east elevation windows are shaded to prevent excessive solar gain. Building services include a biomass boiler, low water consumption and recycling systems, solar water heating, mechanical ventilation with heat recovery and smart energy metering. The living zone is located above the bedrooms which require lower natural lighting levels. The use of double SIP panels with staggered joints was clearly significant in achieving

the necessary airtightness and reduced thermal bridging to fulfil the Code level 6 requirements. Thermal mass is augmented with phase change material within the plasterboard ceilings, and experimentation will determine its efficacy in preventing summertime overheating.

The Barratt 'Green House' incorporates storey-height lightweight concrete panels, joined with thin-layer mortar and hollow concrete intermediate floor slabs to give thermal mass. High levels of insulation afford the required U-values. The SIP butterfly roof has photovoltaic cells on the south-facing aspect and a green sedum roof to the north. High-performance triple-glazed thermally broken timber- frame windows complete the external envelope. Building services include solar water heating, mechanical ventilation with heat recovery, as well as water-saving and rainwater-harvesting systems. External shading is operated by automatic shutters to optimise useful solar gain and prevent summer overheating. The Barratt Green House achieved 92 points, just two over the requirement for an overall Code level 6 rating. The key U-values are listed in Table 18.1.

The University of Nottingham 'Tarmac Homes Project' reflects the different requirements of Code level 4 and Code level 6 housing within a pair of semi-

Fig. 18.1 Code for Sustainable Homes – Code 6 – Sheppard Robson 'Lighthouse', BRE Innovation Park, Watford. *Photograph:* Arthur Lyons

Fig. 18.2 Code for Sustainable Homes – Code 6 – Barratt 'Green House', BRE Innovation Park, Watford. *Photograph:* Arthur Lyons

Table 18.1 Barratt Green House U-values

Building element	U-values (W/m² K)
Roof	0.09
Walls	0.11
Ground floor	0.09
Exposed floor to carport	0.11
Glazing (triple and gas-filled)	0.70
Doors	0.68

Notes:
- Airtightness: 1.0 m³/(hr.m²) @ 50Pa
- Heat loss parameter (HLP): 0.72 W/m² K, c.f. Code 6 HLP 0.8 W/m² K

North Elevation

South elevation

Fig. 18.3 Code for Sustainable Homes – 'Tarmac Homes Project', University of Nottingham. Semi-detached pair. Code 6 (rendered) and Code 4 (brick). *Photographs:* Arthur Lyons

detached homes. Both halves of the pair are built in masonry, one (Code level 6) with solid walls and the other (Code level 4) with cavity construction, but these forms could, in principle, be interchanged with appropriate adjustments to their insulation levels. Both properties incorporate 3 m² solar thermal panels, rainwater harvesting and sun pipes, and they share an individually monitored biomass boiler. The key differences are in the insulation levels associated with the external walls and glazing.

The Code level 4 house of brick and block construction has a 150 mm cavity, partially filled with 100 mm of low-emissivity foil-coated polyisocyanurate (PIR) insulation. The inner leaf is constructed with 100 mm lightweight blockwork and a 12.5 mm dry plasterboard lining, giving an overall wall U-value of 0.20 W/m² K.

The Code level 6 house is constructed from 215 mm large format aircrete blocks using thin joint construction. It is externally insulated with 150 mm of expanded polystyrene and thin-coat render. With an internal finish of thin-coat projection plaster, the overall wall U-value is 0.15 W/m² K.

The Code level 4 house has PVC-U windows with 16 mm argon fill and low-emissivity hard coat, giving an overall U-value of 1.90 W/m² K. The Code level 6 house has timber windows with 16 mm argon fill and low-emissivity soft coat, giving an overall U-value of 1.50 W/m² K and a BFRC rating of band A. The Code level 6 south elevation has a sun space to prevent summer overheating but designed to admit the required solar gain during the cooler months.

The Code level 4 house incorporates a mechanical ventilation system with heat recovery, while the Code level 6 house uses a passive heat-recovery wind cowl ventilation system.

The 72 photovoltaic tiles on the Code level 6 house have a peak generating power of 3.74 kW, which equates to the anticipated peak domestic load of 3.52 kW.

The Caplin Homes 'Zero-Carbon Solar House' in Leicestershire (Fig. 18.4) incorporates an innovative combination of sustainable technologies which may be incorporated into any new-build housing, to collect and store sufficient solar energy for use throughout the year. The project is set to exceed the UK government's 2016 zero carbon target. The house collects enough solar energy to provide heating and hot water, and approximately twice the quantity of electricity required to run the system.

Earth energy
bank

Fig. 18.4 Caplin Homes 'Zero-Carbon Solar House'. *Photograph*: Arthur Lyons. Solar energy storage – Earth energy bank. *Image*: Courtesy of Caplin Homes Ltd

The timber frame house incorporates the principles of passive solar design with appropriate orientation and seasonal shading, low heat losses and airtight construction. Mechanical ventilation incorporates heat recovery from the extracted air. Fresh air is pre-warmed using innovative solar wall technology in which finely perforated metal sheeting warms the external boundary layer of air, which is then drawn through the perforations into the ventilation system. This is more efficient than the standard solar air-heating systems which rely on the solar heat being transmitted through the metal sheeting to the internal layer of air which is then circulated.

The angled solar collectors on the flat roof are hybrid photovoltaic/thermal (PV-T) panels, which have the advantage of cooling the PV system to maintain generating efficiency and producing hot water for both domestic use and, importantly, for charging the 'earth energy bank' under the house.

The earth energy bank (Fig. 18.4) consists of 1.5 m-deep bore holes at 1.5 m centres which have a 32 mm flow and return pipe embedded within thermally conducting bentonite. The pumped water system feeds heat into the bank during the warm months of the year, causing the temperature to rise by about 25°C, and then extracts this heat through a heat pump for underfloor heating during the cold periods. The system has the advantage over standard ground-source installations in that it is contained within the footprint of the building, and does not lose efficiency due to a gradual cooling of the soil over a period of years. In addition, the heat pump is more efficient when extracting heat from a higher temperature source. Heat is retained within the poorly conducting soil by perimeter vertical insulation and 75 mm insulation over the top of the soil. Any excess electrical energy produced by the PV cells and not required for operating the system or for domestic functions is used to generate additional heat to be stored in the energy bank, rather than being sold back to the grid. The heat pump can also extract some night-time heat from the hybrid solar panels warmed by the night air and background infra-red radiation. A sophisticated management system constantly monitors the internal, external and energy bank temperatures, electrical generation and demand, and diverts excess electrical energy either to the domestic hot-water system or to the earth energy bank to optimise overall performance.

Passivhaus

The Passivhaus standard aims to reduce the requirement for space heating and cooling while maintaining good air quality and comfort levels. This requires very high levels of insulation including thermal bridge-free construction and high-performance windows, in addition to an airtight building fabric incorporating mechanical ventilation with full heat recovery. The criteria for Passivhaus certification are based on threshold levels of energy utilisation and air changes with limiting values of thermal performance data, as indicated in Table 18.2. A slightly reduced set of criteria relate to retrofit on existing housing.

The Cartwright Pickard 30-home development in London (Fig. 18.5) includes two blocks built to Passivhaus standards. The Passivhaus units, sympathetic to the Victorian urban environment of Hammersmith, consist of a three-storey terrace of six houses and a five-storey block of 12 flats. Features include adjustable shading to south elevations and light pipes to provide daylight and ventilation to circulation areas.

Table 18.2 Passivhaus criteria

Energy performance targets and air infiltration	
Specific heating demand	≤ 15 kWh/m².Yr
Specific cooling demand	≤ 15 kWh/m².Yr
Specific heating load	≤ 10 W/m²
Specific primary energy demand	≤ 120 kWh/m².Yr
Air changes per hour	≤ 0.6 @ 50 Pa

Typical limiting values normally required to achieve Passivhaus criteria	
Walls, roof and floor U-values	≤ 0.15 W/m² K
Glazing unit	≤ 0.8 W/m² K
Installed glazing	≤ 0.85 W/m² K
Doors	≤ 0.8 W/m² K
Thermal bridging (linear value)	≤ 0.01 W/m K
Heat recovery efficiency	≥ 0.75
Ventilation power utilisation	0.45 Wh/m³
Lighting and appliances	High efficiency

ZERO CARBON NON-DOMESTIC BUILDINGS

The challenge to meet the target of zero carbon non-domestic new buildings by 2019 is complex due to the wide variety and forms of commercial and industrial buildings and their associated business activities. The UK government has set a zero carbon ambition of 2018

Fig. 18.5 Sulgrave Gardens Passivhaus development, Hammersmith, London. *Architects*: Cartwright Pickard Architects. *Client*: Octavia Housing. *Photograph*: Morley Von Sternberg

for all new public sector buildings, with a focus on schools and colleges to achieve the goal by 2016. The Building Regulations Part L (2013 edition) require a 9% fabric energy-efficiency enhancement from the 2010 standard. Only the regulated energy consumption – that is, the energy defined by Part L of the Building Regulations and predicted by the Simplified Building Energy Model (SBEM) – applies. Energy used for industrial processes is classified in this context as unregulated, but it is subject to a range of government incentives aimed at encouraging reductions.

EXISTING BUILDING STOCK

While the current focus is on new building, it may be assumed that at least 70% of the current building stock will still be in use by 2050. The issue of addressing retrofit energy-conservation technologies will therefore increase in relative importance as higher standards become the norm for all new construction. The Building Regulations specify appropriate requirements when building work is carried out on existing buildings.

Energy Performance Certificates (EPCs) are required for all buildings when they are sold, built or rented. Certificates are rated A to G. Currently fewer than half

Table 18.3 Energy Saving Trust best practice targets for the refurbishment of social and private housing

Building element or function	Improvements and target U-values
Walls	Insulation to a maximum U-value of 0.30 W/m²K
Pitched roofs	Insulation between rafters to a U-value of 0.16 W/m²K
Flat roofs	Insulation to a U-value of 0.25 W/m²K or better
Windows	BFRC rating of C or above. Any retained windows to be draught-proofed
Doors	Solid doors to a U-value of 1.0 W/m²K, half-glazed doors to a U-value of 1.5 W/m²K
Space heating	Ideally a domestic wet central heating and hot water system except where electricity is the only option.
Airtightness	A target permeability of 5m³/(m²) @ 50 Pa
Ventilation	A ventilation system to be installed
Light and appliances	When rewiring make 75% of the units for low-energy lamp fittings only. Specify low-energy appliances
Renewable and low -carbon technologies	Encourage the provision of solar hot water heating and other renewable technologies such as photovoltaics

Table 18.4 Cradle to factory gate embodied energy in building materials

Material	Embodied Energy MJ/kg
Aluminium	155
Polypropylene	95
Expanded polystyrene	89
Polyurethane	72
Copper	42
Lead	25
Steel	20
Mineral wool/ Rockwool	17
Glass	15
Timber	10
Plasterboard	6.8
Portland cement CEM1	5.5
Brick	3.0
Marble	2.0
Concrete (25/30 MPa)	0.78
Straw	0.24
Sand	0.08

Notes:
- The embodied energy is the total energy consumed in the processes from cradle to factory gate including extraction, transportation and production.
- The embodied energy is measured in MJ/kg, but different construction systems require widely differing material mass for the equivalent building component (e.g. construction of a wall in concrete, brick, steel, timber or glass).
- The figures are guidelines only as production processes and recycling contents vary considerably.
- The data is based on the BSRIA Guide 'Embodied Carbon. The Inventory of Carbon and Energy' (ICE) 2011.

of the existing housing stock within England and Wales achieves the average D rating. Furthermore, many government buildings fall into the lower categories with respect to energy efficiency.

The Energy Saving Trust has suggested a range of best practice standards (Table 18.3) for the refurbishment of the existing building stock. Clearly, a 60% reduction in carbon is easier to obtain on the poorer quality properties than those already at a higher standard. The best practice represents the optimum balance between environmental performance and practicality.

Specification of materials

As energy considerations have risen up the construction agenda, building material manufacturers have pressed their ecological credentials, often with claims and counter-claims between competing materials and systems. The spin-off from this debate has been a rush to reduce embodied energy in the production processes (Table 18.4), to reduce waste both within the factory and on site, and also to emphasise recyclability.

Metals, such as steel, aluminium and copper, have long had good records of recycling from within the construction industry, but now, for example, significant volumes of PVC-U from the replacement windows industry are recycled. Some plastics products can be co-extruded with a core of recycled material and external layers of virgin material. Plastic bottles may be recycled as loft insulation or into composites as plastic lumber. Low-grade timber waste may be used in non-traditional products such as permanent insulating concrete formwork. These changes significantly affect the balance of ecological ratings traditionally placed on construction materials. Current research on cement is investigating alternative products which have much lower embodied energy and may even be carbon neutral. Excess bottle glass, not required for direct recycling, may be converted into a foamed aggregate for light-weight concrete.

RESPONSIBLE SOURCING OF MATERIALS

Responsible sourcing of materials requires a well-managed supply chain, from the point where the material is mined or harvested through processing and manufacturing to use, reuse, recycling and waste disposal. The BRE Global Certification Scheme BES 6001 provides a framework for responsible and sustainable resourcing of construction products. It also offers a route to gaining credits within the materials section of the Code for Sustainable Homes and BREEAM certification schemes.

Increasingly, manufacturers of building materials and components are following the requirements of BS EN ISO 9001: 2014 and BS EN ISO 9004: 2009 in respect of quality management systems, as well as BS EN ISO 14001: 2004 in respect of environmental management systems.

LIFE CYCLE ASSESSMENT

A full life cycle assessment includes extraction, processing, manufacturing, transportation, construction, maintenance, reuse, recycling and final disposal. This clearly gives a more realistic environmental impact than consideration of embodied energy alone (Table 18.3). For example, steel has an embodied energy of approximately 24 MJ/kg, but it is recycled many times over without any loss of performance. By comparison, concrete, which can normally only be recycled as aggregate, has an embodied energy of approximately 1 MJ/kg. In addition, the quantities used for equivalent structural purposes differ significantly and the energies involved in their deconstruction and recycling are different.

Cradle-to-grave embodied carbon data (Table 18.5) has been collated for some key materials by the Target Zero project which gives guidance to designers, architects and engineers on reducing emissions towards the UK government's 2019 zero carbon target. However, the data on recycling (e.g. re-forming steel), down-cycling (e.g. old concrete used as aggregate) and reuse (e.g. timber beams) is not robust and subject to change as industries strive to raise their ecological credentials. Recent data from Tata Steel suggest that only 5% of concrete waste and 1% of steel from deconstruction sites is sent to landfill sites. TRADA now give the equivalent figure for timber as 25%.

Table 18.5 Life cycle embodied energy in building materials

Material	End-of-life assumption	Total life cycle CO_2 emissions (tCO_2e/t)
Steel sections	99% recycled, 1% landfill	1.01
Steel reinforcement	92% recycled, 6% landfill	0.82
Concrete (C30/37)	72% recycled as aggregate, 23% landfill	0.14
Laminated timber (Glulam)	16% recycled, 4% incineration, 80% landfill	1.1
Plasterboard	20% recycled, 80% landfill	0.15
Aggregate	50% recycled, 50% landfill	0.005
Tarmac	77% recycled, 23% landfill	0.02

Notes:
- tCO_2e/t is total carbon dioxide equivalents per tonne.
- The data was collated by the British Constructional Steelwork Association and is subject to change.
- Different materials require differing masses for equivalent constructions.
- Data on steel was from Tata Steel, 2001.
- Data on concrete, aggregate and tarmac was from the Department for Communities and Local government, 2005.
- Data on plasterboard was from WRAP (Net Waste Tool), 2008.
- Data on timber was from TRADA, 2008. However, TRADA reported in 2012 that only 25% of UK wood waste is now placed in landfill. The majority is used for the manufacture of particleboard or burnt as biomass fuel.

GREEN GUIDE TO SPECIFICATION

Guidance to designers and specifiers on the environmental impact of various building materials and construction systems is given in the BRE *The Green Guide to Specification*. This lists a wide range of construction systems against a standardised set of environmental criteria, giving each criterion a letter grading (A+ to E) together with an overall summary rating (A+ to E). *The Green Guide* is intended for use with whole building assessment tools such as the BRE Environmental Assessment Method (BREEAM) and the Code for Sustainable Homes.

The Green Guide records the relative environmental impacts of the construction materials used in the six generic building types:

Commercial buildings – including offices
Educational buildings – schools, universities, colleges
Healthcare buildings – including hospitals
Retail outlets
Residential buildings
Industrial buildings

Materials and components are arranged within the key building elements:

Ground floors
Upper floors
Roofs
External walls
Windows
Internal walls and partitions
Insulation
Landscaping

Within each category the materials are compared on a like-for-like basis, so that equivalent forms of construction, such as a steel or concrete column, can be directly compared. Where appropriate, materials are compared on a common U-value equating to equivalent heat loss.

The environmental ratings are based on a life cycle assessment which includes greenhouse gas emissions, mineral and water extraction, toxicity of waste and pollution, landfill or incineration, fossil fuel depletion, atmospheric pollution and acid rain. Specifications are based on a 60-year period, with factors included for shorter anticipated periods to replacement. Recycled content is taken into consideration, but this is constantly improving with a greater realisation of its environmental importance.

BIODIVERSITY ISSUES IN PLANNING AND DEVELOPMENT

The British Standard BS 42020: 2013 gives guidance to planners and developers who may encounter biodiversity issues. The Standard sets out to ensure the production of high-quality ecological information to support effective decision-making and compliance with statutory obligations, leading to successful implementation of practical measures during development.

Recycling and deconstruction

Despite the high and increasing landfill tax, approximately 30 million tonnes of construction waste are sent to landfill each year within the UK. The issue of waste on building sites (Fig. 18.6) tends to be a low priority for much of the industry, despite the financial and ecological costs involved. However, some producers such as plasterboard, block and insulation manufacturers will take back clean excess and offcuts for reprocessing.

Fig. 18.6 Building site waste

With the increasing need to recycle materials, it may be anticipated that in the future more attention will be given to design for deconstruction rather than demolition, as this can lead to greater reuse of resources. Deconstruction accounts for typically 6–8% of the whole life cycle energy used within building construction.

References

FURTHER READING

Addis, W. and Schouten, J. 2004: *Principles of design for deconstruction to facilitate reuse and recycling*. London: Construction Industry Research and Information Association.

Barlow, S. 2011: *Guide to BREEAM*. London: RIBA Publishing.

Bere, J. 2013: *Building for the future. An introduction to Passive House*. London: RIBA Publishing.

Berge, B. 2009: *The ecology of building materials*, 2nd edn. Oxford: Architectural Press.

Berry, C. and McCarthy, S. 2011. *Guide to sustainable procurement in construction*. London: CIRIA.

BRE Global. 2009a: *Framework for the responsible sourcing of construction products*. BES 6001 Issue 2.0. Garston: Building Research Establishment.

BRE Global. 2009b: *Methodology for environmental profiles of construction products. Product category rules for Type III environmental product declaration of construction products*. BES 6050 Issue 1.0. Garston: Building Research Establishment.

BREEAM Environmental Assessments. 2007: *Sustainable construction in steel*. Info. Sheet 2. Ascot: Steel Construction Institute.

Concrete Centre. 2011: *Thermal performance. Part L1A. The road to 2016 (zero carbon) fabric and services*. Camberley: The Concrete Centre.

CPA. 2012: *Guide to understanding the embodied impacts of construction products*. London: Construction Products Association.

Edwards, B. 2014: *Rough guide to sustainability. A design primer*, 4th edn. London: RIBA Publishing.

Ellingham, I. and Fawcett, W. 2013: *Whole life sustainability*. London: RIBA Publishing.

Gaze, C. 2009: *Complying with the code for sustainable homes. Lessons learnt on the BRE innovation park*. FB20. Bracknell: IHS BRE Press.

Hadi, M. and Halfhide, C. 2009: *Move to low-carbon design*. FB21. Bracknell: IHS BRE Press.

Halliday, S. 2007: *Sustainable construction*. Oxford: Butterworth Heinemann.

Hammond, G., Jones, C., Lowrie, F. and Tse, P. 2011: *Embodied carbon. The inventory of carbon and energy*. Bracknell: BSRIA.

Institution of Structural Engineers. 2011: *A short guide to embodied carbon in building structures*. London: Institution of Structural Engineers.

Khatib, J. (ed.) 2009: *Sustainability of construction materials*. Abington: Woodhead Publishing Ltd.

Koones, S. 2010: *Prefabulous and sustainable*. New York: Abrams.

Lewry, A. 2012: *Energy management in the built environment. A review of best practice*. Bracknell: IHS BRE Press.

MacKenzie, F., Pout, C., Sharrock, L., Matthews, A. and Henderson, J. 2010: *Energy efficiency in new and existing buildings*. Bracknell: IHS BRE Press.

Menzies, G., Turan, S. and Banfill, P. 2007: *Life-cycle assessment and embodied energy. A review*. Proc. Inst. Civil Engineers. Construction Materials 160. Issue CM4, pp. 135–142. November 2007. Institution of Civil Engineers.

NBS. 2012: *Sustainability bibliography*. Construction Information Service.

NHBC. 2010: *Standards extra*. Technical newsletter 48. Milton Keynes: NHBC.

NHBC Foundation. 2009: *Zero carbon homes. An introductory guide for housebuilders*. NF14. Amersham: NHBC Foundation.

NHBC Foundation. 2012: *Survey of low and zero carbon technologies in new housing*. NF 42. Amersham: NHBC Foundation.

Pelsmakers, S. 2012: *Environmental design pocketbook*. London: RIBA Publishing.

RIBA. 2011: *Guide to the Building Regulations 2011 edition*. London: RIBA Publishing.

RIBA. 2011: *Guide to BREEAM*. London: RIBA Publishing.

Roaf, S., Fuentes, M. and Thomas-Rees, S. 2012: *Ecohouse*, 4th edn. Abingdon: Routledge.

Robust Details. 2008: *Robust details handbook*, 3rd edn (April 2013 update). Milton Keynes: Robust Details Ltd.

SCI. 2009: *Code for sustainable homes. How to satisfy the code using steel technologies*. Publication 386. Ascot: Steel Construction Institute.

Syed, A. 2012: *Advanced building technologies for sustainability*. New Jersey: Wiley.

Target Zero. 2011: *Guidance on the design and construction of low carbon warehouse buildings. Version 2.0*. Target Zero.

Target Zero. 2011: *Guidance on the design and construction of low carbon supermarket buildings. Version 2.0.* Target Zero.

Target Zero. 2011: *Guidance on the design and construction of low carbon office buildings. Version 1.0.* Target Zero.

Target Zero. 2011: *Guidance on the design and construction of low carbon mixed-use buildings. Version 2.0.* Target Zero.

TRADA. 2011: *Green deal. A summary of the consultation.* London: TRADA Technology.

TRADA. 2012: *BREEAM. A summary of changes to the 2011 version.* London: TRADA Technology.

Turrent, D. (ed.) 2007. *Sustainable architecture.* London: RIBA Publishing.

Woolley, T. 2013: *Low impact building. Housing using renewable materials.* Chichester: Wiley Blackwell.

Zero Carbon Hub. 2012: *Informing the Part L 2013 consultation. Fabric energy efficiency for Part L 2013. Classification methodology for different dwelling types.* Milton Keynes: NHBC.

Zero Carbon Hub. 2012: *Informing the Part L 2013 consultation. Fabric energy efficiency for Part L 2013. Worked examples and fabric specifications.* Milton Keynes: NHBC.

Zero Carbon Hub. 2013: *Zero carbon strategies for tomorrow's new homes.* Milton Keynes: NHBC.

STANDARDS

BS 8544: 2013
Guide for life cycle costing of maintenance during the in use phases of buildings.

BS 8900
Managing sustainable development of organizations:

Part 1: 2013	Guide.
Part 2: 2013	Framework for assessment against BS 8900–1. Specification.

BS 8903: 2010
Principles and framework for procuring sustainably. Guide.

BS 8905: 2011
Framework for the assessment of the sustainable use of materials. Guidance.

BS ISO 14067: 2012
Carbon footprints of products. Requirements and guidelines for quantification and communication.

BS ISO 15392: 2008
Sustainability in building construction. General principles.

BS ISO 15686
Building and constructed assets. Service life planning:

Part 1: 2011	General principles.
Part 2: 2012	Service life prediction procedures.
Part 3: 2002	Performance audit and reviews.
Part 4: 2014	Service life planning. Planning using Building Information Modelling.
Part 5: 2008	Life cycle costing.
Part 6: 2004	Procedures for considering environmental impacts.
Part 7: 2006	Performance evaluation for feedback of service life data from practice.
Part 8: 2008	Reference service life and service-life estimation.
Part 9: 2008	Guidance on assessment of service life data.
Part 10: 2010	When to assess functional performance.

BS ISO 16343: 2013
Energy performance of buildings. Methods for expressing energy performance and for energy certification of buildings.

PD ISO/TR 16344: 2012
Energy performance of buildings. Common terms, definitions and symbols for the overall energy performance rating and certification.

BS ISO 16346: 2013
Energy performance of buildings. Assessment of overall energy performance.

pr BS ISO 16745: 2013
Environmental performance of buildings. Carbon metric of a building. Use stage.

BS ISO 16818: 2008
Building environment design. Energy efficiency. Terminology.

BS ISO 21931: 2010
Sustainability in building construction. Framework for methods of assessment of the environmental performance.

BS 42020: 2013
Biodiversity. Code of practice for planning and development.

pr EN ISO 9000: 2014
Quality management systems. Fundamentals and vocabulary.

BS EN ISO 9001: 2014
Quality management systems. Requirements.

BS EN 9004: 2009
Managing for the sustained success of an
organisation. A quality management approach.
PD ISO/TS 12720: 2014
Sustainability in buildings and civil engineering
works. Guidelines on the application of the
general principles in ISO 15392.
BS EN ISO 13370: 2007
Thermal performance of buildings. Heat transfer
via the ground. Calculation methods.
BS EN ISO 14001: 2004
Environmental management systems.
Requirements with guidance for use.
BS EN ISO 14020: 2001
Environmental labels and declarations. General
principles.
BS EN ISO 14024: 2001
Environmental labels and declarations.
Type I Environmental labelling. Principles and
procedures.
BS EN ISO 14025: 2011
Environmental labels and declarations.
Type III Environmental declarations. Principles
and procedures.
BS EN ISO 14040: 2006
Environmental management. Life cycle
assessment. Principles and framework.
BS EN ISO 14044: 2006
Environmental management. Life cycle
assessment. Requirements and guidelines.
BS EN 15232: 2012
Energy performance of buildings. Impact of build-
ing automation, controls and building
management.
BS EN 15241: 2007
Ventilation for buildings. Calculation methods for
energy losses due to ventilation and infiltration in
buildings.
BS EN 15242: 2007
Ventilation for buildings. Calculation methods for
the determination of air flow rates.
BS EN 15243: 2007
Ventilation for buildings. Calculation of room
temperatures and of load and energy for buildings
with room conditioning systems.
pr EN 15603: 2013
Energy performance of buildings. Overarching
standard Energy Performance of Buildings
Directive (EPBD).
BS EN 15643
Sustainability of construction work. Assessment of
buildings:

Part 1: 2010   General framework.
Part 2: 2011   Framework for the assessment
               of environmental performance.
Part 3: 2012   Framework for the assessment
               of social performance.
Part 4: 2012   Framework for the assessment
               of economic performance.
BS EN 15804: 2012
Sustainability of construction works.
Environmental product declarations. Core rules
for the product category of construction products.
PD CEN/TR 15941: 2010
Sustainability of construction works.
Environmental product declarations.
Methodology for selection and use of generic data.
BS EN 15942: 2011
Sustainability of construction works.
Environmental product declarations.
Communication format business-to-business.
BS EN 15978: 2011
Sustainability of construction works. Assessment
of environmental performance of buildings.
Calculation method.
BS EN 16247
Energy audits:
    Part 1: 2012   General requirements.
    pr Part 2: 2012   Buildings.
BS EN 16309: 2014
Sustainability of construction works. Assessment
of social performance of buildings. Calculation
methodology.
pr EN 16627: 2013
Sustainability of construction works. Assessment
of economic performance of buildings.
Calculation method.
pr EN 16751: 2014
Bio-based products. Sustainability criteria.
PD ISO/TR 21932: 2013
Sustainability in buildings and civil engineering
works. A review of terminology.
PD 156865: 2008
Standardized method of life cycle costing for
construction procurement.
PAS 2030: 2012 Ed.2
Improving the energy efficiency of existing
buildings. Specifications for installation process,
process management and service provision.
BIP 2135: 2007
A handbook for sustainable development.
BIP 2203: 2011
The sustainable procurement guide. Procuring
sustainably using BS 8903.

BUILDING RESEARCH ESTABLISHMENT PUBLICATIONS

BRE Digests

BRE Digest 446: 2000
Assessing environmental impacts of construction industry consensus, BREEAM and UK Ecopoints.

BRE Digest 447: 2000
Waste minimisation on a construction site.

BRE Digest 452: 2000
Whole life costing and life-cycle assessment for sustainable building design.

BRE Digest 457: 2001
The Carbon Performance Rating for offices.

BRE Information papers

BRE IP 13/03
Sustainable buildings (Parts 1–4).

BRE IP 4/05
Costing sustainability. How much does it cost to achieve BREEAM and EcoHomes ratings?

BRE IP 15/05
The scope for reducing carbon emissions from housing.

BRE IP 16/05
Domestic energy use and carbon emissions. Scenarios to 2050.

BRE IP 1/06
Assessing the effects of thermal bridging at junctions and around openings.

BRE IP 6/06
Balanced Value for sustainable procurement.

BRE IP 4/07
Environmental weighting. Their use in the environmental assessment of construction products.

BRE IP 1/08
The price of sustainable schools.

BRE IP 2/08
New build and refurbishment in the Sustainable Communities plan.

BRE IP 3/08
Delivering sustainability objectives through planning.

BRE IP 9/08
Part 1. Applying the code for sustainable homes on the BRE Innovation Park. Lessons learnt about building fabric.

BRE IP 9/08
Part 2. Applying the code for sustainable homes on the BRE Innovation Park. Lessons learnt about energy sources, overheating and ventilation.

BRE IP 9/08
Part 3. Applying the code for sustainable homes on the BRE Innovation Park. Lessons learnt about waster use, harvesting, recycling and drainage.

BRE IP 9/08
Part 4. Applying the code for sustainable homes on the BRE Innovation Park. Lessons learnt about architecture, construction and material sourcing.

BRE IP 12/08
An introduction to PassivHaus.

BRE IP 1/09
Performance and service life in the Environmental Profiles Methodology and Green Guide to Specification.

BRE IP 3/09
Lessons learnt from the Barratt Green House. Delivering a zero carbon home using innovative concrete systems.

BRE IP 7/10
SBEM for non-domestic buildings. An introduction.

BRE IP 10/10
SAP for beginners.

BRE IP 11/10
Sustainability in foundations.

BRE IP 14/10
Consumer feedback on low-carbon housing.

BRE IP 18/10
Sustainable housing refurbishment. An update on current guidance information.

BRE IP 2/11
Energy in schools.

BRE IP 3/11
The performance of district heating in new developments.

BRE IP 12/11
Sustainable refurbishment of the BRE Victorian terrace (Parts 1 and 2).

BRE IP 19/11
Assessing the sustainability of office refurbishment with BREEAM. A case study.

BRE IP 3/12
Potential for reducing carbon emissions from commercial and public-sector buildings.

BRE IP 8/12
Sustainability at BRE. Delivering the S-plan.

BRE IP 18/12
Funding renewable energy projects.
BRE IP 19/12
Sustainable information technology.
BRE IP 22/12
Site layout planning for sunlight and solar gain.
BRE IP 23/12
Site layout planning for daylight.
BRE IP 25/12
Making use of carbon emissions in the built environment.
BRE IP 2/13
Designing for future climate change. Lessons learned from developing an adaption strategy.
BRE IP 3/13
Responsible sourcing of materials in construction.
BRE IP 6/13
Surface water run-off in the Code for Sustainable Homes.
BRE IP 7/13
Energy surveys and audits. A guide to best practice.
BRE IP 9/13
Lessons from AIMC4 for cost effective, fabric first, low energy housing.
BRE IP 13/13
Building with confidence using renewable materials.
BRE IP 16/13
Greener homes for Redbridge. Sustainable refurbishment of 19 dwellings.
BRE IP 1/14
Understanding the choices for building controls.
BRE IP 2/14
Operating BEMS. A practical approach to building management systems.

BRE Reports

BR Developer Sheet
EcoHomes 2006. The environmental rating for homes.
BR 390: 2000
The Green Guide to Housing Specification. An environmental profiling system for building materials and components used in housing.
BR 418: 2001
Deconstruction and reuse of construction materials.

BR 487: 2007
Designing quality buildings. A BRE guide.
BR 493: 2007
Creating environmental weightings for construction products. Results of a study.
BR 498: 2008
Sustainability through planning. Local authority use of BREEAM, EcoHomes and the Code for Sustainable Homes.
BR 501: 2009
The Green Guide to Specification (4th edn).
BR 502: 2009
Sustainability in the built environment. An introduction to its definition and measurement.
BR 506: 2009
Smart home systems and the Code for sustainable homes.
EP 99: 2009
Sustainable masonry construction.
EP 103: 2013
Designing resilient cities. A guide to good practice.
FB 24: 2010
A guide to the simplified building energy model (SBEM).
FB 44: 2012
Energy management in the built environment. A review of best practice.

GOVERNMENT PUBLICATIONS

Department for Communities and Local Government. 2014: *Code for sustainable homes and energy performance of buildings: Statistical release.* London: Communities and Local Government.

Department for Communities and Local Government. 2011: *Zero carbon non-domestic buildings. Phase 3. Final report.* London: Communities and Local Government.

Department for Communities and Local Government. 2010: *Code for sustainable homes – technical guide November 2010.* London: Communities and Local Government.

Department for Communities and Local Government. 2010: *Code for sustainable homes – case studies. Volume 2.* London: Communities and Local Government.

Department for Communities and Local Government. 2010: *Summary of the changes to the code for sustainable homes technical guidance. November 2010.* London: Communities and Local Government.

Department for Communities and Local Government. 2009: *Definition of zero carbon homes and non-domestic buildings. Consultation. Summary of responses.* London: Communities and Local Government.

Department for Communities and Local Government. 2008: *Definition of zero carbon homes and non-domestic buildings. Consultation.* London: Communities and Local Government.

HM Government. 2011: *Carbon plan. Delivering our low carbon future. December 2011.* London: HM Government.

UK Green Building Council. 2007: *Report on carbon reductions in new non-domestic buildings.* London: Communities and Local Government.

ADVISORY ORGANISATIONS

Association for the Conservation of Energy, Westgate House, 2a Prebend Street, London N1 8PT (0207 359 8000).

BREEAM Centre, BRE, Garston, Watford, Hertfordshire WD25 9XX (01923 664462).

Centre for Energy and the Environment, The Innovation Centre, University of Exeter, Rennes Drive, Exeter, Devon EX4 4RN (01392 724144).

Energy Saving Trust, 21 Dartmouth Street, London SW1H 9BP (0300 123 1234).

Sustainability Centre, East Meon, Hampshire GU32 1HR (01703 823166).

Sustainable Homes Ltd, Marina House, 17 Marina Place, Hampton Wick, Kingston-upon-Thames, Surrey KT1 4BH (0208 973 0429).

INDEX